Understanding Advanced Statistical Methods

CHAPMAN & HALL/CRC
Texts in Statistical Science Series

Series Editors
Francesca Dominici, *Harvard School of Public Health, USA*
Julian J. Faraway, *University of Bath, UK*
Martin Tanner, *Northwestern University, USA*
Jim Zidek, *University of British Columbia, Canada*

Texts in Statistical Science

Understanding Advanced Statistical Methods

Peter H. Westfall

Information Systems and Quantitative Sciences
Texas Tech University, USA

Kevin S. S. Henning

Department of Economics and International Business
Sam Houston State University, USA

CRC Press
Taylor & Francis Group
Boca Raton   London   New York

CRC Press is an imprint of the
Taylor & Francis Group an **informa** business

A CHAPMAN & HALL BOOK

CRC Press
Taylor & Francis Group
6000 Broken Sound Parkway NW, Suite 300
Boca Raton, FL 33487-2742

© 2013 by Taylor & Francis Group, LLC
CRC Press is an imprint of Taylor & Francis Group, an Informa business

No claim to original U.S. Government works

Printed on acid-free paper
Version Date: 20130227

International Standard Book Number-13: 978-1-4665-1210-8 (Hardback)

Visit the Taylor & Francis Web site at
http://www.taylorandfrancis.com

and the CRC Press Web site at
http://www.crcpress.com

Contents

List of Examples

Preface

We wrote this book because there is a large gap between the elementary statistics course that most people take and the more advanced research methods courses taken by graduate and upper-division students so they can carry out research projects. These advanced courses include difficult topics such as regression, forecasting, structural equations, survival analysis, and categorical data, often analyzed using sophisticated likelihood-based and even Bayesian methods. However, they typically devote little time to helping students understand the fundamental assumptions and machinery behind these methods. Instead, they teach the material like witchcraft: Do this, do that, and voilà—statistics! Consequently, students learn little about what they are doing and why they are doing it. Like trained parrots, they learn how to recite statistical jargon mindlessly. The goal of this book is to make statistics less like witchcraft and to treat students as intelligent humans and not as trained parrots—thus the title, *Understanding Advanced Statistical Methods*.

This book will surprise your students. It will cause them to think differently about things, not only about math and statistics, but also about research, the scientific method, and life in general. It will teach them how to do good modeling—and hence good statistics—from a standpoint of *deep* knowledge rather than *rote* knowledge. It will also provide them with tools to think critically about the claims they see in the popular press and to design their own studies to avoid common errors.

There are plenty of formulas in this book, because to understand advanced statistical methods requires understanding probabilistic models, and probabilistic models are necessarily mathematical. But if your students ever find themselves plugging numbers into formulas mindlessly, make them stop and ask, "Why?" Getting students to ask and answer that question is the main objective of this book. Having them perform mindless calculations is a waste of your time and theirs, unless they understand the *why*. Every formula tells an interesting story, and the story explains the *why*.

Although all statistics books purport to have the goal of making statistics understandable, many try to do so by avoiding math. This book does not shy away from math; rather, it teaches the needed math and probability along with the statistics. Even if your students are math "phobes" they will learn the math and probability theory and hopefully enjoy it, or at least appreciate it.

In particular, statistics is all about unknown, algebraic quantities. What is the probability of a coin landing heads up when flipped? It is not 50%. Instead, it is an unknown algebraic quantity θ that depends on the construction of the coin and on the methods of the coin-flipper. Any book that teaches statistics while avoiding algebra is therefore a book of fiction!

This book uses calculus where needed to help readers understand continuous distributions and optimizations. Students should learn enough calculus to understand the logical arguments concerning these core concepts. But calculus is not a prerequisite. We only assume that students have a familiarity with algebra, functions and graphs, and spreadsheet software such as Microsoft Excel®. The book employs a "just-in-time" approach, introducing mathematical topics, including calculus, where needed. We present mathematical concepts in a concrete way, with the aim of showing students how even the seemingly hard math is really not so hard, as well as showing them how to use math to answer important questions about our world.

As far as probability theory goes, we employ a laser-beam focus on those aspects of probabilistic models that are most useful for statistics. Our discussion therefore focuses more on distributions than on counting formulas or individual probability calculations. For example, we present Bayes' theorem in terms of distributions rather than using the classical two-event form presented in other sources. For another example, we do not emphasize the binomial distribution; instead, we focus on the Bernoulli distribution with independent and identically distributed observations.

This book emphasizes applications; it is not "math for math's sake." We take real data analysis very seriously. We explain the theory and logic behind real data analysis intuitively and gear our presentation toward students who have an interest in science but may have forgotten some math.

Statistics is not a collection of silly rules that students should recite like trained parrots—rules such as $p < 0.05$, $n > 30$, $\rho > 0.3$, etc. We call these *ugly rules of thumb* throughout the book to emphasize that they are mere suggestions and that there is nothing hard-and-fast about any of them. On the other hand, the logic of the mathematics underlying statistics is not ugly at all. Given the assumptions, the mathematical conclusions are 100% true. But the assumptions *themselves* are never quite true. This is the heart and soul of the subject of statistics—how to draw conclusions successfully when the premises are flawed—and this is what your students will learn from this book.

This book is not a "cookbook." Cookbooks tell you all about the *what* but nothing about the *why*. With computers, software, and the Internet readily available, it is easier than ever for students to lose track of the *why* and focus on the *what* instead. This book takes exactly the opposite approach. By enabling your students to answer the *why*, it will help them to figure out the *what* on their own—that is, they will be able to develop their own statistical recipes. This will empower your students to use advanced statistical methods with confidence.

The main challenge for your students is not to understand the math. Rather, it is to understand the statistical point of view, which we present consistently throughout this book as a mantra:

Model Produces Data

More specifically, the *statistical model* is a *recipe* for *producing random data*. This one concept will turn your students' minds around 180°, because most think a statistical model is something *produced by data* rather than a *producer of data*. In our experience, the difficulty in understanding the statistical model as a data-generator is the single most significant barrier to students' learning of statistics. Understanding this point can be a startling epiphany, and your students might find statistics to be fun, and surprisingly easy, once they "get it." So let them have fun!

Along with the presentation of models as *producers of data*, another unique characteristic of this book is that it avoids the overused (and usually misused) "population" terminology. Instead, we define and use the "process" terminology, which is always more correct, generally more applicable, and nearly always more scientific. We discuss populations, of course, but correctly and appropriately. Our point of view is consistent with the one presented in *Statistical Science* (26(1), 1–9, 2011) by Robert E. Kass and several discussants in an article entitled "Statistical inference: The big picture."

Another unique characteristic of this book is that it teaches Bayesian methods *before* classical (frequentist) methods. This sequencing is quite natural given our emphasis on probability models: The flow from probability to likelihood to Bayes is seamless. Placing Bayesian methods before classical methods also allows for more rounded and thoughtful discussion of the convoluted frequentist-based confidence interval and hypothesis testing concepts.

This book has no particular preference for the social and economic sciences, for the biological and medical sciences, or for the physical and engineering sciences. All are useful, and the book provides examples from all these disciplines. The emphasis is on the overarching *statistical* science. When the book gives an example that does not particularly apply to you or your students' fields of study, just change the example! The concepts and methods of statistics apply universally.

The target audience for this book is mainly upper-division undergraduates and graduate students. It can also serve lower-division students to satisfy a mathematics general education requirement. A previous course in statistics is not necessary.

This book is particularly useful as a prerequisite for more advanced study of regression, experimental design, survival analysis, time series analysis, structural equations modeling, categorical data analysis, nonparametric statistics, and multivariate analysis. We introduce regression analysis (ordinary and logistic) in the book, and for this reason, we refer to the data as Y, rather than X as in many other books. We use the variable designation X as well, but mainly as a predictor variable.

The spreadsheet software Microsoft Excel is used to illustrate many of the methods in this book. It is a good idea, but not strictly necessary, to use a dedicated mathematical or statistical software package in addition to the spreadsheet software. However, we hope to convince your students that advanced statistical methods are really not that hard, since one can understand them to a great extent simply by using such commonplace software as Excel.

About Using This Book

- Always get students to ask "Why?" The point of the book is not the *what*; it is the *why*. Always question assumptions and aim to understand how the logical conclusions follow from the assumptions.

- Students should read the book with a pencil and paper nearby, as well as spreadsheet or other software, for checking calculations and satisfying themselves that things make sense.

- Definitions are important and should be memorized. Vocabulary terms are given in **boldface** in the book, and their definitions are summarized at the ends of the chapters. Strive to teach the definitions in the context of your own field of interest, or in the context of your students' fields of interest.

- Some formulas should be memorized, along with the stories they tell. Important formulas are given at the ends of the chapters.

- We often give derivations of important formulas, and we give the reasons for each step in parentheses to the right of the equations. These reasons are often simple, involving basic algebra. The reasons are more important than the formulas themselves. Learn the reasons first!

- The exercises all contain valuable lessons and are essential to understanding. Have your students do as many as possible.

- A companion website http://courses.ttu.edu/isqs5347-westfall/westfall_book.htm includes computer code, sample quizzes, exams and other pedagogical aids.

Acknowledgments

We would like to thank Josh Fredman for his excellent editing and occasional text contributions; students in Dr. Westfall's ISQS 5347 class, including Natascha Israel, Ajay Swain, Jianjun Luo, Chris Starkey, Robert Jordan, and Artem Meshcheryakov for careful reading and feedback; Drs. Jason Rinaldo and D. S. Calkins for careful reading, needling, and occasional text passages; and the production staff at Taylor & Francis Group/CRC Press, including Rachel Holt and Rob Calver, as well as Remya Divakaran of SPi for helpful direction and editing. Most graphics in the book were produced using the SGPLOT and SGPANEL procedures in SAS software.

Authors

Peter H. Westfall has a PhD in statistics and many years of teaching, research, and consulting experience in biostatistics and a variety of other disciplines. He has published over 100 papers in statistical theory and methods, won several teaching awards, and has written several books, one of which won two awards from the Society for Technical Communication. He is former editor of *The American Statistician* and is a Fellow of both the American Statistical Association and of the American Association for the Advancement of Science.

Kevin S. S. Henning has a PhD in business statistics from Texas Tech University and currently teaches business statistics and forecasting in the Department of Economics and International Business in the College of Business at Sam Houston State University.

1

Introduction: Probability, Statistics, and Science

1.1 Reality, Nature, Science, and Models

So, what is reality? Yes, this may be an odd question to start a statistics book. But reality is what science is all about: It is the study of what is real. "What is real?" is a topic that fills volumes of philosophy books, but for our purposes, and for the purposes of science in general, the question of what is real is answered by "That which is *natural* is real." Of course, that raises the question, "What is natural?"

Without delving too far into philosophy, **Nature** is all aspects of past, present, and future existence. Understanding Nature requires common observation—that is, it encompasses those things that we can agree we are observing. As used in this book, Nature includes the physical sciences (e.g., planets, galaxies, gravity), the biological sciences (e.g., DNA, medicine), and the social sciences (e.g., economics, psychology). Nature includes man-made things such as dams, as well as social constructs such as economic activity; we certainly do not limit our definition of Nature to those things that are without human intervention. In fact, most examples involving Nature in this book *do* involve human activity.

Science is the study of Nature. It involves understanding why Nature is the way that it is and using such knowledge to make predictions as to what will happen—or would have happened—under various circumstances.

Personal realities which are not commonly observed or agreed upon—for example, those of a mystical or spiritual quality—are outside the scope of science. Someone may believe that the Earth rests upon a large turtle, and while this point of view may offer comfort and meaning, it is not a common, agreed-upon observation and is therefore not a scientific proposition. The same can be said about major religions: Tenets of faith lacking agreed-upon observation cannot be subjected to measurement and testing and hence are outside the scope of science.

Statistics is the language of science. In its broadest form, statistics concerns the analysis of recorded information or data. Data are commonly observed and subject to common agreement and are therefore more likely to reflect our common reality or Nature. Data offer us a clearer picture of what Nature is and how Nature works, and statistical analyses of data allow us to reverse-engineer natural processes and thus gain scientific knowledge.

To understand Nature, you must construct a model for how Nature works. A **model** helps you to understand Nature and also allows you to make predictions about Nature. There is no right or wrong model; they are all wrong! But some are better than others. The better models are the ones you want to use, and in this book we'll help you identify them.

If you have ever played with toy trains or dolls, you are probably very familiar with the general concept of modeling. Your first toys probably only resembled their real-world counterparts in the most elementary of ways. As you grew older, however, your toys probably became more like the real thing, and hence, they became better models. For example, your first toy train might have been nothing more than a piece of wood sculpted to look like a locomotive, with no working parts, but when you got older, you may well have played with a working toy locomotive that ran on electric tracks and pulled a few miniature cars. This train was a better model because the principles behind its operation were closer to those of real trains. They were still not identical, of course. Real trains have sophisticated throttle and communications equipment and are many orders of magnitude larger than toy trains.

Trains and dolls are *physical models*. The focus of this book will be on another class of models, called *mathematical models*, which are built out of equations rather than materials. As with physical models such as the toy train, these mathematical models are not how Nature really operates, but if they are similar to Nature, they can be very informative. Thus, your model is good if it produces data resembling what Nature would produce. These models are personal: They are mental abstractions that *you* create, and that *you* use. Someone else may create and use a different model.

We will often represent models using graphs. When you see a graph, always ask yourself "What is the information that is provided in this graph?" To answer, look carefully at the axis labels and the numbers on the axes, and be sure you understand what they mean. Also, read the figure legends and the surrounding text. While a picture may be worth 1000 words, it is only worth one equation. But it is a lot more fun to look at than the equation! It is also easier to remember. When you see an equation, ask yourself, "How does that look in a graph?"

Example 1.1: A Model for Driving Time

You will drive x kilometers. How long will it take you? If you typically average 100 km/hour (or 62.1 miles/hour), then your driving time y (in hours) may be given by the model $y = x/100$; Figure 1.1 shows a graph of this equation.

Thus, if your distance is 310 km, then your driving time may be given by 3.10 hours or 3 hours and 6 minutes.

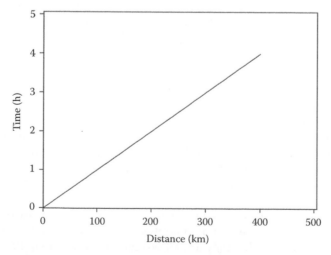

FIGURE 1.1
A model for driving time as a function of distance: $y = x/100$.

Two things you should note about the driving time model: First, a model allows you to make *predictions*, such as 3 hours and 6 minutes. Note that a prediction is not about something that happens in the future (which is called a **forecast**). Rather, a **prediction** is a more general, "what-if" statement about something that might happen in the past, present, future, or not at all. You may never in your life drive to a destination that is precisely 310 km distant, yet still the model will tell you how long it would take if you did.

Second, notice that the *model produces data*. That is, if you state that $x = 310$, then the model produces $y = 3.10$. If you state that $x = 50$, then the model produces $y = 0.50$. This will be true of all models described in this book—they all produce data. This concept, *model produces data*, may be obvious and simple for this example involving driving time, but it is perhaps the most difficult thing to understand when considering statistical models.

Of course, the model $y = x/100$ doesn't produce the data all by itself, it requires someone or something to do the calculations. It will not matter who or what produces the data; the important thing is that the model is a *recipe* that can be used to produce data. In the same way that a recipe for making a chocolate cake does not actually produce the cake, the mathematical model itself does not actually produce the data. Someone or something must carry out the instructions of the recipe to produce the actual cake; likewise, someone or something must carry out the instructions of the model to produce the actual data. But as long as the instructions are carried out correctly, the result will be the chocolate cake, no matter who or what executes the instructions. So you may say that the cake recipe produces the cake, and by the same logic, you may also say that the model produces the data.

A **statistical model** is also a *recipe* for *producing* data. Statistics students usually think, incorrectly, that the data produce the model, and this misconception is what makes statistics a "difficult" subject. The subject is much easier once you come to understand the concept *model produces data*, which throughout this book is an abbreviated phrase for the longer and less catchy phrase, "the model is a recipe for producing data." You can use data to *estimate* models, but that does not change the fact that your model comes first, before you ever see any data. Just like the model $y = x/100$, a statistical model describes how Nature works and how the data from Nature will appear. Nature is already there before you sample any data, and you want your model to mimic Nature. Thus, you will assume that your model produces your data, not the other way around.

A simple example will clarify this fundamental concept, which is absolutely essential for understanding the entire subject of statistics. If you flip a perfectly balanced coin, you think there is a 50% chance that it will land heads up. This is your model for how the data will appear. If you flip the coin 10 times and get 4 heads, would you now think that your coin's Nature has changed so that it will produce 40% heads in the future? Of course not. *Model produces data*. The data do not produce the model.

1.2 Statistical Processes: Nature, Design and Measurement, and Data

Statistical analysis requires data. You might use an experiment, or a survey, or you might query an archived database. Your method of data collection affects your interpretation of the results, but no matter which data collection process you choose, the science of studying Nature via statistics follows the process shown in Figure 1.2.

Notice that Nature produces data but only after humans tap Nature through *design and measurement*.

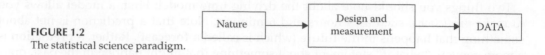

FIGURE 1.2
The statistical science paradigm.

In *confirmatory research*, design and measurement follow your question about Nature. For example, you might have the question, "Does taking vitamin C reduce the length of a cold?" To answer that question you could design a study to obtain **primary data** that specifically addresses that question. In *exploratory research*, by contrast, your question of interest comes to mind after you examine the data that were collected for some other purpose. For example, in a survey of people who had a cold recently, perhaps there was a question about daily vitamin intake. After examining that data, the question "Does taking vitamin C reduce the length of a cold?" may come into your mind. Since the survey was not intended to study the effects of vitamin C on the duration of colds, these data are called **secondary data**. Conclusions based on confirmatory research with primary data are more reliable than conclusions based on exploratory research with secondary data. On the other hand, secondary data are readily available, whereas it is time-consuming and costly to obtain primary data.

Both types of analyses—those based on primary data and those based on secondary data—are useful. Science typically progresses through an iterative sequence of exploratory and confirmatory research. For example, after you notice something interesting in your exploratory analysis of secondary data, you can design a new study to confirm or refute the interesting result.

To understand Figure 1.2, keep in mind that the arrows denote a *sequence*: Nature precedes your design and measurement, which in turn precede your DATA. The capital letters in *DATA* are deliberate, meant to indicate that your data have not yet been observed: They are potential observations at this point and are unknown or random. When we discuss data that are already observed, we will use the lowercase spelling *data*. These data are different, because they are fixed, known, and hence nonrandom.

The uppercase versus lowercase distinction (*DATA* versus *data*) will be extremely important throughout this book. Why? Consider the question "Does vitamin C reduce the length of a cold?" If you design a study to find this out, you will collect lowercase "d" data. These data will say something about the effects of vitamin C on the length of a cold *in this particular study*. However, they are not the only data you could possibly have collected, and they cannot describe with absolute perfection the nature of the effect of vitamin C on the length of cold. Your data might be anomalous or incomplete, suggesting conclusions that differ from the reality of Nature. In contrast, the as-yet unobserved DATA include all possible values. Statistical theory is all about generalizing from data (your sample) to the processes that produce the entirety of DATA that could possibly be observed. With proper statistical analyses, you are less likely to be misled by anomalous data.

In other statistics sources, *DATA* refer to a *population*, and *data* refer to a *sample from the population*. If it helps you to understand the DATA/data distinction, go ahead and think this way for now, but we suggest that you avoid the "population" terminology because it is misleading. You will learn much more about this in Chapter 7.

You will never see all the DATA; generally, it is an infinite set of possible outcomes of everything that could possibly happen. On the other hand, you do get to see your data. A main goal of statistical analysis is to use the data that you have observed to say something accurate about the potential DATA that you have not observed.

Definitions of Terms in Figure 1.2

- **Nature** is the real situation. It might refer to a phenomenon in biology, physics, or human societal interactions. It is there whether you collect data or not.
- **Design** is your plan to collect data. Broadly speaking, design involves deciding how you are going to study Nature. You could directly observe the phenomenon of interest, conduct an experiment, or analyze existing measurements contained in a database; the design refers to the methods you will use to collect your data. Think of design as something that happens before you get the actual numbers.
- **Measurement** refers to the *type* and *units* of the data that you will record and use; for example, a measurement could be height in feet, rounded to the nearest inch. The binary "yes" or "no" choices on a questionnaire is another example of a measurement. A measurement can also be a processed number such as the average of responses to questions one through five on a questionnaire, where each response is 1, 2, 3, 4, or 5.
- **DATA** are the potential data that you might observe. At this point, you should visualize a data set that will be in your computer (e.g., in a spreadsheet), but you don't know what the numbers are.

Example 1.2: The Statistical Science Paradigm for Temperature Observation

"How about the weather today?" is a common elevator topic. Suppose you designed a simple study to measure temperature. In this case:

- *Nature* refers to weather.
- *Design* refers to your plan to get the data. For example, you may plan to go outside and look at your thermometer. Or, you may plan to go online and see what a weather website tells you.
- *Measurement* refers to the type and units of the data you will actually collect. If your thermometer measures temperature in Celsius, then the measurement will be temperature in the Celsius scale. Further, if you plan to report the temperature to the nearest degree, the measurement can further be refined to be temperature in the Celsius scale rounded to the nearest integer.
- *DATA* refer to the actual number you will observe, before you have observed it. It could be any value, so you must represent it algebraically as Y (a capital letter). Once you actually observe a temperature—say, 15°C—then that's your lowercase "d" data, $y = 15°C$ (note the lowercase y).

This example would be more interesting if your design were to collect data over 365 consecutive days, in which case your data set would include 365 numbers instead of just one.

Example 1.3: The Statistical Science Paradigm for Presidential Approval Polling

What do people think about the current president? In this case, the elements are as follows:

- *Nature* is public opinion.
- *Design* is your plan to collect the data. This plan should be much more elaborate than in the weather example, Example 1.2. For instance, you may hire a staff of phone interviewers, obtain a list of randomly selected telephone numbers, write a script for the interviewers to explain what they are doing to the people who answer the phone, decide how many times to follow up

TABLE 1.1

A DATA Set

ID	Response
0001	Y_1
0002	Y_2
0003	Y_3
0004	Y_4
...	...

if no one is home, decide on how many people to call, and decide on how many responses to obtain.

- *Measurement* refers to the type of data that will be collected to measure opinion about the president. If you ask the question "Do you like the president?" then the measurement is simply a yes-or-no outcome. This type of measurement is common in statistics and is sometimes called a **binary response**. Or, you might ask respondents a battery of questions about different aspects of the president's performance, on which they rate their preference according to a 1, 2, 3, 4, 5 scale. In this case, the measurement might be average preference using a collection of questionnaire items, sometimes called a **Likert scale**.
- *DATA* refer to the actual numbers that will be in your spreadsheet or other database. For example, in the simple "yes/no" measurement, the data might look like as shown in Table 1.1.

The DATA values are as-yet unknown, so you have to represent them algebraically as Y_i rather than as specific values. Once you observe the data, they become specific data values such as $y_1 =$ "yes," $y_2 =$ "no," $y_3 =$ "no," $y_4 =$ "no," and so on, assuming the measurement is a binary yes-or-no outcome.

Example 1.4: The Statistical Science Paradigm for Luxury Car Sales

How are luxury car sales these days? In an era of expensive gas prices, people tend to shy away from gas-guzzling luxury cars. If you were studying trends at a dealership, the elements might be defined as follows:

- *Nature* is car purchasing behavior.
- *Design* is your plan to collect data. You may plan to contact people in the car industry and request annual sales figures. You will need to define specifically what is meant by a luxury car first.
- *Measurement* refers to the type of data you will record. In this case, that might be annual U.S. sales (in millions of dollars) of luxury cars. Alternatively, you might decide to measure numbers of cars sold (in thousands). Or, you might decide to measure both dollar sales and car count; this would be called a **bivariate measurement**.
- *DATA* refer to the values you will collect. See Table 1.2.

Prior to observation, the DATA are random, unknown, and hence indicated algebraically as Y_i. Once you collect the data, you can replace the uppercase DATA values Y_i with the actual numbers.

TABLE 1.2

Annual Sales DATA for
Luxury Cars

Year	Annual Sales
2000	Y_1
2001	Y_2
2002	Y_3
...	...

1.3 Models

A statistical model is an abstraction of Figure 1.2. It is a simplification that allows you to both explain how Nature works and make predictions about how Nature works. To explain, and make predictions, the process by which data are produced in Figure 1.2 is represented using the model shown in Figure 1.3.

The simplest case of a probability model $p(y)$ is the coin flip model: $p(\text{heads}) = 0.5$ and $p(\text{tails}) = 0.5$. As a data producer, this model produces the outcomes heads or tails randomly, just like coin flips. It can produce as many random coin flips as you want. Isn't that handy! The model $p(y)$ can be an automatic coin flipper!

Your model $p(y)$ substitutes for both Nature and design and measurement shown in Figure 1.2 and states that the mathematical function $p(y)$ produces the data, as shown in Figure 1.3. Your real DATA are produced from Nature, as tapped through your design and measurement. Your probabilistic model $p(y)$ also produces DATA; you will see examples of this repeatedly throughout the book, where we produce DATA* from models $p(y)$ using computer random number generators. When we use the computer to generate DATA, we call the resulting values DATA*, designated with an asterisk*, to distinguish them from real DATA.

Probabilistic models $p(y)$ are usually *wrong* in one way or another, as they are oversimplifications, just like a toy train is an oversimplification of the real train. But the model is *useful* if it is good, meaning that the DATA* it produces look like your real DATA. The more similar your model's DATA* are to Nature's own DATA—as tapped through your design and measurement—the better your model. By analogy, the model train is *good* if it faithfully represents the real train, but the model train is obviously *wrong* in that it is not the real train.

Just as Figure 1.2 shows how Nature's data are produced, the model shown in Figure 1.3 also *produces* data. Note that the term *model* is used in two senses here: First, Figure 1.3 itself is a model for how Nature works, and second, the function $p(y)$ is called a **probability model**. To summarize these two meanings in a single sentence, your *model* for reality is that your DATA come from a *probability model $p(y)$*. So the statement, *model produces data*, is itself a model—your model—for how your DATA will be produced.

The dual meanings of the word *model* are so important; they need a shout out.

The Dual Meanings of the Term "Model"

Your *model* for Nature is that your DATA come from a *probability model $p(y)$*.

FIGURE 1.3
The model for the statistical science paradigm shown in Figure 1.2.

Such a model $p(y)$ can be used to predict and explain Nature. Again, the term *prediction* here does not necessarily refer to predicting the future, which is called *forecasting*. Rather, a prediction is a guess about unknown events in the past, present, or future or about events that may never happen at all. The best way to understand prediction is to think of what-if scenarios: "What if I invest $100,000 in this mutual fund? How much money would I have at the end of the year?" Or, "What if the state had issued 1000 more hunting licenses last year? What would the deer population be today?" These are examples of predictions, covering the past, present, future, or none of the above.

At this point, the meaning of $p(y)$ may be unclear, especially in relation to the example of driving time of Example 1.1, where the model was $f(x)$. The following sections clarify the distinctions between the deterministic model $y = f(x)$ and the probabilistic model, which is represented by the following expression:

$$Y \sim p(y)$$

The symbol $\sim$ can be read aloud either as "produced by" or "distributed as." In a complete sentence, the mathematical shorthand $Y \sim p(y)$ states that your DATA Y are produced by a probability model having mathematical form $p(y)$. The expression $Y \sim p(y)$ is just a shorthand notation for the graphical model shown in Figure 1.3.

1.4 Deterministic Models

A **deterministic model** is a model where an outcome y is completely determined by an input x. It is a mathematical **function** $y = f(x)$ that allows you to make predictions of y based on x. Here, $f(.)$ is used rather than $p(.)$ as shown in Figure 1.3 to underscore the distinction between deterministic and probabilistic models. The driving time model of Example 1.1 is an example of a deterministic model: There, $f(x) = x/100$. This type of model is deterministic because y is completely determined if you know x: Given $x = 310$, there is one and only one possible value for y according to this model, namely, $y = f(310) = 3.10$.

In case you have forgotten, and since we use a lot of functions in this book, here is a refresher on the meaning of a mathematical *function*. A function is a mapping of values of x to values of y such that, given a particular value in a relevant range of x values, there is one and only one resulting y value. For example, $y = x^2$ maps $x = 4$ to only one value, namely, $y = 16$, and is a function of the form $y = f(x)$. But if $y^2 = x$, then $x = 4$ corresponds to two values of y, namely, $y = 2$ and $y = -2$; hence, $y^2 = x$ is not a function of the form $y = f(x)$. A deterministic model $y = f(x)$ states that for a given x, there can be one and only one possible value of y, namely, $y = f(x)$.

Example 1.5: A Deterministic Model for a Widget Manufacturer's Costs

Suppose that y represents the total yearly cost of your business, x represents how many widgets you will make in your company per year, c is your collective fixed cost, and m is your cost to produce each widget. Then a simple model that relates the number of widgets you make to your total cost y comes from the slope–intercept form of a deterministic straight-line model:

$$y = c + mx.$$

You are probably quite familiar with deterministic models like the slope–intercept equation earlier from your previous math courses. These models are often useful for describing basic relationships between quantities. However, these models have a major weakness in that they do not explicitly account for variability in Nature that we see and experience in every second of our lives. Because there is variability in the real world, deterministic models are obviously wrong. They tell you that things are perfectly predictable, with no variation. While probabilistic models are not exactly correct either, they are more realistic than deterministic models because they produce data that vary from one instance to another, just as you see in Nature. Deterministic models, on the other hand, produce data with no variability whatsoever.

As side note, if you happen to have read something about *chaos theory*, then you know that there are deterministic models that look a lot like probabilistic models. Go ahead and have a look—there are plenty of fun things to discuss about chaos theory, probability, determinism, and free will, perhaps with your colleague Hans while enjoying a pint at the local pub!

1.5 Variability

Do you eat precisely the same food every day? Shop at exactly the same stores? Arrive at work at precisely the same instant? Does Hans prefer the same brand of toothpaste as Claudia? If you saw somebody jump off a cliff, would you do it too? The answer to all these questions is, of course, a resounding "No!" And aren't we lucky! If everything were the same as everything else, imagine what a dull world this would be.

Variability is everywhere. Every time you drive 310 km, it takes a different amount of time. Every day the stock markets go up or down, different from the day before. Hans does not buy the same toothpaste as Claudia. Everybody lives to a different age. If you roll a die 10 times, you won't get the same result every time. One spoonful of vegetable soup is not identical to another spoonful. Variability is so real, you can taste it!

Deterministic models are obviously wrong because the data they produce do not exhibit variability. Every time you plug in $x = 310$ in the equation $y = x/100$, you will always get $y = 3.10$. Try it a few times: Plug $x = 310$ into the equation $y = x/100$ and calculate. Repeat, repeat, repeat. Do you ever get a different y?

You must use **probabilistic** (or **stochastic**) **models** to account for natural variability. In Example 1.1, your actual driving time Y is variable, because your average speed changes depending on variables like road conditions, city versus highway driving, your attitude about speeding, and on your need for bathroom breaks! Thus, your driving time Y is not precisely equal to $x/100$; rather, it deviates from $x/100$ by a variable amount.

Are deterministic models ever true? Perhaps in the physical and engineering sciences? Rarely, if ever! In physics, you will see deterministic models that purport to govern the physical universe, but these models have idealized assumptions that are not precisely true, leading to actual outcomes that vary from the model's predictions. For example, the models used by NASA to predict the location where a Martian rover will land will be wrong every time (although not by much), because of numerous uncontrollable factors. Further, such deterministic models often break down completely at the quantum level, where variability and randomness take over. Finally, experimental validations of physical models of the universe result in measurements that vary from experiment to experiment, again requiring statistical and probabilistic models to analyze the data.

For an engineering example, consider the maximum stress that a dam can bear. This value cannot be predicted perfectly. It depends on many unknown variables, such as the type and preparation of concrete down to the atomic level, the exact quality of the construction, and the behavior of the dam in its environment in real time. It is impossible to characterize this information so completely as to arrive at a deterministic prediction of the maximum stress level that the dam can bear at any given moment. Neither the physicist nor the engineer can tell you precisely what will happen, despite all of their wonderful deterministic mathematical models.

But, truth be told, deterministic models are often at least approximately correct in the physical and engineering sciences and give reasonably accurate predictions in spite of their failings. This happens when the variability (called *noise* in their jargon) is tiny relative to the deterministic component (called the *signal*). Thus, while deterministic models are wrong, they can still be useful.

In the social sciences, on the other hand, deterministic models are usually just plain silly. Can you precisely determine tomorrow's Dow Jones Industrial Average? No. Can you precisely determine how much money Jae Hwa will spend on a particular trip to the market? No. Can you precisely determine what answer Alejandra will enter on a survey, when asked about her coffee preference? No.

Nor are relationships deterministic in the biological and medical sciences. Can you precisely determine whether a diseased patient will survive 5 years? No. Can you precisely determine how many deer will be born in a drought season? No. Can you precisely determine whether a child born of two brown-eyed parents having recessive blue eye color genes will have blue eyes? No. On the other hand, you can predict all of these things very well by using probability models, but only in an aggregate sense—not individually.

In summary, you need probabilistic models in all areas of science. Deterministic models are obviously wrong because the data they produce lack variability, unlike the real data that you see. If the variability is tiny relative to the deterministic component, then you can still use the deterministic model; otherwise, you should use a model that includes a probabilistic component if you want realistic predictions.

Probabilistic models assign likelihoods to the outcomes of interest, rather than assigning a determination that a certain outcome will occur with 100% certainty. And while 100% certain determinations can be more comforting, likelihoods are more realistic and are quite useful for making decisions. For example, if 95% of stage II ovarian cancer patients survive 5 years when given therapy A, and if only 80% of them survive 5 years when given therapy B, then, all other things being equal, you would choose therapy A for treatment. This does not mean that, in hypothetical worlds (also called *counterfactual* worlds) where you could somehow play out your potential futures using either therapy, you would always live longer with therapy A. What it does mean is that you have a better *chance* of living 5 years with therapy A. In these counterfactual worlds, you estimate that in 95% of them you would live 5 years or more with therapy A, while in only 80% of them you would live 5 years or more using therapy B. You decide: Do you want A or B?

You might find probability models challenging because they have a strong conceptual component. Just look at the previous paragraph: While the choice of therapy A seems obvious, the rationale for preferring therapy A involves potential, counterfactual worlds and is therefore quite conceptual. Guess what: Statistics and probability require imagination! You probably didn't think you would have to use your imagination in a statistics class, did you?

Most students learn about Nature in a categorical style that emphasizes "right" and "wrong." For instance, your teachers may have said things like "*Answer A is wrong,*

Answer B is right, *Answer C* is wrong, and *Answer D* is wrong. Therefore, you should fill in the bubble for *Answer B*." Categorical thinking is so natural, fundamental, and well rehearsed for people (and probably animals) that probabilistic thinking may seem unnatural in comparison. Indeed, probabilistic investigation as a science is much more recently developed in human history. It is not well rehearsed in daily life and must be learned through a course of study such as you will find in this book. Although deterministic models can be used in an attempt to assign absolute truths, such as "If I drive 310 km at 100 km/hour, then it will take me precisely 3.10 hours to reach my destination," these kinds of determinations are in fact 100% false! You will *never* arrive *precisely* 3.10 hours later, at least when time is measured precisely, say, by using a stopwatch. Probabilistic models are much more realistic, giving you predictions such as "If I drive 310 km, I will arrive in less than 3.50 hours 90% of the time."

Before discussing probability models more formally, we must introduce the concept of a *parameter*, a concept that applies to both deterministic and probabilistic models.

1.6 Parameters

Whether deterministic or probabilistic, models have parameters that govern their performance. A **parameter** is a numerical characteristic of the data-generating process, one that is usually unknown but often can be estimated using data. For example, suppose you don't know the sales tax rate. A model for the amount you pay is as follows:

$$y = \text{Round}\{(1 + \rho)x\}$$

Here, the variable x is the price of the object before the tax.

The variable ρ is the Greek lowercase letter rho, pronounced "row," and is a parameter of the model; you can estimate it using transaction data. Here and throughout the book, unknown parameters are denoted by Greek letters such as ρ, θ, μ, σ, β, λ, δ, and π.

Note that, even though ρ is unknown, this model still produces the data y. This is the ordinary situation: Models produce data, but they have unknown parameters.

Much of the statistics you will do or see involves **statistical inference**. Statistical inference is the science of using data—produced by Nature as tapped through design and measurement—together with assumptions about the data-generating process (which this book covers), to make defensible conclusions about Nature. The probability that is present in statistical models comprises an essential component of statistical inference, as it allows you to quantify the effects of chance variability on your data and thereby separate the real conclusions from those that are explainable by chance alone. We will devote considerable time in the chapters ahead to deciding whether your statistical results are explainable by chance alone.

Here, we come to the main Mantra that will be repeated throughout this book, again and again. Memorize it now!

Mantra #1:

Model produces data.
Model has unknown parameters.
Data reduce the uncertainty about the unknown parameters.

Here and throughout the book, the Greek lowercase letter theta, θ, denotes a **generic parameter**. Thus, θ could represent a tax rate, a slope, an intercept, or another quantity, depending on the specific application. In our Mantra, there may be more than one parameter (e.g., mean and standard deviation) yet we still call the parameters, collectively, "θ." In the case where θ is comprised of a list of values, it is called a parameter **vector**. It will be clear from context whether θ refers to a single parameter or a list of parameters.

While it may seem abstract to use algebraic symbols like θ to denote parameters, there really is no other way, because *model has unknown parameters*. You can reduce your uncertainty about the values of these parameters, but you cannot eliminate your uncertainty outright. Instead, you need to use probabilistic analysis: You can never make a claim like "The parameter is certainly equal to 4.5," but you will be able to state something like "The parameter is most likely between 4.3 and 4.7." *Data reduce the uncertainty about the unknown parameters.* Data do not eliminate the uncertainty about the unknown parameters.

1.7 Purely Probabilistic Statistical Models

A purely probabilistic statistical model states that a variable quantity Y is generated at random. This statement is represented mathematically as $Y \sim p(y)$, where $p(y)$ is a **probability distribution function (pdf)**, a function that assigns relative likelihoods $p(y)$ to the different observable values of the data y. The function $p(y)$ tells you what kind of numbers you will see: If $p(y_1)$ is large, you will see relatively many values of Y near y_1; if $p(y_2)$ is small, you will see relatively few values of Y near y_2. In Figure 1.4, a model for time it takes to drive 310 km, you can see an example: When the function $p(y)$ is large, for example, when $y = 3.1$, many of your driving times are near $y = 3.1$. When $p(y)$ is small, for example, when $y = 3.2$, few of your driving times are near $y = 3.2$.

NOTE: The expression $Y \sim p(y)$ is quite different from the equation $Y = p(y)$. The expression $Y \sim p(y)$ states that Y is *produced by the function $p(y)$*, while the expression $Y = p(y)$ states that

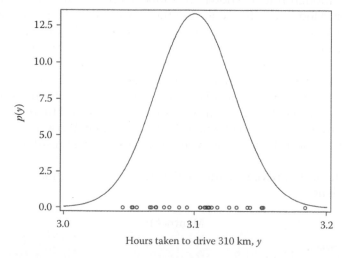

Hours taken to drive 310 km, y

FIGURE 1.4
A model that produces data ($p(y)$, solid curve), and a sample of data produced by that model (circles).

Y is *equal to the function p(y)*. In Figure 1.4, the data values of Y appear on the horizontal axis, while the function $p(y)$ is the curve. In this book, you will *never* see the expression $Y = p(y)$.

The parameters of the model are never known, because *model has unknown parameters.* Figure 1.4 shows a precise curve $p(y)$ that produces Y, but in practice you will never know this precise curve. A main goal of statistical analysis is to estimate the parameters that make this curve what it is (*data reduce the uncertainty about the unknown parameters*). A statistical model, then, is a statement that your data are produced by a model with unknown parameters. In the purely probabilistic case, the definition is as follows:

Definition of a Purely Probabilistic Statistical Model

A purely probabilistic statistical model states that a variable Y is produced by a pdf having unknown parameters. In symbolic shorthand, the model is given as $Y \sim p(y|\theta)$.

Note the distinction between Y and y in the expression $Y \sim p(y|\theta)$. Capital Y refers to a single random outcome, and lower case y refers to fixed realization of Y. Earlier in the discussion, we referred to DATA in uppercase letters, and this upper case Y is equivalent to DATA, because it refers to the case where the data are not yet observed. The circles on the horizontal axis of Figure 1.4 are observed and therefore constitute lowercase "d" data.

This distinction between uppercase "D" and lowercase "d" is extremely important for your understanding of probability models and for your understanding of how to think in probabilistic terms. But what *is* probability, anyway? You often see percentages used to communicate probabilities, and this is indeed a good way to think about them. If the probability of A is 40%, then in (roughly) 40 out of 100 instances, A will occur, and in the other 60 instances, A will not occur. For example, if the probability of a die showing 1 is 1/6, or 17%, then in (roughly) 17 out of 100 rolls of the die, you will see a 1. You can also see this using the computer and a spreadsheet program: If you produce DATA* from a model where $p(1) = 1/6$, then roughly 17 out of 100 Y*s will have the value 1.

As with any model, the probability model is a *mental* conception. With the die, you *imagine* that about 17 out of 100 rolls will produce a 1, but this is only based on your mental assumption that the die is fair. What if the die is a trick die, with no 1 on it? Or what if the die is loaded so that the 1 rarely comes up? Then your mental model is wrong. A more believable mental model would be one that states that the probability of seeing a 1 is an unknown parameter θ. You can never know the precise numerical value of this parameter θ, but you can estimate it using data (*data reduce the uncertainty about the unknown parameters*).

In some cases, the 100 instances (e.g., rolls of the die) that you can use to understand probability are completely in your mind, as opposed to being real-world actions such as physically rolling the die. For example, what is the probability that the best football team in the league will beat the worst one in tomorrow's game? Here, the 100 instances would have to be repeated plays of the game under identical circumstances, much like rolls of a die. But it is impossible to play the game over and over with exactly the same people, weather, fan support, etc. Instead, you have to imagine *potential futures*: In 100 *potential future* plays of the game that you can imagine, how many times will the best team win? The number of wins in 100 potential futures depends on your personal judgment. So, what do you think? You might understand what you think a little better by putting some money on the line! If you are willing to bet 1 (dollar, euro, pound, etc.) in hopes of winning 10 (so your net earnings is $10 - 1 = 9$), then you think the probability is 10% that the underdog will win: In 10 out of 100 of your potential futures, you will net 9, for a total of 90 won, and in the remaining 90 out of 100 of your potential futures, you will lose 1, for a total of 90 lost. Thus, over all the potential futures that you can imagine, you will come out even. This type of balancing of payouts is the way that professional oddsmakers assign probabilities.

A probability model $p(y)$ does not care whether the 100 instances correspond to physical or mental reality. It's just a model for how the future data will appear, no matter whether the futures are potential or actual. Either way, $p(y)$ will produce data for you when you use the computer—for example, you can use the computer to play out a future football game repeatedly under identical conditions, getting different outcomes from one potential future to the next.

The probability model allows you to make what-if predictions as to the value of Y, but unlike the deterministic model, it does not presume to know what the precise value of Y will be. For example, in the car driving time example, a probability model would not produce $y = 3.10$ (hours) when $x = 310$ (km); rather, it would produce random values in a neighborhood of 3.10, such as 3.09, 3.14, 3.14, 3.08, 3.19, 3.13, 3.12, …, as shown in Figure 1.4. This model is much more realistic than the deterministic model, because in repeated driving of the 310 km distance, your driving times will vary similarly.

Example 1.6: A Probability Model for Car Color Choice

Suppose you wish to predict whether the next customer will buy a red car, a gray car, or a green car. The possible values of Y are *red, gray,* and *green,* and the distribution $p(y)$ might have the form shown in Table 1.3.

Probability distributions are best understood using graphs. Figure 1.5 shows a **needle plot** of the distribution. A **bar chart**, possibly called a *column chart* by your spreadsheet software, is another similar, commonly used graph to depict a probability distribution.

In Figure 1.5, the vertical lines (the "needles") are in place to make the graph easier to read and are not technically part of the function $p(y)$. The pdf could have simply been depicted using only the solid dots on top of the lines, with no vertical lines.

The model of Table 1.3 does not tell you precisely what the next customer will do; the model simply says it is random: Y could be red, gray, or green. However, the model does allow you to make aggregate what-if predictions as follows: "If I sold cars to the next 100 customers, then about 35 of them would buy a red car, 40 would buy a gray car, and 25 of them would buy a green car." You should say *"about* 35" because the actual number is unknown. However, the *law of large numbers*, covered in Chapter 8, states that the sample proportion from actual data gets closer to the true probability from the model as the sample size increases.

TABLE 1.3

Probability Distribution
of Color Choice

Color Choice, y	$p(y)$
Red	0.35
Gray	0.40
Green	0.25
Total	1.00

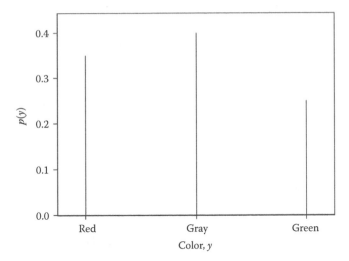

FIGURE 1.5
Graph (needle plot) of the probability distribution of Y.

Again, this prediction does not necessarily concern an event in the past, future, or even the present. It is simply a hypothetical, what-if statement about what would be likely to happen in a given scenario.

This model contains a very consequential assumption about reality: It assumes that only three possible choices of car color are possible. This implies that, in the universe described by this model, no customer will ever choose blue, orange, brown, white, or any other color for their car. Is this a good **assumption**? It might be, if you model the sales of cars at a dealership that sells only red, gray, and green cars. But it is a bad assumption if you model a dealership that offers more variety and a completely useless one if you model a dealership that only sells black and white cars.

This is a good time to restate the basic concept for evaluating the quality of a model. A model is good if the data it produces (recall that this is denoted as DATA* when generated by a computer) look like the data (denoted as DATA previously) produced by Nature. In this example, if you go through the sales records and notice that a brown car was sold on Thursday, then you would question the usefulness of the model, because the only DATA* you get from the model will be red, gray, and green.

The model of Table 1.3 is also bad if a sample of sales records data show drastically different percentages for the various colors, such as 10%, 10%, and 80%, rather than the 35%, 40%, and 25% anticipated by your model. This is a distinction between probability models and statistical models. Probability models assume specific values of the parameters, statistical models do not. Instead, in statistical models, the probabilities are always unknown parameters. (*Model has unknown parameters.*) This makes statistical models more believable in that the probabilities could be any numbers. If you think about it, how could you possibly know what the percentages in Table 1.3 really are? You can't. These percentages are always algebraic unknowns; we'll call them θ_1, θ_2, and θ_3 rather than 35%, 40%, and 25%; or 10%, 10%, and 80%; or anything else. It is believable that the true percentages are *some* numbers θ_1, θ_2, and θ_3, but it is not believable that the percentages are specific values like 35%, 40%, and 25%.

Thus, the requirement for a good statistical model is that the DATA* produced by the model look like the actual DATA for *some settings of the parameters.* You do not have to know what those parameter values are.

1.8 Statistical Models with Both Deterministic and Probabilistic Components

The model with both deterministic and probabilistic components is a **regression model**, which is a model for how the distributions of Y change for different X values. The regression model is represented as follows:

$$Y \sim p(y \mid x)$$

The symbol $p(y|x)$ is read aloud as "the probability distribution of Y given a particular X." The symbol $|$ is shorthand for "given" or "given that." The model $Y \sim p(y|x)$ reads, in words, as follows:

> For a given $X = x$, Y is generated at random from a probability distribution whose mathematical form is $p(y|x)$.

While more cumbersome, the following notation is a more specific and more correct shorthand to represent the regression model:

$$Y \mid X = x \sim p(y \mid x)$$

In the example with Y = driving time and X = distance, the model states that "For a given distance $X = x$, driving time Y is generated at random from a probability distribution that depends on $X = x$, whose mathematical form is $p(y|x)$." In other words, there is a different distribution of possible driving times when $X = 100$ km than when $X = 310$ km (shown in Figure 1.4). This makes sense: While the relationship between Y and X is not deterministic, it is certainly the case that the time Y will tend to be much longer when $X = 310$ km than when $X = 100$ km; hence, the distributions of Y differ for these two cases.

In the regression case, the parameters of the model are also never known. Hence, the definition of the statistical model is as follows:

Definition of Statistical Model with Both Deterministic and Probabilistic Components

> This model states that, given the value of a variable $X = x$, a variable Y is produced by a pdf that depends on x and on unknown parameters. In symbolic shorthand, the model is given as $Y \mid X = x \sim p(y|x, \theta)$.

This model also allows you to make what-if predictions as to the value of Y. Like the deterministic model, these predictions will depend on the specific value of X. However, since it is also a probabilistic model, it does not allow you to say precisely what the value of Y will be; as shown in the previous example, probabilistic models only allow you to make what-if predictions in the aggregate.

Take the car example previously. If X = age of customer, then the distribution of color preference will depend on X. For example, when $X = 20$ years, your distribution might be as shown in Table 1.4 and graphed in Figure 1.6.

But when $X = 60$ years, your distribution might be as shown in Table 1.5 and graphed in Figure 1.7.

The model does not tell you precisely what the next customer will do, but does allow aggregate what-if predictions of the following type: "If I sold cars to the next 100

TABLE 1.4

Probability Distribution of Color
Choice for 20-Year-Old Customers

y	$p(y \mid X = 20)$
Red	0.50
Gray	0.20
Green	0.30
Total	1.00

TABLE 1.5

Probability Distribution of
Color Choice for 60-Year-Old
Customers

y	$p(y \mid X = 60)$
Red	0.20
Gray	0.40
Green	0.40
Total	1.00

20-year-old customers, then about 50 would buy a red car, 20 would buy a gray car, and 30 would buy a green car." Similarly, you can say "If I sold cars to the next 100 60-year-old customers, then about 20 would buy a red car, 40 would buy a gray car, and 40 would buy a green car."

There are so many models to learn—probabilistic, deterministic, and the combination of the two. But really, it's easier than you might think: Just memorize the combination model of this section. The purely probabilistic model is a special case of it, one where the distribution of Y does not depend on X. And the deterministic models that the physicists and engineers use so much are also special cases, ones where the distributions have no variability.

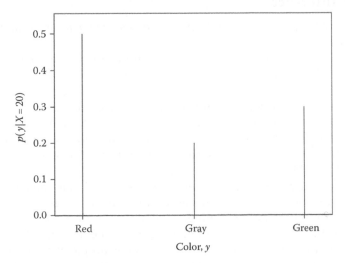

FIGURE 1.6
Graph of the probability distribution of Y when $X = 20$ years.

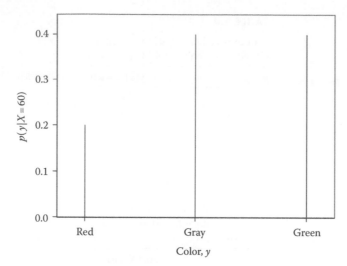

FIGURE 1.7
Graph of the probability distribution of Y when $X = 60$ years.

For example, the deterministic driving time model states that when $x = 310$, then y is equal to 3.10 with 100% probability, and there is no variability. For another deterministic example, a model that states that 100% of 60-year-olds buy gray cars also has no variability in the distribution. And these models are obviously incorrect, like all deterministic models! No matter which models you like, they are all special cases of the combination model. Thus, you can say that the model with both deterministic and probabilistic components is the *mother of all models*.

1.9 Statistical Inference

Recall the Mantra: *Model produces data. Model has unknown parameters. Data reduce the uncertainty about the unknown parameters.*

How does this work? The simple coin toss thought exercise provides the entire mental framework you need to understand even the most complex of statistical models. So if you toss a coin, it's a 50% heads and 50% tails, right? Wrong! It's close to 50–50, but due to slight imperfections in every coin—resulting in one side being just a tad heavier than the other—the probability of "heads" is not exactly 0.500000000000000000000000000000000000 000000000000000000000000000000000000000000... with the zeroes continuing ad infinitum, but instead something slightly different, such as 0.500000000000000000000000000000000000 000000000000000032021324534222234200788...

For all intents and purposes, this number is so close to 0.5 that you could go ahead and assume 0.5, but while highly accurate, this assumption is not actually true.

Let's make the example more interesting. Take a pair of pliers and bend the coin. Now what is the probability of heads? You can no longer assume it's 0.5. All you can say is that it is simply π, some unknown probability. This π is an example of an unknown parameter in the Mantra *Model has unknown parameters* and is also an example of the generic θ mentioned previously. The model that produces the data is therefore as given in Table 1.6.

TABLE 1.6

Probability Distribution
for a Bent Coin

Outcome, y	$p(y)$
Tails	$1-\pi$
Heads	π
Total	1.00

We apologize for the abuse of notation here, as the Greek letter π is more commonly used as the famous trigonometric constant $\pi = 3.14159\ldots$. We will use π that way later when we discuss the normal pdf—the famous bell curve graphed in Figure 1.4. Meanwhile, in this coin toss example, π is simply a number between 0 and 1, the unknown probability of getting heads when you flip the bent coin.

How can you learn about this model? The simple answer is "Collect some data!" (*Data reduce the uncertainty about the unknown parameters.*) Flip the bent coin many times, and count how many tosses turn up heads. If the proportion is 3 out of 10, or 30%, you now have a better idea about π: It is somewhere near 0.30. Your uncertainty about the unknown parameter π is reduced when you have data.

However, you are still uncertain: The true π is not 0.30; it is still the same unknown value that it was before. By analogy, if you flip a fair coin 10 times and get three heads, you shouldn't think the probability is 0.30; you should still think it is 0.50 (or darn close to 0.50 as discussed previously).

The model still produces the data. The data do not produce the model. If you think the data produce the model, then you would think, based on 10 flips and three heads, that suddenly the coin's Nature has changed so that it now will give heads in 30% of the subsequent flips. The true π is not 0.30 anymore than it is 0.50 for a fair coin; it is still the same unknown value that it was before. The data you have collected only suggest that the probability of getting heads is *near* 0.30, not that it is *equal to* 0.30.

Now, how to apply the lowly coin toss example to something that resembles typical research? Simple. Refer to Table 1.3, the example of car color choice. The statistical model looks, in reality, as shown in Table 1.7.

Here, the numbers π_1, π_2, and π_3 are the unknown model parameters, again an example of a generic parameter vector $\theta = (\pi_1, \pi_2, \pi_3)$. The model is good in that DATA* produced by the model will look like the DATA that you actually see, *for some settings of the parameter* $\theta = (\pi_1, \pi_2, \pi_3)$. You do not have to know what the parameter values are to know that the model is good.

By collecting data, you can easily reduce your uncertainty about the parameters π_1, π_2, and π_3, although you can never determine them precisely.

TABLE 1.7

Probability Distribution
of Color Choice

y	$p(y)$
Red	π_1
Gray	π_2
Green	π_3
Total	1.00

Model produces data. Data do not produce the model. Instead, data reduce your uncertainty about the unknown model parameters. The reduction in uncertainty about model parameters that you achieve when you collect data is called **statistical inference**.

A note on notation: While Table 1.7 shows the model as $p(y)$, we sometimes represent the model as $p(y|\theta)$ to emphasize that the model depends on the unknown parameter(s) θ. Usually, $p(y|\theta)$ is the more correct notation. We often use the notation $p(y)$ rather than $p(y|\theta)$, just for the sake of simplicity.

1.10 Good and Bad Models

Compare Figures 1.2 and 1.3. The model of Figure 1.3 is "good" if, for some parameter settings, the DATA* produced by the model "look like" the DATA that you see in reality (Figure 1.2). But why the quotes around the words *look like*? What does that mean, specifically?

To answer, make Figure 1.3 specific to the coin toss case. Also, assume a fair coin whose probability is exactly 0.5. (This example is hypothetical, since such a coin does not exist!) A model for this process is $Y \sim p(y)$, where Y can be either *heads* or *tails* and where $p(y)$ is given as in Table 1.8.

This distribution is closely related to a special distribution called the **Bernoulli distribution**. (In Chapter 2, we cover this distribution and others in more detail.) The Bernoulli distribution produces 0s and 1s instead of heads and tails, but you can easily recode a 0 as tails and 1 as heads to arrive at the distribution in Table 1.8. You can do this in Microsoft Excel, after adding in the Data Analysis toolpack. Once you select "Random Number Generation" from the "Data Analysis" menu, the screenshot should look something like Figure 1.8.

Click OK and the result looks as shown in Figure 1.9. (Note that your numbers may differ due to randomness.)

You can recode the zeroes and ones to tails and heads as shown in Figure 1.10.

So the result is the sequence of heads, tails, heads, heads, tails, tails, heads, heads, heads, and tails. This is an example of the DATA* that can be produced by computer random number generators. However, since the data are in hand, as opposed to being in a potential future, you should call them data* instead of DATA*.

Figure 1.10 shows an example of DATA* for the coin toss case. But what do the actual DATA look like? Well, the DATA are what you would get if you actually flipped the coin 10 times. For example, a real sequence of coin flips—from actual coin tossing, not from computer generation—might be heads, heads, tails, tails, tails, tails, tails, heads, heads, and tails. This is an example of what is meant by DATA, and since these values are now in hand (as opposed to being in a potential future), you should call them data.

TABLE 1.8

Probability Distribution
for a Fair Coin

y	$p(y)$
Tails	0.5
Heads	0.5
Total	1.0

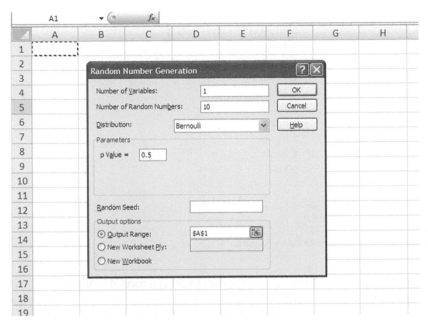

FIGURE 1.8
Generating Bernoulli random numbers using Microsoft Excel®.

	G8		f_x	
	A	B	C	D
1	1			
2	0			
3	1			
4	1			
5	0			
6	0			
7	1			
8	1			
9	1			
10	0			
11				

FIGURE 1.9
A sample of $n = 10$ observations produced by the Bernoulli(0.5) distribution.

The model is good if the DATA* produced by the model look like the real DATA. But if you compare the two actual sequences of computer-generated data* and the actual coin-tossed data, they won't match, flip for flip. So is the model still "good"? Yes! In fact it is an excellent model.

Definition of a Good Model

A model is **good** if:

    a. For some parameter settings, the *set* of possible outcomes produced by the model well matches the set of possible outcomes produced by Nature, design, and measurement.

    b. For some parameter settings, the *frequencies of occurrences* of the specific outcomes, as well as successive combinations of outcomes, well match the *frequencies of occurrences* of the specific outcomes and successive combinations of outcomes produced by Nature, design, and measurement.

	A	B	C	D	E
1	1	=IF(A1=1,"Heads", "Tails")			
2	0	Tails			
3	1	Heads			
4	1	Heads			
5	0	Tails			
6	0	Tails			
7	1	Heads			
8	1	Heads			
9	1	Heads			
10	0	Tails			
11					
12					

FIGURE 1.10
Recoding the Bernoulli data to create coin toss data.

The Bernoulli(0.5) random number generation in Excel is a model that passes on both counts: (a) the set of possible outcomes is {heads, tails}, exactly the same as that in Nature, and (b) the frequencies of occurrences are reasonably similar—both near 50% heads. Note that with more data and data*, these frequencies can be ascertained better; in the previous example, there are simply not enough data to make a firm judgment. Nevertheless, the model does a very good job of meeting our criteria (a) and (b) for a good model, and it doesn't really matter that it consistently fails to produce exactly the same sequence of heads and tails that you would get if you manually tossed the coins. It would actually be kind of creepy if that happened, right?

What does a "bad" model look like? Here are two examples of "bad" models for the coin toss process.

Bad Model #1: For toss i, where $i = 1, 2, 3, \ldots$, the outcome is heads if i is odd and tails if i is even. The sequence is thus alternately heads, tails, heads, tails, heads, and so forth. This model seems okay at first: The set of values produced is {Heads, Tails}, just like in Nature, and the frequency of heads is 0.5 as it should be. Where it fails is in the frequencies of occurrences of *successive* outcomes. The successive outcome "heads followed by heads" is impossible with this model, but very frequent in reality: In 25% of adjacent flips, both will be heads.

Bad Model #2: The Bernoulli distribution does a very good job of modeling coin flips. What about another distribution? The *normal* distribution is the most commonly assumed distribution in all of statistics. How does it work here? You can use the normal random number generator and produce some values as shown in Figure 1.11.

Figure 1.12 shows a sample from the normal distribution. Your numbers may vary due to randomness.

The numbers shown in Figure 1.12 are another example of DATA*—that is, data produced by a computer's random number generator. Is this model good? Clearly not since the set of outcomes produced consists of numbers filling a continuum between approximately −3 and +3, which do not at all match the discrete, whole integer outcomes {0, 1}.

Figure 1.4 shows another example of the normal distribution. It has a famous "bell curve" shape, producing DATA* values in the middle of the curve more often and DATA* at the extremes less often.

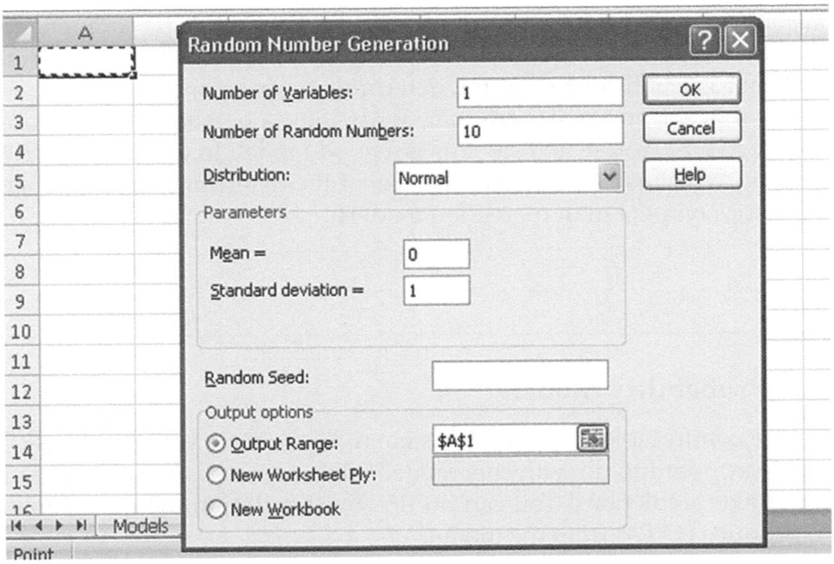

FIGURE 1.11
Generating values from a normal distribution.

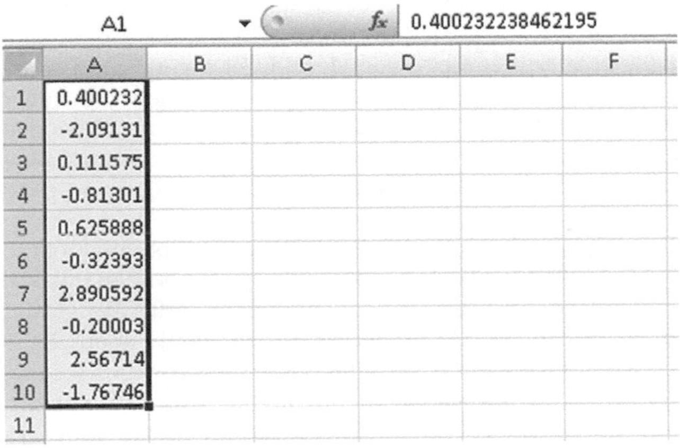

FIGURE 1.12
A sample of data* produced by a normal distribution.

   You do not have to know the model's parameter values to know that it is a good model. That is a relief, because *model has unknown parameters* anyway. For example, a good model for a bent coin is the Bernoulli(π) model, since the Bernoulli(π) model produces 0s and 1s that look like the bent coin results (heads = 1, tails = 0) for some values of π between 0 and 1. For example, you could specify the parameter settings π = 0.20, 0.25, and 0.30 and have the computer produce Bernoulli data for each of these settings. The resulting DATA* would look like the results of flipping a coin with a particular kind of bend. Thus, the criterion for a model being good is that, for *some parameter settings*, the DATA* produced by the model look like the DATA that are actually observed. You don't have to know the actual parameter values; that's what you use the DATA for: *data reduce the uncertainty about the unknown parameters.*

In cases where the model has both deterministic and probabilistic components, there is an additional criterion that is sometimes used: A model may be called "good" if the probabilistic component is small relative to the deterministic component. Again, imagine you have a model that is able to predict precisely what color a person would choose for their car with 100% certainty: You would say this is a good model! In Chapter 17, we define the *R-squared* statistic, which is a measure of the size of the deterministic component relative to the probabilistic component in regression models.

1.11 Uses of Probability Models

Suppose you are comfortable that a model is good. "So what?" you should ask. "What in the world am I supposed to do with this model?" The answer is simple and very important: You can make predictions! You can do this by **simulation**, which means using the computer to produce DATA* from the model.

Example 1.7: Estimating the Probability of Getting 50% Heads in 10 Flips

If you flip a coin 10 times, you should get heads 5 times, right? Wrong! To test this, you could flip coins over and over again, generating DATA, and note how often you get 5 heads out of 10 flips. If you repeated the process 1000 times—more than 1000 would be even better for greater accuracy—you should get a very good estimate. But that would be tedious! Instead, you can let the computer do the work, generating DATA* instead of DATA: Create 10 columns and 1000 rows of Bernoulli values with $p = 0.5$ to simulate 1000 instances of flipping the coin 10 times. Then count how many of the rows, out of 1000, yield exactly 5 heads. Figure 1.13 shows how to generate the data in Excel.

Figure 1.14 shows the tallying of the number of heads in Column K of the spreadsheet. The command " = COUNTIF(K:K, " = 5")/1000 counts how many of the 1000 samples have exactly 5 heads and divides that number by 1000. We got 0.228, but your number may differ slightly due to randomness. Thus, only about 22.8% of the time will you get exactly 5 heads out of 10 flips. The true probability can be calculated here to be 24.6% using the *binomial distribution*, which we will not discuss. As this example shows, simulation provides an excellent and useful approximation to the true probability.

It may be surprising that, even with a fair coin, the probability of getting exactly 50% heads is somewhat low—since the probability is only 24.6%, most of the time you will *not* get 50% heads. Challenge question: What is the probability of seeing 50% heads when there are 100 flips?

Example 1.8: Choosing an Optimal Trading Strategy

Who cares about flipping coins? Let's *earn* some coins! You have heard the phrase, "Buy low, sell high," right? Makes sense! Suppose you buy shares of stock in a company that looks promising. When do you sell? When do you buy? Suppose you debate two strategies for buying and selling this particular stock over the next 250 trading days (roughly one calendar year).

Strategy 1: Buy and hold.

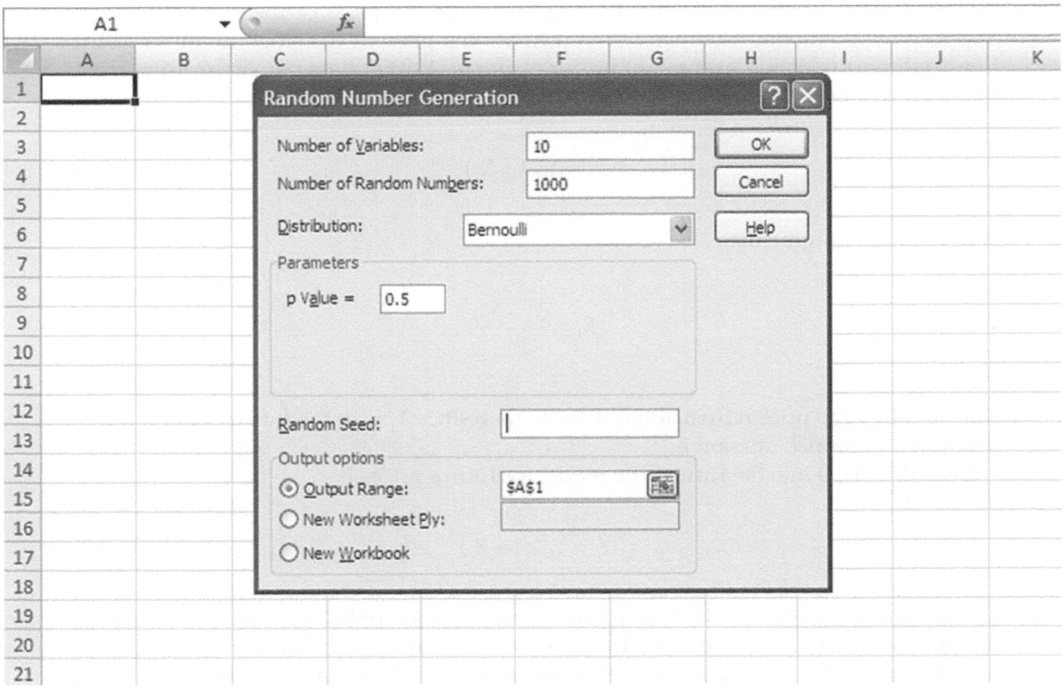

FIGURE 1.13
Generating 1000 samples of $n = 10$ Bernoulli observations per sample.

	A	B	C	D	E	F	G	H	I	J	K
1	0	1	0	1	0	1	0	1	0	1	=SUM(A1:J1)
2	1	1	0	1	1	0	1	0	0	0	5
3	0	1	1	1	0	0	1	0	1	1	6
4	1	0	1	1	0	1	1	1	1	0	7
5	0	1	0	0	0	0	1	1	0	1	4
6	0	0	1	0	1	0	1	0	1	1	5
7	1	0	0	0	1	0	0	1	1	0	4

FIGURE 1.14
Counting the number of heads per sample.

Strategy 2: Sell when there are three consecutive days where the stock price rises. Buy when there are three consecutive days where it drops. Otherwise, hold.

The rationale behind strategy 2 is your gut feeling that "what goes up must come down," and also, in the case of stock prices, "what goes down must come back up."

But is your gut feeling right? Which strategy is better? To determine the best strategy, you can create a realistic model to simulate possible future values of stock prices. You can then simulate 1000 potential futures, each containing the results from 250 consecutive trading days. Then you can try both strategies and compare your earnings after 250 trading days using each strategy. You will have 1000 values of earnings when using strategy 1 and 1000 values of earnings when using strategy 2. Then you can compare the two to find out which works better, on average, and pick the winner.

Here are the mathematical details. Let Y_t denote the price of the stock at time t. Suppose $t = 0$ is today, 6:00 p.m., so you know the price of the stock today from the financial reports. It is Y_0 and might be, for example, 23.32 dollars per share. Tomorrow, the price will be as follows:

$$Y_1 = Y_0(1 + R_1) \tag{1.1}$$

A little algebra shows that

$$R_1 = \frac{Y_1 - Y_0}{Y_0} \tag{1.2}$$

This is called the price **return** at day 1. Note that, since Y_1 is in the future and therefore unknown, R_1 is also unknown.

Extending (1.2) into the future, the model for future prices is

$$Y_t = Y_{t-1}(1 + R_t) \tag{1.3}$$

Thus, you can generate all the future price values Y_t if you only knew the future returns R_t. Of course, you don't know the future R_t values, but financial theory says that they behave remarkably like the coin tosses generated by Excel—except that, instead of being produced by a Bernoulli distribution, the return DATA look more like DATA* produced by a normal distribution, such as the one shown in Figure 1.12.

Figure 1.15 shows how to generate return DATA* and hence possible future trajectories of the stock price.

In Figure 1.15, you will notice that the normal distribution depends on two parameters, the mean and the standard deviation. These are very important statistical parameters and will be described in much greater detail later, starting with Chapter 9.

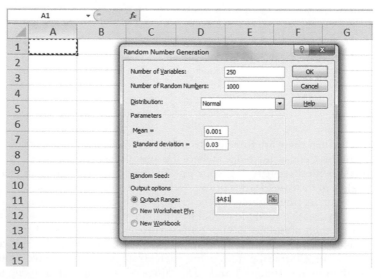

FIGURE 1.15
Generating 1000 potential future return sequences for the next 250 trading days.

For now, just suppose they are parameters that govern the particular normal distribution that you assume to produce your data: If you pick different parameter values, you get a different normal distribution. Here, θ = (mean, standard deviation) is an example of a parameter *vector*.

In Figure 1.15, the mean and standard deviation are set to 0.001 and 0.03, but these are just hypothetical values. Those parameters are never truly known, not by all the Economics Nobel Prize winners, not by billionaire financier Warren Buffett, and not even by the musician Jimmy Buffett, because *model has unknown parameters*. On the other hand, *data reduce the uncertainty about the unknown parameters*, so you can use historical data on stock returns to suggest *ranges* of plausible values of θ = (mean, standard deviation) and then perform multiple analyses for the parameters within those ranges, which is also called **sensitivity analysis**.

The simulated returns look as shown in Figure 1.16; your numbers may vary due to randomness.

Supposing today's price (at $t = 0$) is $y_0 = 23.32$, the potential prices are calculated as shown in Equation 1.3: $y_1 = 23.32(1 + r_1)$, $y_2 = y_1(1 + r_2)$, $y_3 = y_2(1 + r_3)$, etc. You can do this in another tab of the spreadsheet as shown in Figure 1.17.

At the end of the analysis, each row shown in Figure 1.17 is a potential future trajectory of the stock prices over the next 250 trading days.

	A	B	C	D	E	F	G	H	I	J	
A1			f_x	-0.00800696477401652							
1	-0.00801	-0.03733	0.00833	0.03929	0.03695	0.05299	-0.06451	-0.00603	0.03385	-0.0316	-0
2	0.0152	0.02853	-0.00108	-0.00062	0.00744	-0.00956	0.02884	-0.01383	0.0713	0.02332	0
3	0.02219	-0.02165	0.01033	-0.03678	0.06951	0.02584	-0.03274	0.04577	-0.00717	-0.04164	-0
4	-0.00953	-0.01672	-0.00996	0.02639	-0.00277	0.05546	-0.01222	-0.03397	0.02017	-0.00207	
5	-0.01993	0.01759	0.02893	-0.02836	0.0408	-0.00226	0.02206	0.01128	-0.01662	-0.02543	-0
6	0.0254	0.05137	0.01859	0.01544	-0.0047	0.0024	-0.02688	-0.01987	0.03148	-0.00959	-0
7	0.03706	0.05832	0.01709	0.028	-0.02405	-0.00335	-0.01138	0.03185	-0.04959	0.0272	-0
8	-0.01813	-0.01812	0.05204	0.0054	0.04013	-0.01544	0.01625	0.02259	-0.01519	-0.01753	0
9	0.00441	0.05627	-0.0138	0.01377	-0.01423	-0.02384	0.02206	0.01765	-0.00524	0.03535	-0
10	-0.03806	0.02397	-0.025	0.05071	-0.0606	-0.02172	0.00522	0.01634	-0.00123	-0.01651	

FIGURE 1.16
Potential future return trajectories.

	A	B	C	D	E	F	G	H
SUM			f_x	=A2*(1+Sheet1!A1)				
1	0	1	2	3	4	5	6	
2	23.32	=A2*(1+Sheet1!A1)	22.4552	23.3375	24.1998	25.4823	23.8.	
3	23.32	23.6745	24.35	24.3235	24.3086	24.4894	24.2553	24.9
4	23.32	23.8374	23.3213	23.5623	22.6956	24.2731	24.9004	24.0.
5	23.32	23.0977	22.7114	22.4852	23.0785	23.0145	24.291	23.9!
6	23.32	22.8551	23.2573	23.9302	23.2514	24.2001	24.1453	24.!
7	23.32	23.9123	25.1407	25.6079	26.0032	25.881	25.9431	25.2.
8	23.32	24.1843	25.5948	26.0323	26.7611	26.1175	26.0299	25.7.
9	23.32	22.8972	22.4822	23.6522	23.7799	24.7342	24.3523	24.7.
10	23.32	23.4228	24.7408	24.3994	24.7353	24.3833	23.802	24.3

FIGURE 1.17
Potential future price trajectories.

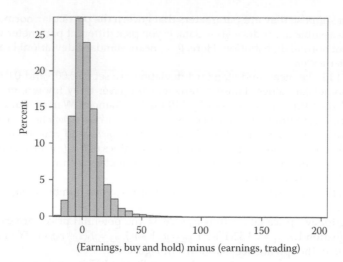

FIGURE 1.18
Histogram of earnings differences between buy and hold versus trading strategies.

With these future trajectories, you can try out each trading strategy to see which one nets you the most cash. For some potential futures, strategy 1 will work better, and for other potential futures, strategy 2 will work better. You want to pick the one that gives you more money on average, *over all potential futures*. Figure 1.18 in the following shows the distribution of the difference of your earnings, over 1000 potential futures, using the buy and hold versus trading strategy.

Figure 1.18 shows the **histogram**—an **estimate** of the probability distribution $p(y)$; see Chapter 4 for further details—of the difference between potential future earnings using strategies 1 and 2. For any particular future, the difference may be positive, meaning strategy 1 is preferred, or negative, meaning strategy 2 is preferred. Contrary to what your intuition might say, it seems that there are greater possibilities for much higher earnings with strategy 1—buying and holding—since Figure 1.18 extends much farther to the right of zero than to the left of zero. In fact, the average difference calculated from the 1000 potential futures is 3.71, meaning that you earn 3.71 more on average using strategy 1 than using strategy 2. So, on average, strategy 1 earns more. However, this does not guarantee that you will be better off using strategy 1 for the next 250 trading days. It only means that strategy 1 is better on average, over all potential futures, based on this model.

Example 1.9: Predicting a U.S. Presidential Election Based on Opinion Polls

Another example where simulation is very useful is in predicting the results of elections. In the United States, we do not directly elect our president. Rather, there is an *electoral college*, a system by which each of the 50 states and the District of Columbia contributes a certain number of votes based on its population. For example, at the time of writing of this book, California, the most populous state, contributes 55 electoral votes out of 538 total votes in the electoral college, whereas Delaware, a small state, contributes 3 votes. A presidential candidate needs to win a total of 270 electoral votes—a simple majority of 538—in order to win the election. Electoral votes are awarded based on the popular vote in each state. When a candidate wins the popular vote in a state, that state awards all of its electoral votes to that candidate. The other candidates get nothing. (Maine and

Nebraska use a different system, but this little detail has never altered the outcome of a presidential election.)

It can happen—and has happened—that a presidential candidate loses the overall popular vote but wins the election. Winning the presidency comes down to winning 270 or more electoral votes by winning the popular vote in any combination of states whose electoral votes total at least 270. This is why political analysts often talk about "battleground" states, where the vote is likely to be close and where the candidates would most benefit from spending their time and money. But how do the analysts suspect which states will have a close vote? Opinion polls!

Opinion polls provide a measure of the likelihood of a candidate winning a state. Using Bayesian calculations, as shown in Chapter 13, you can calculate the probability of winning based on the estimated proportion and the margin of error. For now, don't worry about too many details—we're just introducing the main ideas. Suppose there are two candidates and an opinion poll states that 47% of registered voters favor candidate A while 53% favor candidate B, with a 4% margin of error. In that case, assuming the opinion poll is accurate, then the probability that candidate A will win is 6.7%. In 6.7% of potential future scenarios, candidate A will win, and in the other 93.3% of the same scenarios, candidate B will win.

When you consider the combinations of all such future scenarios, with each state having different win probabilities and with different numbers of electoral votes at stake, along with the resulting potential range of future electoral college vote totals, the math seems daunting. But by using simulation, you can easily generate, say, 10,000 potential future scenarios, based on polling data for each state, and make an informed estimate as to who is likely to win. Each scenario gives a different electoral college tally. Figure 1.19 shows a histogram of 10,000 plausible values of the electoral college outcome, based on opinion polling data.

This simulation-based methodology is explained in much more detail by Christensen and Florence in their article "Predicting Presidential and Other Multistage Election Outcomes Using State-Level Pre-Election Polls," published in the journal *The American Statistician* in February, 2008.

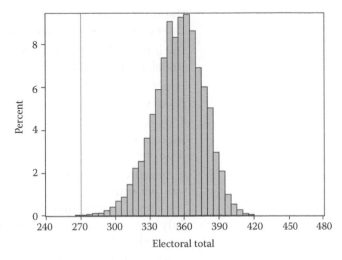

FIGURE 1.19

Histogram of the number of potential electoral votes for a hypothetical candidate, based on polling data. If the polls are accurate, the candidate is almost certain to win, since the number of votes will most likely be greater than the required 270, represented by the vertical line.

Vocabulary and Formula Summaries

Vocabulary

Nature	What is, was, will be, or might have been.
Statistics	The study of Nature using data.
Model	A mathematical representation of the outcomes of the processes of Nature, design, and measurement.
Prediction	A statement about something that might happen in Nature, be it in the past, present, future, or not at all.
Forecast	A statement about what will happen in the definite future.
Statistical model	A probabilistic recipe for how data are produced, one that depends on unknown parameters.
Primary data	Data you collected for a stated purpose.
Secondary data	Data collected for a different purpose; see *primary data*.
Design	A plan to collect data.
Measurement	The type of data to be collected.
DATA	As-yet unseen information produced from Nature, design, and measurement, also called Y.
data	The information after they are collected, also called y.
DATA*	The information to be produced by the model, also called Y^*. (See *simulation* in the following.)
data*	The information that has been produced by the model, also called y^*.
Binary response	Data that are dichotomous, such as 0 or 1 and yes or no.
Likert scale	A response scale used in surveys to indicate degrees of preference, typically comprised of items measured on a 1, 2, 3, 4, 5 scale.
Bivariate measurement	A measurement that consists of two numbers simultaneously.
Probabilistic model	The mathematical function called a pdf, typically written as $p(y)$; also a statement that DATA are produced by such a model.
Function	A mapping of values of x to values of y such that, given a particular value in a relevant range of x values, there is one and only one resulting y value.
Deterministic model	A model that always produces the same output, given the same inputs, typically written as $y = f(x)$.

Stochastic model	A *statistical model*, a *probability model*. Typically discussed in the context of time sequence data.
Parameter	A numerical characteristic of a natural process or model, usually fixed and unknown, indicated using the generic symbol θ.
Statistical inference	The method by which you learn about unknown parameters using data.
Generic parameter	Any parameter, denoted by the symbol θ.
Vector	A list of values.
Probability distribution function (pdf)	A function that assigns relative likelihoods to the different observable values of the data, typically written as $p(y)$.
Needle plot	A graph used to depict a discrete pdf. See also Chapter 2.
Bar chart	A graph used to depict a discrete pdf; see also *needle plot*.
Assumption	Something you stipulate about the model that you assume to produce your DATA.
Regression model	A model for how the distributions of Y change for different X values, written as $p(y\|x)$ or $p(y\|x, \theta)$.
Statistical inference	The reduction in uncertainty about your model parameters that you experience after you collect and analyze your data.
Bernoulli distribution	A probability distribution that produces the values 0 or 1.
Good model	A model where (a) the set of possible outcomes produced by the model well matches the set of possible outcomes produced by Nature, design, and measurement and (b) the frequencies of occurrences of the specific outcomes, and successive combinations of outcomes, well match the frequencies of occurrences of the specific outcomes produced by Nature, design, and measurement.
Bad model	One that is not good.
Simulation	Using the computer to produce DATA* from the model.
Return	The relative change from one time period to the next.
Parameter vector	A vector whose values are all parameters.
Sensitivity analysis	Multiple analyses with different plausible values of the parameters.
Estimate	A guess at the value of some entity.
Histogram	An estimate of a pdf.

Key Formulas and Descriptions

$y = c + mx$	The equation of a straight line with slope m and intercept c.
$y = f(x)$	A statement that y is produced as a deterministic function of x.
$Y \sim p(y)$	A statement that the data Y are produced by a pdf $p(y)$.
$p(y) = 1/2$, for y = tails, or y = heads.	The model for how data produced by flipping a fair coin will look; see *Bernoulli* distribution.
$p(y) = 1/6$, for $y = 1, 2, 3, 4, 5$, and 6.	The model for how data produced by rolling a fair die will look; a *discrete uniform* distribution.
$Y \sim p(y\mid\theta)$	A statement that the data Y are produced by a pdf that depends on an unknown parameter θ; a purely probabilistic statistical model.
$Y\mid X = x \sim p(y\mid x,\theta)$	A statement that the data Y are produced by a pdf that depends on an unknown parameter θ and a known value x; a *regression model*; a statistical model that has both deterministic and probabilistic components; the mother of all models.
$r_t = (y_t - y_{t-1})/y_{t-1}$	The relative change from time $t-1$ to time t is called the *return* at time t.

Exercises

For all exercises here and elsewhere in the book, take care to write in proper sentences. Look back in this chapter for examples of how to incorporate mathematical and statistical terms into text so that your answers read clearly.

1.1 Demonstrate that Equation 1.2 follows logically from Equation 1.1.

1.2 Show that Equation 1.3 is true for the case where $t = 10$.

1.3 A small automobile dealership has a variety of cars, from compact to midsize to luxury. They wish to model how many top-end, gas-guzzling luxury cars they sell in a given day. This particular dealership sells luxury cars infrequently, and people aren't buying many gas-guzzlers lately, so most days there are no luxury cars sold. Occasionally, however, there are days where more than one luxury cars are sold, so the dealership wants to plan appropriately for its inventory.

    Here are the instructions using Microsoft Excel, but it should be easy to do this using many other software packages. Go to Data Analysis → Random Number Generation (you might have to add it in first), and select the Poisson distribution. Enter 1213 in the *random seed* box. (A random seed initializes the stream of random numbers that are generated. When you repeat an analysis using the same seed, you get the same stream of random numbers. If you want different random numbers, use a different seed.) Select five different values of λ (lambda), the theoretical mean number of sales per day, and generate 100 days worth of sales data at random, one for each value of λ that you try, so that you have generated 500 numbers in total. Note that λ can be less than 1.0 since it is the average number of sales per day.

Remember that a probability model is "good" if the data it produces match the real data in a frequency sense, if not in an exact, number-for-number sense, for some parameter settings. Look at your different sets of sales numbers (one set of 100 numbers for each λ), and suggest a value of λ that looks reasonable. Do not compute summary statistics like means or percentages, just look at the actual data, and remember that they are supposed to represent number of cars sold in a given day. Explain in words, with reference to the numbers that you generated, why you think the λ you chose is reasonable and why some of the other λ values you tried are not reasonable. You should have tried values of λ that are clearly too large, as well as values of λ that are clearly too small.

1.4 Recall the empirical science paradigm: Reality is studied by Nature → Design and Measurement → DATA, while the statistical science model for reality is Probability Model → DATA. You can see what kinds of data the model produces by simulating DATA* using particular parameter settings. Answer the following in the context of the auto dealership case study in Exercise 1.3.

A. There are five concepts here: (i) Nature, (ii) design and measurement, (iii) DATA, (iv) probability model, and (v) DATA*. Identify all of them as explicitly as possible.

B. Explain, in terms of this case study, what it means for the model to be "good." Refer to your generated data sets in Exercise 1.3 as part of your answer.

C. Be sure that you distinguish reality from the model. Which of (i), (ii), (iii), (iv), and (v) of Exercise 1.4A belong to reality, and which belong to the statistical model for reality?

1.5 Use the car sales case study of Exercises 1.3 and 1.4. Suppose you are doing confirmatory research so that your question of interest precedes the design and measurement. What question of interest might be answerable using the given design, measurement, and data?

1.6 Using the car sales case study of Exercises 1.3 and 1.4, and the descriptions of bad models given in the chapter, pick a probability model that is obviously bad for this car sales case, in the sense that the DATA* produced by the model are not qualitatively similar to the DATA produced by Nature, design, and measurement, for any parameter settings. Explain why the model is bad by discussing some data* that you simulate from your bad model.

1.7 Model the up or down movement of a stock price using simulation. If using Excel, select Tools → Data Analysis → Random Number Generation, and then select the Bernoulli distribution. Now generate 1000 days of up or down data when the probability of the price going up is $\pi = 0.570$. Enter 2115 as a random seed.

A. Format the 1000 data points so that they indicate "Day" (1, 2, ..., 1000) in one column, "Bernoulli Outcome" in the next column, and "Result" (up or down) in the third column (use the Excel "IF" function to calculate), handing in only the first page.

B. Identify (i) Nature, (ii) design and measurement, (iii) DATA, (iv) probability model, and (v) DATA* in this stock price application. As in Exercise 1.4C, be sure to distinguish which of (i) through (v) belong to reality, and which belong to the model for reality.

C. How many days out of the 1000 did you expect the stock to go up? Explain.

D. Using the concept of *randomness*, explain why there is a difference from the number of up days you observed in Exercise 1.7A and what you expected in 1.7C. Use, as an analogy, the following scenario in your answer: You will flip a coin 10 times. How many heads do you *expect* to get? (5). How many heads *will* you get? What if you flipped it 1000 times?

1.8 Use the methodology of Example 1.7 to estimate the probability of seeing 50% heads when there are 100 flips.

1.9 Redo the analysis of Example 1.8 using a mean return of −0.001 instead of +0.001. How do the results change? What is the practical implication to you as an investor?

1.10 Death and taxes are certain. In a hospital, there will be a certain number of deaths daily, from all causes. Suppose the Poisson distribution with $\lambda = 0.5$ is a good model for the number of deaths in a day, in the sense that the DATA* produced by this Poisson model look like the real deaths (DATA). Use simulation to answer the following:

A. On average, how many total deaths are there in a 7 day week?

B. Over 52 weeks, what is your guess of the worst week, in terms of number of deaths?

1.11 Death and taxes can be a lot like car sales! Refer to Exercise 1.3. Suppose luxury shipments come to the dealership every week and the dealership is open 7 days a week. Use a simulation study to suggest how many cars the dealer should request every week. Looking at the simulated DATA*, discuss (i) what are the consequences if the dealer requests too few and (ii) what are the consequences if the dealer requests too many?

1.12 Probability is a measure of degree of belief, from 0 (complete impossibility) to 1 (complete certainty). A probability of 0.5 means that an event is as likely to occur as it is not to occur. A probability of 0.25 means that the event occurs one time out of four, on average. A probability of 0.0001 means that event occurs once every 10,000 times, and a probability of 10^{-12} (the notation 10^{-12} means $1/10^{12}$, or 1 divided by 1,000,000,000,000) means that the event is a one-in-a-trillion occurrence. Give your own subjective probability (just one number) for each of the following, with written explanations of your logic. The fact that these are subjective means that there are no exactly correct answers. However, there are errors of logic, so please consider your explanations carefully, and write thoughtfully. Do not attempt to answer by collecting data or by calculation. Instead, use your logic, and imagine your potential futures. Be careful about answering 0.0 or 1.0, as these answers convey absolute certainty, which you usually cannot assume about your potential futures.

A. A coin will show heads when tossed.

B. A coin will land on its edge (balanced perfectly, i.e., neither heads nor tails) when tossed.

C. Your immediate family members will all live past 50 years.

D. At least one of your immediate family members will live past 50 years.

E. Your immediate family members' first names all have fewer than 6 letters.

F. The next roll of toilet paper you will see will have the paper coming from the "wall side" of the roll, not over the top.

G. The next dozen toilet paper rolls you will see will all have the paper coming from the "wall side" of the roll, not over the top.

H. You will see a living zebra tomorrow, in person (not on video or television).

I. You will see a living zebra sometime in the next 5 years, in person.

J. You will be involved in an automobile accident tomorrow.

K. You will be involved in an automobile accident sometime in the next 5 years.

L. It will rain tomorrow.

M. The Dow Jones Industrial Average will be higher tomorrow (or the next trading day) than it is today (or the most recent trading day).

N. The Dow Jones Industrial Average will be more than 100 points higher tomorrow (or the next trading day) than it is today (or the most recent trading day).

O. The Dow Jones Industrial Average will be more than 500 points higher tomorrow (or the next trading day) than it is today (or the most recent trading day).

1.13 Describe Nature, design, measurement, and DATA for a study wherein the probability in Exercise 1.12F is estimated. (You don't have to collect any data.)

1.14 Describe Nature, design, measurement, and DATA for a study wherein the probability in Exercise 1.12J is estimated. (You don't have to collect any data.)

1.15 The *model produces data* concept is the essence of probabilistic simulation. Search the web using the key words "probabilistic simulation in _____," where "_____" is your field of interest (e.g., medicine). Explore a few websites and write a paragraph explaining how probabilistic simulation is used in your field of interest. In your answer, discuss why probability is needed for the study.

1.16 Let Y be the time it takes you to get to school (or work); that is, Y is your commute time. Explain why your Y is better explained by a probabilistic model than by a deterministic model.

1.17 A jet airliner crashes hard into a rugged mountainside. You can use the flip of a fair coin to model whether a person on board lives or dies (e.g., heads = lives, tails = dies). Using the definition of a good model, explain why this model is not good.

1.18 Let Y be the number of credit cards owned by a generic person (0, 1, 2, ...). Following the structure and notation of Table 1.7, define a model that you think could produce these Y data. Also, explain why it is better to use π's for the probabilities rather than actual numbers in your model.

1.19 Use the definition of a good model in Section 1.10. Why is the deterministic model for driving time when $x = 310$ *not* a good model, according to that definition?

2

Random Variables and Their Probability Distributions

2.1 Introduction

In Chapter 1, you might have wondered about all the capital and lowercase letters. When do you use one versus the other? The answer is simple: Uppercase letters denote **random variables**, and lowercase letters denote **fixed quantities**. Think of what you will see, or could see, or might have seen versus what you actually do see. If you don't actually see something, it is variable in your mind and hence modeled as a random variable. If you actually see it, it is fixed or nonrandom.

Example 2.1: Rolling Dice

Suppose you plan to roll a die. The outcome will be Y, a random variable. The variable Y can be any of the numbers 1, 2, 3, 4, 5, or 6. Now suppose you go ahead and actually roll the die, and it shows "1." Then you have observed $y = 1$.

Example 2.2: Measuring Height

Suppose you plan to select a student at random from the class. His or her height will be Y, a random variable. The variable Y will likely be between 135 and 210 cm. Suppose you select the student, conduct the measurement, and get the result 143.1 cm. Then you have observed $y = 143.1$.

While *future* versus *present* can be a useful way to distinguish between random DATA versus fixed data, a better distinction is *prediction* versus *observation*. Recall that prediction does not necessarily refer to the future; rather, it refers to any what-if scenario that could happen in the past, present, future, or not at all. Uppercase DATA refer to prediction of unknown information; lowercase data refer to existing information that you have.

2.2 Types of Random Variables: Nominal, Ordinal, and Continuous

The "measurement" component of the statistical science paradigm introduced in Chapter 1 (Figure 1.2) defines what kind of data you will see. There are broad classes of measurements known as *nominal*, *ordinal*, and *continuous*, each of which requires different classes of models. The appropriate classification is the first thing you should identify about your measurements. You would not want to use a model that produces nominal DATA* when your DATA are in fact continuous, because that would be a bad model!

Types of DATA

- **Nominal DATA**: These are DATA whose possible values are essentially labels (or, as the word *nominal* suggests, names), with no numerical value. As an example, the color of car chosen (the red, gray, or green example of Chapter 1) is a nominal variable. Other examples include eye color, choice of political party affiliation, choice of religious affiliation, and job title.
- **Continuous DATA**: These are numerical DATA whose possible values lie in a **continuum**, or in a continuous range. For example, the time you have to wait for a call center to answer your call is a continuous random variable whose values lie in the continuous range from 0 to 60 minutes, assuming you will slam down the phone in disgust after waiting an hour! In this example, the variable "waiting time" is continuous because you could imagine there being an infinite number of decimal-valued numbers between, say, 10 and 11 minutes.
- **Ordinal DATA**: These types of DATA are intermediate between nominal and continuous. Unlike nominal data, which can be numbers without intrinsic order such as 1 = male and 2 = female, ordinal DATA are numbers that reflect an intrinsic *order*, hence, the name *ordinal*. Examples of ordinal DATA include the number of different computers you use in a week (i.e., at home, school); the number of siblings you have; your preference for coffee on a scale of 1, 2, 3, 4, and 5; and your education level.

Both nominal DATA and ordinal DATA are examples of **discrete DATA** or DATA whose possible values can be listed. These are quite different from continuous DATA, whose possible values lie in a continuum and therefore cannot be listed.

Some statistics sources classify DATA types further as *ratio, interval, ordinal, and nominal.* Often, these sources give rules of thumb for different ways to analyze the data, depending on their typology; see Velleman, Paul F. and Wilkinson, Leland (1993), Nominal, Ordinal, Interval, and Ratio Typologies Are Misleading, *The American Statistician* (47(1): 65–72). We won't discuss these typologies. Instead, the methods that we present in this book cover the ratio, interval, ordinal, and nominal types: Ratio and interval DATA are special cases of the continuous type, while ordinal and nominal DATA are special cases of the discrete type.

While it is easy to distinguish nominal DATA from the other types, students have a hard time distinguishing between the continuous and discrete ordinal types. The first thing to remember here is that the question of discrete versus continuous is about the DATA that you *might see*, not about the data that you *have collected*. The data you have collected are always discrete, since you can easily list the values and they can never fill any continuous range. For example, no matter how many times you call the call center, the data will never come close to filling the entire continuum from 0 to 60 minutes. Continuous DATA fill the continuum, but observed data can never fill a continuum.

In the earlier discussion, the phrase *possible values* appears repeatedly. This is an important phrase! The set of possible values of a random variable Y is called the **sample space**. When choosing a model, your first consideration should be that the sample space of the model matches (or nearly matches) the sample space of your DATA; this was criterion (a) for a good model given in Chapter 1 (Section 1.10).

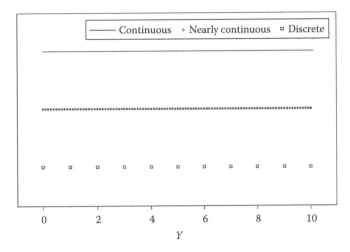

FIGURE 2.1
Continuous, discrete but nearly continuous, and discrete measurements. In the continuous case, the DATA can occur anywhere between 0 and 10. In the discrete cases, the data can occur only where there are circles or squares.

For a simple example, consider the Bernoulli coin flip where the outcome is either 0 or 1 (tails or heads). Here the sample space is $S = \{0, 1\}$. This is the most discrete that an outcome can possibly be, except for the degenerate case where the outcome is always 0 and the sample space is $S = \{0\}$. (Imagine a coin with both sides being tails.)

While we spend much time discussing continuous DATA, you cannot observe it in practice. Even if the quantity that you *wish to* measure is continuous, such as the time you spend waiting on hold to talk to customer service, the actual quantity that you *can* measure is discrete because whatever measuring device you will use must round the DATA off to some decimal. With an extremely accurate measure of time, your measured wait times (in minutes) might be 3.2123, 4.3299, 10.0023, 0.0231, etc. While these numbers certainly fill the continuum better than values rounded to the nearest minute (3, 4, 10, and 0, respectively), they cannot fill gaps in the continuum, such as the one between 3.00000 and 3.00001. Therefore, they are not truly continuous. Any random variable which is perfectly continuous—such as your true waiting time—is always a **latent variable** or one that you cannot measure directly. Instead, the best you can do is to observe a rounded-off version of the true latent quantity.

Figure 2.1 illustrates the difference between perfectly continuous, discrete but nearly continuous, and discrete, as discussed in the call center example given earlier.

So, you ask, "If the real DATA that I can actually analyze is always discrete at some level, then why bother using models that produce continuous DATA at all?" Great question! There are several reasons for using such continuous models.

Reasons to Use Models that Produce Continuous DATA, Even Though Observable DATA Are Always Discrete

1. Even though the actual DATA might be discrete, the averages computed from such DATA are often more nearly continuous because of the central limit theorem, discussed in Chapter 10. This famous theorem states that the distribution of averages and sums is approximately normal distribution. The normal distribution is the most important distribution in all of statistics, and it is a continuous model (i.e., one that produces continuous DATA).

2. When the level of measurement is precise enough to appear nearly continuous, there are so many possible values for a random outcome that it becomes unwieldy to assign every single value a probability. It is much easier to specify a continuous model.
3. A continuous model can easily be discretized, that is, rounded off, at whatever decimal level you wish to produce DATA* that closely match the decimal level of your DATA.
4. Your model does not have to be perfect. The DATA* produced by your model do not have to match DATA perfectly, only well enough for the practical purpose of resolving your statistical question of interest.
5. There is the matter of *parsimony*: Estimates using simpler models tend to give better results than estimates using more complex models, even when the simpler model is *wrong* and the more complex model is *right*. Continuous models tend to be simpler than highly discrete models. You will see more on this in Chapter 17.

So, when is a continuous approximation of a discrete distribution good enough? The answer depends on context as well as intent. There is no single, black and white answer—as in all of statistics, the correct answer involves shades of gray. One answer is that the better the discrete DATA fill the continuum, the better the continuous model is as an approximation. While "shades of gray" answers are often not very helpful, they are at least factual, as the black and white answers are nearly always wrong, to one degree or another.

But if you absolutely must have a black and white type of answer, here is one.

Ugly Rule of Thumb 2.1

If the set of possible discrete outcomes is 10 or more, then a continuous model may provide an adequate approximation to the distribution of the discrete random variable.

Throughout this book, we provide rules of thumb such as this one. As with all rules of thumb provided in this book, we call this one "ugly" because you should not take it too seriously. Does something magical happen when the number of outcomes changes from 9 to 10? Of course not. That's why rules of thumb are ugly. Unlike mathematical facts, rules of thumb are not logically precise. But we do intend them to be useful, so please use them freely!

Ugly rules of thumb presented in this book are just crude guidelines, usually based on the authors' expertise and experience. If you cite our rules of thumb in your research, please don't use the term "ugly." We use the term "ugly" to emphasize that there is a great distinction between *crude guideline* and *logical fact*.

2.3 Discrete Probability Distribution Functions

The model for a random variable is called a **probability distribution function**, or pdf, which is defined as a function $p(y)$ that you assume produces your DATA, Y. The function $p(y)$ tells you (1) the *set* of possible Y values that will be produced and (2) the frequencies of occurrence of the different Y values.

TABLE 2.1

List Form of a Discrete pdf

y	$p(y)$
y_1	$p(y_1)$
y_2	$p(y_2)$
...	...
Total	1.0

If your random variable is discrete, then you should use a discrete pdf for your model. If continuous, then you should use a continuous pdf. The acronym *pdf* (this is not "portable document format"!) will be used for either case in this book; it will be clear from context whether the pdf is discrete or continuous. In some sources, including texts and online documents, a discrete pdf may be called a *probability mass function*, while a continuous pdf may be called a *probability density function*.

While on the subject of acronyms, we will abbreviate *random variable* as *RV*. (This is not "recreational vehicle"!) Unlike the acronym pdf, RV is capitalized to remind you that random quantities are denoted with capital letters and fixed quantities with lowercase letters.

Recall that a discrete RV has potential values that can be listed. A discrete pdf is simply a listing of these values, y_1, y_2, ..., often arranged in numerical order if ordinal and in lexicographical order if nominal, along with their associated probabilities $p(y_1)$, $p(y_2)$, ... This gives you the **list form** of a discrete pdf shown in Table 2.1.

There are infinitely many possible discrete pdfs $p(y)$. All you need for $p(y)$ to be a discrete pdf are the following two conditions.

Requirements for a Discrete pdf

1. $p(y) \geq 0$, for $y = y_1, y = y_2, \ldots$
2. $p(y_1) + p(y_2) + \cdots = 1$

Requirement (1) states that probabilities are never negative, and requirement (2) states that the total probability is 1.0 or 100%. To put it in a different way, (2) tells you that when you observe an RV Y, it will certainly be one of the values in its sample space $S = \{y_1, y_2, \ldots\}$.

Notice that requirement (1) does not say that the values of Y must be greater than 0; it only says that the pdf *function* $p(y)$ must be greater than or equal to 0 for all values (whether negative or positive) of y. RVs Y are often negative: For example, if Y is defined as the increase in the gross national product (gnp), then Y is negative when the gnp drops.

Introduction to Summation and Set Notation

The requirements for a discrete pdf are equivalently written as follows:

1. $p(y) \geq 0$, for $y \in S$
2. $\sum_{y \in S} p(y) = 1$

The relational symbol "$\in$" reads aloud as "an element of," and the symbol "Σ" is the uppercase Greek letter sigma, which you read aloud as "summation." The subscript of Σ, $y \in S$, tells you the values that you are summing up. So (2) says that the sum of the probability function $p(y)$, over all of the possible values of Y, equals 1.

The function $p(y)$ can be specified in the **list form** shown in Table 2.1 or in **function form**. Sometimes it is easier to specify $p(y)$ in list form, sometimes in function form. The following examples will clarify.

Example 2.3: The Bernoulli Distribution

The Bernoulli distribution in Chapter 1 is the most discrete of discrete distributions because there are only two possibilities. Recall the Ugly Rule of Thumb 2.1: The more possible values of the RV, the closer to continuous it becomes. Since there are only two possible values of the Bernoulli RV, it isn't at all close to continuous. In list form, the distribution is specified as shown in Table 2.2, a repeat of Table 1.4.

The list form can easily be confused with a data set that you might have collected in your spreadsheet. After all, your data are also a list of data values. Do not confuse the list form of the distribution with your data set! To help avoid confusion, note that the pdf describes the DATA you *will see*, not the data that you *have already seen*. Also, for a discrete pdf, the list is much shorter than in your observed set since you do not list repeats in your pdf. For example, if your data set has a list of $n = 100$ zeroes and ones—for instance, yes/no responses—then the data list will span 100 rows of your spreadsheet, whereas the pdf that produces these values will have only 2 rows in the list, namely, the 0 and the 1.

While you can always write discrete distributions in list form, the function form is more useful in many cases. In function form, the Bernoulli pdf can be written as

$$p(y|\pi) = \begin{cases} 1 - \pi, & \text{for } y = 0 \\ \pi, & \text{for } y = 1 \end{cases} \tag{2.1}$$

Even more cleverly, you can write the Bernoulli pdf as

$$p(y|\pi) = \pi^y (1 - \pi)^{1-y}, \quad \text{for } y \in \{0,1\} \tag{2.2}$$

This would be a good time to browse the Internet to learn about the Bernoulli distribution. There you will see all of the same math, but the symbols used are different. You may as well get used to it. Symbols are almost always different from one source to the next. An important rule:

Don't get too hung up on the particular symbols that different sources use. Instead, strive to understand what the symbols *mean*.

TABLE 2.2

List Form of the Bernoulli Distribution

y	$p(y)$
0	$1 - \pi$
1	π
Total	1.0

Example 2.4: The Car Color Choice Distribution

We're pulling your leg. Unlike the Bernoulli distribution, there really is no distribution named the "car color choice distribution." Rather, it's just a generic discrete distribution, which also can be called a **multinomial distribution**. The point is that in the nominal case (as in color choice), the pdf is a discrete pdf, one where the set of possible outcomes $S = \{y_1, y_2, \dots\}$ happen to be labels rather than numbers. In list form, the generic car color choice pdf shown in Chapter 1 is given by Table 2.3, a repeat of Table 1.5.

Here, $\theta = (\pi_1, \pi_2, \pi_3)$ is the unknown parameter vector.

This is an example where list form is preferable to function form for simplicity's sake. Here is the function form of the car color choice distribution:

$$p(y|\theta) = \pi_1^{I(y=\text{red})} \pi_2^{I(y=\text{gray})} \pi_3^{I(y=\text{green})}, \quad \text{for } y \in \{\text{red}, \text{gray}, \text{green}\}$$

In this expression, $I(\text{condition})$ is the **indicator function**, returning the value 1 if the condition is true and 0 if the condition is false. For instance, $I(y = \text{red}) = 1$ when $y = \text{red}$, but $I(y = \text{red}) = 0$ when $y = \text{gray}$.

Example 2.5: The Poisson Distribution

Introduced in the Chapter 1 exercises as a model for both the number of car sales at a dealership and for the number of deaths at a hospital, the Poisson distribution is a discrete pdf that is best displayed in function form. That form is

$$p(y|\lambda) = \frac{\lambda^y e^{-\lambda}}{y!}, \quad \text{for } y = 0, 1, 2, \dots$$

Now we have some explaining to do. First, the symbol λ is the Greek lowercase letter lambda and represents the theoretical average. Second, the letter e used here and throughout this book refers to one of the most famous special numbers in math, called **Euler's constant**. (*Euler* is pronounced "oiler.") The number appears everywhere. Along with the famous $\pi = 3.14159\dots$, Nature itself tells us many things that involve Euler's constant. Its numerical value is

$$e = 2.71828\dots$$

Finally, the term $y!$ does not mean "indeed, most emphatically, y"; rather, it reads aloud as "y factorial" and is defined as

$$y! = 1 \times 2 \times 3 \times \cdots \times y$$

TABLE 2.3

List Form of the Car Color Choice Distribution

y	$p(y|\theta)$
Red	π_1
Gray	π_2
Green	π_3
Total	1.00

TABLE 2.4

List Form of the
Poisson Distribution

y	$p(y \mid \lambda)$
0	$e^{-\lambda}$
1	$\lambda e^{-\lambda}$
2	$\lambda^2 e^{-\lambda}/2$
3	$\lambda^3 e^{-\lambda}/6$
4	$\lambda^4 e^{-\lambda}/24$
...	...
Total	1.00

Thus, $1! = 1$, $2! = 1 \times 2 = 2$, $3! = 1 \times 2 \times 3 = 6$, and so on. Notice, however, that y can be 0 in the Poisson model. In this case, $0!$ is simply defined as the number 1. Yes, that seems weird, but it works, in that the definition $0! = 1$ makes the probability function behave as required. (Review the "requirements for a discrete pdf" given earlier.)

The list form of the Poisson distribution is somewhat more cumbersome, but still instructive. Plugging in $y = 0, 1, 2, \ldots$ successively into the Poisson function form gives you the list form shown in Table 2.4.

The list is infinite, so any list that you may create is incomplete—hence, the need for the "…" symbol.

This would be a good time to browse the Internet to learn about the Poisson distribution. Compare the notation you see elsewhere with the notation you see here, and convince yourself that there is no difference, even though you may see different symbols in different sources.

How can such a weird-looking function such as the Poisson distribution possibly work in practice? As it turns out, there are physical laws describing the universe that give rise to the Poisson model for certain observable quantities, for example, number of photons arriving to a telescope. The Poisson distribution also arises as a simple approximation to other discrete pdfs such as the binomial distribution. But, most importantly, whether by physics, by approximation of other distributions, or just by plain luck, the Poisson model simply works well in a variety of cases. In such cases, for some choices of λ, the DATA* produced by a Poisson model look like DATA you will actually observe. Since DATA* produced by the Poisson model are the numbers 0, 1, 2, 3, …—that is, data with sample space $S = \{0, 1, 2, \ldots\}$— your DATA must also be 0, 1, 2, 3, … if you are considering using a Poisson model. In particular, if your data will be numbers like 5.3, 5.7, 10.4, …, then you can rule out the Poisson model.

There are many named discrete distributions, and there are many more that are not named. Common named discrete pdfs include Bernoulli, Poisson, geometric, hypergeometric, binomial, and negative binomial. Take a moment to search for these distributions.

2.4 Continuous Probability Distribution Functions

Whoa, here we go. Continuous functions! Limits! Infinitesimals! Calculus! Accckkk! But, seriously, there is no way to talk about continuous distributions without calculus. It is the common, accepted language used to describe continuous functions. We could develop a whole new language for describing these distributions, but what would be the point? A perfectly good language already exists—calculus—and it has stood the test of time for nearly half a millennium. Further, the most important distribution in our universe is a continuous distribution. It is the famous *normal* distribution introduced in Figure 1.4.

It is impossible to make sense of the normal distribution without talking about the area under its curve. The area under a curve of a function is the *definite integral* of a function, a fundamental concept of calculus. Okay, so let's learn some calculus!

Example 2.6: Diabetes, Body Mass Index, and Weight

An example will help you to ease into the high-minded mathematics. No matter what your field of interest, you probably know someone with diabetes. Diabetes has become a health crisis, with great personal and societal costs. Obesity is associated with diabetes, so many studies of diabetes also involve weight measurement, particularly adjusted for height. The body mass index (BMI) is a measure of a person's weight relative to their height, defined as the following ratio:

$$\text{BMI} = \frac{\{\text{Weight (in kg)}\}}{\{\text{Height (in m)}\}^2}$$

A person may be classified as obese when $\text{BMI} \geq 30$.

While we are not medical doctors, we can confidently say that this is another "ugly rule of thumb." Nothing magical happens between 29.9 and 30.1 BMI. Still, you might be interested to know how this threshold translates to weight. What is the obesity cutoff, in kilograms, for a person of typical height, say 1.7 m (or 5 ft, 7 in.)?

Here is the math. In it, you will see the relational symbol $\Leftrightarrow$, which reads aloud as "is equivalent to." Here and throughout this book, take a little time to understand the mathematical logic. Take particular note of the parenthetical explanations: A goal of this book is that you *understand* advanced statistical methods! So please follow carefully; it's not hard.

$\text{BMI} \geq 30$	(Criterion for obesity)
$\Leftrightarrow \{\text{Weight (in kg)}\}/\{\text{height (in m)}\}^2 \geq 30$	(By substitution)
$\Leftrightarrow \{\text{Weight (in kg)}\}/1.7^2 \geq 30$	(By height assumption)
$\Leftrightarrow \{\text{Weight (in kg)}\} \geq 30 \times 1.7^2$	(By algebra)
$\Leftrightarrow \{\text{Weight (in kg)}\} \geq 86.7 \text{ kg}$	(By arithmetic)

So a person 1.7 m tall would be classified as obese if he or she weighs 86.7 kg or more. For those who like pounds instead of kilograms, that makes 191 lb the obesity threshold. Again, this is just an ugly rule of thumb. Consult your doctor, not a statistics book, if you want more information about obesity and diabetes!

How many people meet the criterion for obesity? Imagine measuring the weights of many people who are 1.7 m tall. Here, the *design* is to identify a group of people who are 1.7 m tall and plop them all down on scales that measure in kilograms, and the *measurement* is weight, rounded to the nearest whole kilogram. You will get DATA when you do this.

What kind of model produces DATA* that look like these DATA?

If the measured DATA will be numbers like 0, 1, 2, ..., 90, 91, ..., all in kilograms, then the distribution is discrete. The distribution may look as shown in the needle plot displayed in Figure 2.2. For the purposes of the discussion, it is best not to worry too much how we constructed this distribution. The main point is that you will assume that the DATA are produced by a similar distribution (*model produces data*), one that you do not know precisely (*model has unknown parameters*). If you actually collected data on weights of people who are 1.7 m tall, you would have a better idea what this distribution really looks like (*data reduce the uncertainty about the unknown parameters*).

Figure 2.3 is a zoomed-in version of Figure 2.2, one that shows the area of concern between 85 and 90 kg.

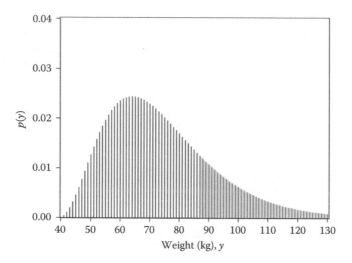

FIGURE 2.2
A plausible distribution for producing weight (roundest to the nearest kilogram) for people who are 1.7 m tall. The area of concern for obesity is shown in the shaded portion of the graph.

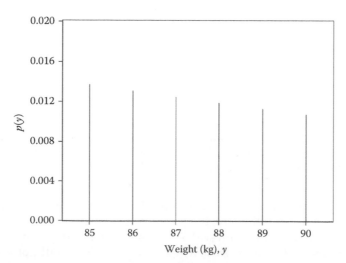

FIGURE 2.3
The portion of the discrete distribution shown in Figure 2.2 from 85 to 90 kg.

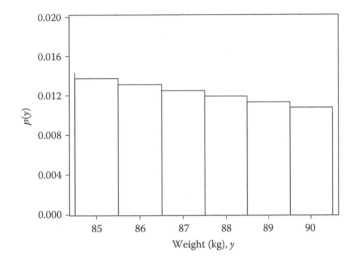

FIGURE 2.4
The discrete distribution of weight showing probabilities as rectangular areas. The tops of the rectangles form a distribution function for the continuous measurements.

Of course, there are weights between 85 and 86 kg, between 86 and 87 kg, etc. All the weights between 85.5 and 86.5 are rounded to 86, so the probability that the weight is between 85.5 and 86.5 is, from inspection of the graph, about 0.013. In other words, if this distribution is true, then about 13 out of 1000 people who are 1.7 m tall weigh between 85.5 and 86.5 kg. A simple and natural way to model *all* the continuous weights that are in between the discrete observed weights is to use a step function (or rectangular function) rather than discrete needles, where the **area under the function** is equal to the probability. Such a function is shown in Figure 2.4.

In Figure 2.4, notice that the **area of the rectangle** above 85.5 and 86.5 is given by

$$\text{Area of rectangle} = \text{Base} \times \text{Height} = (86.5 - 85.5) \times 0.013 = 0.013$$

In the rectangular model, the probability of the ±0.5 kg range of continuous measurements surrounding 86.0 is exactly the same as the discrete probability of observing 86.0. This is reasonable because all values in the ±0.5 kg range of 86.0 all are rounded to 86.0.

Seems reasonable? Good! You're now hooked! The function that connects the tops of the rectangles in Figure 2.4 is an example of a **continuous distribution**. Its function form is given as follows:

$$\tilde{p}(y) = p(y_i), \quad \text{for } y_i - 0.5 \le y < y_i + 0.5 \tag{2.3}$$

Explanation of Equation 2.3

- The mark ~ over the p is called a *tilde*; it is simply there to distinguish the continuous pdf $\tilde{p}(y)$ from the discrete pdf $p(y_i)$.
- The terms y_i are the possible discrete values. Here they are $y_0 = 0, y_1 = 1, y_2 = 2, \ldots,$ $y_{90} = 90, \ldots.$
- Unlike the discrete pdf $p(y_i)$, the continuous pdf $\tilde{p}(y)$ is defined to be a positive number for *all* $y > 0$ and not just for the discrete observable values y_i.

The function $\tilde{p}(y)$ is a continuous pdf because it provides a model for all the weights in the continuum, not because the function itself is continuous. Actually, the function $\tilde{p}(y)$ itself is discontinuous, being a step function. Technically, a pdf is a continuous pdf if its *cumulative distribution function* (or cdf) is a continuous function; the cdf is defined later in this chapter. The cdf corresponding to the pdf $\tilde{p}(y)$ is indeed continuous, even though $\tilde{p}(y)$ itself is discontinuous.

- The range $y_i - 0.5 \le y < y_i + 0.5$ refers to the way that numbers are rounded off. For example, all numbers y such that $50 - 0.5 \le y < 50 + 0.5$ are rounded off to 50.

- Perhaps most importantly, the continuous pdf $\tilde{p}(y)$ *does not give you probabilities.* For example, $\tilde{p}(86.23290116) = 0.013$ *does not mean* that there is a 0.013 probability that a person will weigh precisely 86.23290116 kg. When using continuous pdfs, probabilities can *only* be determined as areas under the curve. Thus, the continuous pdf $\tilde{p}(y)$ tells you that the probability *between* 85.5 and 86.5 is 0.013. Also, by the area of a rectangle formula, the continuous pdf $\tilde{p}(y)$ tells you that the probability between 86.0 and 86.5 is $(86.5 - 86.0) \times (0.013) = 0.0065$. The number $\tilde{p}(86.23290116) = 0.013$ can be interpreted as the "relative likelihood" in the sense that 86.23290116 is relatively more likely than 90.12538544, for which, by visual inspection of Figure 2.4, $\tilde{p}(90.12538544) = 0.011$ or so. Again, $\tilde{p}(86.23290116) = 0.013$ does *not* mean that 13 out of 1000 people weigh precisely 86.23290116 kg. In fact, it is quite unlikely that *anyone* in a sample of 1000 people would weigh precisely 86.23290116 kg.

- Since the areas of the rectangles are equal to the discrete probabilities, and because the discrete probabilities sum to 1.0, it follows that the total area under the continuous pdf $\tilde{p}(y)$ is also equal to 1.0.

The pdf $\tilde{p}(y)$ is not the true pdf; it is an *approximation* that assumes all numbers within a ±0.5 range of any integer are equally likely. This seems like a bad assumption: For the weight values considered in Figure 2.4, the higher numbers in the interval ranges should be less likely. Also, the values of $\tilde{p}(y)$ jump discontinuously at 85.5, 86.5, etc., and there is no natural reason why this should happen.

One way to improve the rectangular approximation is to use a finer measurement, say rounded to the nearest 10th of a kilogram. In this case, the zoomed-in area of interest might look as shown in Figure 2.5.

Note that the probabilities shown in Figure 2.5 are approximately one-tenth the size of those shown in Figure 2.3. For example, the probability of 86.0 shown in Figure 2.3 is about 0.013, whereas in Figure 2.5, it is much smaller, about 0.0013. This makes perfect sense when you realize that in Figure 2.3, the number 0.013 is the probability that weight is in the interval 85.5–86.5 kg, whereas in Figure 2.5, the number 0.0013 is the probability that weight is in the interval 85.95–86.05 kg. The interval in Figure 2.5 has one-tenth the size of the interval in Figure 2.3, so it makes sense that one-tenth as many people will be in the narrower interval.

To represent all of the intermediate weights that round off to tenths of a kilogram using the rectangle-based distribution function, you can draw rectangles as in Figure 2.4, obtaining Figure 2.6.

But in Figure 2.6, the base × height formula no longer works to find the probability of the measurements that round to 86.0, since from Figure 2.6:

$$\text{Area of rectangle} = \text{Base} \times \text{Height} = (86.05 - 85.95) \times 0.0013 = 0.00013$$

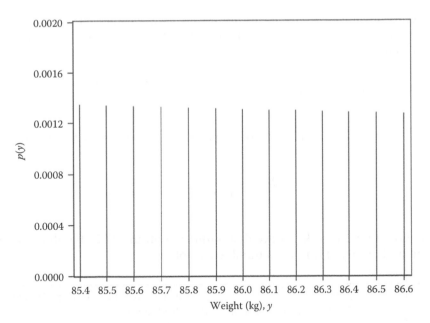

FIGURE 2.5
A zoomed-in portion of the discrete distribution of weight when measured to the nearest tenth of a kilogram.

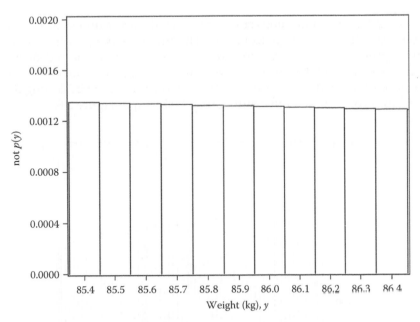

FIGURE 2.6
Weights, rounded to nearest tenth of a kilogram, showing rectangles to capture continuous measurements.

The number 0.00013 is 10 times too small. That's why the label of the vertical axis in Figure 2.6 is labeled "not $p(y)$" instead of $p(y)$. If you want the areas of the rectangles to correspond to probabilities, you have to adjust the heights in Figure 2.6 by multiplying by 10. In this case, the vertical axis values will be closer to those shown in Figure 2.4. In general, if the numbers y_i are rounded to the nearest Δ (the uppercase Greek letter delta), where $\Delta = 0.1$ represents tenths, $\Delta = 0.01$ represents hundredths, and so on, then the height of the bar chart which makes the areas of the rectangles correspond to probabilities of the continuous measurement is

$$\text{Height of rectangle above } y_i = \frac{(\text{discrete probability of } y_i)}{\Delta}$$

This gives you the following revision of Equation 2.3 in case the data are rounded to the nearest Δ:

$$\tilde{p}(y) = \frac{p(y_i)}{\Delta}, \quad \text{for } y_i - \frac{\Delta}{2} \le y < y_i + \frac{\Delta}{2} \tag{2.4}$$

To see how Equation 2.4 works, notice that numbers within $\pm\Delta/2$ of y_i all round to y_i. You can find the probability of this set of numbers as follows:

Area = Base × Height	(The area of a rectangle, area = probability)
$= \{(y_i + \Delta/2) - (y_i - \Delta/2)\} \times \{p(y_i)/\Delta\}$	(By substitution)
$= \Delta \times \{p(y_i)/\Delta\}$	(By algebra)
$= p(y_i)$	(By algebra)

Thus, Equation 2.4 gives a continuous curve for which the areas under the rectangles equal the probabilities of the discrete outcomes y_i. The numbers $p(y_i)/\Delta$ typically approach a smooth curve $p(y)$ as Δ shrinks to zero; such a function $p(y)$ is the pdf of the true (not rounded off) measurement. Figure 2.7 shows how these functions can converge.

The continuous limit of the curves $\tilde{p}(y) = p(y_i)/\Delta$ shown in Figure 2.7, letting the roundoff range Δ tend to zero, is the smooth continuous pdf shown in Figure 2.8.

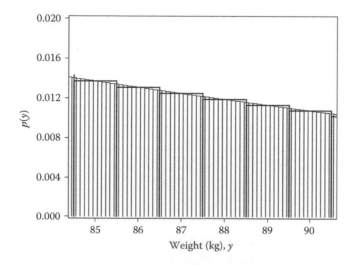

FIGURE 2.7
Rectangular continuous pdfs corresponding to rounding weight to the nearest kilogram ($\Delta = 1$, darker) and to the nearest tenth of a kilogram ($\Delta = 0.1$, lighter). With smaller Δ, the rectangle distribution is closer to a smooth distribution.

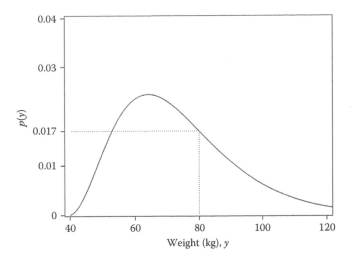

FIGURE 2.8
Graph of a smooth continuous pdf $p(y)$.

How do you interpret the pdf $p(y)$? Comparing Figure 2.8 with Figure 2.2, you can see that the values of $p(y)$ are approximately the probability that the RV Y lies within a ± 0.5 range of a particular y. But you can't always count on this correspondence. In cases where the range of data is less than 1 unit—for instance, if all of the weights were measured in thousands of kilograms (or *metric tons*)—many values of the pdf $p(y)$ would have to be greater than 1.0 in order for the total area under the curve to be 1.0 (see Figure 2.9).

Notice the vertical axis of Figure 2.9 has to go much higher than 1.0 to account for the narrow range on the horizontal axis. If the height of the curve were less than 1.0, then rectangular area calculation would tell you that the total area would be less than $(0.12 - 0.04) \times 1.0 = 0.08$. But the area under a continuous pdf is equal to 1.0, so the height must be more than 1.0 when the data range is less than 1.0. The bottom line is that with a continuous distribution, the numbers $p(y)$ on the vertical are clearly not probabilities, since they

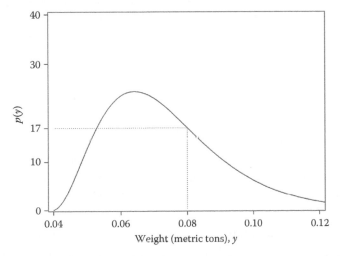

FIGURE 2.9
The pdf $p(y)$ for weight measured in thousands of kilograms (metric tons).

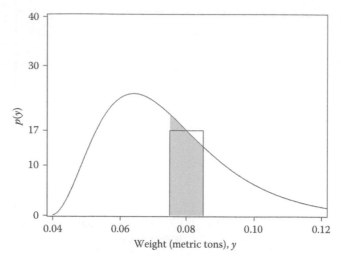

FIGURE 2.10
The probability that weight (in thousands of kilograms) lies between 0.075 and 0.085 is the shaded area. The rectangular approximation $p(0.080) \times 0.01 = 17 \times 0.01$ is also shown.

can be more than 1.0. On the other hand, with discrete distributions, the numbers $p(y)$ *are* probabilities.

Here is the correct way to interpret $p(y)$ for continuous pdfs.

Interpretation of $p(y)$ in Terms of Probability When $p(y)$ Is a Continuous pdf

- The probability that Y lies within $\pm\Delta/2$ of y is approximately equal to $p(y)\Delta$.
- The approximation is better when Δ is small.

This interpretation is understood by approximating the area under a curve over a Δ range using a rectangle. For example, see Figure 2.10.

Then the probability of observing a weight (in metric tons) in the range 0.08 ± 0.005 is approximately $17 \times 0.01 = 0.17$. Or in other words, about 17 out of 100 people will weigh between 0.075 and 0.085 metric tons, equivalently, between 75 and 85 kg. Clearly, this is a somewhat crude approximation, since the rectangle doesn't match the shaded area in Figure 2.10 very well. But the approximation will be better as the interval width Δ decreases.

What is the probability of someone's weight being exactly 80 kg? Not 80 rounded to the nearest kilogram or 80.0 rounded to the nearest tenth of a kilograms, but precisely 80.00 00000000000000000000000000000000000000000000000000000000000000000000000000000... kilograms, with infinite precision? Well, from the rectangle logic, the area under the pdf in the $\pm\Delta/2$ range of 80 is approximately $p(y)\Delta$, so the probability that the measurement is close to 80 can be approximated as shown in Table 2.5.

These probabilities get closer and closer to 0 as Δ decreases, and the probability of seeing exactly 80.00000000000000000000000000000000000000000000000000000000000000000000000000 00000000000... is smaller than all of them! So the only thing that makes sense is that the probability that a continuous RV equals some fixed constant is exactly zero. Let's set that off so you can find it easily later.

There is 0 probability that a continuous RV Y equals some specific value y.

This seems counterintuitive. After all, someone who weighs about 81.0 kg now *had to weigh exactly* 80.00000000000000000000000000000000000... kg at some point in time before

TABLE 2.5

Approximate Probabilities of Observing
a Weight within a Range around 80 kg

Weight Range	Approximate Probability
80 ± 0.05	$0.017 \times 0.1 = 0.0017$
80 ± 0.005	$0.017 \times 0.01 = 0.00017$
80 ± 0.0005	$0.017 \times 0.001 = 0.000017$
80 ± 0.00005	$0.017 \times 0.0001 = 0.0000017$

they weighed 81.0 kg. There are large treatises in math and even philosophy concerning conundrums like this one, all having to do with infinitesimals. The resolution is that the RV *can* equal particular values, but the set of situations where this can happen is relatively so small that the probability must be defined as zero. To think of it another way, imagine that there are 100 different ways that $Y = 80.000000000000000000000000000\ldots$ but also that there are infinitely many (∞'ly many) other equally likely possible values of Y. Then you would have to take the probability of seeing $Y = 80.000000000000000000000\ldots$ to be $100/\infty$ or 0.

As another way to understand this seemingly counterintuitive concept, recall that probabilities are areas under the continuous curve. This means the probability that $Y = 80$ is the area under the curve above 80. That's $80.000000000000000000000000000000\ldots$ exactly, not 80 rounded to the nearest integer. In other words, the shape whose area you want is a vertical line. And the width of a vertical line is 0, so the area is base $\times$ height $= 0 \times p(80) = 0$.

This would be a good time to introduce the **population** concept and give a reason we don't use it. In many statistics sources, probability is defined as follows:

Probability of an Outcome Using the Population Definition

$$\Pr(\text{outcome}) = \frac{\text{No. of elements in population where the outcome occurs}}{\text{Total no. of elements in population}}$$

Here and throughout this book, the Pr(.) symbol refers to probability of a specific outcome or set of outcomes. This differs slightly from the pdf $p(y)$ in that the pdf $p(y)$ is an entire function, not just a single number like Pr(.). Also, the pdf $p(y)$ is never a probability for a set of outcomes; it only shows likelihoods for individual outcomes.

For example, if the outcome is "someone weighs between 79.5 and 80.5 kg," and if a person is randomly sampled from a population of 1000 people, 4 of whose weights are indeed between 79.5 and 80.5, then the probability will be $4/1000 = 0.004$, using the population definition. While this is true, it is neither scientific nor useful. Is the universe completely described by these 1000 people at this particular point in time? No, there are other possibilities. So 0.004 could be considered at best, an estimate of the probability that weight is between 79.5 and 80.5. Further, this population interpretation can be bizarre: Suppose no one of the 1000 happens to weigh in that range. In that case, the probability is $0/1000 = 0.0$. But are you then to conclude that it is *impossible* for someone to weigh between 79.5 and 80.5 kg? This is clearly illogical and unscientific: All you can say is that no one in this particular group of people at this particular point in time has weight between 79.5 and 80.5 kg. This is not a general scientific truth about human physiology, so the population definition of probability is not scientific.

Recall also that in the weight example discussed earlier, the person's height is 1.7 m. When using the population definition of probability, you would have to restrict your attention to the subpopulation whose height is exactly 1.7 m and then define probability using that subpopulation. But with a precise measurement of height, with many decimals such as 1.70000000000, there will not be *anyone* in that subpopulation. The population-based definition of probability cannot even provide an answer to the question, "What is the probability that a person who is 1.7 meters tall will weigh between 79.5 and 80.5 kg?" because there is a "0" in the denominator and you can't divide by zero. Again, the population definition of probability is not scientific: Certainly it is possible, physiologically, that a person who is 1.7 m tall will weigh between 79.5 and 80.5 kg. The reason people weigh what they do is not a function of the population they belong to; it is a function of the processes that produce their weight, including genetics, diet, and exercise.

These are just some of the reasons we avoid the population terminology. Statistics sources use the population terminology frequently, so you should learn it. But you should also learn why it is wrong. We discuss population sampling and the *process* versus *population* interpretation of probability, in much more detail in Chapter 7.

As with discrete pdfs, there are infinitely many possible continuous pdfs. All you need is to satisfy two requirements. These are essentially the same as the discrete pdf requirements, with the exception that the summation symbol Σ is replaced by the integration symbol $\int$.

Requirements for a Continuous pdf

1. $p(y) \geq 0$, for all $y \in S$

2. $\int_{y \in S} p(y)dy = 1$

The first requirement states that the pdf values are never negative. The second introduces the calculus notion of an **integral** and states simply that the area under the pdf curve is 1.0. This corresponds to the requirement for a discrete distribution that the sum of all probabilities is 1.0 and is a natural requirement since area under the curve is probability. The second requirement simply states that, as in the case of the discrete distribution, the total probability is 1.0% or 100%.

One way to understand the expression $\int_{y \in S} p(y)dy = 1$ is by the rectangle logic. You can approximate the area under the curve by creating a list of rounded-off continuous measurements y_i, say, 100 or so rounded off to the nearest Δ and then add up all of the rectangles' areas. This is the method of *Riemann sums*, given by the following formula:

$$\int_{y \in S} p(y)dy \cong \sum_i p(y_i)\Delta \qquad (2.5)$$

Example 2.7: The Normal pdf

To illustrate approximation (2.5), let us introduce the most common continuous pdf, the **normal distribution**, also known as the famous (sometimes called *infamous*) "bell curve." Its function form is

$$p(y \mid \mu, \sigma^2) = \frac{1}{\sqrt{2\pi}\sigma} \exp\left\{\frac{-(y-\mu)^2}{2\sigma^2}\right\}, \quad \text{for } -\infty < y < \infty$$

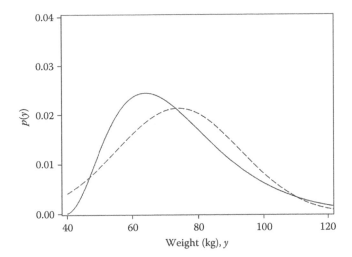

FIGURE 2.11
The normal distribution (dashed curve) as an approximate weight producer. The actual distribution of weight (solid curve) is right-skewed, while the normal approximation is symmetric.

The parameter μ is the Greek lowercase letter mu, pronounced "mew," and serves to locate the center of the distribution. The parameter σ is the Greek lowercase letter sigma and serves as a measure of the horizontal variability of the distribution. For the body weight example, approximate values of μ and σ are $\mu = 74.0$ and $\sigma = 18.7$. (There will be much more on these parameters later.) Figure 2.11 shows the resulting normal distribution superimposed over the weight distribution shown in Figure 2.8.

The normal distribution shown in Figure 2.11 is not quite right as a data producer. The reason is that the normal distribution is **symmetric**—that is, it predicts that you are just as likely to see that someone's weight is 40 kg less than the center as you are to see that their weight is 40 kg more than the center. From a human physiology perspective, this is illogical: People who have such low weight are likely to be very ill or dead. On the other hand, there are many people who are perfectly healthy despite being on the heavy side. Therefore, you should expect that the tail of the true weight distribution will be longer on the right side than on the left. This is a characteristic of what is called a **right-skew distribution**. (A **left-skew distribution**, on the other hand, has a left tail that is longer than the right tail.)

In practice, the normal distribution is never precisely correct because there is always some degree of skewness in DATA produced by Nature, design, and measurement. But the normal distribution is often a convenient and simple approximation, one that gives answers that are "good enough." Much of the remainder of this book is devoted to clarifying the meaning of "good enough," so don't worry that you don't know precisely what that means. But as a hint, look at Figure 2.11. The normal distribution is not "good enough" if you want to predict how often a hospital will see extremely obese patients: Since the extreme upper tail of the normal distribution is too small, the normal distribution model predicts too few people in this range.

The function form of the normal distribution may look puzzling:

$$p(y|\mu,\sigma^2) = \frac{1}{\sqrt{2\pi}\sigma}\exp\left\{\frac{-(y-\mu)^2}{2\sigma^2}\right\}, \quad \text{for } -\infty < y < \infty$$

Let's unpack it a little.

First, $p(y|\mu,\sigma^2)$ is a function of y, so y is a variable. The parameter vector $\theta = (\mu,\sigma^2)$ can also take on different values, but you should assume, for now, that μ and σ^2 are constants. (Recall that the $|$ symbol means *given that* or *assuming constant*.) So the form of the function is

$$p(y|\theta) = \text{Constant} \times g(y)$$

where

$$\text{Constant} = \frac{1}{\sqrt{2\pi}\sigma}$$

$$g(y) = \exp\left\{\frac{-(y-\mu)^2}{2\sigma^2}\right\}$$

There are similar multiplicative constants in many pdfs, and they all do the same thing: They ensure that the area under the pdf is 1.0. It then follows that the area under the curve

$$g(y) = \exp\left\{\frac{-(y-\mu)^2}{2\sigma^2}\right\}$$

is equal to $\sqrt{2\pi}\sigma$. That's how you get the constants: They are 1/(area under the curve *without the multiplicative constant*); this ensures that the area under the curve *with the constant* is equal to 1.0.

A curve such as $g(y)$ without the multiplicative constant is the *kernel* of the function; the kernel will appear again in Chapter 13 about Bayesian analysis, where it simplifies the calculations.

The normal distribution appears over and over in statistics. So to understand it better, you should draw its graph. Here are instructions for drawing graphs in general using Excel.

Instructions for Drawing Graphs of pdfs *p(y)*

1. In column A, put y in the first row. Then create a list of y values, such as 0.0, 0.1, 0.2, ..., 10.0 in the rows below the first row. Choose the minimum, maximum, and increment (0, 10, 0.1, e.g.) based on what you know about $p(y)$. You may need to modify them later if the graph does not show enough detail or if there is wasted space in the graph.
2. In column B, put the label $p(y)$ in the first row. Then enter the function formula for $p(y)$ in the rows below the first row, referring to the values in column A.
3. Select columns A and B simultaneously. Select "Insert," "Scatter," and "Scatter with smooth lines." You should now see a graph of the function.
4. If the graph does not cover enough range, or if it looks "choppy," change the minimum, maximum, and/or increment in step 1. Repeat step 3 and look at your graph. This may require several passes until the graph looks just right.

5. Label the horizontal axis "y" and the vertical axis "$p(y)$," or use more descriptive names.
6. Make cosmetic changes to the graph as needed: Change the axes increments or minima and maxima shown in the graph if they don't make sense. Remove the legend. Software is great but sometimes the defaults are not ideal!

Example 2.8: Verifying That the Area under the Normal Distribution Function Equals 1.0

Let's check that the area under the normal distribution curve is in fact equal to 1.0. Pick $\mu = 74.0$ and $\sigma = 18.7$, as in the weight example. You can draw the curve in Excel using the instructions shown earlier; it should look like the dashed line curve in Figure 2.11. To find the approximate area under this curve, you can use the Riemann sum approximation of Equation 2.5 with $\Delta = 0.1$. Figure 2.12 shows how the calculations look in Excel and also illustrate some of the steps that you can use to draw graphs of functions.

The formulas in the cells of Figure 2.12 are as follows:

Cell B2: = 1/(SQRT(2*3.14159265)*18.7)

Cell C2: = EXP(−0.5*(A2 − 74^2/18.7^2)

Cell D2: = B2*C2

Cell E2: = D2*0.1

Cell F2: = SUM(E2:E802)

Note that the end result, 0.958788, is not exactly 1.0. You could make the number closer to 1.0 by expanding the range. In Figure 2.11, the normal distribution seems to extend well beyond the 40–120 range, which, from the calculation shown in Figure 2.12, captures only around 96% of the total area.

	A	B	C	D	E	F	G
		S1	▾	f_x			
	A	B	C	D	E	F	G
1	y	Constant	Kernel	$p(y)$	$p(y)\Delta$	$\Sigma p(y)\Delta$	
2	40	0.021334	0.191495	0.004085	0.0004085	0.958788	
3	40.1	0.021334	0.193363	0.004125	0.0004125		
4	40.2	0.021334	0.195244	0.004165	0.0004165		
5	40.3	0.021334	0.197138	0.004206	0.0004206		
6	40.4	0.021334	0.199044	0.004246	0.0004246		
7	40.5	0.021334	0.200963	0.004287	0.0004287		
8	40.6	0.021334	0.202894	0.004329	0.0004329		
9	40.7	0.021334	0.204839	0.00437	0.000437		
10	40.8	0.021334	0.206796	0.004412	0.0004412		
11	40.9	0.021334	0.208765	0.004454	0.0004454		

FIGURE 2.12
Excel screenshot showing area calculation via the method of Riemann sums.

At this point, "So what?" is a good question to ask again. Why should you care about the normal distribution? As it turns out, most statistical methods assume that a normal distribution is in fact the data producer. This assumption is nearly always questionable, but as we will show throughout this book, it is often okay to use the normal distribution model, even when it is wrong. Further, this distribution is often a very good model for statistical estimates, such as averages, even when the data themselves are clearly non-normal. So if you want to understand statistical methods, you must understand the normal distribution very well.

As in the case of discrete distributions, there are many named continuous distributions. The most famous of these is the normal distribution, but other famous ones include the uniform, exponential, beta, gamma, Cauchy, Weibull, chi-squared, F, and t-distributions. There are infinitely many others without names that you can use as well. As always, the question "Which distribution should I use?" is answered by "a distribution that produces DATA* similar to your DATA, for some value(s) of the parameters." We give additional guidance in Chapter 4.

2.5 Some Calculus—Derivatives and Least Squares

Now that we've introduced the integral, we'd better explain more about it. You already know the single most important thing about an integral, as far as statistics goes: The integral of a pdf is simply the area under the pdf curve. Further, you can approximate this area easily using the rectangle approach, which is also important because it gives you an intuitive way to understand the integral. If you have already taken calculus, you know techniques for finding these integrals. If you have not had calculus, we'll present just enough calculus techniques so that you can understand advanced statistical methods.

Before integrals, you need to understand derivatives. For a generic function $f(x)$, the **derivative** of $f(x)$ at a point $x = x_0$ is the *slope of the tangent line* (assuming such a line exists) to $f(x)$ at the point $x = x_0$ (see Figure 2.13). Informally, a line is tangent to the curve at a point

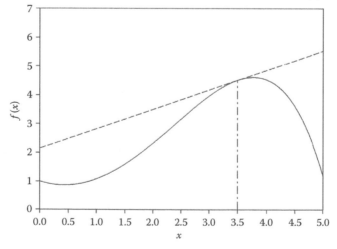

FIGURE 2.13
A generic function $f(x)$ (solid curve) showing the tangent line (dashed line) at $x = 3.5$.

if it "just touches" the curve there, in a sort of glancing blow. In Figure 2.13, the line "just touches" the curve at $x = 3.5$.

What is this mysterious $f(x)$ in Figure 2.13? If you can't stand the suspense, read ahead, but for now, it's not important that you know what it is. In fact, we'd rather you *didn't* know yet, because it will distract you from the more important points regarding the definition of a derivative that follow.

The slope of the tangent line is the derivative. Recall the following formula:

$$\text{Slope} = \frac{\text{Rise}}{\text{Run}}$$

This formula states that the slope is equal to the increase in the vertical axis variable per unit increase in the horizontal axis variable. Examining Figure 2.13, the rise of the tangent line over the entire range graph is $5.537 - 2.153 = 3.384$. The run is $5.0 - 0.0 = 5.0$, so rise/run $= 3.384/5.0 = 0.68$ is the slope. The derivative of the function shown in Figure 2.13 at $x = 3.5$ is therefore 0.68; this is written as $f'(3.5) = 0.68$.

You have a different derivative $f'(x)$ for every x. When $x = 4.5$, the tangent line is sloped downward; therefore, $f'(x) < 0$ when $x = 4.5$. Figure 2.14 shows the graph of the derivative function $f'(x)$ over x in the interval $0 \leq x \leq 5$.

Correspondences between the Derivative Function $f'(x)$ Shown in Figure 2.14 and the Original Function $f(x)$ Shown in Figure 2.13

- The derivative is positive between $x = 0.4$ and $x = 3.8$, and negative outside this interval. These cases correspond to locations where the original function is increasing and decreasing, respectively.
- The derivative is equal to zero twice in the range $0 \leq x \leq 5$, at about $x = 0.4$ and $x = 3.8$. These points correspond to a local minimum and a local maximum, respectively, in the original function.

It is important to know when the derivatives are positive and when they are negative. For example, if the relationship between $y =$ cotton yield and $x =$ fertilizer is increasing

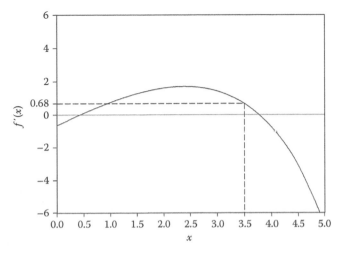

FIGURE 2.14
The derivative function $f'(x)$ corresponding to the function $f(x)$ shown in Figure 2.13, showing that $f'(3.5) = 0.68$.

when $x = 10$, then you should use more fertilizer to increase yield. But if the relationship is decreasing when $x = 10$, then you should use less fertilizer to increase yield. At the optimal fertilizer level, the derivative is zero. The case where the derivative is zero is also used to estimate parameters via least squares, covered later in this section, and via maximum likelihood, covered in Chapter 12.

How can you calculate a derivative, when the rise/run formula requires *two* x values and their corresponding y values, whereas the derivative just involves a *single* x value? It's a mystery all right! Isaac Newton and Gottfried Leibniz solved it hundreds of years ago by using the method of *successive approximation*. The idea is as follows: The slope of the tangent line to $f(x)$ at $x = 3.5$ is approximately the slope obtained using the points $x = 2.5$ and $x = 3.5$:

$$f'(3.5) \cong \frac{f(3.5) - f(2.5)}{3.5 - 2.5}$$

See Figure 2.15.

In Figure 2.15, the approximating slope is 1.35, quite a bit different—twice as much, in fact—from the actual slope of the tangent line, which is 0.68.

This is a good time to bring up an important point that you will see repeatedly throughout this book: The words *approximately equal* are **weasel words**. The term *weasel words* is an informal slang indicating words that are vague, often used with the intent to mislead. The problem with the words *approximately equal*, and the reason we call them weasel words, is that you could say anything is "approximately equal" to anything else. No one could argue, no matter how far apart the items are. You could say that the approximating slope 1.35 is approximately equal to 0.68, since after all, it's a lot closer to 0.68 than is the number 1,000,000. But you could also say that 1,000,000 is approximately equal to 0.68 as well. After all, 1,000,000 is a lot closer to 0.68 than the number 1,000,000,000,000,000,000,000,000.

If you disagree that 1,000,000 is approximately equal to 0.68, think of it this way: Measured in billions of units, the number 1,000,000 is equal to 0.001 and 0.68 is

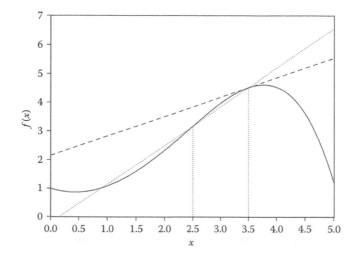

FIGURE 2.15
The original curve $f(x)$ (solid curve), the tangent line at $x = 3.5$ (dashed line), and an approximation to the tangent line using the points $x = 2.5$ and $x = 3.5$ (dotted line).

equal to 0.00000000068. Both are pretty close to zero and hence are "approximately equal," especially as compared to 1,000,000,000,000,000,000,000,000,000, which is equal to 1,000,000,000,000,000 billions. For an analogy, when viewed from the Andromeda galaxy, our Earth and Sun are in approximately the same location in space. However, from our standpoint, we are not in approximately the same location in space as the Sun at all. If we were, then we would burn to a crisp. So, depending on your point of view, you can indeed claim that $1,000,000,000 \cong 0.68$. See how weasely the words "approximately equal to" can be?

Much of the subject of statistics concerns approximations. The Mantra *data reduce the uncertainty about the unknown parameters* is itself a statement that the true parameters are approximated by using their estimated values from the data. A main concern in the subject of statistics is to identify how good these approximations are and how to make them better.

Outside the realm of numbers, a biologist could tell you "A cow is approximately a sunflower," and he would be right. A physicist might say "I am approximately a light bulb," and he would be right. Your colleague Hans might tell you, "I am approximately a brilliant researcher," and he would be right, too! There's just no quality control for the word "approximate." So, whenever you see that word or the symbol $\cong$, you should immediately think "Danger! Someone is trying to fool me!"

Never think the symbols $\cong$ and $=$ mean the same thing or that the terms *approximately equal* and *equal* are the same. Whenever you see the word *approximately* or the $\cong$ symbol, you should ask "How good is the approximation?" and "How can the approximation be made better?"

In the example where the approximate slope is 1.35 as shown in Figure 2.15, how can you make the approximation better? Simple: Just make the two points closer to 3.5. Figure 2.16 shows the approximation using $x = 3.4$ and $x = 3.5$.

As you can see in Figure 2.16, the closer the two x points are to the number 3.5, the closer the approximating slope is to the slope of the tangent, that is, to the derivative.

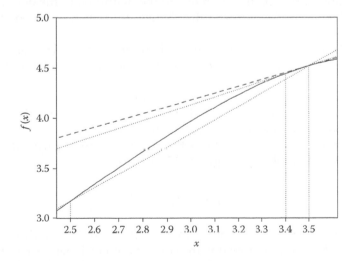

FIGURE 2.16
A zoomed-in version of Figure 2.15, comparing the approximations (dotted lines) to the tangent line slope (solid dashed line) of the curve (solid curve) at $x = 3.5$.

Thus, the conundrum regarding how to define the derivative when there is only one x value at the point of the tangent line is resolved by defining the derivative as the limit of successive approximations.

Definition of Derivative of a Function $f(x)$ at the Point Where $x = x_0$

$$f'(x_0) = \lim_{x \to x_0} \frac{f(x_0) - f(x)}{x_0 - x} \tag{2.6}$$

The symbols $x \to x_0$ are read "as x tends toward x_0."

A technical point: Equation 2.6 assumes that the limit exists and that it is the same, no matter whether x approaches x_0 from the left or from the right. For non-smooth functions, the limit may be undefined at certain points x_0.

The understanding of a derivative as a slope, and its relation to the graphs shown earlier, is the most important thing you need to know. If you have taken calculus, you may have learned a bunch of formulas for how to calculate derivatives. Those are useful skills, but without the understanding of what the formulas mean, your knowledge of derivative formulas is like that of a trained parrot. In statistics it is more important that you understand what a derivative *is* than how to calculate it. Even so, there are a few derivative formulas that you should commit to memory as long as you first understand what the formula tells you in terms of graphs of functions and slopes of tangent lines.

First, a notation convention. Sometimes it is more convenient to indicate the derivative of a function $f(x)$ using the notation $f'(x)$ as mentioned earlier, and sometimes it is more convenient to use the following notation:

$$\frac{\partial f(x)}{\partial x} = f'(x)$$

The ∂ symbol means "differential" and refers to the definition of the derivative as shown in Equation 2.6. The term $\partial f(x)$ refers to the change in $f(x)$ that corresponds to the small change ∂x in x.

The ∂ form of the derivative is most handy when the function $f(x)$ has a specified form like x^2. In such cases an expression like $\partial x^2 / \partial x$ is easier to understand than the expression $(x^2)'$.

Table 2.6 summarizes the formulas that you will need to understand this book. In the table, the symbol a or n denotes constants that can take just one value (like 3.4 or 5), while the symbol x denotes a variable that can take many values.

Now it's time for the big reveal! What was the mysterious function $f(x)$ that was graphed back in Figure 2.13 and in subsequent graphs? Here it is:

$$f(x) = 2 + x^2 - e^{0.65x}$$

Thus, the derivative function $f'(x)$ that is graphed in Figure 2.14 is given by $f'(x) = 2x - 0.65e^{0.65x}$, which can be found by applying the rules shown in Table 2.6. Pay attention to the reasons—they are the most important thing that you should learn here. The answer itself is not what you should try to learn—don't be a trained parrot!

TABLE 2.6

Essential Derivative Formulas and Their Stories

Label	Formula	Words and Stories
D1	$\dfrac{\partial a}{\partial x} = 0$	The derivative of a constant is 0.
D2	$\dfrac{\partial af(x)}{\partial x} = a\dfrac{\partial f(x)}{\partial x}$	The derivative of a constant times a function is equal to the constant times the derivative of the function.
D3	$\dfrac{\partial\{f(x)+g(x)\}}{\partial x} = \dfrac{\partial f(x)}{\partial x} + \dfrac{\partial g(x)}{\partial x}$	The derivative of a sum is the sum of the derivatives.
D4	$\dfrac{\partial x^n}{\partial x} = nx^{n-1}$	The derivative of x to an exponent is the exponent times x raised to the exponent minus one.
D5	$\dfrac{\partial \ln(x)}{\partial x} = \dfrac{1}{x}$	The derivative of the natural logarithm of x is equal to $1/x$.
D6	$\dfrac{\partial e^x}{\partial x} = e^x$	The derivative of the exponential function (base e) is equal to the exponential function itself. (This is rather remarkable and is one reason that the number $e = 2.718\ldots$ is so important.)
D7	$\dfrac{\partial e^{ax}}{\partial x} = ae^{ax}$	This formula generalizes D6; plugging $a = 1$ in gives you D6 as a special case.
D8	$\dfrac{\partial (a-x)^2}{\partial x} = 2(x-a)$	This formula is used in model fitting via least squares.
D9	$\dfrac{\partial g\{f(x)\}}{\partial x} = g'\{f(x)\} \times f'(x)$	This formula is called the *chain rule*. It's a little more complicated than the others, so we'll try not to use it too much. If you stare at them long enough, you'll recognize formulas D7 and D8 as applications of this rule.

$f'(x)$

$$= \frac{\partial(2 + x^2 - e^{0.65x})}{\partial x} \qquad \text{(By definition)}$$

$$= \frac{\partial\{2 + x^2 + (-1)e^{0.65x}\}}{\partial x} \qquad \text{(By algebra, the derivative formulas earlier mention sums but not differences, so this step is needed to write terms as sums)}$$

$$= \frac{\partial(2)}{\partial x} + \frac{\partial x^2}{\partial x} + \frac{\partial\{(-1)e^{0.65x}\}}{\partial x} \qquad \text{(By the formula labeled D3 in Table 2.6)}$$

$$= 0 + \frac{\partial x^2}{\partial x} + \frac{\partial\{(-1)e^{0.65x}\}}{\partial x} \qquad \text{(By D1)}$$

$$= 2x + \frac{\partial\{(-1)e^{0.65x}\}}{\partial x} \qquad \text{(By D4, letting } n = 2)$$

$$= 2x + (-1)\frac{\partial\{e^{0.65x}\}}{\partial x} \qquad \text{(By D2, where } a = -1)$$

$$= 2x - 0.65e^{0.65x} \qquad \text{(By D7, where } a = 0.65)$$

Example 2.9: Obtaining the Sample Mean from the Calculus of Least Squares

A common application of calculus in statistics is in finding the x that is "as close as possible" to a data set such as {3.2, 5.6, 1.0, 1.5}. Figure 2.17 shows a **dot plot** of the $n = 4$ data points in this data set.

What single number x comes closest to all of these data values? To answer, you need to know precisely what *close* means. The best x using one definition of *close* will not be the best x when you use a different definition.

A common definition of closeness is the *least squares* criterion. You can say that x is *close* to a data set if the *sum of squared deviations* is small. That is to say, an x for which $f(x) = (3.2 - x)^2 + (5.6 - x)^2 + (1.0 - x)^2 + (1.5 - x)^2$ is small is an x that is close to the data set. And the x that makes $f(x)$ as small as possible is called the **least squares estimate**.

Figure 2.18 shows a graph of the sum of squares function $f(x)$.

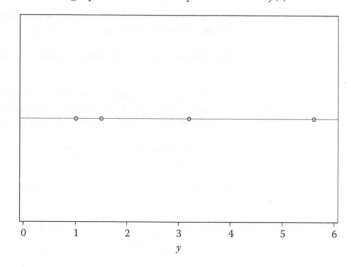

FIGURE 2.17
A dot plot of the data values {3.2, 5.6, 1.0, 1.5}.

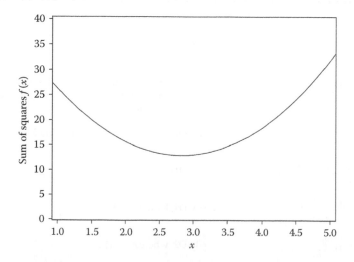

FIGURE 2.18
The sum of squares function $f(x)$.

The x that is "closest" to the data is the one that makes $f(x)$ a minimum; from Figure 2.18, this looks to be some number between 2.5 and 3.0. Note that, at the minimum, the tangent line is flat, that is, the derivative is 0. To locate the value of the "closest" x precisely, you can find the derivative function $f'(x)$ and then solve $f'(x) = 0$ for x. So first let's find the derivative function $f'(x)$.

$f'(x)$

$$= \frac{\partial\{(3.2-x)^2 + (5.6-x)^2 + (1.0-x)^2 + (1.5-x)^2\}}{\partial x} \quad \text{(By definition)}$$

$$= \frac{\partial(3.2-x)^2}{\partial x} + \frac{\partial(5.6-x)^2}{\partial x} + \frac{\partial(1.0-x)^2}{\partial x} + \frac{\partial(1.5-x)^2}{\partial x} \quad \text{(By D3)}$$

$$= 2(x-3.2) + 2(x-5.6) + 2(x-1.0) + 2(x-1.5) \quad \text{(By D8)}$$

$$= 8x - 22.6 \quad \text{(By algebra and arithmetic)}$$

When the derivative function $f'(x) = 8x - 22.6$ is positive, the function $f(x)$ graphed in Figure 2.18 is increasing. When $f'(x)$ is negative, the function is decreasing. And when $f'(x) = 0$, the function is neither increasing nor decreasing; that is, it is at the minimum. The location of the minimum is easy to find here:

$f'(x_{min}) = 0$ (The slope is zero at the minimum)

$\Leftrightarrow 8x_{min} - 22.6 = 0$ (By substitution, using the calculated derivative given earlier)

$\Leftrightarrow x_{min} = 22.6/8 = 2.825$ (By algebra)

Again, the symbol $\Leftrightarrow$ means, in words, "is equivalent to." When you work with a formula, as we just did, the formula takes a different form in each step of your work. Each of these forms is equivalent to the others. Do not confuse the phrase "is equivalent to" with the phrase "is equal to." For example, going from the second to third lines in the derivation, a mistranslation of "$\Leftrightarrow$" as "is equal to" would tell you that $0 = 2.825$, which is nonsense.

It is no accident that the least squares estimate $x_{min} = 2.825$ is the same as the ordinary average of the data, denoted by the symbol "$\bar{x}$" and calculated as

$$\bar{x} = \frac{3.2 + 5.6 + 1.0 + 1.5}{4} = 2.825$$

The least squares criterion provides a justification for using the average $\bar{x}$ as a representative value of a data set.

2.6 More Calculus—Integrals and Cumulative Distribution Functions

Albert Einstein showed us that everything is relative; it all depends on your point of view. Whether you are watching the train go by, or whether you are on the train watching the land go by, the system is the same, but the point of view differs. Similarly, integrals are part of the same system as derivatives, but the point of view differs. To introduce this new point of view, consider the weight distribution example of Section 2.4 and a new concept called the **cumulative distribution function**, or cdf for short. The cdf is the function $P(y)$ which gives the probability that a person's weight is less than or equal to y.

In general, the cdf is defined as follows, for both discrete and continuous pdfs.

Definition of Cumulative Distribution Function

$$P(y) = \Pr(Y \le y)$$

Note that the capital letter P in the cdf $P(y)$ is used to distinguish it from the lower case p in the pdf $p(y)$; it has nothing to do with the fixed/random distinction of the RV Y versus the fixed quantity y.

As y increases—for example, as weight y changes from, say, 60–70 kg—more people are included in the set of people whose weight is less than or equal to y; hence, the cdf $P(y)$ increases. Also, as y gets large without bound, *all* people are included, and hence, $P(y)$ approaches 1.0 as y increases without bound. Finally, unlike a continuous pdf $p(y)$, which is *never* interpreted directly as a probability, a cdf $P(y)$ is *always* interpreted directly as a probability, no matter whether discrete or continuous.

Figure 2.19 is the graph of the cdf $P(y)$ that corresponds to the weight (in kg) pdf $p(y)$ graphed in Figure 2.8.

How is $P(y)$ related to $p(y)$? Pretty simple! For a continuous $p(y)$

$$P'(y) = p(y)$$

In words,

> The derivative of the cumulative distribution function (cdf) of a continuous random variable is equal to the probability distribution function (pdf).

The equation $P'(y) = p(y)$ has applications beyond statistics; in fact, it is a very famous equation, having the highfalutin designation as **the Fundamental Theorem of Calculus**. Despite its highfalutin designation, it's rather easy to see why $P'(y) = p(y)$ is true. From the derivative definition shown in Equation 2.6

$$P'(y_0) = \lim_{y \to y_0} \frac{P(y_0) - P(y)}{y_0 - y}$$

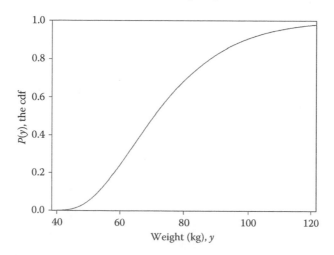

FIGURE 2.19
The cdf of weight.

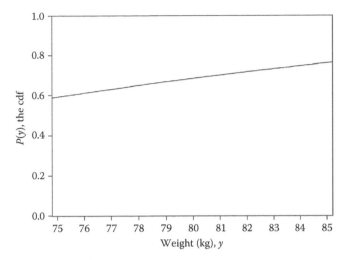

FIGURE 2.20
The cdf $P(y)$ in the {75–85 kg} range.

Suppose you are interested in $y_0 = 80$ kg. Figure 2.20 is Figure 2.19 but zoomed-in on the {75–85 kg} range.

Now, pick a y close to 80, like 79. Then $P(y_0) - P(y) = P(80) - P(79)$, which is the cumulative probability of weight up to 80 kg, minus the cumulative probability of weight up to 79 kg. This difference is just the probability of someone's weight being in the range from 79 to 80. It can be shown as area under the curve of the pdf $p(y)$ and approximated using a rectangular region as shown in Figure 2.21.

In Figure 2.21, the shaded area is exactly $P(80) - P(79)$. The rectangular approximation shows that

$$P(80) - P(79) \cong \text{Base} \times \text{Height} = (1.0) \times p(80)$$

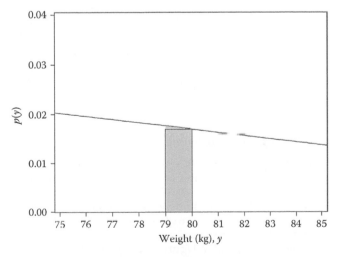

FIGURE 2.21
The probability between 79 and 80 kg (shaded) and the approximating rectangle.

Weasel word alert! Here you see the "approximately equals" symbol, $\cong$, again. This approximation can be made better with a smaller Δ:

$$P(80) - P(80 - \Delta) \cong \text{Base} \times \text{Height} = \Delta \times p(80)$$

Let $\Delta = 80 - y$. Then

$$\frac{P(80) - P(y)}{80 - y} \cong \frac{(80 - y)p(80)}{80 - y} = p(80)$$

The approximation becomes better when y is closer to 80. In other words, the derivative of the cdf is equal to the pdf.

Our demonstration that $P'(y) = p(y)$ is not mathematically rigorous. The intuitive idea is simple, as shown earlier, but it took the most brilliant mathematical minds in the history of our world hundreds of years to rigorously remove the "weaseliness" from the phrase *approximately equal to* in the proof of the Fundamental Theorem of Calculus.

With the knowledge that $P'(y) = p(y)$, you now know the most important thing you can possibly know about integral calculus. You are also subject to all kinds of devious calculations your instructor might throw at you!

Here is some calculus notation. The expression $\int_a^b p(y)dy$ is called a *definite integral* and refers to the *area under the curve $p(y)$* between constants a and b. For example, if $a = 70$ kg and $b = 80$ kg, then $\int_{70}^{80} p(y)dy$ is the area under the weight pdf curve from 70 to 80 kg. This area is also the probability that a person weighs between 70 and 80 kg, which is the cumulative probability up to 80 kg, minus the cumulative probability up to 70 kg. Thus, $\int_{70}^{80} p(y)dy = P(80) - P(70)$. In general

$$\int_a^b p(y)dy = P(b) - P(a) \tag{2.7}$$

All you need now is $P(y)$. But you know that because you know that $P'(y) = p(y)$. And now you know the most important parts of the calculus story!

The rest are technical details, which are useful to gain practice and experience. But if you haven't understood the forgoing discussion, please go back and read it again. The technical calculus details are worthless without understanding why they are true. Please, please, don't be a trained parrot!

Example 2.10: The Triangular Distribution

To get started simply, suppose you will observe a measurement Y produced by the pdf $p(y)$ shown in Figure 2.22, an example of the distribution known as the **triangular distribution**.

This pdf might be used to model student grades that are between 0 and 100, for example. It produces grades that are more often higher than 50 than lower than 50. No grades are more than 100 or less than 0.

The equation of the pdf graphed in Figure 2.22 is $p(y) = 0.0002y$, for $0 \leq y \leq 100$, and $p(y) = 0$ otherwise. The vertical dotted line in Figure 2.22 is not part of the function; it is just there to help you see that $p(y) = 0$ when $y > 100$.

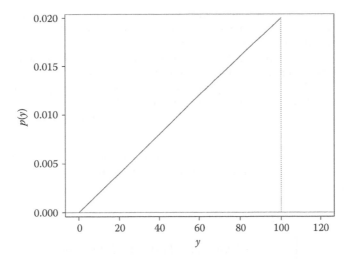

FIGURE 2.22
A triangular pdf.

Now, why 0.0002? Notice that the pdf is a triangle and that the height of the triangle is 0.02. The formula for area of a triangle is (1/2) × base × height, which is here equal to (1/2) × 100 × 0.02 = 1.0, as required of a pdf. Thus, the odd-looking constant 0.0002 is simply what is required to make the area under the pdf equal to 1.0.

Suppose this triangle distribution *does* produce the grades. What proportion of the grades will fall between 70 and 90, which, in some grading systems, translates to a grade of a "B" or "C"? This is the area under the pdf between 70 and 90, represented as follows:

$$\int_a^b p(y)dy = P(b) - P(a)$$

or

$$\int_{70}^{90} (0.0002y)dy = P(90) - P(70)$$

What are $P(90)$ and $P(70)$? It's actually easier to answer the more general question "What is $P(y)$?" The Fundamental Theorem of Calculus tells you that $P'(y) = p(y)$ or here, $P'(y) = 0.0002y$. So all you need is a function $P(y)$, such that when you take its derivative you get $0.0002y$. Let's play around a little.

The Hunt-and-Peck Method for Finding an Integral

- Is $P(y) = y$? No, since then you would have $P'(y) = 1$, and you want $P'(y) = 0.0002y$.
- Is $P(y) = 0.0002y$? No, since then you would have $P'(y) = 0.0002$, and you want $P'(y) = 0.0002y$.

- Is $P(y) = y^2$? No, since then you would have $P'(y) = 2y$, and you want $P'(y) = 0.0002y$. But you're getting closer, since there is a "y" in the derivative, and you need that. So keep the "y^2" and keep trying.
- Is $P(y) = 0.0002y^2$? No, since then you would have $P'(y) = 0.0004y$, and you want $P'(y) = 0.0002y$. But now you have almost figured it out.
- Is $P(y) = 0.0001y^2$? Well it certainly could be because then you would have $P'(y) = 0.0002y$, as you wanted. You're done right?
- Oh no! Could it also be that $P(y) = 0.0001y^2 + 1$? Then you would also have $P'(y) = 0.0002y$.

Actually, the equation $P'(y) = 0.0002y$ doesn't uniquely determine $P(y)$, since $P(y)$ can be $0.0001y^2 + c$, where c is any constant. Fortunately, you don't have to worry about the constant since it cancels out:

$$\int_a^b p(y)dy = \{P(b)+c\} - \{P(a)+c\} = P(b) - P(a)$$

While the "hunt-and-peck method" described earlier is not the way to go about solving this kind of problem in practice, it is useful to understand so that you remember the big picture, which is $P'(y) = p(y)$, and also so that you can check your work. You may find what you think to be the right $P(y)$ from somewhere (maybe the Internet), but if a calculation of $P'(y)$ does not yield $p(y)$, then you know there was a mistake.

Now you can write the following equations:

$$\int_{70}^{90} (0.0002y)dy = P(90) - P(70)$$
$$= 0.0001(90)^2 - 0.0001(70)^2$$
$$= 0.81 - 0.49 = 0.32$$

Thus, 32% of the grades will be between 70 and 90 (or Bs and Cs), according to this model.

There is commonplace shorthand for showing the evaluations of integrals. It is useful, but unfortunately, it takes the important $P(y)$ out of the picture. It looks like this:

$$\int_{70}^{90} (0.0002y)dy = (0.0001)y^2 \Big|_{70}^{90}$$
$$= 0.0001(90)^2 - 0.0001(70)^2$$
$$= 0.81 - 0.49 = 0.32$$

You get the same answer either way. The term $(0.0001)y^2 \Big|_{70}^{90}$ is the shorthand for $P(90) - P(70)$. We suggest you use the shorthand $P(y)\Big|_a^b$ in place of $P(b) - P(a)$ because it is standard notation. But we also suggest that you think about the meaning of cumulative area $P(y)$, rather than apply the shorthand blindly.

You could have also found the answer 0.32 by calculating areas of the triangles: The triangle with base from 0 to 90 has area (1/2) × base × height = (1/2) × (90) × (0.0002) × (90), and the triangle with base from 0 to 70 has area (1/2) × base × height = (1/2) × (70) × (0.0002) × (70). Taking the difference gives 0.0001(90)² − 0.0001(70)² = 0.32, the same result. It's nice when there are alternative ways to calculate areas, such as the rectangle or triangle formulas, as this gives you a simple way to check your answers. More importantly, simple rectangular and triangular formulas give you a deeper intuitive understanding of integral calculus. However, most of the interesting distributions in statistics are not rectangular or triangular. For such distributions, the simple formulas are still useful to provide quick approximations, but you need formal calculus to get the precise answers. The following is a realistic example of such a distribution.

Example 2.11: Waiting Times and the Exponential Distribution

Do you hate being put on hold when you call technical support for help? You're not alone! Companies realize that this is irritating to customers and that customer loyalty can be adversely affected by long waiting times, so they monitor their waiting time data carefully.

Let Y denote the time you have to wait on your next call. One model for producing DATA* that look like Y is the **exponential distribution**, defined as follows:

$$p(y|\lambda) = \lambda e^{-\lambda y}, \quad \text{for } y \geq 0$$

It may go without saying, but $p(y|\lambda) = 0$ for $y < 0$, because you can't have a negative waiting time!

The term λ is another one of those parameters you keep hearing about (*model has unknown parameters*). For the exponential distribution, the parameter λ is the reciprocal of the mean. For instance, if the average waiting time is 2.0 minutes, then $\lambda = 0.5$.

Suppose indeed that $\lambda = 0.5$. Then the distribution of waiting time is as shown in Figure 2.23.

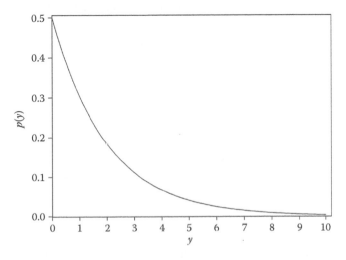

FIGURE 2.23
The exponential distribution with $\lambda = 0.5$.

If this model is reasonable, how often will customers have to wait between 1 and 5 minutes? The calculation is as follows:

$$\int_1^5 0.5e^{-0.5y}dy = P(5) - P(1)$$

Here, $P(y)$ is a function such that $P'(y) = 0.5e^{-0.5y}$. Noting from rule D7 in Table 2.6 that $\partial e^{-0.5y}/\partial y = -0.5e^{-0.5y}$, you can determine $P(y) = -e^{-0.5y} + c$. Again, the constant c cancels, so you can write the integral as follows:

$$\int_1^5 0.5e^{-0.5y}dy = P(5) - P(1) = (-e^{-0.5(5)}) - (-e^{-0.5(1)})$$

$$= e^{-0.5} - e^{-2.5} = 0.6065 - 0.0821 = 0.524$$

Thus, the wait times for 52.4% of the calls will be between 1 and 5 minutes if this model is valid.

To make the calculation of the integrals such as this one easier, you should memorize the integral formulas shown in Table 2.7. This table gives you *indefinite integrals*; when you place the limits of integration on the integral, such as the range of 1–5 minutes, the result is called a *definite integral*. As in Table 2.6, the symbols a and n denote constants, while the symbol y denotes a variable. Notice how the integral formulas I1 through I7 in Table 2.7 correspond to the derivative formulas D1 through D7 in Table 2.6.

TABLE 2.7

Some Essential Indefinite Integral Formulas (the Term "+c" Is Omitted from All) and Their Stories

Label	Formula	Words and Stories
I1	$\int ady = ay$	The integral of a constant a is the linear function with slope $= a$.
I2	$\int af(y)dy = a\int f(y)dy$	The integral of a constant times a function is equal to the constant times the integral of the function.
I3	$\int \{f(y) + g(y)\}dy = \int f(y)dy + \int g(y)dy$	The integral of a sum is the sum of the integrals.
I4	$\int y^n dy = \dfrac{y^{n+1}}{n+1}$, when $n \neq -1$.	This corresponds to D4 in Table 2.6, but is the inverse operation. Note that $n = -1$ is not allowed since then there would be a division by zero.
I5	$\int y^{-1}dy = \ln(y)$	This takes care of the $n = -1$ case in I4.
I6	$\int e^y dy = e^y$	The integral of the exponential function (base e) is equal to the exponential function. Again, this is a remarkable fact about the number $e = 2.718\ldots$.
I7	$\int e^{ay}dy = \dfrac{e^{ay}}{a}$, when $a \neq 0$.	This corresponds to D7 in Table 2.6, but the inverse operation.

To see how to use the indefinite integral formulas in Table 2.7, consider the triangular distribution $p(y) = 0.0002y$, for $0 \le y \le 100$. Here

$$\int p(y)dy$$

$$= \int (0.0002y)dy \qquad \text{(By substitution)}$$

$$= (0.0002)\int ydy \qquad \text{(By I2, using } f(y) = y)$$

$$= (0.0002)\frac{y^2}{2} \qquad \text{(By I4, using } n = 1)$$

$$= 0.0001y^2 \qquad \text{(By algebra)}$$

For another example, consider the waiting time distribution $p(y) = 0.5e^{-0.5y}$. Here

$$\int p(y)dy$$

$$= \int 0.5e^{-0.5y}dy \qquad \text{(By substitution)}$$

$$= 0.5\int e^{-0.5y}dy \qquad \text{(By I2, using } f(y) = e^{-0.5}y)$$

$$= 0.5\frac{e^{-0.5y}}{-0.5} \qquad \text{(By I7, using } a = -0.5)$$

$$= -e^{-0.5y} \qquad \text{(By algebra)}$$

The waiting time example is one where the usually omitted constant c of the indefinite integral has special importance. Specifically, the cumulative probability distribution is most definitely *not* given as $P(y) = -e^{-0.5y}$; after all, this would produce negative probabilities. It turns out in this case that the constant $c = +1$ gives the cdf. To see why this is true, suppose that you want the cumulative probability that the waiting time is less than 1 minute. This probability is represented as follows:

$$P(1) = \int_0^1 0.5e^{-0.5y}dy = -e^{-0.5y}\Big|_0^1 = (-e^{-0.5(1)}) - (-e^{-0.5(0)}) = -e^{-0.5} + 1$$

The cdf, or cumulative probability up to a number y, is represented as

$$P(y) = \int_0^y 0.5e^{-0.5t}dt = -e^{-0.5t}\Big|_0^y = (-e^{-0.5(y)}) - (-e^{-0.5(0)}) = -e^{-0.5y} + 1$$

Where did the t come from? Well, any variable other than y would work. The expression $P(y) = \int^y p(y)dy$ is simply nonsense, since the same letter y is used to denote two different things, (i) a constant upper limit and (ii) a variable whose value can be anything less than the upper limit. So the expressions $P(y) = \int^y p(t)dt$, $P(y) = \int^y p(u)du$, $P(y) = \int^y p(x)dx$,

$P(y) = \int^y p(a)da$, and even $P(y) = \int^y p(\text{dog})d\,\text{dog}$ are all correct, but the expression $P(y) = \int^y p(y)dy$ is incorrect.

Sometimes infinity (represented by the symbol ∞) is a limit of integration. For example, the cdf formula itself is given as

$$P(y) = \Pr(Y \le y) = \int_{-\infty}^{y} p(t)dt$$

In the case of the exponential waiting time pdf, this expression is evaluated as

$$P(y) = \int_{-\infty}^{y} 0.5e^{-0.5t}dt = \int_{-\infty}^{0} 0\,dt + \int_{0}^{y} 0.5e^{-0.5t}dt = 0 + (-e^{-0.5y} + 1)$$

Note that the area under a function with zero height is also zero.

But in many cases, the function is not zero, and there is an infinite limit of integration. For example, the probability that your waiting time will be more than 5 minutes is written as

$$\Pr(Y > 5) = \int_{5}^{\infty} 0.5e^{-0.5y}dy = -e^{-0.5y}\Big|_{5}^{\infty} = -e^{-0.5(\infty)} - (-e^{-0.5(5)}) = -e^{-0.5(\infty)} + e^{-2.5}$$

How to interpret an expression like $e^{-0.5(\infty)}$? The trick is to replace ∞ by something really, really big, like 1,000,000. But infinity is larger than 1,000,000, so then replace it by something even bigger, like 10,000,000. Repeat with even larger numbers. A pattern may emerge so that you can see that the numbers converge to some value (like zero); even more importantly, you will understand the logic of the convergence. If the numbers do converge, then you can take an expression involving an ∞ to be that converged value.

For instance, replacing ∞ with 1,000,000, you get $e^{-0.5(1,000,000)} = e^{-500,000} = 1/e^{500,000}$. This is an extremely small number, very close to zero. You can also see logically that the numbers get smaller, closer and closer to zero, as you replace ∞ with 10,000,000, then 100,000,000, and so on; hence, you can take the value of $-e^{-0.5(\infty)}$ to be 0.0.

Vocabulary and Formula Summaries

Vocabulary

Random variable	An entity that can take one of a set of possible values, and whose particular manifestation is uncertain.
Fixed quantity	An entity that is not random.
Continuum	A continuous range of numbers.

Sample space	The set of possible values of a RV Y, abbreviated S.
Discrete data	Data whose sample space can be listed.
Continuous data	Data whose sample space is a continuum.
Nominal data	Data whose values are labels, not numbers.
Ordinal data	Discrete data that have numerical value.
Probability distribution function (pdf)	A function $p(y)$ that tells you (1) the *set* of possible data values that will be produced and (2) the frequencies of occurrences of the different data values that will be produced.
List form (of a discrete pdf)	A representation of the function $p(y)$ using the list of possible values (y_1, y_2, ...) in one column, and their associated probabilities ($p(y_1)$, $p(y_2)$, ...) in the next column.
Function form (of any pdf)	A representation of the pdf $p(y)$ as a mathematical function of y; for example, $p(y) = \lambda e^{-\lambda y}$.
Multinomial distribution	A probability distribution used for nominal RVs.
Indicator function	A function whose value is 1 when a condition is true, and 0 when it is false.
Euler's constant	The famous number $e = 2.718...$.
Area under the function	Also called an *integral*, used to model continuous probability.
Area of a rectangle	The formula base × height, often used to approximate the area under a curve.
Discrete distribution	A distribution function $p(y)$ that assigns probability to different discrete y values, for which the sum of all $p(y)$ values is 1.0.
Continuous distribution	A distribution function $p(y)$ that assigns relative likelihood to y values falling in a continuum, for which the total area under the curve $p(y)$ is 1.0.
Latent variable	A variable, such as the true waiting time measured to infinite decimals, that cannot be observed directly.
Population	A collection of items. In statistics, it is often discussed in the context of a smaller subset (called a *sample*) taken from the population.
Process	The collective mechanisms that produce the data.
Integral	A mathematical operation describing the area under a curve, when the curve is positive.
Normal distribution	An example of a continuous symmetric distribution, the most famous distribution in statistics.

Symmetric distribution	A distribution whose left tail, when reflected about the center of the distribution, is identical to the right tail.
Right-skew distribution	A distribution whose right tail produces more extreme values than does the left tail.
Left-skew distribution	A distribution whose left tail produces more extreme values than does the right tail.
Derivative	A mathematical operation describing the slope of the line that is tangent to a function.
Weasel words	Words that are vague, often used with the intent to mislead.
Dot plot	A plot where data are indicated by dots on a number line.
Least squares estimate	A quantity that minimizes sum of squared deviations.
Cumulative distribution function (cdf)	The function $P(y) = \Pr(Y \le y)$, giving the cumulative probability that an RV Y is at most the value y.
Fundamental Theorem of Calculus	A theorem stating that the derivative of the integral of a function is equal to the function being integrated.
Triangular distribution	A pdf $p(y)$ whose graph looks like a triangle.
Exponential distribution	The pdf given by $p(y) = \lambda \exp(-\lambda y)$, for $y > 0$.

Key Formulas and Descriptions

$\sum_{y \in S} p(y) = 1.0$	The sum of discrete probabilities is 1.0.
$\int_{y \in S} p(y)dy = 1$	The integral of a continuous pdf is 1.0.
$p(y \mid \pi) = \pi^y (1 - \pi)^{1-y}$, for $y \in \{0,1\}$.	The Bernoulli pdf.
$y! = 1 \times 2 \times 3 \times \cdots \times y$, $0! = 1$.	The factorial function.
$p(y \mid \lambda) = \lambda^y e^{-\lambda}/y!$, for $y = 0,1,2,\dots$.	The Poisson probability distribution function.
$\tilde{p}(y) = p(y_i)/\Delta$, for $y_i - \Delta/2 \le y < y_i + \Delta/2$.	The rectangular approximation to a continuous pdf, based on a discrete pdf $p(y_i)$ with increments Δ.
$p(y \mid \mu, \sigma^2) = \dfrac{1}{\sqrt{2\pi}\sigma} \exp\left\{\dfrac{-(y-\mu)^2}{2\sigma^2}\right\}$, for $-\infty < y < \infty$.	The normal pdf.
$f'(x_0) = \lim_{x \to x_0} \dfrac{f(x_0) - f(x)}{x_0 - x}$	The definition of the derivative of a function $f(x)$ at the point $x = x_0$.

$$f'(x) = \frac{\partial f(x)}{\partial x}$$
Alternative expressions for the derivative.

Specific derivative formulas
See D1–D9 in Table 2.6.

$P'(y) = p(y)$
The derivative of the cdf equals the pdf.

$$P(y) = \int_{-\infty}^{y} p(t)dt$$
The area under the pdf to the left of y equals the cdf.

Specific integral formulas
See I1–I7 in Table 2.7.

Exercises

2.1 Using software, draw a graph of the pdf $p(y) = e^{-y}$, using the range $0 \le y \le 10$, where $e = 2.7183$ is the famous number called Euler's constant. Put y on the horizontal axis and $p(y)$ on the vertical axis. Make the graph a smoothly connected curve with no points or dots.

2.2 Suppose the wingspan of peregrine falcons is normally distributed with mean 1 m and standard deviation 0.1 m. Draw a graph of this pdf using the built-in normal distribution function in the computer (in Excel, it is called NORM.DIST). Put "Wingspan, y" on the horizontal axis, and "$p(y)$" on the vertical axis. Make the graph a smooth connected curve with no points or dots.

2.3 The *chi-squared distribution* is a famous one that is used to model variances. One example of a chi-squared distribution (there are many others) is $p(y) = (1/16)y^2e^{-y/2}$, for $y \ge 0$. Using the computer, draw a graph of this pdf that extends far enough to the right so that the essential range of the distribution is visible, but not so far that there is wasted space in the graph.

2.4 The *Cauchy distribution* is also famous and is used to model outlier-prone processes. It is given by $p(y) = (1/\pi)\{1/(1 + y^2)\}$ for $-\infty < y < \infty$. Using the computer, draw a graph of this pdf that extends far enough to the right and left so that the essential range of the distribution is visible, but not so far that there is wasted space in the graph.

2.5 Without data, pick examples of your own choosing involving a variable quantity Y. Examples of such a Y include survey response, financial outcome, height, weight, time until events, number of wolves in an area, etc. Without any data, identify examples where the distribution $p(y)$ of Y is (1) discrete and roughly symmetric, (2) discrete and right-skewed, (3) discrete and left-skewed, (4) continuous and roughly symmetric, (5) continuous and right skewed, and (6) continuous and left-skewed. Create hand-drawn graphs of how the pdfs of such a Y would look in all cases, labeling and numbering the both axes. Explain why, in terms of the subject matter, you think that the pdf looks like it does.

2.6 Specify $p(y)$ in both list and function forms when Y has the Bernoulli distribution with $\pi = 0.3$. (**NOTE:** When the parameter θ is known or specified to be some particular value like 0.3 in this exercise, then the "$|\theta$" term in the function form "$p(y|\theta)$" can be dropped, leaving just $p(y)$.)

2.7   Look up the Bernoulli distribution on the Internet. Make a table showing how the algebraic terms described in this chapter concerning the Bernoulli distribution correspond to terms introduced on that Internet page.

2.8   Show how the function form of the car color distribution in Example 2.4 gives you the list form shown in Table 2.3. Do this by plugging in y = red, then y = gray, and then y = green into the function form and simplifying. Provide explanations for each step.

2.9   Construct the list form of the Poisson distribution, with actual numbers given as rounded to three decimals, when $\lambda = 0.5$. Draw a graph of the resulting distribution, labeling and numbering the axes. Repeat when $\lambda = 1.0$ and describe how the data produced by these two models are (1) similar and (2) different.

2.10   Search the Internet for "negative binomial distribution." How is this distribution similar to the Poisson distribution? How is this distribution different from the Poisson distribution?

2.11   *Model produces data*. For each of the pdfs $p(y)$ graphed in the following, produce, *from your mind*, a list of $n = 10$ randomly selected observations. For example, if the pdf was Bernoulli, your list might look like 0, 0, 0, 1, 0, 1, 1, 0, 0, 0.

A.

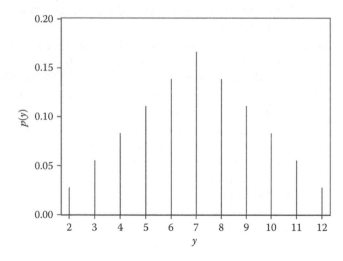

B.

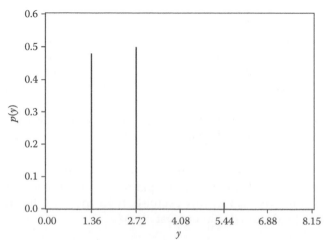

C.

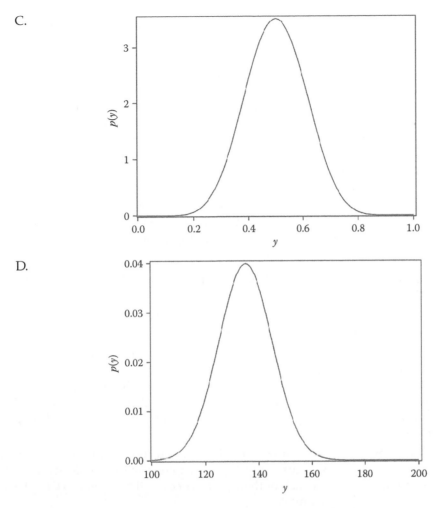

D.

2.12 A biologist counts the number of pupfish found in each of 100 water scoops. Results are as follows:

Pupfish	Water Scoops
0	35
1	34
2	24
3	6
4	1
Total	100

A. Do the data look like they could have come from a Poisson distribution with $\lambda = 1$? Generate 10 tables as shown earlier, each having 100 scoops, using this distribution. Then make a qualitative conclusion regarding whether the model is "good."

B. Assume the model with $\lambda = 1$ is good. Since models are assumed to produce the observed data, they allow you to generalize to other data. Using the 10 repeated samples of 100 scoops generated from the Poisson distribution with $\lambda = 1$, how often (out of the 10 repeats) did you see more than four pupfish?

2.13 A quality control engineer records widths of $n = 30$ computer chips produced in the factory; they are 311, 304, 316, 308, 312, 308, 314, 307, 302, 311, 308, 300, 316, 304, 316, 306, 314, 310, 311, 309, 311, 306, 311, 309, 311, 305, 304, 303, 307, and 316.

   A. The quality control engineer states that the process is running smoothly when the data behave as if produced from a normal distribution with $\mu = 310$ (the target width) and $\sigma = 4.5$ (the natural process variation). Do the data look like they could have been produced by normal distribution with $\mu = 310$ and $\sigma = 4.5$ and then rounded to the nearest integer? Generate 10 samples of $n = 30$ from this distribution, compare the resulting rounded-off numbers to those given, and make a qualitative conclusion regarding whether the model is good.

   B. Assume the model with $\mu = 310$ and $\sigma = 4.5$ is good—that is, that the process is running smoothly. Use the 10 repeated samples of 30 observations generated from the given normal distribution to answer the question, "Even when the process is running smoothly, how far can the chip width be from the target?"

2.14 Consider the function $f(x) = (2 - x)^2 + (10 - x)^2$.

   A. Using the computer, draw a graph of this function that shows its minimum clearly.

   B. Using calculus, find the minimum. Give reasons for every step.

   C. Relate your answer in Exercise 2.14B to least squares as discussed in this chapter.

2.15 Your colleague Hans claims that product preference is related to product complexity with an inverted "U" shape: When a product has too few features, people don't like it because it does not do what they want it to do. When the product has too many features, people don't like it because it does more than they want. His deterministic model is Preference = $6 + $ Complexity $- 0.03$ Complexity^2, where the Complexity measure lies between 0 and 30.

   A. Explain why Hans' deterministic model is a bad model. To do so, consider 10 different people who evaluate a particular product with complexity 15.0. What does the model predict each of their preferences will be? Refer to Chapter 1 for the definition of a good model.

   B. Using the computer, draw a graph of the deterministic model that shows its maximum clearly. Label the axes "Complexity" and "Preference."

   C. Using calculus, find the value of Complexity that produces maximum Preference when using this model. Give reasons for every step.

2.16 A pdf is $p(y) = 0.01$, for $0 < y < 100$. (This is an example of a *uniform distribution*.)

   A. Is this a discrete pdf or a continuous pdf? How can you tell?

   B. Using the computer, draw a graph of this function.

   C. Referring to Exercise 2.16B, what kind of DATA* is produced by this distribution? Describe how a sample of such DATA* would look, say, if in your spreadsheet. (See Exercise 2.13 earlier, for example.)

2.17 See Example 2.8. Using the same methods shown there, show how to use the method of Riemann sums to see that the area under the normal pdf is much closer to 1.0 than the reported value 0.958788.

2.18 Which of the following functions $f(y)$ are pdfs? First, identify whether discrete or continuous, and then check the necessary conditions. In all cases, assume $f(y) = 0$ for values of y not given.

A. $f(y) = 1/6$, for $y = 1, 2, ..., 6$

B. $f(y) = 1/6$, for $y = 1, 2, ..., 10$

C. $f(y) = 1/k$, for $y = 1, 2, ..., k$

D. $f(y) = y - 5.4$, for $y = 1, 2, ..., 10$

E. $f(y) = (\frac{1}{2})^y$, for $y = 1, 2, ...$

F. $f(y) = 0.25$, for $y \in \{-4.5, -4.2, 0.2, 100.1\}$

G. $f(y) = 1/6$, for $1 \le y \le 6$

H. $f(y) = 1/6$, for $0 \le y \le 6$

I. $f(y) = 1/6$, for $20 \le y \le 26$

J. $f(y) = y^2$, for $0 < y < 1$

K. $f(y) = 1/y^2$, for $0 < y < 1$

L. $f(y) = 1/y^2$, for $1 < y < \infty$

2.19 A pdf is $p(y) = a + e^{-y}$, for $0 < y < 2$; $p(y) = 0$ otherwise, where a is a constant.

A. Find a.

B. Using the computer, graph the pdf.

2.20 A pdf is $p(y) = a/y$, for $1 < y < 10$; $p(y) = 0$ otherwise, where a is a constant.

A. Find a.

B. Using the computer, graph the pdf.

3

Probability Calculation and Simulation

3.1 Introduction

Probabilities are useful for prediction. How often will you win at solitaire? How often will you wreck your car on the way to work? How often will your research study turn out the way you want? Probabilities can't predict these outcomes *individually*, but they can give you *aggregate* predictions, and these predictions are very useful for making decisions. For example, in your potential futures, you will die early more frequently if you don't wear your seat belt than if you do wear it. This knowledge gives you a rational basis for making the decision to buckle your seat belt. But, it is entirely possible that you could die of natural causes after a long life of never wearing a seat belt; that is one of your potential futures. Similarly, a chain-smoker might live a long life cancer-free; that is one of the chain-smoker's potential futures.

It is very common for people who are unfamiliar with the notion of probability to use *anecdotal claims*. An anecdotal claim is based on a person's individual experience and is very limited in its generality. You may have heard someone say that a particular scientific study is flawed because they know of a case where the study's findings did not hold; this is an example of an anecdotal claim. Anecdotal claims use only one potential future path; there are many others that are possible. While no one can say for certain what will happen to you individually—what your ultimate future path will be—statistical models allow you to make strong claims about, for example, what will happen to a large number of people who are like you in some regard (e.g., gender, age, occupation). This information provides a rational basis for predicting how your particular collection of plausible future paths might look. While anecdotal claims consider only one possible future path, statistical models consider them all. Thus, claims based on statistical models are generally more valid than anecdotal claims.

This chapter covers probabilistic foundations of statistical models, showing how you can calculate probabilities associated with them. These probabilities can help you to establish that results are generalizable rather than anecdotal. Many of the calculations we present will be hypothetical because they will assume that you know the model's parameters. Recall the Mantra: *Model produces data; model has **unknown** parameters*. The true probabilities are examples of unknown parameters. You never know the parameters with perfect precision, but in some cases, you can know what they are well enough. The most famous cases involve games of chance like craps, roulette, blackjack, and slots. These cases resemble coin flips, where the parameter of interest is known to be $\pi = 0.5$—or is at least so close to 0.5 that the assumption of $\pi = 0.5$ provides excellent predictions. In the casino games, the probabilities are also known well enough. They aren't 0.5, like the coin flip, but they are values that the casinos know will make you lose.

While you can't know the probabilities in practice, outside of casino games, coin flipping, die tossing, etc., hypothetical calculations based on known probabilities are not useless! They form the basis for *hypothesis testing*, a commonly used method for assessing whether your data can be explained by chance alone, discussed further in Chapter 15.

3.2 Analytic Calculations, Discrete and Continuous Cases

Example 3.1: Auto Fatalities

Imagine a discrete random variable (RV) Y—say the number of fatal automobile accidents in a given day in a large city. The sample space of possible values for Y is $S = \{0, 1, 2, \ldots\}$. One prediction of interest to the traffic management authorities is whether there will be any fatalities on a given day. This eventuality is called, logically enough, an **event** and is represented by the set $\{1, 2, \ldots\}$, which is the same as the entire sample space with the $\{0\}$ excluded.

Generically, an event is a *subset* of the sample space. Call the subset A; then $A \subset S$, which is read aloud as "the event A is a subset of the sample space S."

More notation: Recall from Chapter 2 that the symbol Pr means "probability." The probability that the RV Y will be in the set A is denoted by the symbolic expression $\Pr(Y \in A)$. In the example given earlier, where $A = \{1, 2, \ldots\}$, you can also write this probability as $\Pr(Y \geq 1)$.

Let $p(y)$ be the probability distribution function (pdf) for a discrete RV Y. Probabilities for events are calculated as follows.

Probability of an Event A for a Discrete RV

$$\Pr(Y \in A) = \sum_{y \in A} p(y)$$

In words, this equation tells you that "the probability that the discrete RV Y is in the set A is equal to the sum of the probabilities of the individual outcomes that are in the set A."

While it is important to remember that the discrete case differs from the continuous case, the formulas are fortunately very similar, so you don't have to remember twice as many. Just remember that summation Σ in the discrete case becomes an integral $\int$ in the continuous case, and you'll be fine.

Probability of an Event A for a Continuous RV

$$\Pr(Y \in A) = \int_{y \in A} p(y)dy$$

Back to car fatalities: Suppose, hypothetically, that the number of fatal accidents in a day looks as if produced by a Poisson distribution with parameter $\lambda = 0.5$. You can find the probability of one or more fatal accidents on a given day as follows:

$\sum_{y \in A} p(y)$	(Since the event "one or more" means that Y is in the set $A = \{1, 2, \ldots\}$)
$= \sum_{y \in \{1,2,\ldots\}} p(y)$	
$= \sum_{y \in \{1,2,\ldots\}} \dfrac{e^{-0.5} 0.5^y}{y!}$	(By substituting the Poisson pdf with $\lambda = 0.5$)
$= \dfrac{e^{-0.5} 0.5^1}{1!} + \dfrac{e^{-0.5} 0.5^2}{2!} + \dfrac{e^{-0.5} 0.5^3}{3!} + \dfrac{e^{-0.5} 0.5^4}{4!} + \dfrac{e^{-0.5} 0.5^5}{5!} + \cdots$	$\left(\text{By definition of summation, } \sum \right)$
$= 0.3033 + 0.0758 + 0.0126 + 0.0016 + 0.0002 + \cdots$	(By arithmetic)
$= 0.3935 + \cdots$	(By arithmetic)

What to do with the dots ... at the last step? Since the terms get so small, this term can probably be ignored here. But this is another case of the weasely $\cong$, so you should be careful. Fortunately, it is easy to check the size of the omitted terms using the computer. In Excel, for example, there is a function POISSON.DIST(y, 0.5, FALSE) that returns the summands indicated earlier. You can easily add them up for values 6, 7, 8, ..., as far as you want and see that the remainder ... is tiny. So you can take $\Pr(Y \geq 1)$ to be 0.3935, meaning that in about 40 out of every 100 days, there is one or more auto fatality.

At this point, it is worth mentioning another type of weasely $\cong$ that you might worry about. Mathematically, you should write $1/3 \cong 0.3333$, since these two numbers differ. But this approximation is minor compared to all the other approximations in statistics, and anyway you can easily make the approximation better by adding some more threes. So as long as there are enough significant digits in the decimal representation, you can write $1/3 = 0.3333$ and not worry about the $\cong$ symbol. In general, you should report decimal values to at least three significant digits (0.3333 has four significant digits, as does 45.03). When calculating, however, you should avoid roundoff—just let the computer do it all. The computer also rounds off, but somewhere around the 15th significant digit, so the effects of roundoff are negligible.

How about a continuous example? We've already done a few in Chapter 2. What's different here is just some notation. For example, in the waiting time example, let Y be the time you have to wait. If the question of interest is how often waiting time exceeds 2 minutes, the set of values is $Y > 2$ or, in set notation, $Y \in A$, where $A = \{y; 2 < y < \infty\}$. The set A so described includes everything on the number line to the right of the number 2.00000.... Then you can unpack the integral probability formula $\Pr(Y \in A) = \int_{y \in A} p(y)dy$ as follows:

$\Pr(Y \in A)$

$= \int_{y \in A} p(y)d$ (By definition)

$= \int_2^\infty p(y)dy$ (By substituting $A = \{y; 2 < y < \infty\}$)

$= \int_2^\infty 0.5e^{-0.5y}dy$ (By substituting the exponential pdf for $p(y)$)

$= -e^{-0.5y}\big|_2^\infty$ (By integral properties I2 and I7 in Table 2.7)

$= (-e^{-0.5(\infty)}) - (-e^{-0.5(2)}) = 0 + e^{-1} = 0.368$ (By method of dealing with infinite integral limits shown in Section 2.6 and by arithmetic)

In words, around 37 out of every 100 callers will have to wait more than 2 minutes.

Many continuous and discrete probability functions are readily available in software; in these cases, you don't need to use calculus. For example, in Excel, you can calculate cumulative probabilities from the normal distribution using the NORM.DIST function: If Y is produced by the normal distribution with parameters μ and σ, or in shorthand if $Y \sim N(\mu, \sigma^2)$, then $P(y) = $ NORM.DIST(y, μ, σ, TRUE). The last argument, TRUE, tells the NORM.DIST function to return the normal cumulative distribution function (cdf), when FALSE it returns the normal pdf.

3.3 Simulation-Based Approximation

Real systems are complex. Research is never as simple as saying "Let's find the probability that there are one or more auto fatalities in a given day" or "Let's find the probability that a customer's waiting time exceeds two minutes." Consider the call center operation in more detail: There are myriad factors that affect waiting time, including other callers; number of telephone operators available, who is on break and who is sick; the time of day; the day of the week; the state of the Internet connections; and so on. You could construct a model incorporating random variation from many of these factors, giving you something much more complex than the simple exponential distribution, and attempt to answer the question of how often will waiting time be longer than 2 minutes using an integral as shown earlier. However, it is likely that even the most brilliant team of mathematicians would not be able to do the calculus. Instead, complex systems are analyzed using simulation or the method of producing DATA* introduced in Chapter 1, a method you will see throughout this book. But don't dismiss the calculus! The analytic calculation using summation (in the discrete case) or integration (in the continuous case) is the gold standard because it gives you the precisely correct answer. Simulation gives you a good approximation, but not the precisely correct result.

Simulation is very useful in statistics. Seemingly simple statistical questions, such as "What happens when the distribution is different than what I assumed?" are notoriously difficult to answer unless you use simulation, in which case, the answer is relatively easy. Simulation is also used for data analysis. Two commonly used methods of analyzing data, one called *bootstrapping* and the other called *Markov Chain Monte Carlo*, both discussed in later chapters, utilize simulation. Simulation also makes an excellent pedagogical tool because of its transparency. Using simulation, you can estimate probabilities exactly as you would estimate the probability of heads using 100 coin flips—just count number of heads and divide by 100. The calculus-based analyses, in comparison, might seem more mysterious. Finally, what better way to understand the ubiquitous concept *model produces data* than simulation? A simulation analysis is, after all, based on DATA* that you have produced from your model.

The following algorithm shows how to estimate $\Pr(Y \in A)$, where Y is from the pdf $p(y)$, no matter whether $p(y)$ is discrete (either ordinal or nominal) or continuous.

Estimating Probabilities via Simulation

1. Generate, via simulation, $Y_1^*, Y_2^*, Y_3^*, \ldots, Y_{NSIM}^*$ from $p(y)$.
2. For each simulated Y_i^*, check whether $Y_i^* \in A$.
3. Estimate $\Pr(Y \in A)$ using $\dfrac{\#Y_i^* \in A}{NSIM}$. (This is mathematical shorthand that means "count how many Y_i^* are in the set A and divide by the number of simulations *NSIM*.")

This is just like estimating the probability of a bent coin landing on heads: Flip it 1000 times (that's your *NSIM*), count how many times it turns up heads (the set A is just the set comprised of the single element, $A = \{\text{heads}\}$), and divide the number of heads by 1000.

However, this is also a weasely method because it produces an approximation—an estimate. You can make the estimate better (i.e., closer to $\Pr(Y \in A)$) by choosing a larger *NSIM*. Intuitively, this should make sense: If you flip the bent coin 1,000,000 times, your estimate of $\Pr(Y = \text{heads})$ should be better than your estimate based on 1,000 flips.

You can use simulation for the auto fatality calculation given earlier with the Poisson distribution, where $\lambda = 0.5$ yielding $\Pr(Y \geq 1) = 0.3935$ using any random number generator. One simulation yields $y_1^* = 2$, $y_2^* = 0$, $y_3^* = 1, \ldots, y_{1000}^* = 0$; note the switch from capital Y^* to lower case y^*. Of these 1000 y^* values, 397 are greater than or equal to 1, so the estimate of $\Pr(Y \geq 1)$ is $397/1000 = 0.3970$. Which number is right, 0.3970 or 0.3935? It is easy to remember that the simulation-based estimate 0.3970 is wrong, because the simulation is based on random numbers. Generate another sample and you get a different estimate, like 0.3840. On the other hand, you always get the same 0.3935 when you use the Poisson probability function, and that answer is the correct one. Using larger *NSIM*, for example, *NSIM* = 1,000,000, you might get an estimate like 0.3931—still wrong but much closer to the true value 0.3935.

3.4 Generating Random Numbers

In Excel and other software, you will find built-in random number functions for Poisson, uniform, normal, and other common distributions. However, you might not find a particular distribution, such as the triangular one mentioned in Chapter 2. What to do? Simple. Use the **inverse cdf method**, of course!

Okay, maybe we should tell you what is this thing called the "inverse cdf method." The acronym "cdf" you may remember from Chapter 2 stands for "cumulative distribution function." For any RV Y, it is defined as $P(y) = \Pr(Y \leq y)$; for a continuous RV, it is defined as $P(y) = \int_{(-\infty, y)} p(t)dt$. The inverse cdf method only applies to the continuous case; it does not work for the discrete case.

Another thing you need to know about the inverse cdf method: It is based on random numbers that are simulated from the **uniform distribution**, that is, the distribution that makes all values between 0 and 1 equally likely. The pdf of the uniform distribution is $p(y) = 1$, for $0 < y < 1$. The graph of the distribution is just a flat line with height 1 and that the area under the function is 1.0 by the base × height formula. It turns out that the uniform distribution is very useful, not only for simulating other distributions—we call it the *mother of all distributions* for this reason—but also because the p value that is used for testing statistical hypotheses has this distribution (see Chapter 15).

The **inverse of a function** $y = f(x)$ is the function $x = f^{-1}(y)$, if it is defined. Recall the definition of a function: It is a mathematical relationship between two sets of numbers such that if you input a number x, you get a unique number y. An inverse function simply flips the direction of the relationship: Instead of starting with an x and ending with y, you start with y and end with x.

For example, if $y = f(x) = x^2$, for $x > 0$, then $x = f^{-1}(y) = y^{1/2}$ is the inverse function. Note that an inverse function may not always be defined. For example, if you consider all x, positive or negative, in the function $f(x) = x^2$, then there is no inverse function: If $y = 4$, there are two possible values of x that could be returned, namely, $x = -2$ and $x = +2$. Such ambiguity is not allowed in the definition of a function, so you sometimes have to constrain the set of x values to have a well-defined inverse function.

To find the inverse cdf, set $p = P(y)$, and solve for $y = P^{-1}(p)$. The function $P^{-1}(p)$ is the inverse cdf function, also called the *quantile function* discussed in Chapter 4.

You can generate *NSIM* data values $Y_1^*, Y_2^*, Y_3^*, \ldots, Y_{NSIM}^*$ from any continuous pdf $p(y)$ using the following simple steps.

Generating a Random Sample from Any Continuous pdf

1. Generate $U_1^*, U_2^*, U_3^*, \ldots, U_{NSIM}^*$ from the U(0, 1) distribution (the uniform distribution on the (0,1) range).
2. Let $Y_1^* = P^{-1}(U_1^*), Y_2^* = P^{-1}(U_2^*), Y_3^* = P^{-1}(U_3^*), \ldots, Y_{NSIM}^* = P^{-1}(U_{NSIM}^*)$.

For the triangular distribution in Chapter 2, $p(y) = 0.0002y$, for $0 \le y \le 100$, and $P(y) = \int_{\{-\infty, y\}} p(t)\, dt = \int_{\{0,y\}} 0.0002t\, dt = 0.0001t^2 \big|_0^y = 0.0001y^2 - 0.0001(0^2) = 0.0001y^2$. You can obtain the inverse function by solving $p = 0.0001y^2$ for y, obtaining $y = P^{-1}(p) = (10{,}000p)^{1/2}$ as follows:

$p = P(y)$	(Start with the cdf of the triangular
$\quad = 0.0001y^2$	distribution; call it p)
$\Rightarrow p/0.0001 = y^2$	(By algebra)
$\Rightarrow 10{,}000p = y^2$	(By arithmetic)
$\Rightarrow y = \sqrt{10{,}000p} = (10{,}000p)^{1/2}$	(By algebra)

Note that the solution is unique since $y \ge 0$.

You can generate a sample from this triangular distribution as $Y_i^* = \left(10{,}000U_i^*\right)^{1/2}$, for $i = 1, 2, \ldots, NSIM$. Figure 3.1 illustrates the concept.

In Figure 3.1, you see the cdf $P(y) = 0.0001y^2$ and a randomly generated $u^* = 0.731$ selected from the U(0, 1) distribution. The function is then used *backward* to find the

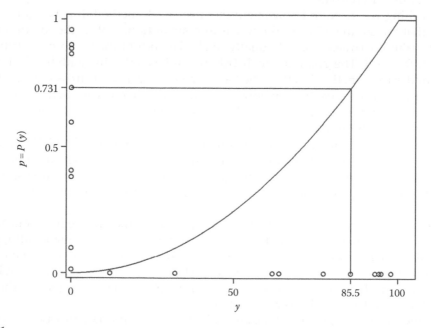

FIGURE 3.1
The cdf of the triangle distribution and a value $y^* = 85.5$ randomly generated from this distribution by using the randomly generated U(0, 1) value $u^* = 0.731$, as well as nine other values.

value $y^* = 85.5$ via $y^* = \{10{,}000(0.731)\}^{1/2}$. You will get different values y^* from the triangular distribution when you sample different u^* from the U(0, 1) distribution. For example, a random sample of $NSIM = 10$ values from the U(0, 1) distribution might be 0.731, 0.382, 0.101, 0.596, 0.899, 0.885, 0.958, 0.014, 0.407, and 0.863, shown on the vertical axis of Figure 3.1; these give the corresponding random sample of $NSIM = 10$ values 85.5, 61.8, 31.8, 77.2, 94.8, 94.1, 97.9, 11.8, 63.8, and 92.9, respectively, from the triangular distribution, shown on the horizontal axis.

It's pretty easy to see why this method works: Look carefully at Figure 3.1. The uniform U(0, 1) random numbers u^* fall on the vertical axis, and the resulting values y^* fall on the horizontal axis. Because the distribution is uniform, you know that 73.1% of the u^*s will be less than 0.731. But whenever u^* is less than 0.731, the resulting value y^* will also be less than 85.5. Thus, the method produces y^* values so that 73.1% are less than 85.5. This is exactly what you want because 85.5 is the 0.731 quantile. The same argument works for any quantile, so the method gives you data that appear in the correct frequencies.

In the discrete case, you can also generate random numbers from the uniform distribution. Just recode the U^* values as the discrete values so that the probabilities are right. For example, consider the car color choice distribution from Table 1.3, shown again here as Table 3.1.

Here, you can generate a U^*, and recode it to red, gray, and green as follows:

If $0 \leq U^* < 0.35$, then $Y^* =$ "red."

If $0.35 \leq U^* < 0.75$, then $Y^* =$ "gray."

If $0.75 \leq U^* \leq 1.00$, then $Y^* =$ "green."

See Figure 3.2.

Using the same uniform random numbers shown in Figure 3.1, Figure 3.2 shows two red, four gray, and four green simulated values.

Let's think a bit about why this works. Recall that for continuous distributions, the area under the curve is probability. In the case of the U(0, 1) distribution, the curve forms a perfect 1×1 square. Probabilities are thus very easy to calculate for a uniform distribution: Simply find the area using base × height of the rectangle whose base is found as the difference $b - a$ between two points (a, b) on the horizontal axis. For example, the probability that a U(0, 1) RV U is between 0 and 0.35 is found as $(0.35 - 0) \times 1 = 0.35$, which is the probability of selling a red car in the example given earlier. This is just what you want! Similarly, the probability between 0.35 and 0.75 is found as $(0.75 - 0.35) \times 1 = 0.40$, the probability of selling a gray car.

So, generating random numbers from a discrete distribution boils down to partitioning the interval (0, 1) into rectangles corresponding to the probabilities of the events

TABLE 3.1

Probability Distribution of Color Choice

y	$p(y)$
Red	0.35
Gray	0.40
Green	0.25
Total	1.00

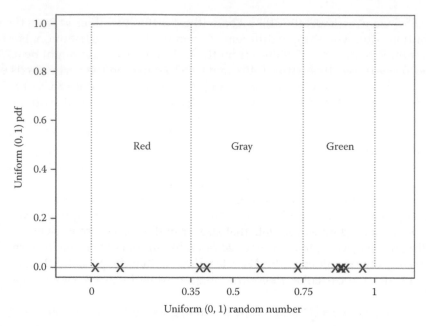

FIGURE 3.2
Recoding the uniform RV as discrete values red, gray, or green. The uniform random numbers on the vertical axis of Figure 3.1 are shown with "*x*" marks.

you are interested in. The U(0, 1) distribution can thus be called the **mother of all distributions**, since all distributions, whether discrete or continuous, can be constructed from the U(0, 1) pdf.

But how do you simulate values U^* from the uniform U(0, 1) distribution to begin with? Sure, you can assume that any software worth its salt will simulate uniforms, but how do *they* do it? There is a long and lively literature on this subject, with mathematicians, statisticians, and computer scientists developing their own and criticizing others' algorithms. While interesting, much of this research is tangential to our main points, so you can just assume that the U(0, 1) random numbers generated by whatever software you use are adequate.

Vocabulary and Formula Summaries

Vocabulary

Sample space   The set of possible values of the RV Y.

Event   A particular outcome or set of outcomes for the RV Y, a subset of the sample space.

Inverse of a function   The function $x = f^{-1}(y)$ obtained by solving $y = f(x)$ for x.

Inverse cdf method   A method for generating random numbers having any continuous distribution.

Uniform distribution   The continuous distribution that makes all values between 0 and 1 equally likely, abbreviated U(0, 1).

Mother of all distributions   The U(0, 1) distribution.

Key Formulas and Descriptions

$$\Pr(Y \in A) = \sum_{y \in A} p(y)$$ The probability that a discrete RV Y lies in the set A.

$$\Pr(Y \in A) = \int_{y \in A} p(y)dy$$ The probability that a continuous RV Y lies in the set A.

$$\Pr(Y \in A) \cong \frac{\#\{Y_i^* \in A\}}{NSIM}$$ The approximate probability that an RV Y lies in the set A, obtained via simulation.

$x = f^{-1}(y)$ The inverse of the function $y = f(x)$.

$p(y) = 1$, for $0 < y < 1$ The uniform pdf over the range from zero to one, abbreviated $U(0, 1)$.

$Y_i^* = P^{-1}(U_i^*)$ Random variable having continuous distribution $p(y)$ obtained using the inverse cdf method.

Exercises

3.1 After the normal distribution, the uniform distribution over the (0, 1) range (abbreviated as the U(0, 1) distribution) is perhaps the most important in all of statistics. What characterizes the DATA* that are produced by the uniform distribution? Answer this question by experimenting with the uniform random number generator, and provide a qualitative (not quantitative) discussion. Refer to your experimentation in your answer.

3.2 The U(0, 1) distribution has the function form $p(y) = 1.0$, for $0 \le y \le 1$; $p(y) = 0$ otherwise.

   A. Draw a graph of this pdf using the computer.

   B. Discuss the appearance of the graph in relation to the type of DATA* it produces, as you have seen in Exercise 3.1.

   Express the following as integrals, and compute their values.

   C. The probability that a U(0, 1) RV is less than 0.5.

   D. The probability that a U(0, 1) RV is more than 0.2.

   E. The probability that a U(0, 1) RV is between 0.2 and 0.5.

3.3 An example of a triangular distribution on the (0, 1) range distribution has the function form $p(y) = 2y$, for $0 \le y \le 1$, and $p(y) = 0$ otherwise.

   A. Draw a graph of this pdf using the computer.

   B. Show that it is a valid pdf by checking the necessary conditions.

   C. Discuss the appearance of the graph in Exercise 3.3A in relation to the type of DATA* it will produce. Compare and contrast the type of DATA* produced by this distribution with the DATA* produced by the U(0, 1) distribution.

   D. What does $p(0.75) = 1.5$ tell you? Explain in a couple of sentences. Refer to Chapter 2 for how to interpret $p(y)$ for a continuous pdf.

   Express the following as integrals, and compute their values.

   E. The probability that an RV with the given triangular distribution is less than 0.5.

F. The probability that an RV with the given triangular distribution is more than 0.2.

G. The probability that an RV with the given triangular distribution is between 0.2 and 0.5.

3.4 Use a random number generator in the following problems. Write your answers professionally, tutorial style; don't just give the final answer. Estimate the probability that

A. A U(0, 1) RV is less than 0.5. Compare and contrast with Exercise 3.2C.

B. A U(0, 1) RV is more than 0.2. Compare and contrast with Exercise 3.2D.

C. A U(0, 1) RV is between 0.2 and 0.5. Compare and contrast with Exercise 3.2E.

3.5 Use a random number generator, along with the inverse cdf method, in the following problems. Write your answers professionally, tutorial style; don't just give an answer. Estimate the probability that

A. An RV having the triangular distribution of Exercise 3.3 is less than 0.5. Compare and contrast with Exercise 3.3E.

B. An RV having the triangular distribution of Exercise 3.3 is more than 0.2. Compare and contrast with Exercise 3.3F.

C. An RV having the triangular distribution of Exercise 3.3 is between 0.2 and 0.5. Compare and contrast with Exercise 3.3G.

3.6 Answers to the survey question, "How much do you like coffee?" are given on a 1, 2, 3, 4, and 5 scale. Suppose the following distribution, in list form, is a reasonable model for producing the data that you might see:

y	$p(y)$
1	0.25
2	0.15
3	0.20
4	0.10
5	0.30
Total	1.00

A. Show how to write $Pr(Y > 3)$ using the summation and set notation formula; then show how to calculate that probability using that formula.

B. Show how to write $Pr(Y \geq 3)$ using the summation and set notation formula; then show how to calculate that probability using that formula.

C. Use simulation to estimate the probability in Exercise 3.6B. Compare and contrast your simulation-based result with your answer to Exercise 3.6B.

3.7 Consider the following function $p(y)$:

y	$p(y)$
0.00	0.50
0.50	0.00
1.20	0.10
1.70	0.20
1.90	0.15
1.95	0.05
Total	1.00

    A. Draw a graph of this function using the computer.

    B. Show that this is a valid pdf.

    C. Find $\Pr(Y < 1.70)$.

    D. Find $\Pr(Y = 1.95)$.

    E. Find $\Pr(0.50 < Y < 1.90)$.

    F. Find $\Pr(0.50 \leq Y \leq 1.90)$.

    G. Find $\Pr(|Y - 1.2| < 0.5)$.

    H. Find $\Pr(|Y - 1.2| > 0.5)$.

3.8 Hans is investigating effectiveness of banner ads. He asks the participants in the study to mark their intention to click on the banner ad by moving their mouse cursor to a position between 0 and 100 on a specially designed web page. Suppose that the distribution of the continuous responses is $p(y) = 0.0002y$, for $0 \leq y \leq 100$, as in Example 2.10.

    A. Show how to write $\Pr(Y > 50)$ using the integral formula; then show how to calculate that probability using the integral formula.

    B. Show how to calculate $\Pr(Y > 50)$ as the difference between the areas of two triangles.

3.9 A pdf is $p(y) = 0.01$, for $0 < y < 100$.

    A. Use calculus to help answer the following question: If 1000 DATA* values were produced by this pdf, about how many of them would be more than 90? Explain your logic.

    B. Repeat A but using the area of a rectangle formula.

3.10 A pdf is $p(y) = a + e^{-y}$, for $0 < y < 2$; $p(y) = 0$ otherwise, for some constant a.

    A. Find $\Pr(Y > 1)$ and show how this probability appears in the graph.

    B. Find $\Pr(Y \geq 1)$.

    C. Find $\Pr(0.5 < Y < 1.5)$.

3.11 A pdf is $p(y) = a/y$, for $1 < y < 10$; $p(y) = 0$ otherwise, for some constant a.

    A. Find $\Pr(Y > 2)$ and show how this probability appears in the graph.

    B. Find $\Pr(Y \geq 2)$.

    C. Find $\Pr(1.5 < Y < 2.5)$.

3.12 Consider the function $p(y) = \log_{10}(1 + y^{-1})$, for $y = 1, 2,..., 9$. (This is the famous probability distribution that defines *Benford's law*.) Repeat Exercise 3.7A through H using this distribution.

3.13 Consider the function $p(y) = y/2$, for $0 < y < 2$; $p(y) = 0$ otherwise. Repeat Exercise 3.7A through H using this distribution.

3.14 Consider the function $p(y) = 1/y$, for $1 < y < e$, $p(y) = 0$ otherwise. Repeat Exercise 3.7A through H using this distribution.

3.15 Express the following probabilities as integrals, and calculate their values using the NORM.DIST function of Excel or other computer-based normal cdf calculator.

    A. $\Pr(Y > 10)$, where $Y \sim N(20, 5^2)$

    B. $\Pr(Y < 10)$, where $Y \sim N(20, 5^2)$

    C. $\Pr(Y > 20)$, where $Y \sim N(20, 25)$

    D. $\Pr(Y \leq 10)$, where $Y \sim N(10, 5)$

    E. $\Pr(Y \geq 10)$, where $Y \sim N(20, 100)$

4

Identifying Distributions

4.1 Introduction

Researchers often ask, "Which distribution should I use?" The answer is simple and hopefully familiar by now: Use a distribution that, for some particular parameter settings, produces DATA* that look like your DATA.

While simple in principle, the answer is more difficult in practice. There are myriad distributions from which to choose and usually no right answer. After all, as the Mantra tells you, *model has unknown parameters*. The probability distribution $p(y)$ is unknown, since it consists of these unknown parameters. But the next part of the Mantra offers the hope and resolution: *Data reduce the uncertainty about the unknown parameters*. Using data, you can identify models that are reasonable, and you can rule out models that are unreasonable.

Usually, even before collecting data, (i.e., when you are contemplating your future DATA), you have some ideas what models might be appropriate. For example, you know the types of measurements you will obtain, whether discrete or continuous, the range of their possible values, and you should have some ideas about symmetry or asymmetry. Using these bits of information, you can narrow the list of possible distributions $p(y)$ that produce DATA* similar to your DATA.

In some cases, you can determine the distribution $p(y)$ purely from theory, but more often the choice is not a precise science. Rather it is an art guided by your knowledge of historical data and by what you can anticipate about theoretical future DATA.

In many cases, it is not particularly important that you identify a named form for the distribution $p(y)$—for example, normal, Poisson, uniform, and Bernoulli. Models where you do not have to assume such a form for $p(y)$ are called *nonparametric models*. These models will be discussed later in the book. In this chapter, we explain why you might want to pick a particular distribution form, and we offer guidance on how to do so.

Do not expect to find "The Answer" to the question "Which distribution should I use?" in this chapter. The process of distribution selection, and of statistical model selection in general, is not one of picking the right model (which is impossible), but rather a process of ruling out bad models and then selecting an expedient one from what remains. You can assume that your selection will be wrong, in one way or another. But that's okay—as famous statistician George Box said, roughly, "All models are wrong but some are useful." This chapter describes methods for ruling out bad models and identifying useful ones.

4.2 Identifying Distributions from Theory Alone

Example 4.1: The Distribution of a Bent Coin

The easiest case for identifying a probability distribution function (pdf) is the familiar and lowly coin toss. Let's make it more interesting by bending the coin with pliers. Although it's still a pretty simple example, it provides a great way to understand many advanced concepts in statistics.

There are only two possible outcomes for Y, so there is only one possibility for a pdf $p(y)$: the Bernoulli. No fancy theory is needed here. The answer is Bernoulli(π); end of story. You do not need to know the value of π; the answer is still Bernoulli(π). If you plug in any specific number for π, your model is wrong—for example, the model Bernoulli(0.5) is *wrong*, the model Bernoulli(0.4) is *wrong*, and the model Bernoulli(0.77777) is *wrong*, but the model Bernoulli(π) is *right*.

The previous example is a rare exception to our admonishment that all models are wrong. If there is just one coin toss, with an unknown outcome, then the model Bernoulli(π) is right. With multiple tosses, however, there are assumptions that you must make concerning identical distributions and independence that are usually wrong, to more or less of a degree; these issues will be discussed later.

Most gambling games are similar to a bent coin. If you play red on Roulette, for example, your outcome Y will be either win or loss, with a $\pi = 18/38$ chance of a win, assuming a perfectly balanced American roulette wheel. (Your chance of winning is slightly better in European roulette, with $\pi = 18/37$.)

Simply knowing that the sample space of DATA* values ({0, 1}) matches the sample space of your DATA (also {0, 1} or suitably recoded) tells you that the Bernoulli model for a single observation is valid. But with more than two possible outcomes for your DATA, the form of the distribution is more difficult to determine, because most named distributions have **constraints**. These constraints imply that the probabilities follow certain function forms and therefore limit their applicability. Even when your DATA will be in the set {0, 1, 2, ...}, as required for the Poisson model, the Poisson model might not be reasonable, as the following example shows.

Example 4.2: The Distribution of a Number of Insects Caught in a Trap

Suppose your study involves counting insects caught in traps. Around 50% of your traps produce no insects, and the remainder produce a range from 1 to 500, with many over 100. Is the Poisson model a good model? It has the right sample space, $S = \{0, 1, 2, ...\}$, so it satisfies the first criterion for a good model given in Section 1.10. The second criterion is that it must produce DATA* that look like the DATA for some parameter settings. But you can never coax the Poisson model to produce DATA* that look like the DATA—the percentage of zeroes in your DATA* will be too small when λ is large, and the percentage of DATA* values over 100 will be too small when λ is small. So, while the choice of parameter λ allows some flexibility, the particular constraints imposed by the Poisson distribution imply that the model is inappropriate for this application, even though it has the correct sample space.

As an alternative to the Poisson distribution, you might consider the *negative binomial distribution*. It has the same {0, 1, 2, ...} sample space but is more flexible (less constrained) than the Poisson and is often used when the Poisson model fails.

A **generic distribution** is one that is completely flexible, with no constraints. Such a distribution is always applicable for a single observation. In the trapped insect case of Example 4.2, you could simply say that the distribution is $p(y)$, or that $Y \sim p(y)$, without stating anything else. No one will argue with you! Since you would not have assumed anything about the form of the distribution $p(y)$, it could be anything. And certainly, DATA* produced by $p(y)$ will look like the real DATA you observe for some $p(y)$, when you allow yourself the flexibility to say that $p(y)$ can be any distribution whatsoever, so the model is correct, by the definition in Chapter 1, at least for a single observation.

A benefit of stating distributions generically is that you give the reviewers of your work less to complain about. If your research conclusions depend strongly on the assumption of a Poisson model, then a reviewer can criticize your conclusions based on the potential invalidity of the Poisson model. If you make no such assumption, asserting only that the distribution is some generic $p(y)$, and can still arrive at the same essential conclusion, then you give the reviewer one less thing to criticize, and your conclusions are more defensible.

Generic distributions (e.g., as used in nonparametric methods) are advantageous because they make your life easier. Why not always use generic distributions then? Because, as you might expect, there are trade-offs. When you identify the distribution more specifically (as normal, Bernoulli, Poisson, etc.), you are telling a more complete story about Nature; that is, you are doing better research. Also, if you identify the distribution specifically, you can more easily make predictions about the future DATA, since you can simulate DATA* from a known distribution—see Example 1.8 concerning trading strategies. Another benefit of using specific distributions is that they provide optimal statistical procedures, as we discuss in later chapters. Finally, some of the more complex research tools, such as models to identify unobserved human thought processes, require specific distribution forms such as the normal distribution to tease out subtle effects from your observed data.

But it is not a one-sided argument in either case. Sometimes research questions are better answered without identifying the specific form of $p(y)$ (normal, Poisson, etc.), and sometimes they are better answered by identifying the form. In this book, we usually suggest that you try to identify the form of $p(y)$ as specifically as possible, because it helps you identify appropriate statistical methods, because it helps you diagnose process variation, and because it is simply good science.

In physics, the motion of particles is well modeled in terms of normal distributions (see *Brownian motion*). Without knowing this distribution form, the physicist's predictions of particle motion would be less precise. The physicist also describes radioactive decay using the exponential distribution; again, without knowing this distribution form, the physicist would not be able to make as accurate predictions. Knowing the distribution form allows you to make more accurate predictions.

Example 4.3: The Stoplight Case

Like the Bernoulli distribution for the coin toss, this example is a case where you can identify the specific form of the distribution $p(y)$ from theory alone. Suppose you drive to work every day. At one particular intersection, there is a stop sign, and you will turn right from there. At the intersection, there is a large bush that obstructs your view until you inch forward a bit. Once you inch forward a bit, you can see a stoplight a few blocks away, through which you must pass to get to work. See Figure 4.1.

Suppose the stoplight signal is green when you see it. How long will it stay green? Let Y denote the time it stays green. You could have arrived at the stop sign intersection at any time in the stoplight's cycle, and there is no reason to assume that you are more

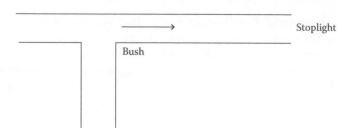

FIGURE 4.1
Diagram of the stoplight example.

likely to arrive in the middle of the cycle than at the beginning or end. Thus, if the light stays green for a total length of θ seconds, your time Y is uniformly distributed between 0 and θ and so has the following distribution:

$$p(y) = \frac{1}{\theta} \quad \text{for } 0 < y < \theta; \quad p(y) = 0 \quad \text{otherwise}$$

If you happen to know that the light is preprogrammed to stay green for a total of $\theta = 2$ minutes, then the time you see it stay green is distributed as uniform from 0 to 2 minutes; or in shorthand, $Y \sim U(0, 2)$. See Figure 4.2.

This example is unusual in that the distribution form $p(y)$ is known purely from the theory of the context. Even so, you can't be 100% sure that the uniform distribution is correct, because it depends on your behavior. You could possibly game the system to make the distribution nonuniform. Supposing that the light turned green at precisely 7:30 a.m. every day, you could plan to arrive just a second after 7:30 a.m. to make the time longer. But barring this kind of gaming of the system, the $U(0, \theta)$ distribution is the correct model.

You could be generic in the stoplight case by stating the distribution is just $p(y)$, without stating anything else, and you would be right. But the stoplight example is one where identifying the actual form of $p(y)$ is beneficial. If you say the distribution is $U(0, \theta)$, then

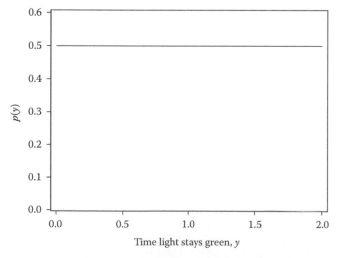

FIGURE 4.2
Distribution of time the light stays green from when you first see it, assuming a cycle time of 2 min.

you are communicating much more information than if you say simply that the distribution is some generic $p(y)$. Specifically, you are communicating that the time the light stays green has a distribution that looks like that shown in Figure 4.2, rather than, say, a normal distribution. If you state that the distribution is just $p(y)$, then your audience can reasonably conclude that it might be anything—normal, skewed, discrete, etc., even including a distribution that only produces negative numbers. Hence, you are better off stating specifically what the distribution is, when you know, rather than being completely generic. It's just more scientific.

4.3 Using Data: Estimating Distributions via the Histogram

Recall the Mantra: *Data reduce the uncertainty about the unknown parameters.* In most cases, you cannot identify the distribution $p(y)$ purely from theory. But if you collect some data (lowercase d), or if you have access to relevant historical data, you can use it to suggest types of distributions that are plausible producers of your DATA (uppercase D), and you can rule out other distributions that are clearly wrong.

Again, don't think that data can tell you the answer to your question, "Which distribution produced the data?" There are many plausible candidates for a $p(y)$ that you can assume to have produced your data, and you will not be able to tell which one is right and which ones are wrong. To be specific, suppose you collect the following data on the time in minutes until a phone call to a call center is answered: $y_1 = 0.3$, $y_2 = 2.3$, $y_3 = 1.0$, $y_4 = 0.1$, $y_5 = 3.9$, and $y_6 = 0.8$. What distribution $p(y)$ might have produced these values? There are many possibilities. Figure 4.3 shows four distributions that are plausible. You cannot tell

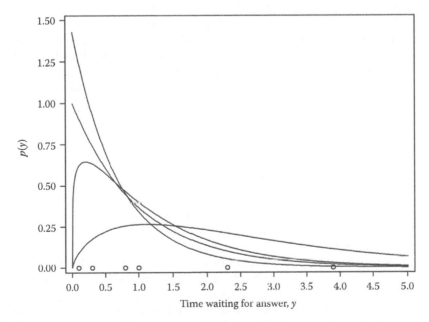

FIGURE 4.3
Four distributions $p(y)$ (solid curves) that could have produced the observed data (circles).

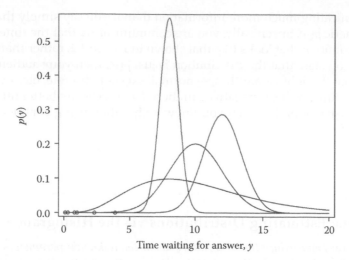

FIGURE 4.4
Four distributions $p(y)$ (solid curves) that are unlikely to have produced the observed data (circles).

from the data which one of these four distributions is the true one or whether some other distribution not shown in Figure 4.3 is true. You can't use data to *rule in* any particular distribution.

On the other hand, you can use data to *rule out* many distributions. Figure 4.4 shows four distributions that are essentially *ruled out* by the same six data values, at least beyond a reasonable doubt. It is theoretically possible that the data could have been produced by these distributions, in other words, it is possible that the positions of such data values are *explainable by chance alone,* but the likelihood is so small that you can rationally discount these cases. The notion of whether results are *explainable by chance alone* is formalized by using probability calculations, covered in Chapter 15 and on.

Although you cannot possibly identify the distribution using data, you can use the data to form an *estimate* (weasel word alert!) of the distribution that produced your data. There are several ways to estimate distributions, and much of the statistical literature is devoted to this very topic. One such estimate of the distribution $p(y)$ is the **histogram**. While the histogram cannot tell you whether any particular named distribution (such as the normal, Poisson, exponential, etc.) produced your data, it can help you identify generic properties of that distribution, such as *symmetry, skewness,* and *bimodality.*

The rectangular pdf graphs in Chapter 2 look a lot like histograms. Those graphs were purely theory based, however, and did not use any data set. Histograms, on the other hand, use a data set to estimate probabilities in the rectangular regions.

For example, consider the previous data, with $y_1 = 0.3$, $y_2 = 2.3$, $y_3 = 1.0$, $y_4 = 0.1$, $y_5 = 3.9$, and $y_6 = 0.8$. Suppose you wish to construct a rectangular approximation to the true pdf. How would you estimate the probabilities in the different intervals? Table 4.1 shows how.

The logic for the calculations in the *Estimated Probability* column of Table 4.1 is the same as the logic for estimating the probability of heads for a bent coin. What is the best guess of heads? Just count the total number of heads, and divide by the total number of flips. What is the guess of the probability of the range $0.0 \leq y < 0.5$? Just count the number of times that $0.0 \leq y < 0.5$ occurs, and divide by the total number of data points. There are two out of the six data values in the range $0.0 \leq y < 0.5$; namely, $y_1 = 0.3$ and $y_4 = 0.1$, and hence, the estimated probability of the range is $2/6 = 0.333$.

TABLE 4.1

Counts of Call Lengths in the Different Interval Ranges and Estimated pdf

Interval (or Bin)	Midpoint y_i	Count	Estimated Probability $\hat{p}(y_i)$	Estimated pdf $\tilde{p}(y)$
$0.0 \leq y < 0.5$	0.25	2	$2/6 = 0.333$	$(2/6)/(1/2) = 0.667$
$0.5 \leq y < 1.0$	0.75	1	$1/6 = 0.167$	$(1/6)/(1/2) = 0.333$
$1.0 \leq y < 1.5$	1.25	1	$1/6 = 0.167$	$(1/6)/(1/2) = 0.333$
$1.5 \leq y < 2.0$	1.75	0	$0/6 = 0.000$	$(0/6)/(1/2) = 0.000$
$2.0 \leq y < 2.5$	2.25	1	$1/6 = 0.167$	$(1/6)/(1/2) = 0.333$
$2.5 \leq y < 3.0$	2.75	0	$0/6 = 0.000$	$(0/6)/(1/2) = 0.000$
$3.0 \leq y < 3.5$	3.25	0	$0/6 = 0.000$	$(0/6)/(1/2) = 0.000$
$3.5 \leq y < 4.0$	3.75	1	$1/6 = 0.167$	$(1/6)/(1/2) = 0.333$
	Totals	6	1.00	

The *Estimated pdf* column of Table 4.1 comes from the area of a rectangle formula and Equation 2.4: This equation states that the rectangular approximation to the distribution is given by

$$\tilde{p}(y) = \frac{p(y_i)}{\Delta}, \quad \text{for } y_i - \frac{\Delta}{2} < y < y_i + \frac{\Delta}{2}$$

Here, $p(y_i)$ is the probability within the $\pm\Delta/2$ range of y_i, the midpoint of the interval. For the area under the rectangle to be a probability, the height of the rectangle must be the probability divided by the interval width Δ, with $\Delta = 0.5$ in this example. Figure 4.5 shows the resulting approximation.

Figure 4.5 is clearly an approximation, as you would expect that the true distribution looks more like one of the distributions shown in Figure 4.4—smooth and without stretches of zero likelihood. There are two approximations at work here. First, just like with the toss of a bent coin, you get better estimates with more coin tosses. Here there are only six observations—by analogy, six coin tosses—so the estimated probabilities are

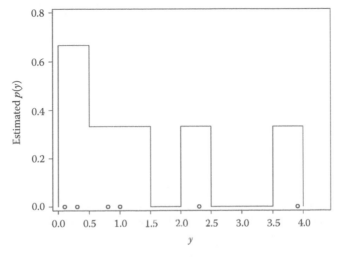

FIGURE 4.5

Histogram approximation to $p(y)$. The six data values used to construct the histogram are also shown as circles.

not very accurate. Second, as discussed in Chapter 2, the rectangular approximations are better with narrower intervals. You would like narrower intervals, but with small sample sizes, you will have too few observations in the intervals to estimate the probabilities. Many of the intervals will be empty, given the grossly incorrect estimate of 0.0 for a probability. So the number of intervals you can use for a histogram depends on the sample size: obtain a larger sample size and you can have more (and narrower) intervals.

If you wish to construct a histogram "by hand," as shown in Figure 4.5, you'll need to use some trial and error to produce a nice-looking graph. Or, you can let your statistical software select the intervals; the defaults are often adequate. Most often, you'll need to use a combination of software defaults and your own post-processing to clean up the axes and to label things appropriately as needed.

How many intervals should you pick in a histogram? There is no one right answer, but with a larger sample size, you can pick more intervals. Here is a good place to give an ugly rule of thumb.

Ugly Rule of Thumb 4.1

If n is your sample size, pick $n^{1/2}$ intervals to use for your histogram.

This is not the formula used by most software, but it serves okay as an ugly rule of thumb. In particular, it shows you that with larger n, you can use more intervals. Table 4.1 shows eight intervals, and $n = 6$, so we violated the rule of thumb: Since $n^{1/2} = 6^{1/2} = 2.4$, the ugly rule of thumb suggests using two or three intervals instead of the eight that we chose. That's ok; it's just an ugly rule of thumb! And anyway, Table 4.1 would have been pretty darn boring with just two intervals.

Figure 4.6 shows a default histogram produced by software for 1000 call times. Notice that the resulting graph looks much more like one of the continuous pdfs shown in Figure 4.3.

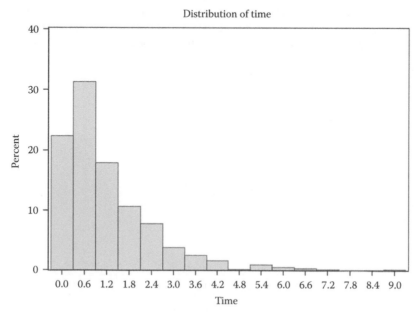

FIGURE 4.6
Histogram of 1000 wait times.

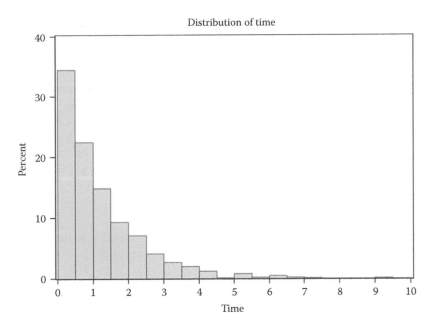

FIGURE 4.7
Revised Figure 4.6 to ensure no negative endpoints.

Notice in Figure 4.6 that the leftmost bar seems to suggest that some of the wait times are less than zero. The software indeed chose a negative leftmost endpoint. Software can be ignorant like that! You might have to instruct the software take zero as the lower endpoint, as shown in Figure 4.7.

There is a general principle at work here, and we set it off in bold so you can find later:

> Don't blindly accept the software defaults. Make adjustments so that the software presents the data accurately and in an aesthetically pleasing way.

Hmmm... something still doesn't seem right about the computer-generated histograms in Figures 4.6 and 4.7: The vertical axes are labeled as "Percent" rather than "Estimated pdf" or something similar. In fact, those histograms are not valid pdfs because the area under the curves is not 1.0. To make Figures 4.6 and 4.7 bona fide pdfs, you'd have to divide the percentages on the vertical axes by the interval width or by 0.5 in the case of Figure 4.7. Such a change would not change the essential *appearance* of the graph, however: It would look exactly the same, but with the maximum value on the vertical axis near 0.70 (70%) rather than near 0.35 (35%) as shown in Figure 4.7. So you can use the computer-generated histograms to suggest the shape and location of the underlying pdf, even though the values on the vertical axis are not on the pdf scale.

Specifically, what is graphed in Figure 4.7 is $\hat{p}(y_i)$ rather than $\hat{\tilde{p}}(y) = \hat{p}(y_i)/\Delta$. But these two differ only by the constant of proportionality $1/\Delta$, so the graphs will have the same appearance, other than a different vertical axis scale. In either case, the histogram shows what you want to know: It shows the distribution's degree of symmetry, asymmetry, range, and peak locations. All of these are equally visible, no matter whether you look at $\hat{p}(y_i)$ or $\hat{\tilde{p}}(y) = \hat{p}(y_i)/\Delta$.

The histogram is a nonparametric tool. It estimates a generic $p(y)$ without making any assumption about any named form (normal, Poisson, etc.) of the distribution. Sometimes it is handy to compare the histogram estimate of $p(y)$ with a distribution having a particular named form, to see whether such a named model would be reasonable. If the histogram and the named distribution are very similar in appearance, then you can be more confident in using the named distribution as a model.

As you can see from comparing the previous histograms with $n = 6$ versus $n = 1000$ observations, the histograms look a lot better with larger n. This is no surprise, since probability estimates are better with larger n—just think of flipping a bent coin $n = 6$ times versus $n = 1000$ times. So, larger n is better, but how large is "large enough?" Here is another ugly rule of thumb:

Ugly Rule of Thumb 4.2

With a sample size of $n \geq 30$, the histogram is an adequate estimate of the distribution $p(y)$ that produced your data. Larger n provides better accuracy.

Interpreting Histograms

1. Notice the range (on the horizontal axis) of the observed values.
2. Notice the approximate center of the distribution of the observed values.
3. The histogram is an estimate of the pdf. Look for indications of symmetry or asymmetry or any special unusual features such as bimodality (i.e., two distinct peaks).
4. Often the histogram is used to assess the adequacy of the normal model. If so, look for rough symmetry and bell shape of histogram. Do not look for perfection though: Even data* produced by a normal distribution will not have a perfectly bell-shaped histogram due to randomness.
5. Discuss the sample size. With larger sample sizes, histograms become better approximations to the distribution that produced the data.

If you use the histogram—or the *quantile–quantile plot* discussed later in this chapter—to assess a particular distribution form, like normality, you should *never* conclude "the data are normally distributed" or "the data are not normally distributed." Both statements are meaningless, as data can never be normally distributed, or have any other continuous distribution, for that matter. They are just discrete values, even if measured to many decimals. So any claim that the *data* are or are not normally distributed is purely nonsense.

The question of normality, or of any other distribution form such as the exponential, Poisson, etc., is a question about the *process* that produced your data. (*Model produces data.*) It is not a question about your data set. The data do shed light on your question, "Which distribution produced my data?" but they do not answer it. *Data reduce the uncertainty about the unknown parameters.* They do not eliminate your uncertainty.

Example 4.4: Estimating the Distribution of Stock Market Returns via the Histogram

In Chapter 1, we introduced the concept of the financial *return*, which is the relative price change from one day to the next. We showed in Example 1.7 how you can create potential future stock price trajectories when you know the distribution of the returns. Example 1.7 assumed a normal distribution. Is that a reasonable assumption? One way to check whether you can assume a normal distribution to produce returns is to examine historical return data, draw the histogram, and superimpose (via software) a normal distribution. Data from the Dow Jones Industrial Average (DJIA) are freely

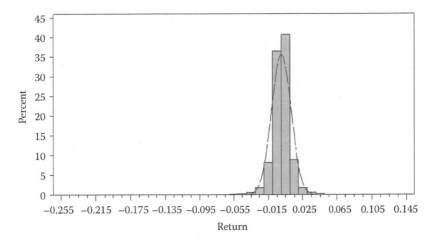

FIGURE 4.8
Histogram of $n = 18{,}834$ daily DJIA returns, with superimposed approximating normal distribution.

available from a variety of Internet sites; consider a data set from January 1, 1930, to December 31, 2004. If you calculate the daily returns and draw a histogram using software, along with the approximating normal pdf curve, you will see a graph as shown in Figure 4.8.

From the graph of Figure 4.8, the normal approximation seems okay, right? The bell curve seems to hug the histogram well, suggesting a good approximation, right? Well, that seems reasonable and is the right logic, but it turns out to be wrong in this case! But it's not your fault that you are wrong. The fault lies with the histogram: The tail behavior of the data is not visually apparent. For a better comparison of the true distribution with an assumed model, one that shows the tail behavior more clearly, you'll need to use the quantile–quantile plot.

4.4 Quantiles: Theoretical and Data-Based Estimates

The histogram is a great tool for looking at your data, and you should use it whenever you analyze data. However, for diagnosing tail behavior, a histogram is of limited use because the heights of the tails are so small as to be barely visible. Yet the tail behavior is often the most important aspect of a distribution. In his book *The Black Swan: The Impact of the Highly Improbable*, author Nassim Nicholas Taleb argues that it is the unusual events—those that are in the tails of the distribution, what he calls *black swans*—that have the largest impact on financial markets and on society in general. To diagnose tail behavior, a graphical tool called the **quantile–quantile plot** (or *q–q* plot) outperforms the histogram. The tool uses (obviously!) **quantiles**, which are interesting in their own right.

We introduced quantiles in Chapter 3, although not by name. Now, what exactly is a quantile?

Definition of the Quantile

If $P(y_p) = p$, then y_p is the p quantile of the distribution $p(y)$.

In other words, if proportion p (out of 1.0) of the data Y produced by $p(y)$ is less than or equal to y_p, then y_p is called the p quantile of the distribution.

Often these are called **percentiles**: If $(100 \times p)\%$ of the data produced by $p(y)$ are less than or equal to y_p, then y_p is called the $(100 \times p)$th percentile of the distribution. For example, if your college entrance exam score was 640, and if they told you that your score was at the 80th percentile, then $y_{0.80} = 640$.

Some percentiles are so famous they have special names: The 50th percentile is called the **median**; the 25th, 50th, and 75th percentiles are collectively called **quartiles**; the 20th, 40th, 60th, and 80th percentiles are collectively called **quintiles**; and the 10th, 20th, …, 90th percentiles are collectively called **deciles**.

Have a look again at Figure 3.1, the cumulative distribution function (cdf) of the triangular distribution, a continuous pdf. For any continuous distribution, the p quantile is unambiguously defined as follows:

$$y_p = P^{-1}(p)$$

This is the inverse cdf evaluated at p. For example, using the triangular pdf $p(y) = 0.0002y$ from Chapter 3, the cdf is $p = P(y) = 0.0001y^2$, and the inverse cdf calculation yields $y_p = (10{,}000p)^{1/2}$. Hence, the median is $y_{0.5} = \{10{,}000(0.5)\}^{1/2} = 70.71$, and the 10th percentile is $y_{0.1} = \{10{,}000(0.1)\}^{1/2} = 31.62$.

The discrete case is ambiguous since there is often no y for which $P(y) = p$. For example, what is the 0.75 quantile (or upper quartile) of the Poisson distribution with $\lambda = 0.5$? You can calculate $P(0) = e^{-0.5}0.5^0/0! = 0.6065$, and $P(1) = e^{-0.5}0.5^0/0! + e^{-0.5}0.5^1/1! = 0.6065 + 0.3033 = 0.9098$, but there is no value y for which $P(y) = 0.75$ since there is no possible y between 0 and 1 for the Poisson distribution.

The same kind of ambiguity occurs when you estimate quantiles using data. For example, what is your best guess of the median of the call center data set $y_1 = 0.3$, $y_2 = 2.3$, $y_3 = 1.0$, $y_4 = 0.1$, $y_5 = 3.9$, and $y_6 = 0.8$? For any number between 0.8 and 1.0, half of the data values are lower (0.3, 0.1, and 0.8) and half are higher (2.3, 1.0, and 3.9). Hence, there are infinitely many candidates for the estimate of the median. When your software calculates a median or any other estimated quantile, it must decide how to handle such ambiguous cases. In the case of the median, an obvious choice is to take the average $(0.8 + 1.0)/2 = 0.9$ as the estimated median, although this definition does not necessarily provide the best estimate, particularly if the distribution that produces the data is skewed.

Another source of ambiguity is how to interpret the smallest data value, for example, $y_4 = 0.1$ in the set data set. Is this an estimate of the $1/6$ quantile, since there are $n = 6$ data values and 0.1 is the smallest of the six? This *seems* logical…let's continue with that thought. Then the second smallest, 0.3, estimates the $2/6$ quantile, and 0.8 estimates the $3/6$ quantile… uh oh. We already decided that the median (the $3/6$ quantile) should be estimated by 0.9, not 0.8. So the logic that led us to saying 0.1 is an estimate of the $1/6$ quantile fails. But it gets worse! Continuing, 1.0, 2.3, and 3.9 would be estimates of the $4/6$, $5/6$, and $6/6$ quantiles. But do you really think that $6/6$ (or 100%) of the DATA will be less than 3.9? Of course not…you could easily be on hold longer than 3.9 minutes. The number 3.9 is just the largest in the data set. It's not the largest possible call wait time in general.

There are many suggestions about how to estimate quantiles using the data to solve these dilemmas. Here is one popular way to do it. First, define the **order statistics** using the symbol $y_{(i)}$, with parentheses in the subscript, as the ordered values:

$$y_{(1)} \leq y_{(2)} \leq \cdots \leq y_{(n)}$$

Thus, in the call center data, $y_{(1)} = 0.1$, $y_{(2)} = 0.3$, $y_{(3)} = 0.8$, $y_{(4)} = 1.0$, $y_{(6)} = 2.3$, and $y_{(6)} = 3.9$. Now, here is a common assignment of quantile estimates to the order statistics:

$$\hat{y}_{(i-0.5)/n} = y_{(i)} \tag{4.1}$$

Let's unpack Equation 4.1 a little. First, the "∧" over y denotes "estimate of." Without the ∧, the term $y_{(i-0.5)/n}$ refers to the actual $(i - 0.5)/n$ quantile of the distribution, rather than the estimate from the data. The right-hand side is the ith ordered value of the data. Putting it all together, Equation 4.1 reads as follows:

> The $(i - 0.5)/n$ quantile of the distribution is estimated by the ith ordered value of the data.

The "−0.5" term in Equation 4.1 solves the problems with median and 100th percentile definition as noted previously. Table 4.2 shows how this works with the call center data.

You can see that the definition of Equation 4.1 states that the third ordered value, 0.8, is not an estimate of the median, but rather an estimate of the 0.417 quantile. This is sensible, because the estimate of the median should be larger than 0.8. Further, the largest value, 3.9, is not an estimate of the 100th percentile, but rather of the 91.7th percentile. This is also sensible, because values larger than 3.9 minutes are possible. Without the "−0.5" term, the number 3.9 would be an estimate of the 6/6 quantile or the 100th percentile.

To see another way that the "−0.5" term works, suppose there were only $n = 5$ data points in the call center data, with all but the largest value. Then the middle value is 0.8: Two values are higher, and two are lower. It is hard to imagine any other estimate of the median. Equation 4.1 gives you $\hat{y}_{(3-0.5)/5} = y_{(3)}$, or $\hat{y}_{0.5} = 0.8$. Without the "−0.5" term, you would say that 0.8 is an estimate of the 0.6 quantile.

Finally, we reiterate that the word *estimate* is a weasel word. The word *estimate* means that the number obtained is *wrong* because it is *not equal to the true value*; the hat (∧) alerts you to this problem. The lowly coin toss is a good concept to revisit: If you toss a bent coin six times, the proportion of heads in the six flips is just an *estimate* of the probability of heads. By the same token, the value 0.3 in the call center data is just an *estimate* of the 25th percentile of the call center distribution. With sample sizes larger than $n = 6$, estimates become more accurate, no matter whether an estimate of the proportion of heads in coin flips or the estimate of a quantile from call center data.

TABLE 4.2

Applying Equation 4.1 with the Call Center Data

i	$y_{(i)}$	$(i - 0.5)/n$	$\hat{y}_{(i-0.5)/n} = y_{(i)}$
1	0.1	$(1 - 0.5)/6 = 0.083$	$\hat{y}_{0.083} = 0.1$
2	0.3	$(2 - 0.5)/6 = 0.250$	$\hat{y}_{0.250} = 0.3$
3	0.8	$(3 - 0.5)/6 = 0.417$	$\hat{y}_{0.417} = 0.8$
4	1.0	$(4 - 0.5)/6 = 0.583$	$\hat{y}_{0.583} = 1.0$
5	2.3	$(5 - 0.5)/6 = 0.750$	$\hat{y}_{0.750} = 2.3$
6	3.9	$(6 - 0.5)/6 = 0.917$	$\hat{y}_{0.917} = 3.9$

4.5 Using Data: Comparing Distributions via the Quantile–Quantile Plot

Look again at Equation 4.1, which states $\hat{y}_{(i-0.5)/n} = y_{(i)}$. This is a completely nonparametric statement, making no assumption about distribution form. It simply states that the $(i - 0.5)/n$ quantile of $p(y)$ is estimated by the ith ordered data value.

You can also estimate the quantile using a particular assumed form of a distribution. For example, if you estimate the parameters (μ, σ) of the normal distribution as $(\hat{\mu}, \hat{\sigma})$, then you can estimate the p quantile of the distribution as $P^{-1}(p \mid \hat{\mu}, \hat{\sigma})$. In general, letting $\hat{x}_p$ denote the estimated p quantile of a distribution having a particular assumed form (such as the normal), the q–q plot is the plot of the values $(\hat{x}_{(i-0.5)/n}, \hat{y}_{(i-0.5)/n})$. If the assumed distribution form is the true distribution, then these two numbers should be approximately equal (although not exactly equal because of random variation), leading to the approximate appearance of a straight line. If the assumed form of the distribution is much different from the true distribution, then the graph will have a pronounced curvature.

Example 4.5: Investigating Normality of Stock Market Returns via the q–q Plot

Revisit Figure 4.8, which shows a histogram of the Dow Jones data. Figure 4.9 shows a q–q plot comparing the actual quantiles of the Dow Jones return data $\hat{y}_{(i-0.5)/n}$ to quantiles $\hat{x}_{(i-0.5)/n}$ estimated from a normal distribution. The reference line $(\hat{x}_{(i-0.5)/n}, \hat{x}_{(i-0.5)/n})$ is also shown.

Unlike Figure 4.8, the discrepancy from normality is readily apparent in Figure 4.9 since the values do not fall near the straight line. The extreme returns are much farther out in the tail of the return distribution than would be expected, had the return distribution been normal. For example, the bottom leftmost point where the return is around −0.25 (the DJIA lost 25% of its value on October 19, 1987), but if the data were from a normal distribution, you would expect this number to be near the line, somewhere around −0.05. Financial analysts know that return distributions have heavier tails than the normal, so this conclusion is no surprise. Again, you can see the discrepancy from normality clearly in the q–q plot of Figure 4.9, but not in the histogram of Figure 4.8.

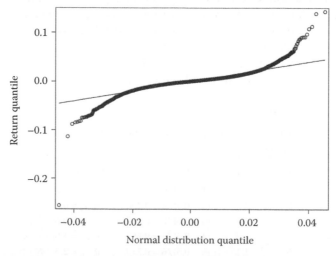

FIGURE 4.9
Quantile–quantile plot of the DJIA return data.

In a q–q plot given by your software, you may see the horizontal axes displayed with different units. It's no problem. The main point is still to compare the plot to the straight line.

Example 4.6: Investigating the Normality of the Call Center Data-Generating Process via the q–q Plot

To understand the q–q plot more clearly, let's walk through the calculations using the call center data $y_1 = 0.3$, $y_2 = 2.3$, $y_3 = 1.0$, $y_4 = 0.1$, $y_5 = 3.9$, and $y_6 = 0.8$. The question is could these data values have come from a normal distribution? Estimates of the mean and standard deviation (obtained using any software; these estimates will be discussed in much more detail in later chapters) are $\hat{\mu} = 1.40$ and $\hat{\sigma} - 1.45$. You can then construct the q–q plot "by hand," as shown in Table 4.3 using Microsoft Excel, whose inverse cdf is invoked as NORM.INV (p, μ, σ).

The q–q plot then looks like as shown in Figure 4.10.

In Figure 4.10, there are not enough data points to make a firm determination about deviation from the line, as some of the deviations can easily be explained by randomness alone. However, the fact that the lower quantiles of the normal distribution extend into negative numbers is a cause for concern, because of course waiting times cannot be negative. In addition, the largest call time is larger than you would expect, had the normal distribution produced these data. Thus, based on the graph *and on subject matter considerations*, it appears that the normal distribution is not a plausible model for producing these waiting time data.

TABLE 4.3

Calculations for Normal q–q Plot Using Microsoft Excel®

i	$y_{(i)}$	$p = (i - 0.5)/n$	$\hat{y}_{(i-0.5)/n} = y_{(i)}$	NORM.INV $(p, 1.40, 1.45)$
1	0.1	$(1 - 0.5)/6 = 0.083$	$\hat{y}_{0.083} = 0.1$	−0.605
2	0.3	$(2 - 0.5)/6 = 0.250$	$\hat{y}_{0.250} = 0.3$	0.422
3	0.8	$(3 - 0.5)/6 = 0.417$	$\hat{y}_{0.417} = 0.8$	1.095
4	1.0	$(4 - 0.5)/6 = 0.583$	$\hat{y}_{0.583} = 1.0$	1.705
5	2.3	$(5 - 0.5)/6 = 0.750$	$\hat{y}_{0.750} = 2.3$	2.378
6	3.9	$(6 - 0.5)/6 = 0.917$	$\hat{y}_{0.917} = 3.9$	3.405

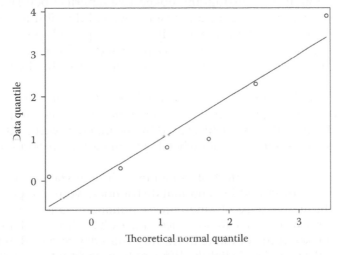

FIGURE 4.10
Quantile–quantile plot of the wait time data.

To be consistent with the ugly rule of thumb regarding sample size for the histogram, here is a similar one for the q–q plot.

Ugly Rule of Thumb 4.3

With a sample size of $n \geq 30$, the quantiles are adequate estimates, making the q–q plot reasonably trustworthy. Larger n provides better accuracy.

Interpreting q–q Plots

1. Compare the data to the straight line. The closer they are to the line, the more they appear as if produced by the assumed model.
2. Data points higher than the line in the upper right suggest that the pdf that produced the data tends to give more extreme values in the upper tail than does the assumed distribution. Data points lower than the line in the lower left suggest that the pdf that produced the data tends to give more extreme values in the lower tail than does the assumed distribution.
3. Data points lower than the line in the upper right suggest that the pdf that produced the data tends to give less extreme values in the upper tail than does the assumed distribution. Data points higher than the line in the lower left suggest that the pdf that produced the data tends to give less extreme values in the lower tail than does the assumed distribution.
4. Do not look for absolute perfection. Even if the data are produced by the assumed distribution, the points will not fall exactly on the line due to randomness.
5. Discuss sample size. With larger sample sizes, the estimated quantiles are more accurate, and hence, the q–q plot more trustworthy.
6. Pronounced horizontal lines are evidence of discreteness.

4.6 Effect of Randomness on Histograms and q–q Plots

We all love perfection! If your histogram looked like a perfect bell shape, you would probably like to say that your data came from a normal distribution. If your normal q–q plot had all of the points falling perfectly on the expected line, you would probably be happy to say that the data came from a normal distribution. And if you flipped a coin 1000 times and got exactly 500 heads, you would probably be happy to call it a fair coin.

We hate to spoil your happiness, but perfection does not exist in the real world, so none of your claims would be right. Your data never perfectly reflect the process that produced your data. There is always doubt, caused by the inherent variability in DATA. One sample is not the same as another, even when both samples are produced from the same identical process.

Thus, you must consider the effect of randomness in any statistical analysis. Even if your data were, in fact, produced by a normal distribution, your q–q plot would still not be a perfect straight line because of randomness. Randomness makes 10 flips of a coin result in the number 3 instead of 5. Randomness makes 1000 flips result in 511 rather than 500. Randomness makes your commute times vary from day to day. Randomness makes the last chocolate cake you baked slightly different from the next one that you will bake. Randomness is so real you can taste it!

It is easy to assess the effect of randomness on the histogram and *q–q* plots by using simulation. Suppose you use these plots to assess normality. Your question is "Could my data have been produced by a normal distribution?" When you look at your histogram, it's not a perfectly smooth and symmetric bell shape, and when you look at your *q–q* plot the data points do not line up perfectly on a line. So you are uncertain. But if the deviations from perfection are far more than what is explainable by chance alone, then you can confidently claim that your data are not produced by a normal distribution. And if the deviations from perfection are within the range of deviations that are explainable by chance alone, you could not rule out (based on data only) that the distribution producing the data was normal. But you cannot *prove* normality ever, using data: Even if the plots are very close to what is expected for the normal distribution, some other distribution that looks nearly normal, but with perhaps some slight asymmetry, could have produced your data. In summary,

> You cannot prove normality, or any other distribution form, using data.
> You can, however, disprove normality or other distributional forms using data.

Example 4.7: Investigating the Effect of Randomness in the Interpretation of the *q–q* Plot of Stock Market Returns

Consider the *q–q* plot of *n* = 18,834 stock market returns shown in Figure 4.9. The estimated mean return is 0.00017076 and the estimated standard deviation is 0.01121524. One random generation of 18,834 data* values from the normal distribution with this mean and standard deviation gives the *q–q* plot shown in Figure 4.11.

You can see in Figure 4.11 that there is no perfection, even though you can say that the distribution that produced the data* *is in fact a normal distribution*. You can say the distribution *is normal* in this case because that's how the data* were produced: They were simulated from a normal distribution. So the deviations you see from a straight line are purely chance deviations, just like coin toss deviations such as 3/10 instead of 5/10 and 511/1000 instead of 500/1000.

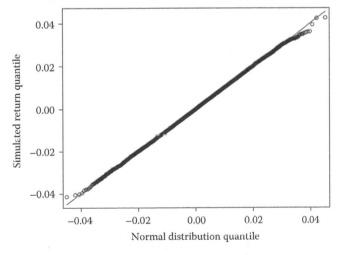

FIGURE 4.11
Quantile–quantile plot of *n* = 18,834 data* values produced by a normal distribution.

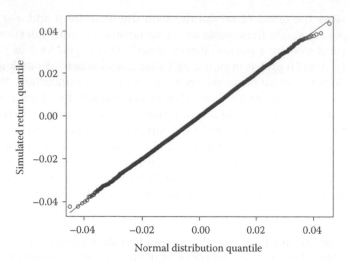

FIGURE 4.12
Quantile–quantile plot of another n = 18,834 data* values produced by the same normal distribution that produced the data* in Figure 4.11.

Comparing Figure 4.11 with Figure 4.9, it appears that the deviations from the line with the original data in Figure 4.9 are far more than the deviations from the line with the simulated data* in Figure 4.11. Hence, the deviations in the original data do not appear explainable by chance alone. The story does not end here, though: Figure 4.11 is just one simulation. Maybe in other simulations from the normal distribution, some deviations as large as those seen in Figure 4.9 could appear? Figure 4.12 shows another generation of n = 18,834 data* values from the same normal distribution that produced the data* in Figure 4.11. The pattern is slightly different, due to randomness, from that shown in Figure 4.11.

Using only two simulations as shown in Figures 4.11 and 4.12 does not suffice to show the potential deviations caused by chance alone; you need to perform more simulations. But more simulations (try some yourself!) would show the same thing: The chance deviations from the line with data* produced by a normal distribution are much less than the deviations from the line shown in the original data. The deviations in the original data are not explainable by chance alone; instead, you must conclude that some distribution other than the normal distribution produced the stock return data.

Once you understand the essential concept illustrated by Figures 4.9, 4.11, and 4.12, the entire subject of statistical hypothesis testing becomes conceptually simple. The formalities are given in Chapter 15.

Example 4.8: Investigating the Effect of Randomness in the Interpretation of the q–q Plot of Call Center Data

The case of the wait time data with the much smaller sample size (n = 6) provides the opposite story. If you simulate data* from a normal distribution and construct the q–q plots, you will see that the deviations shown in Figure 4.10 are entirely explainable by chance alone. It is the subject matter—wait time can't be less than zero; occasionally very large wait times are expected—that tells you that the distribution is not normal in this case. You could not prove non-normality from the graph in Figure 4.10 alone, since the deviations from the line are easily explainable by chance alone.

As a general rule, randomness has a larger effect with small samples than with larger samples. In 10 flips of a fair coin, 30% heads (3 out of 10) is quite likely. In 1000 flips, 30% heads (300 out of 1000) is virtually impossible. That's why larger samples are desirable in statistical analyses—with larger sample sizes you can more easily rule out chance as an explanation for your results.

Vocabulary and Formula Summaries

Vocabulary

Constraints	Restrictions on the model that may or may not be realistic. See also *assumptions* (Chapter 1).
Generic distribution	A distribution $p(y)$ that is not assumed to be constrained to any particular function form.
Histogram	An estimate of the probability distribution function $p(y)$.
Quantile	The value y_p such that $\Pr(Y \leq y_p) = p$.
Quantile–quantile (q–q) plot	A plot of estimated data quantiles against quantiles estimated using a particular distribution.
Percentile	Same as quantile, but phrased in the percentage terminology of "$100p$ percent."
Median	The $p = 0.50$ quantile or 50th percentile.
First quartile	The $p = 0.25$ quantile or 25th percentile.
Second quartile	The $p = 0.50$ quantile or the median.
Third quartile	The $p = 0.75$ quantile or 75th percentile.
Quintiles	The 0.2, 0.4, 0.6, and 0.8 quantiles.
Deciles	The 0.1, 0.2,..., and 0.9 quantiles.
Order statistics	The values of the data when ordered form smallest to largest.

Key Formulas and Descriptions

$p(y) = 1/\theta$, for $0 < y < \theta$	The pdf of the uniform distribution between 0 and θ, abbreviated the U(0, θ) distribution.
$P(y_p) = p$	The equation defining the p quantile of a continuous pdf.
$y_p = P^{-1}(p)$	The solution for the p quantile of a continuous pdf.
$y_{(1)} \leq y_{(2)} \leq \cdots \leq y_{(n)}$	The order statistics from a data set.
$\hat{y}_{(i-0.5)/n} = y_{(i)}$	The $(i - 0.5)/n$ quantile is estimated by the ith order statistic.
$\hat{x}_p = \hat{P}^{-1}(p)$	An estimate of the p quantile of a distribution when a particular distribution is assumed, and its parameters are estimated from the data.

Exercises

4.1 Data from surveys are often recorded on a 1, 2, 3, 4, 5 scale. Here are some responses from $n = 10$ people surveyed by an automobile retailer regarding their level of customer satisfaction: 3, 4, 3, 5, 5, 5, 5, 5, 5, and 4.

   A. Using the computer, draw three distinct distributions that *could have* produced these data. Try not to make the distributions too similar to one another. Be realistic: It is possible that there are dissatisfied customers. You just didn't happen to see any in this particular data set.

   B. Using the computer, draw three distinct distributions that most likely *could not have* produced these data.

   C. Write down the generic distribution $p(y)$ for these data in table form, in terms of unknown parameters, and explain why this distribution is more believable than any of the distributions you drew in Exercise 4.1A or B.

4.2 In Example 4.5, you saw that the stock market returns are not produced by a normal distribution. State the generic model for how these returns are produced, and explain why the generic model is believable.

4.3 In Example 4.6, you read that the call center wait times are not produced by a normal distribution. State the generic model for how the wait times are produced, and explain why the generic model is believable.

4.4 Check that the area under the function in Figure 4.5 is 1.0. What would be the area under the curve if you didn't divide by Δ to obtain the function?

4.5 See Example 4.8. Follow the method shown in Example 4.7 to show that the deviations from the line with the call center data are easily explainable by chance alone.

4.6 See Figure 4.8. Is the difference between the histogram and the normal approximation curve explainable by chance alone? Simulate data from the normal distribution, construct the histogram and the normal approximation curve, and compare the result with Figure 4.8.

4.7 The following data were obtained from the Internet *data and story library*, or DASL, with URL http://lib.stat.cmu.edu/DASL/ at the time of writing this book. The data are scores given by taste testers on cheese, as follows: 12.3, 20.9, 39, 47.9, 5.6, 25.9, 37.3, 21.9, 18.1, 21, 34.9, 57.2, 0.7, 25.9, 54.9, 40.9, 15.9, 6.4, 18, 38.9, 14, 15.2, 32, 56.7, 16.8, 11.6, 26.5, 0.7, 13.4, and 5.5.

   A. Construct and interpret the histogram of the data.

   B. Create a table like Table 4.3 for these data.

   C. Construct and interpret the normal q–q plot of the data.

   D. Simulate a data* set from a normal distribution having the same mean, standard deviation, and sample size as for the taste variable. Construct the histogram and q–q plot of the simulated data*. Repeat nine times, getting a total of 10 histograms and 10 q–q plots. You now have an idea of the effect of chance variation on the histograms and q–q plots. Without doing any formal test, do the histogram and q–q plot of the original data differ from normal by amounts that are explainable by chance alone? Discuss.

4.8 Another data set from DASL involves industrial plant *waste run-up*, a measure of waste where higher values are bad and lower values (especially negative numbers) indicate exceptional performance. The data are 1.2, 16.4, 12.1, 11.5, 24.0, 10.1, −6.0, 9.7, 10.2, −3.7, −2.0, −11.6, 7.4, 3.8, 8.2, 1.5, −1.3, −2.1, 8.3, 9.2, −3.0, 4.0, 10.1, 6.6, −9.3, −0.7, 17.0, 4.7, 10.2, 8.0, 3.2, 3.8, 4.6, 8.8, 15.8, 2.7, 4.3, 3.9, 2.7, 2, 2.3, −3.2, 10.4, 3.6, 5.1, 3.1, −1.7, 4.2, 9.6, 11.2, 16.8, 2.4, 8.5, 9.8, 5.9, 11.3, 0.3, 6.3, 6.5, 13.0, 12.3, 3.5, 9.0, 5.7, 6.8, 16.9, −0.8, 7.1, 5.1, 14.5, 19.4, 4.3, 3.4, 5.2, 2.8, 19.7, −0.8, 7.3, 13.0, 3.0, −3.9, 7.1, 42.7, 7.6, 0.9, 3.4, 1.4, 70.2, 1.5, 0.7, 3.0, 8.5, 2.4, 6.0, 1.3, and 2.9.

   A. Construct and interpret the histogram of the data.

   B. Construct a table like Table 4.3 for these data.

   C. Construct and interpret the normal q–q plot of the data.

   D. Simulate a data* set from a normal distribution having the same mean, standard deviation, and sample size as for the "run-up" variable. Construct the histogram and q–q plot of the simulated data*. Repeat nine times, getting a total of 10 histograms and 10 q–q plots. You now have an idea of the effect of chance variation on the histograms and q–q plots. Without doing any formal test, do the histogram and q–q plot of the original data differ from normal by amounts that are explainable by chance alone? Discuss.

4.9 The following data are wait times for customer service (in minutes): 0.48, 1.15, 0.26, 0.05, 0.06, 0.02, 2.12, 0.45, 0.07, 0.99, 0.70, 1.55, 1.72, 0.90, 0.76, 2.03, 0.63, 0.53, 0.30, 0.51, 0.49, 0.52, 0.05, 0.38, 0.43, 0.60, 0.01, 0.11, 0.00, 0.68, 0.02, 1.46, 0.17, 0.10, 0.01, 0.38, 0.60, 0.14, 0.52, 0.13, 1.30, 0.81, 1.37, 0.51, 0.36, 0.34, 0.49, 0.01, 1.60, 0.73, 2.65, 0.04, 1.15, 0.68, 0.13, 0.19, 0.11, 0.16, 1.23, and 1.01. Construct and interpret the q–q plot for checking whether the data might have come from an exponential distribution.

4.10 Hans spins a top and lets it fall and settle. He had marked a location on the top in indelible ink; if the top settles with the mark perfectly at the peak, then Hans records $Y = 0$. If the top settles with the mark perfectly at the bottom, then he records $Y = 180$. Otherwise, he records Y, the angle in degrees from the peak, where the mark appears. The range of Y is $0 \leq Y < 360°$.

   A. Explain why the pdf $p(y)$ that produces Y might be the U(0, 360) distribution, and draw a graph of $p(y)$. Label and number all axes.

   B. Explain a circumstance where the U(0, 360) pdf might be wrong, and draw a graph of a possible $p(y)$ that might be more correct in this circumstance. Label and number all axes.

   C. Suppose you decide to check whether the U(0, 360) model is reasonable using a q–q plot. Describe Nature, design, measurement, and DATA for this study.

   D. Suppose Hans' data set from Exercise 4.10C is 149, 174, 309, 1, 82, 9, 218, 231, 49, 76, 0, 219, 215, 119, 148, 187, 231, 7, 2, 3, 10, 25, and 86, all degrees from the peak. Construct the q–q plot for checking whether the U(0, 360) model might have produced Hans' data and discuss.

4.11 In Example 4.2, it is stated that around 50% of the traps produce no insects, and the remainder of the traps produce a range from 1 to 500, with a large proportion over 100. Using different values of λ, try to produce 100 data* values from the Poisson distribution that look like this. Summarize your results and explain why the Poisson model is not valid.

5

Conditional Distributions and Independence

5.1 Introduction

You will experience a sense of déjà vu when you read these words: A **conditional distribution** is denoted by $p(y|x)$, which is read aloud as "the probability distribution of Y given a particular X." You should have a sense of déjà vu because this is precisely the definition of a regression model—a statistical model that has both deterministic and probabilistic components—given in Chapter 1.

You can use conditional distributions for making predictions. In addition to the primary mantra (*model produces data...*), here's a second mantra you'll also see repeatedly in this book.

Mantra #2

Use what you know ($X = x$) to predict what you don't know (Y).

For example, suppose you are a doctor faced with the unpleasant task of informing your patient that he has cancer. He asks, "How bad is it? Should I get my affairs in order, or just carry on as usual?" It all depends on what you know about his condition—the type and stage of cancer, his age, health, etc. If he has an early stage and mild form of cancer and is young and otherwise in excellent health, then his chance of surviving 5 years is very good. But if he has late stage and aggressive form of cancer and is elderly and otherwise in poor health, then his chance of surviving 5 years is not very good. The unknown Y in either case is binary—your patient lives or dies, coded as $Y = 1$ or $Y = 0$. There are many more potential futures where $Y = 1$ (the patient lives) in the scenario with early stage cancer in a young and healthy patient than in the scenario with late stage cancer in an elderly and unhealthy patient. In other words, the conditional distributions of Y differ, because what you know (the X) differs in the two cases.

Because conditional distributions are so useful for making predictions, empirical research involves the concept of conditional distributions more than it involves anything else. Conditional distributions come into play whenever you see the phrase *effect of* in empirical research. Dissertation titles and research papers typically include this very phrase, such as "The Effect of Ambiguity on Business Decisions," "The Effect of Communication on Cancer Beliefs," "The Effect of Nicotine Patches on Smoking Cessation," and on and on. Translated, these titles address the following questions: Are business decisions generally better with less ambiguity? Is a patient's knowledge about their cancer generally better when the physician communicates more clearly? Do people generally quit more when they use a smoking patch? Even if the exact phrase *effect of* does not appear in the title, empirical research typically involves something similar, namely, addressing how two variables relate to each other.

Notice that the word *generally* appears in all three translations of the dissertation titles. This is because none of these relationships is deterministic. There is a *distribution* of successes and failures (the binary *Y*s) in business decisions when ambiguity is low, abbreviated $p(y|X = \text{Low})$, and there is a potentially different *distribution* of successes and failures when ambiguity is high, abbreviated $p(y|X = \text{High})$. There is a *distribution* of understanding of cancer (say, as measured by a questionnaire, the *Y*s) among patients whose physicians communicate poorly, abbreviated $p(y|X = \text{Poorly})$, and there is a potentially different *distribution* of understanding of cancer among patients whose physicians communicate well, abbreviated $p(y|X = \text{Well})$. There is a *distribution* of smoking behavior (say, as measured by packs per week, the *Y*s) among people who use the nicotine patch, abbreviated $p(y|X = \text{Patch})$, and there is a potentially different *distribution* of smoking behavior among people who do not use the nicotine patch, abbreviated $p(y|X = \text{No Patch})$.

Notice also that the term *potentially different* appears in all three examples. If the distributions do not differ at all, then the *X* variable (e.g., ambiguity, communication, nicotine patch use) has no effect on *Y*. If the distributions differ at all, then the *X* variable has an effect on *Y*. The statement that *X* has an effect on *Y* is a typical **research hypothesis**; the *effect itself* is an example of the unknown parameter θ that is in our Mantra, *model has unknown parameters*. The effect parameter θ can take a variety of forms, some of which will be explained in later chapters, including the difference between means, the correlation coefficient, the odds ratio, the hazard ratio, and the regression coefficient, depending on the type of distributions $p(y|x)$ that you use. Data reduce your uncertainty about the unknown effect parameter θ, allowing you to make more certain claims about whether an effect exists in reality.

So, what is *X* and what is *Y*? It's easy to remember: *X* is an "explanatory" variable. That is, *X* *explains* some of the variation in *Y*. The variable you call *X* goes by other names too: *predictor variable, exogenous variable*, and *descriptor variable*, among other names. The *Y* variable is also called a *response variable*, an *endogenous variable*, and a *target variable*, among other names.

Sometimes the *X* variable is called an *independent variable*, and the *Y* variable is called a *dependent variable*. While the "dependent" moniker for *Y* is reasonable, the "independent" designation for *X* is too easily confused with a different concept called *statistical independence*, and therefore, we suggest that you *not* use the dependent variable/independent variable designations for your *Y* and *X* variables.

When we say "*X* explains *Y*," we do not necessarily mean that *X* *causes* *Y*, but that's another way to think about what should be *X* and what should be *Y*. If there is a causal connection, then pick *X* to be the variable that causes *Y*. For example, a change in *X* = smoking behavior causes a change in *Y* = lung cancer occurrence. Researchers often want to make causal claims, but rigorous claims of causality usually require carefully controlled experiments (see Section 16.7 of Chapter 16). On the other hand, if you find that *X* explains *Y* very well, you should investigate further to see whether there might be a causal connection between the two. If so, then you have the ability to change *Y* if you can control *X*.

Your *Y* variable should be something that is important. For example, the time until you die (*Y*) is probably interesting to you. Some *X* variables that might influence your *Y* are your genetic makeup and your lifestyle choices such as smoking behavior, drinking behavior, seat-belt-wearing behavior, and illicit drug use behavior. How does the conditional distribution of your remaining lifetime, *Y*, depend on those *X* variables? This chapter gives you tools you can use to answer this question.

5.2 Conditional Discrete Distributions

More déjà vu: In Chapter 1, Tables 1.4 and 1.5, you saw examples of discrete conditional distributions. Arranged in a table to show the dependence of the probabilities on both the color choice and on age, you can construct a single two-way table as shown in Table 5.1. In Table 5.1, the rows add to 1.0, so you can see that the distributions are in the rows, not columns as in Tables 1.4 and 1.5.

The notation $p(y|x)$ refers to numbers such as $p(\text{Red}|\text{Younger}) = 0.50$ and $p(\text{Green}|\text{Older}) = 0.40$. For any fixed X (or fixed row), the numbers sum to 1.0, as is true for any discrete distribution:

$$\sum_{\text{all } y} p(y|x) = 1.0, \quad \text{for every } x$$

For instance, when $X = $ Older, the summation formula simply states that $0.20 + 0.40 + 0.40 = 1.0$.

Table 5.1 is, of course, only hypothetical. Recall the Mantra: *Model produces data, model has unknown parameters*. The probabilities listed in Table 5.1 are not from data; they are just hypothetical values that we have made up to make things concrete. Really, the true data-producing distributions are abstract, algebraic quantities as shown in Table 5.2. They have to be abstract, because *model has unknown parameters*. (Remember?)

The unknown parameters are $\pi_{i|j}$, the probability that a person in age category $X = j$ selects car color $Y = i$, which you can write as $\Pr(Y = i | X = j) = \pi_{i|j}$.

Table 5.2 shows conditional distributions for Y (color) given two discrete X values (younger, older). There are also conditional distributions of Y for continuous X values: $\Pr(Y = i | X = x) = \pi_{i|x}$ where for each x, the numbers $\pi_{1|x}, \pi_{2|x}$, and $\pi_{3|x}$ are nonnegative and sum to 1.0. While these π's are unknown (*model has unknown parameters!*), it is sensible to assume that they are continuous functions of x. In other words, the car color choice distribution should not differ much for a 20.0-year-old person versus a 20.1-year-old person, and they should not differ much for a 60.0-year-old person versus a 60.1-year-old person. A popular advanced statistical

TABLE 5.1

Hypothetical Conditional Distributions $p(y|x)$ for the Car Color Choice Example

		Red	Gray	Green	Total
			Y		
X	Younger	0.50	0.20	0.30	1.00
	Older	0.20	0.40	0.40	1.00

TABLE 5.2

Actual Conditional Distributions $p(y|x)$ for the Car Color Choice Example

		Red	Gray	Green	Total
			Y		
X	Younger	$\pi_{1|1}$	$\pi_{2|1}$	$\pi_{3|1}$	1.00
	Older	$\pi_{1|2}$	$\pi_{2|2}$	$\pi_{3|2}$	1.00

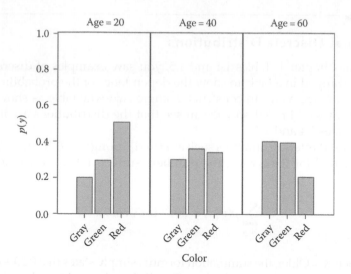

FIGURE 5.1
Potential conditional distributions of color choice (Y) when age (X) is 20, 40, and 60, shown as bar charts.

model that allows the probabilities to vary continuously is the *multinomial logistic regression model*. We will discuss the closely related *logistic regression model* in later chapters.

Figure 5.1 displays bar charts that show how these conditional distributions might look for X = 20, 40, and 60.

Think of the concept of **morphing** to understand the concept of distributions varying continuously as x changes. Figure 5.2 shows another look at Figure 5.1, with the probabilities continuously changing with different x, but always adding to 1.0.

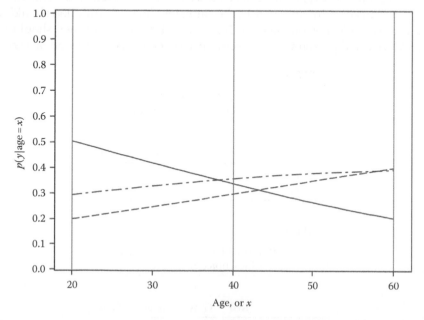

FIGURE 5.2
Potential conditional distributions of color choice as continuous functions of age in the range $20 \le x \le 60$. Solid curve: $p(\text{Red}|\text{Age} = x)$, dashed curve: $p(\text{Gray}|\text{Age} = x)$, and dash-dot curve: $p(\text{Green}|\text{Age} = x)$.

The graphs shown in Figure 5.2 make an assumption that the distributions vary continuously. Any assumption that you make implies a constraint on the class of models that you specify. Always question assumptions! The best way to critique assumptions is to try to identify cases where the assumption is wrong and to logically evaluate whether those cases are plausible. If they are plausible, then your assumptions are not good. How can the assumption of continuously morphing probabilities be wrong? Well, suppose that in the country Greenland, everyone were required to buy a green car after the age of 40. In that case, at $X = 40$, the distributions will shift dramatically, with probability on green jumping to near 1.0. Only the criminals will have red or gray cars!

The farcical example of Greenland shows how unusual things must be for the distributions to change discontinuously. Hence, the continuity assumption seems reasonable, since it would take such a bizarre case to make it untrue. Nevertheless, if you are still uncertain about the continuity assumption, you can always reduce your uncertainty about this or any other assumption using data. *Data reduce uncertainty about the unknown parameters.*

There are occasionally cases where discontinuities occur. For example, if a certain tax break kicks in when people earn less than $60,000, then the distribution of tax paid will jump discontinuously at income = $60,000. If you want to entertain the notion of conditional distributions varying discontinuously, you'll need a strong reason, such as in the tax break example. But usually the following, third and final mantra of this book, is a safe bet.

Mantra #3

Nature favors continuity over discontinuity.

Always remember that the distributions, conditional or otherwise, are not from data. Nor are the distributions from populations. The terms $\Pr(Y = i | X = x) = \pi_{i|x}$ refer to probabilities that are part of a process model that you assume to produce the DATA that you might observe. To see why these probabilities cannot possibly be calculated from sample or population data, imagine the case where $X = 20.000000000000000000\ldots$ years old. This is a person who is precisely, down to the infinitesimal decimal of a second, 20 years old. How many of these people are there in the world at the precise instant when you finish reading this sentence? None! So the probabilities cannot possibly refer to counts out of a total, such as in the statement "50 out of 100 20-year-olds chose the red car," simply because there is not one single 20.0000000000… year old person on the entire planet Earth (let alone 100 of them) at this precise instant.

Understanding models as *producers of data* helps you to understand what they really mean and to avoid making silly, vacuous claims about populations or silly misinterpretations of data. It also helps you to understand the logical foundation for making assumptions such as continuity—it's all about the processes, all about Nature. It is logical to assume that the behavioral processes that lead to car color choice for people who are precisely 20.000000… years old do not differ much from the behavioral processes that lead to car color choice for people who are precisely 20.000000032 years old (20 years and 1 s). Have a look at Figure 5.2 again—do you see the continuity? Does it make sense?

5.3 Estimating Conditional Discrete Distributions

As in the case of estimating probability distribution functions (pdfs) as shown in Chapter 4 via histograms, you often have to use the *binning* method, where you identify ranges of values (or *bins*) within which there are sufficient numbers of observations. In the car

TABLE 5.3

Data from a Sample of 20-Year-Olds

Color Choice	Count	Percent
Red	1271	55.00
Gray	347	15.00
Green	694	30.00
Total	2312	100.00

color choice example, there is no one in the world who is precisely 20.0000000... years old, so if you want to estimate $p(y|X = 20)$, you'll have to settle for an estimate that includes people who are in a range of 20 years old. If you have a huge sample size with thousands of people who are between 20 and 21 years old (so they call themselves 20 years old), then you can use that data directly to estimate the probabilities. Arranged in a **frequency table**, such data might look as shown in Table 5.3.

But data do not produce the model; rather, *model produces data*. Instead, *data reduce the uncertainty about the unknown parameters.* Here, you have **estimates** (which are merely educated guesses) for the values $\pi_{Red|20}$, $\pi_{Gray|20}$, and $\pi_{Green|20}$. You need a hat ^ on them to emphasize that they are not the same as the true values, as follows:

$$\hat{\pi}_{Red|20} = 0.55, \quad \hat{\pi}_{Gray|20} = 0.15, \quad \hat{\pi}_{Green|20} = 0.30$$

If your sample size were much smaller than 2312, you would need to expand the range of X values to estimate the conditional probabilities. For example, if in Table 5.3 the counts for the 20-year-olds were 3, 0, and 1 ($n = 4$) instead of 1271, 347, and 694 ($n = 2312$), then the resulting probability estimates $\hat{\pi}_{Red|20} = 0.75$, $\hat{\pi}_{Gray|20} = 0.00$, $\hat{\pi}_{Green|20} = 0.25$ would clearly lack precision. To increase the precision, you would have to include more people, say 16–25-year-olds. As always in statistics, there is a trade-off here: While the resulting probabilities will be estimated more precisely, they no longer refer specifically to 20-year-olds but instead to a broader group of young people. This could be problematic, for example, if you were designing a marketing campaign directed specifically at 20-year-olds.

How should you create these tables? This is a good time for another ugly rule of thumb.

Ugly Rule of Thumb 5.1

When creating frequency tables, aim for at least five in each category to obtain reliable estimates of the true percentages.

In Table 5.3, this ugly rule of thumb is easily met, since there are 1271, 347, and 694 in the three categories, all of which are easily more than five. On the other hand, in the subsequent discussion, the data set with counts of 3, 0, and 1 does not satisfy Ugly Rule of Thumb 5.1.

5.4 Conditional Continuous Distributions

Why do you want to make money? It's not a silly question. Some people are perfectly content with very little. But most like the comforts that money affords. For example, people who make more money generally buy more expensive homes. This is not a *deterministic* relationship: Just because Joe Smith makes more money than Mary Jones does not tell you that Joe will have a more expensive home than Mary. Instead, the relationship is probabilistic. You cannot say for certain what any individual like Joe will do. But you can say that "Joe-like people" tend to have more expensive homes than "Mary-like people." In the language of conditional probability distributions, the conditional distribution of house value for "Joe-like people" is *morphed to the right* of the conditional distribution of house value for "Mary-like" people.

As with all data that we can measure with our finite human capabilities, *price of home* is technically a discrete variable. In particular, *selling price* typically takes values that look like data rounded to the nearest $1,000, or for larger homes, to the nearest $10,000, such as $245,000 and $510,000. It would be rare to see a home sell for a price like $189,543.12! However, you can assume the house price variable is nearly continuous since its values reasonably fill a continuum.

A conditional continuous pdf is simply a continuous pdf that depends on the particular $X = x$. If $Y =$ home value and $X =$ annual income, there is a potentially different continuous pdf of home value for every different value of annual income. These pdfs are just like ordinary continuous pdfs in that they are nonnegative functions whose area under the curve is 1.0. The only difference is that there are many of them, one for each x, so you need to indicate the dependence on x using the notation $p(y|x)$. You can use the same notation $p(y|x)$ for both discrete and continuous conditional pdfs; the difference is that for any given x, the integral, rather than the sum, must be equal to 1.0 for continuous distributions:

$$\int_{all\ y} p(y|x)dy = 1.0, \quad \text{for every } x$$

As in the discrete car color choice example in the previous sections, you should assume that these distributions vary continuously with X. Specifically, while the distribution of home value among people making $1,000,000 should differ appreciably from the distribution of home value among people making $50,000, there should be very little difference between the distributions for people making $50,000.00 and $50,000.01. You can logically assume a continuous morphing of these conditional distributions, as shown in Figure 5.3. *Nature favors continuity over discontinuity.*

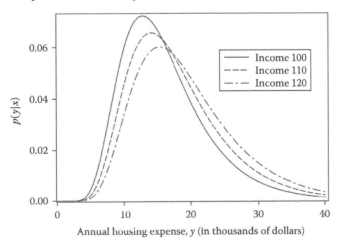

FIGURE 5.3
Possible conditional distributions of housing expense for annual income levels of $100K, $110K, and $120K.

5.5 Estimating Conditional Continuous Distributions

As you saw in Chapter 4, the primary tool for estimating a continuous distribution is a histogram. The same applies for conditional continuous distributions. The only wrinkle is that conditional distributions refer to subsets of potential DATA defined by a particular $X = x$, and there might not be enough data in a given subset to adequately estimate the histogram. Recall Ugly Rule of Thumb 4.2, which suggests that you need $n \geq 30$ for an adequate histogram estimate. You will need to group the X data judiciously to arrive at sensible ranges that include adequate data. Some judgment is needed here and some art.

In set notation, you will need to estimate the conditional distributions such as $p(y|X \in A)$, for judiciously chosen set A of X values, rather than the conditional distributions $p(y|x)$ (which is the same as $p(y|X \in A)$, with $A = \{x\}$), because there are more data where X can take on a range of values (in A) than there are where X takes only one particular value (x).

You might divide the X data into two groups—say separated by median—or perhaps into four groups, separated by quartiles. But the medians and quartiles might not be easy numbers to display: For example, if the quartiles are 13,212.33, 23,991.21, 59,567.36, and 121,544.95, a more aesthetic display might use the values 15K, 25K, 60K, and 120K instead.

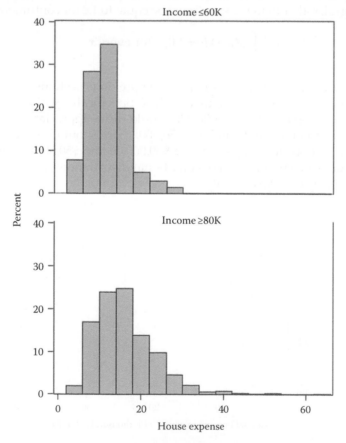

FIGURE 5.4
Estimated conditional pdfs (i.e., histograms) of house expense for income ≤60K and income ≥80K.

There is no Ugly Rule of Thumb here, although there is an overriding guiding principle. We won't call this one a Mantra; instead it's just good common sense:

Be aesthetically pleasing in your presentations of tables and graphs.

In many cases, your audience would rather see the numbers 15K, 25K, 60K, and 120K and be told that they are roughly quartiles, rather than see the actual quartiles themselves. An addendum to the rule "Be aesthetically pleasing" is the rule "Know your audience." One audience has different needs from another. For example, if you are intending to publish in a journal where precise quartile splits are commonly used, despite the ugliness of the resulting numbers, then by all means, use the ugly quartile splits.

Figure 5.4 displays two estimated conditional pdfs (i.e., histograms) of housing expense data using income ranges of $\leq$60K and $\geq$80K, corresponding to sets $A_1 = \{x; x \leq 60\}$ and $A_2 = \{x; x \geq 80\}$, respectively.

Notice that the histograms displayed in Figure 5.4 show the expected morphing appearance suggested by the model of continuously changing distributions shown in Figure 5.3.

Note another interesting feature: There appears to be more variability in house expense among the $\geq$80K income group than among the $\leq$60K income group. Why? One explanation is that having a larger income gives people more freedom: Some will pinch pennies, and others will spend lavishly. With less income, people have fewer choices, and there is correspondingly less variability in their data.

5.6 Independence

Most research questions involve a question about the effect of an X variable on a Y variable. This is a question about Nature, not data, but you can use data to learn about Nature (*data reduce the uncertainty about the unknown parameters*).

Maybe at the fundamental level of Nature, there is no relationship at all between X and Y. Maybe the distribution of successes and failures in business decisions is completely unrelated to ambiguity. Maybe patients' understanding of cancer is completely unrelated to their physician's communication skill. Maybe smoking behavior is completely unrelated to wearing or not wearing a nicotine patch.

Definition of Independence between Random Variables (RVs) X and Y

If $p(y|X \in A_1) = p(y|X \in A_2)$ for *all* sets A_1 and A_2, then Y and X are independent RVs.

Restated, the definition says that X and Y are independent if the *conditional distribution* of Y is *unchanged*, no matter what, or where, X happens to be.

Definition of Dependence between RVs X and Y

If $p(y|X \in A_1) \neq p(y|X \in A_2)$ for *some* sets A_1 and A_2, then Y and X are dependent RVs.

Restated, the definition says that X and Y are dependent if the *conditional distribution* of Y is *different*, depending on the particular X value or set of X values.

Have a look again at the morphing shown in Figures 5.1 through 5.3. If the variables were independent, you would see no such morphing. The distributions of car color choice

TABLE 5.4

Summarizing the Successive Tosses
of a Coin

		Next Toss		
		Heads	Tails	Total
First toss	Heads	2	2	4
	Tails	3	2	5

would be exactly the same, for all ages, and the distributions of housing expense would be the same for all incomes. Based on common sense alone, without any data, you should suspect that the variables are dependent in both cases. In the first case, car color preferences should change with age, and in the second case, housing preferences should change with income.

Independence is a property of Nature, not of data. Your data can easily steer you wrong here: Almost always, your data will show some evidence of relationship between X and Y, even when there is none in reality. Take the lowly coin toss as an example. Flip the coin 10 times, and you may get a sequence like this: T, H, T, T, T, H, H, H, T, H, where T = tails and H = heads. Is there a relationship between the outcome of one toss and the next one? Table 5.4 summarizes the results.

So, can you infer that if you toss a coin and get tails, then the next outcome is more likely to be heads? In other words, is Pr(Head|Tail previously) higher (3/5 = 60%) than Pr(Head|Head previously) (2/4 = 50%)? While the data may suggest this, recall this: *Data reduce the uncertainty about the unknown parameters*. They do not eliminate the uncertainty; that is, the numbers from the data are not the same as the model parameters. They are just estimates and not particularly good ones in this case because the sample size is so small.

Because independence refers to the data-generating process, and not the data, here is a general rule that you can count on:

> You cannot prove independence using data

Based on data, you can say that variables *appear* independent, or are *close to independent*, or that *independence is a reasonable model*, if the *estimated* distributions $\hat{p}(y|X \in A_1)$ and $\hat{p}(y|X \in A_2)$ are reasonably similar for various choices of A_1 and A_2. One way to translate *reasonably similar* is "within chance variation," a topic covered further in Chapter 15. A different and perhaps better way to translate *reasonably similar* is as "practically insignificant" or "close enough so that there is no practical difference."

Either translation involves actual distributions, not estimated distributions. It takes some judgment and experience to understand the nature of chance variation, but always think back to the coin flip: You expect 50%, but you get something else. The difference between what you get and what you expect is precisely explained by chance variation. For example, suppose you flip a coin 1000 times and get 509 heads. The difference between 500 and 509 is explained by chance variation alone.

Conversely, based on the data, you can say that variables *appear dependent* if the estimated distributions $\hat{p}(y|X \in A_1)$ and $\hat{p}(y|X \in A_2)$ are grossly different for some choices of A_1 and A_2. One translation of *grossly different* is "outside the range of chance variation," which is another term for "statistically significant," or "not easily explainable by chance alone." A different, perhaps better way to translate *grossly different* is "practically significant" or "different enough to make a practical difference." Again, these translations involve actual distributions, not estimated distributions. It is usually easy to establish that certain differences are obviously outside the realm of chance. You will find formal probability calculations in Chapter 15; at this point, we are just asking you to develop some common sense intuition. Again, the lowly coin toss example helps: If you flip a coin 1000 times and get 100 heads, is this "outside the realm of chance variation when the coin is fair?" If you can't answer the question immediately as "Yes, of course!" then just simulate 1000 Bernoulli RVs with $\pi = 0.5$, count the 1s, and repeat a couple of times. You'll get it very quickly.

You can usually assume that successive coin flips are independent. Your chance of landing heads on a subsequent flip is the same, regardless of whether the previous flip is heads or tails. However, it is possible to force strong dependence between successive coin flips by trickery: Let the coin fall to the floor, then pick it up, just *one millimeter*, no more, and drop it. It will land on whatever face it showed on the first flip. Pick it up 10 mm and it will likely land on the same face. Even when the coin is flipped in the usual way, there is slight dependence between successive coin flips, for the same reason: The coin flipper picked it up face up from the last toss and then tossed it again. But, just like the Bernoulli(0.5) model is an excellent model albeit not precisely true to the infinite decimal, the independence model is also an excellent model albeit not precisely true to the infinite decimal, when coins are tossed in the usual way. Thus, any suggestion of dependence in a collection of successive coin tosses, as shown in Table 5.4, is usually explained by chance alone.

Example 5.1: Investigating the Independence of Consecutive Market Returns

Can you make money by watching the stock market go up or down? Unlike the case study of comparing trading strategies in Chapter 1, this example will use real stock market data, the same Dow Jones Industrial Average (DJIA) data set analyzed in Chapter 4. The question is whether today's stock price movement is related to yesterday's. An investor would like to know because if the stocks are more likely to rise after a day where they fell, then the investor will want to invest after a down day.

Let Y = today's return on the DJIA, and let X = yesterday's return. Figure 5.5 displays the estimated conditional distributions of Y, separated by the conditions $X \le 0$ and $X > 0$.

Figure 5.5 shows very little difference in the estimated distributions. It seems not to matter whether yesterday's DJIA average was up or down, today's distribution is approximately the same. There *are* slight differences in the histograms—the second-largest bar in the lower graph is a little smaller than the corresponding bar in the upper graph—but these minor discrepancies seem explainable by chance alone, much like tosses of coin. In particular, the centers of the distributions seem about the same, so it appears that you can't make more money on average by putting your money in the market the day following a down day than you can by putting your money in the market in the day following an up day.

The analysis of Figure 5.5 has not proven independence of today's DJIA return and yesterday's return; you can never prove that. In fact, financial researchers argue that there *are* subtle dependencies. Maybe you can see the dependencies by being more thorough in checking the assumption. After all, the definition of dependence is that $p(y|X \in A_1) \ne p(y|X \in A_2)$ for *some* sets A_1 and A_2. Figure 5.5 only looked at the sets $A_1 = \{x; x \le 0\}$ and $A_2 = \{x; x > 0\}$. To make the analysis more complete, you should technically check *all possible* sets A_1 and A_2, but this is clearly not feasible since there are infinitely many of them. But you can check other pairs.

Perhaps you, as a keen investor, believe there is little difference between earnings when yesterday's stock market is down versus when it is up. But you might wonder, "What if there is a *big drop* in the market today? Surely there must be some 'rebound' effect where tomorrow's

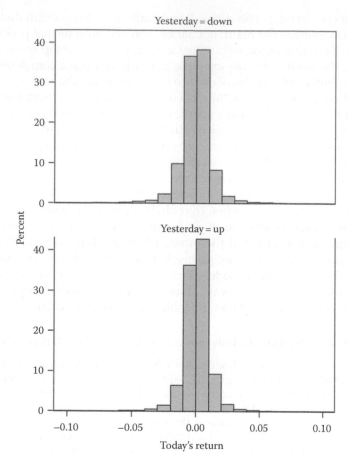

FIGURE 5.5
Conditional histograms of DJIA return, classified by either down or up on the previous day. Occasional returns more than 0.10 and less than −0.10 are not shown.

return will most likely be higher?" And conversely, you might wonder whether there is a reverse rebound if today's market has a big increase. Over the history of the DJIA, the market experienced a drop of 1% or more (Return ≤ −0.01) in about 10% of the trading days. Likewise, the market experienced a rise of 1% or more (Return ≥ +0.01) about 10% of the trading days. Figure 5.6 recreates Figure 5.5 but using the sets $A_1 = \{x; x \leq -0.01\}$ and $A_2 = \{x; x \geq +0.01\}$.

The two conditional distributions in Figure 5.6 still look similar, but there is perhaps a greater, more noticeable discrepancy than what is shown in Figure 5.5: The highest and second highest frequency categories are reversed, suggesting typically *higher* returns on days following a rise in the index. But the distributions still look remarkably similar in shape, center, and spread. Whether the discrepancy is explainable by chance is taken up later in the discussion of hypothesis testing in Chapter 15. For now it is interesting to note that, if there is a difference, it may well run counter to your intuition, if your intuition told you that days with big changes would be followed by a rebound in the other direction. Instead, days following extreme "up" days tend to have *higher* returns, and days following extreme "down" days tend to have *lower* returns.

You were uncertain about the rebound hypothesis, weren't you? And now, you should be uncertain about whether the effect is exactly the opposite of the rebound hypothesis. Data

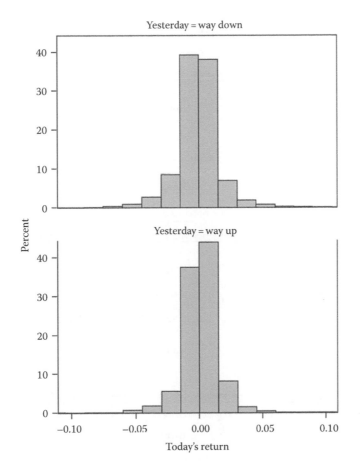

FIGURE 5.6
Conditional histograms of today's DJIA return, classified by either "way down" (a drop of 1% or more on the previous day) or "way up" (a rise of 1% or more on the previous day). Occasional returns more than 0.10 and less than −0.10 are not shown.

reduce uncertainty, but they do not eliminate it altogether. One thing should be fairly certain, though: If there is dependence between yesterday's and today's returns, it is slight. You can be confident in this conclusion because the sample sizes leading to the conditional histograms $\hat{p}(y \mid X \in A)$ shown in Figures 5.5 and 5.6 are very large, meaning that they are accurate estimates of the true distributions $p(y \mid X \in A)$. And since the histograms differ only slightly, it is logical to infer that the corresponding true distributions $p(y \mid X \in A_1)$ and $p(y \mid X \in A_2)$ differ only slightly as well. *Data reduce the uncertainty about the unknown parameters.*

The previous example is one where the independence model seems reasonable. Here is an example where there is clearly dependence.

Example 5.2: Evaluating Independence of Responses on a Survey

A survey was filled out by $n = 33$ faculty and staff concerning the desirability of various speakers who might be invited to a large southwestern university. Two of the potential speakers were George H.W. Bush, the 41st President of the United States, and his wife Barbara Bush. (Their son George W. Bush was the 43rd president.) The data are as follows, with each pair indicating a particular person's rating of the (George Barbara) combination: (1 2), (4 3), (4 3), (1 1), (4 4), (1 1), (3 1), (4 4), (4 4), (4 4), (4 3), (3 3), (3 2), (5 2),

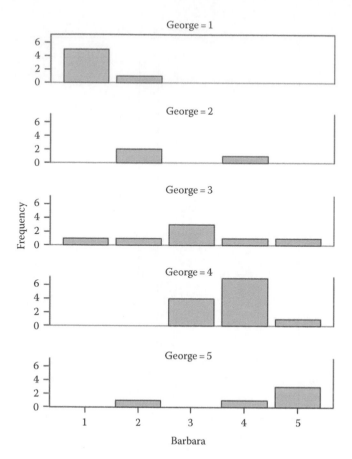

FIGURE 5.7
Estimated conditional distributions of Barbara Bush support for different values of G.H.W. Bush support.

(2 2), (2 2), (2 4), (1 1), (5 4), (1 1), (5 5), (1 1), (4 4), (3 3), (4 4), (4 5), (3 4), (4 4), (4 3), (3 3), (3 5), (5 5), and (5 5). One thing you should notice is that there are many cases where people rated both the same, such as (1 1), indicating lowest ratings for both, and (5 5), indicating highest ratings for both. If you let Y = Barbara Bush rating and X = G.H.W. Bush rating, it appears that there is dependence: The distribution of Y changes for different X. Figure 5.7 displays the estimated conditional distributions.

The morphing of distributions that characterizes dependence is clear in Figure 5.7: The distributions of Barbara Bush support shift to the right, as G.H.W. Bush's support increases. This makes perfect sense and is likely driven by politics: Since G.H.W. Bush was a republican president, republicans are likely to rate both of them highly, and democrats are likely to rate both of them lowly.

Note also that, unlike Figure 5.6 with the DJIA return data, the dependence is obvious here since the centers of the distributions obviously move. However, while the dependence is clear in the graphs, and corroborated by common sense, a critical reviewer of this analysis can still play devil's advocate and suggest that the patterns in Figure 5.7 might be explainable by chance alone. This criticism is addressed by producing data like the aforementioned, except where the variables are independent. Figure 5.8 shows the results of one such simulation.

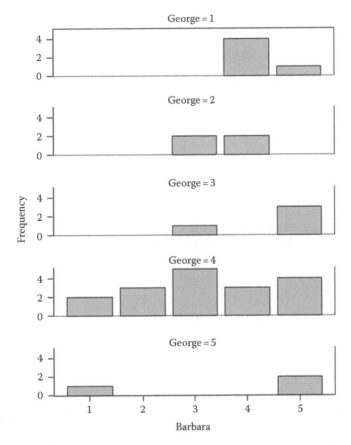

FIGURE 5.8
Appearance of estimated conditional distributions of Barbara Bush support for different values of G.H.W. Bush, when data are simulated independently.

In the simulation model for the data shown in Figure 5.8, the five true distributions $p(y|x)$ are exactly the same: $p(y|X=1) = p(y|X=2) = p(y|X=3) = p(y|X=4) = p(y|X=5)$. The common distribution used for all is simply the sample distribution of Barbara Bush responses in the original data set shown in Example 5.2 (see Table 5.5).

The Barbara Bush ratings shown in Figure 5.8 are simulated from the discrete distribution in Table 5.5; see Chapter 3, Figure 3.2, to recall how to do this. The George H.W. Bush data are simulated similarly from the sample distribution of George H.W. Bush responses in the original data set. While these are not the true distributions, they provide good estimates according to Ugly Rule of Thumb 5.1. Sensitivity analysis using different distributions would show the same kinds of random patterns shown in Figure 5.8.

Even though the distributions $p(y|X=1)$, $p(y|X=2)$, $p(y|X=3)$, $p(y|X=4)$, and $p(y|X=5)$ that produced the data shown in Figure 5.8 are exactly the same—namely, the distribution shown in Table 5.5—the estimated distributions $\hat{p}(y|X=1)$, $\hat{p}(y|X=2)$, $\hat{p}(y|X=3)$, $\hat{p}(y|X=4)$, and $\hat{p}(y|X=5)$ shown in Figure 5.8 are clearly different, and these differences are therefore caused purely by chance variation. (Again, think of a coin: Flip it 10 times, and you might get three heads.) So, you can see that with a sample size, this small chance variation has a big effect. Nevertheless, the trends seen in Figure 5.7 are not seen in Figure 5.8. In addition, if you create many more simulated versions of Figure 5.8, they would all be different,

TABLE 5.5

Sample Distribution of Barbara
Bush Ratings

Rating	Count	Percent
1	6	18.18
2	5	15.15
3	7	21.21
4	10	30.30
5	5	15.15
Total	33	100.0

but—importantly—you probably wouldn't see trends as pronounced as in Figure 5.7. So the trends of Figure 5.7 appear not to be explainable by chance alone, despite the fact that chance alone has a large effect with such a small sample size.

Vocabulary and Formula Summaries

Vocabulary

Conditional distribution   A distribution of RV Y when an RV X is equal to a particular value or lies in a particular set of values.

Research hypothesis   A hypothesis about the state of Nature that can be addressed using data.

Morph   To change continuously from one shape to another.

Frequency table   A table showing counts of outcomes of a discrete variable Y.

Estimate   As a noun, a guess based on data. As a verb, the act of using data to produce an estimate.

Independent   If the *distribution* of Y is the same, no matter what is X, then X and Y are independent.

Dependent   If the *distribution* of Y changes, depending on X, then X and Y are dependent.

Key Formulas and Descriptions

$p(y|x)$ The conditional probability distribution of Y given that X equals the value x.

$\sum_{all\ y} p(y|x) = 1.0$, for every x The sum of the conditional discrete probabilities is 1.0.

$\Pr(Y = i|X = x) = \pi_{i|x}$ The conditional probability that a discrete RV $Y = i$, given that an RV $X = x$, is denoted by the symbol $\pi_{i|x}$.

$\int_{all\ y} p(y|x)dy = 1.0$, for every x — The integral of the conditional continuous pdf is 1.0.

$p(y|X \in A_1)$ — The conditional probability distribution of Y given that X lies in the set A_1.

$\hat{p}(y|X \in A_1)$ — The estimated conditional probability distribution of Y given that X lies in the set A_1.

$A = \{x; x \le 0\}$ — Notation indicating a particular set A of x values.

$p(y|X \in A_1) = p(y|X \in A_2)$ for all sets A_1 and A_2 — The definition of independence of Y and X.

$p(y|X \in A_1) \ne p(y|X \in A_2)$ for some sets A_1 and A_2 — The definition of dependence between Y and X.

Exercises

5.1 Perform an Internet search for "effect of ____ on ____," where the blanks are topics that interest you. Find a page of interest and read enough to help you understand.

   A. Explain the meaning of $p(y|x)$ in that context. As part of your answer, explain why there is more than one y value (i.e., there is a distribution of y values) for a given x value.

   B. If the researcher's X affects the researcher's Y, what does that tell you about the distributions $p(y|x)$?

   C. If the researcher's Y is independent of the researcher's X, what does that tell you about the distributions $p(y|x)$?

5.2 Let Y be speed that a car is observed to be traveling on a highway, say as measured by an automated device placed on the side of the highway, and let X be the age of car in years. Draw (subjectively) graphs of $p(y|X = 1)$ and $p(y|X = 20)$. Put numbers and labels on both axes of both graphs. Remember that the area under the curve is 1.0 when deciding numbers to put on the vertical axes. Explain why you showed a difference between your two graphs. The fact that speed is a behavioral choice of the driver should be a part of your explanation.

5.3 On any given night, there will be cars making trips from place to place. On a given trip, the driver will either be sober ($X = 0$) or drunk ($X = 1$). And, on a given trip, the ride will result in either a successful outcome ($Y = 0$) or a fatal accident ($Y - 1$).

   A. Explain the meaning of the probability distribution $p(y|x)$ when $X = 0$, and write down your guess of what this distribution will be, with actual numbers, in list form. Repeat for $X = 1$.

   B. Explain the meaning of the probability distribution $p(x|y)$ when $Y = 0$, and write down your guess of what this distribution will be, with actual numbers, in list form. Repeat for $Y = 1$.

   C. Describe Nature, design, measurement, and DATA from a study that will allow you to estimate the distributions in Exercise 5.3A and B.

5.4 In the discussion of Figure 5.8, it is mentioned that "the common distribution used for all is simply the sample distribution of Barbara Bush responses in the data set," shown in Table 5.5.

A. Use the data set given in Example 5.2 to find a similar distribution of the George H.W. Bush responses.

B. Using the distributions in Exercise 5.4A and Table 5.5, simulate many 10 more data sets, each having the same number of observation pairs ($n = 33$) as in the original data set given in Example 5.2, but where the Barbara Bush and George H.W. Bush responses are independent. Do this by simulating a Y from the distribution of Table 5.5 and an X from your distribution of Exercise 5.4A, with $n = 33$ such pairs in each simulated data set. These data will be independent, because random number generators are designed so that all data values are generated independently. For each of the 10 simulated data sets having $n = 33$ pairs in each, construct a graph such as the one shown in Figure 5.8. Among the 10 graphs so constructed from independent responses, answer the question, "How often do you see a trend as pronounced as that shown in Figure 5.7?" Using your answer to that question, answer the question, "Does the trend in Figure 5.7 appear to be explainable by chance alone?" (Comment: The p-value, discussed in Chapter 15 and onward, is a measure of how often you see patterns such as those in Figure 5.7 by chance alone; i.e., the p-value measures how often you see patterns such as those in Figure 5.7 in repeated simulations from a process where the variables are independent, such as your many Figure 5.8 replications.)

5.5 Consider Figure 5.3. How would this graph look if Y and X were independent? Explain your answer in terms of the definition $p(y|X \in A_1) = p(y|X \in A_2)$ for all sets A_1 and A_2. Identify all terms explicitly: What is Y? What is X? What is A_1? What is A_2? What is $p(y|X \in A_1)$? What is $p(y|X \in A_2)$?

5.6 Suppose Hans rolls two six-sided dice, with faces numbered 1–6. One is red, yielding the number Y, and one is green, yielding the number X. It is commonly assumed that X and Y are independent.

A. How could Hans make X and Y dependent? Consider some trickery or mechanical device that Hans might employ.

B. Explain how the dependence in Exercise 5.6A is manifested by the definition $p(y|X \in A_1) \neq p(y|X \in A_2)$ for some sets A_1 and A_2. Identify all terms explicitly: What is Y? What is X? What is A_1? What is A_2? What is $p(y|X \in A_1)$? What is $p(y|X \in A_2)$?

5.7 Barring trickery, you can assume that X and Y in Exercise 5.6 are independent.

A. How could you estimate $p(y|X \in A_1)$, where $A_1 = \{1\}$, by using hundreds of rolls of the two dice? Call the resulting estimate $\hat{p}(y|X = 1)$.

B. Now explain why the following sentence is true: $p(y|X = 1) = p(y|X = 2)$, but $\hat{p}(y|X = 1) \neq \hat{p}(y|X = 2)$.

5.8 Suppose X has the Poisson distribution with $\lambda = 0.3$. When $X = 0$, Y has the Bernoulli distribution with $\pi = 0.6$. When $X > 0$, Y has the distribution given in the following table:

y	$p(y)$
0	0.4
1	0.6
Total	1.0

Are X and Y independent? Apply the definition of independence or dependence to explain.

5.9 Suppose X has a Bernoulli distribution. When $X = 0$, then $Y \sim U(0, 1)$. When $X = 1$, then $Y \sim U(0, 1.00001)$. Are X and Y independent? Apply the definition of independence or dependence to explain.

5.10 Suppose X has a normal distribution. Suppose that, given $X = x$, Y has the Poisson distribution with mean $\lambda = e^{1.0-0.2x}$. Are X and Y independent? Apply the definition of independence or dependence to explain.

5.11 In Table 5.4 summarizing the successive tosses of a coin, suppose your data from nine successive tosses looked like this:

		Next Toss		
		Heads	Tails	Total
First toss	Heads	2	2	4
	Tails	2	2	4

A. Write down a sequence of nine tosses (H, T, T, ...) that gives you this table.

B. What in the table *suggests* independence of the current toss and the next toss?

C. Based only on the data in this table, could you *conclude* that the next toss is independent of the first toss? Explain.

5.12 Suppose that any outcome of your coin toss is independent of all other outcomes. Suppose also that you flip a fair coin (meaning 0.50 probability of landing heads), getting heads 9999 times in a row.

A. What is the pdf of the outcome for the 10,000th flip? Don't just give an answer. First, apply the definition of independence, and then give the answer.

B. It is virtually impossible to flip a fair coin independently and get 9999 heads in a row. Explain how this might happen for a fair coin with dependent tosses.

C. It is virtually impossible to flip a fair coin independently and get 9999 heads in a row. Explain how this might happen for independent tosses of a coin that is not fair.

5.13 A pharmaceutical company evaluates whether a new drug has more side effects than a placebo. They arrive at the following cross-classification table summarizing the experiences of 160 patients who enrolled in a clinical trial.

	No Adverse Event	Adverse Event
Placebo	72	8
New drug	55	25

A. Estimate the conditional distribution of adverse event outcome when drug = Placebo.

B. Estimate the conditional distribution of adverse event outcome when drug = New drug.

C. Compare the estimated conditional distributions found in Exercise 5.13A and B, from the standpoint of the management of the pharmaceutical company.

D. Is Ugly Rule of Thumb 5.1 met in Exercise 5.13A and B?

5.14 Generate 1000 pairs (U_1, U_2) of uniform $(0, 1)$ RVs using the computer. By computer default, they will be independent. Now, for each of the 1000 pairs (U_1, U_2), create pairs (Y_1, Y_2), where $Y_1 = -\ln(U_1)$ and $Y_2 = -\ln(U_1) - \ln(U_2)$. Is Y_2 independent of Y_1? To answer, construct two histograms such as shown in Figure 5.5, using the subsets $A_1 = \{y_1; y_1 \le 1\}$ and $A_2 = \{y_1; y_1 > 1\}$. Discuss these two histograms in terms of the definition of dependence.

6

Marginal Distributions, Joint Distributions, Independence, and Bayes' Theorem

6.1 Introduction

A sign on the autoroute says "40% of auto fatalities involved a drunk driver." So, if you drink and drive, then you have a 40% chance of dying or killing someone, right? In 40 out of 100 of your potential future car trips where you drink and drive, you or someone else will die, right? Right?

No, that does not seem right at all. If it were, you and others would be taken off this planet after just a few times driving drunk. *All* drunk drivers would depart this planet very quickly. That would certainly solve the problem of drunken driving!

What's going on here? There is nothing wrong with the statistic "40% of auto fatalities involved a drunk driver." But there is a big difference in the percentage of fatalities that involve drunken driving, versus the percentage of drunken driving excursions that end in fatality. In other words, there is a big difference between $p(y|x)$ and $p(x|y)$. It is very important to understand the difference between these terms, as the interpretations differ dramatically.

As you probably know—hopefully intuitively and not from direct experience!—the percentage of drunken driving excursions that end in fatality is actually pretty small, likely much less than 1% and certainly not anywhere close to 40%. This is not to say you should go ahead and drink and drive, though. It is also a fact that the percentage of drunken driving excursions that end in fatality, while small, is *much higher* than the percentage of sober driving excursions that end in fatality. So, if you want to stay alive—and keep others alive—don't drink and drive!

Which is more relevant, $p(y|x)$ or $p(x|y)$? Remember the Mantra of Chapter 5: *Use what you know to predict what you don't know*. So, if you are planning to drive drunk, and if you want to predict your likelihood of dying or killing someone, you want to know Pr(death on the car trip|I drive drunk) and not Pr(I drive drunk|death on the car trip). The conditional probability Pr(death|drunk) is very small, much less than 1% (but still *much higher* than Pr(death|sober)). The conditional probability Pr(drunk|death) is estimated to be 40% by the roadside sign, but it is not relevant for predicting your outcome of auto fatality when you drive drunk. Do not confuse $p(y|x)$ with $p(x|y)$. This can lead to extremely inaccurate numbers and gross misinterpretations.

For a less dramatic example, consider music and math. While the structures of musical theory are appealing to those who like math, the structures of mathematical theory are not necessarily appealing to those who like music. So, in a room with 100 people, all musicians, you are not likely to find any mathematicians. On the other hand, in a

TABLE 6.1

Hypothetical Conditional Distributions
$p(y|x)$ for Y = Car Color Choice and X = Age

		Y			
		Red	Gray	Green	Total
X	Younger	0.50	0.20	0.30	1.00
	Older	0.20	0.40	0.40	1.00

room with 100 people, all mathematicians, you are likely to find several musicians. In other words, Pr(mathematician|musician) is very low—likely less than 1%—while Pr(musician|mathematician) is much higher, likely more than 10%.

For another example, suppose in the familiar car color choice case, you know that a customer purchased a red car. What is the probability that that person was younger? Table 5.1 showing conditional distributions is repeated here as Table 6.1.

Knowing that the car is red, is the probability of the customer being younger equal to 0.50, as Table 6.1 seems to suggest? Consider that for a second. The same logic would tell you that the probability of the customer being older is 0.20. But there are only two possibilities, younger and older, and their probabilities must add to 1.0. Thus, the probabilities can't possibly be 0.50 and 0.20.

Another way to see that 0.50 can't be right involves the meaning of the numbers themselves. The number 0.50 is the probability of a customer choosing red, given that the customer is younger, or Pr(red|younger). But here you are given that the selection is red, so the number you want is Pr(younger|red), not Pr(red|younger). The term Pr(red|younger) can be understood, loosely, in terms of how many red car purchases there are *out of 100 cars purchased by younger people*. The term Pr(younger|red) can be understood, loosely, in terms of how many younger purchasers there are *out of 100 red cars purchased*.

It's important to recognize here that, in general, $p(x|y)$ is neither equal nor reciprocal to $p(y|x)$. (As we will show later, however, they *are* related.) Here is another look at it: Imagine that you have a slip of paper with two pieces of information about a person: Whether they are older or younger, and what color of car they purchased—red, gray, or green. Now imagine that you put all of the slips of paper belonging to the younger people in a bag, and draw one of the slips. Loosely, the probability that the car color you draw will be red is Pr(red|younger). This is "the probability of red among the younger people." Pr(younger|red) can be found in a similar way: Put all of the slips of paper for people who bought a red car in a bag and draw one. Then, loosely, Pr(younger|red) is the probability that the age of the person whose slip you draw is younger. This is "the probability of younger customers among the people who purchased red cars." Note that the number of slips in the first drawing, for younger people who bought a car, will probably be different from the number of slips of in the second drawing, for red cars bought.

To see how to calculate Pr(Younger|Red), assume you have a data set that is consistent with the probabilities in Table 6.1. Table 6.2 shows n = 300 hypothetical car sales, cross-classified by color choice and age, and is called a **contingency table** or also a **cross-classification table**. The estimated conditional probabilities are precisely the same as the true conditional probabilities in Table 6.1. Don't ever expect this to happen in practice; this is just a pedagogical illustration.

You can estimate Pr(red|younger) as 100/200 = 0.50, as shown in Table 6.1. But how would you estimate Pr(younger|red)? Simple. There are 120 sales of red cars, 100 of which

TABLE 6.2

Hypothetical Contingency Table Giving
the Conditional Distributions of Table 6.1

		Y			
		Red	Gray	Green	Total
X	Younger	100	40	60	200
	Older	20	40	40	100
	Total	120	80	100	300

TABLE 6.3

Another Hypothetical Table Giving the
Conditional Distributions of Table 6.1

		Y			
		Red	Gray	Green	Total
X	Younger	50	20	30	100
	Older	200	400	400	1000
	Total	250	420	430	1100

involved younger customers, so your estimate of Pr(Younger|Red) is $100/120 = 0.833$. This is quite different from 0.50!

The calculation of Pr(younger|red) depends greatly on the distribution (younger, older) among purchasers. Table 6.3 gives the same conditional probabilities shown in Table 6.1 but with a dramatically different age distribution, having many more older purchasers.

Using the data in Table 6.3, Pr(younger|red) is estimated as $50/250 = 0.20$, a far cry from both 0.50 and 0.833. The distribution of age clearly matters!

6.2 Joint and Marginal Distributions

When there are two variables present, say X and Y (e.g., customer age and color), the distribution of one of them, irrespective of the value of other, is called its **marginal distribution**. For example, in Table 6.3, the estimated marginal distribution of age is given in a list format as shown in Table 6.4.

The estimated marginal distribution of color choice in Table 6.3 is given similarly as shown in Table 6.5.

TABLE 6.4

Estimated Marginal Distribution
of Age Using Data from Table 6.3

Age	Estimated Probability
Younger	$100/1100 = 0.091$
Older	$1000/1100 = 0.909$
Total	1.000

TABLE 6.5

Estimated Marginal Distribution
of Color Using Data from Table 6.3

Color	Estimated Probability
Red	$250/1100 = 0.227$
Gray	$420/1100 = 0.382$
Green	$430/1100 = 0.391$
Total	1.000

TABLE 6.6

Estimated Joint Distribution of (Age, Color)
Combinations, Using Data from Table 6.3

		Y		
		Red	Gray	Green
X	Younger	0.045	0.018	0.027
	Older	0.182	0.364	0.364
				1.00

It is easy to remember why they are called *marginal distributions*—it's because they are found in the *margins* of the cross-classification table!

The distribution of the *combination* of values (X, Y) is called the **joint distribution**. In Table 6.3, there are $2 \times 3 = 6$ combinations of (age, color), so the joint distribution is a discrete distribution on those values, and the joint probabilities all add up to 1.0. Table 6.6 displays the estimated joint distribution, found by dividing all of the cells in Table 6.3 by the total of 1100.

Notice that *all six* of the numbers inside Table 6.6 add to 1.0; but the numbers do not add to 1.0 in either the rows or the columns. This is a dead giveaway that you are looking at a joint distribution, and not a collection of conditional distributions. You can interpret the joint distribution probabilities easily; for example, in 4.5% of all car sales, the car is red and the purchaser is younger. Note that 0.045 is *not* a conditional probability; it is an estimate of Pr(younger *and* red) or, equivalently, Pr(red *and* younger). To visualize this, continue the analogy discussed previously with the slips of paper. Imagine putting all the slips of paper into a bag and drawing one slip out. Then the probability that you draw out a slip with the combination *younger, red* on it is estimated to be 0.045. The key difference between this joint probability and the conditional probabilities discussed earlier is that it refers to the entire group of buyers rather than a subset of the buyers.

Appending the row and column totals in Table 6.6 gives the marginal distributions shown in Tables 6.4 and 6.5 (see Table 6.7).

Of course, the numbers shown in Table 6.7 are not really correct: Since they are based on data, they are just estimates. The true probabilities are unknown—*model has unknown parameters*. Table 6.8 shows the more correct representation and introduces some new symbols to represent marginal and joint probabilities.

The "." subscript in the margins indicates that the probability is obtained through summation over the "dotted" index. For examples

$$\pi_{1.} = \pi_{11} + \pi_{12} + \pi_{13}$$

$$\pi_{.3} = \pi_{13} + \pi_{23}$$

TABLE 6.7

Estimated Joint and Marginal Distributions
of (Age, Color), Using Data from Table 6.3

		Y			
		Red	Gray	Green	Total
X	Younger	0.045	0.018	0.027	0.091
	Older	0.182	0.364	0.364	0.909
	Total	0.227	0.382	0.391	1.000

TABLE 6.8

True Joint and Marginal Distributions
of (Age, Color)

		Y			
		Red	Gray	Green	Total
X	Younger	π_{11}	π_{12}	π_{13}	$\pi_{1.}$
	Older	π_{21}	π_{22}	π_{23}	$\pi_{2.}$
	Total	$\pi_{.1}$	$\pi_{.2}$	$\pi_{.3}$	1.00

These formulas generalize to other discrete cases. Let $p(x, y)$ be a discrete probability distribution function (pdf) that assigns probabilities to all combinations (x, y). Let $p(y)$ and $p(x)$ denote the marginal distributions. The following formulas are important and should be memorized.

Obtaining Marginal Distributions from the Joint Distribution, Discrete Case

$$p(y) = \sum_{all\ x} p(x, y)$$

$$p(x) = \sum_{all\ y} p(x, y)$$

(6.1)

In other words, the marginal distribution of Y is calculated by summing the joint probabilities over all of the possible values of X. Similarly, the marginal distribution of X is calculated by summing the joint probabilities over all of the possible values of Y.

A side note about notation: The notation $p(y)$ for the marginal distribution of Y and $p(x)$ for the marginal distribution of X is potentially confusing and is called an *abuse of notation*. The functions $p(x)$ and $p(y)$ are technically the same function. For example, if $p(x) = x^2$, then $p(2) = 2^2$, $p(10) = 10^2$, $p(\text{dog}) = \text{dog}^2$, and $p(y) = y^2$. Thus, $p(x)$ and $p(y)$ are identical, namely, a simple quadratic function. So the notations $p(y)$ and $p(x)$ seem to imply that the marginal distributions of X and Y are the same. This is not intended—in general, we assume that a variable Y has a different distribution than another variable X. The alternative, more correct notation is to label the functions according to the particular random variable (RV), as in $p_X(x)$ and $p_Y(y)$, which indicates two potentially different distribution functions, $p_X(.)$ and $p_Y(.)$. While preferable, this notation is more cumbersome and might obscure other points, so we usually adopt the abusive notation $p(x)$ and $p(y)$ and ask you to view these as different functions. When absolutely necessary for clarity, we will use the more correct notations $p_X(x)$ and $p_Y(y)$.

Back to the main story: Note that the joint probabilities sum to 1.0 over all (x, y) combinations, as shown in Tables 6.6 through 6.8. Mathematically, this is expressed using the double summation formula as follows:

$$\sum_{all\,x} \sum_{all\,y} p(x, y) = 1.0$$

In the continuous case, sums become integrals. The logic involves the same kind of rectangular approximations described in Chapter 2, with the widths of the rectangles tending to zero. The formulas relating marginal to joint distributions in the continuous case are similar to those in the discrete case.

Obtaining Marginal Distributions from the Joint Distribution, Continuous Case

$$p(y) = \int_{all\,x} p(x, y)dx$$

(6.2)

$$p(x) = \int_{all\,y} p(x, y)dy$$

But wait, we haven't even told you what is a continuous joint distribution! Well, like the discrete case it assigns relative likelihoods to combinations (x, y), and like the single-variable or *univariate* case, these numbers $p(x, y)$ are not probabilities since the probability of a specific value is zero. Instead, the values $p(x, y)$ tell you something about the probability in a *small neighborhood* of the point (x, y). Specifically, if $p(x_1, y_1) > p(x_2, y_2)$, then the probability of observing an (X, Y) pair in a *small neighborhood* around (x_1, y_1) is higher than the probability of observing an (X, Y) pair in the same sized *small neighborhood* around (x_2, y_2). And like the univariate case, the area under the curve is 1.0; however, in the two-variable or *bivariate* case area becomes volume and is represented as a double integral:

$$\int_{all\,x} \int_{all\,y} p(x, y)\,dy\,dx = 1.0$$

A picture can help. Let's switch from car color and age to the continuous example from Chapter 5 where X and Y are annual income and housing expense, respectively. Then the joint pdf $p(x, y)$ is a 3-D object depicting the triplets (x, y, z), where $z = p(x, y)$. It is difficult to draw on 2-D paper or a computer screen. Fortunately, there are software tools to help (see Figure 6.1).

The surface shown in Figure 6.1 is the continuous joint pdf, which shows likelihoods of pairs (income, housing expense). The higher the likelihood of a particular point (x, y), the more likely it is to see homeowners who have (income, housing expense) nearby that particular (x, y) combination.

The joint distribution shown in Figure 6.1 is not quite realistic because the income distribution should have a more pronounced right skew—there are occasional rich people with incomes 200K, 800K, 10,000K, etc., but no one with gross income less than 0K. You should use Figure 6.1 to understand the general concept of a joint distribution, but you should also understand that it is not a precise model for real incomes and housing expenses.

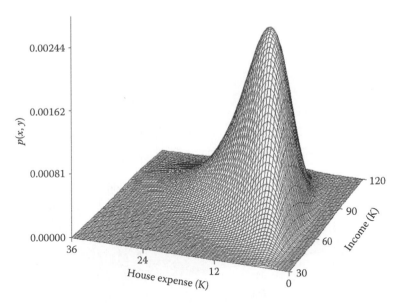

FIGURE 6.1
Hypothetical joint probability distribution of income (*X*) and housing expense (*Y*).

You can obtain the marginal distributions of income and housing expense from the joint distribution shown in Figure 6.1 as $p(x) = \int_{all\,y} p(x,y)dy$ and $p(y) = \int_{all\,x} p(x,y)dx$. For example, the height of the marginal distribution of income when income = 60 is $p_X(60) = \int_{all\,y} p(60,y)dy$, which is the area of the cutout slice of the joint distribution shown in Figure 6.2.

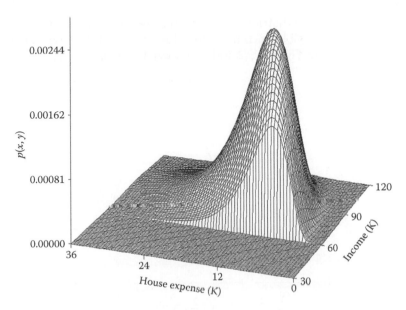

FIGURE 6.2
Joint probability distribution of income (*X*) and housing expense (*Y*), showing the slice where income = 60, namely, $p(60, y)$. The area of the slice is the height of the marginal distribution function of income where income = 60.

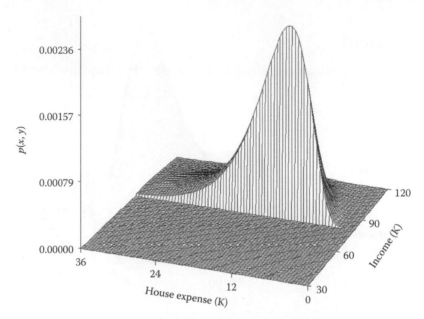

FIGURE 6.3
Joint probability distribution of income (X) and housing expense (Y), showing the slice where income = 80, namely, $p(80, y)$. The area of the slice is the height of the marginal distribution function of income, where income = 80.

While the slice in Figure 6.2 looks like a pdf, it isn't, because the area of the slice is not 1. You will see shortly how to make this slice a pdf (but you may already have an idea…).

Further, the height of the marginal distribution of income when income = 80 is $p_X(80) = \int_{all\ y} p(80, y)\, dy$, which is the area of the cutout slice of the joint distribution shown in Figure 6.3.

The heights of the marginal distribution of income are the areas as shown in Figures 6.2 and 6.3, one for every possible income slice. This marginal distribution is shown in Figure 6.4, with the areas in Figures 6.2 and 6.3 shown by dots.

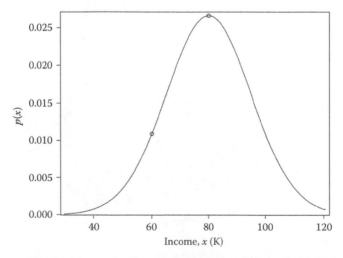

FIGURE 6.4
Marginal distribution of income, with areas under curve slices in Figures 6.2 and 6.3 shown as dots.

6.3 Estimating and Visualizing Joint Distributions

Have a look at the joint distribution of Figure 6.1. How do you estimate such a thing using data? The first tool to understand how to do this is called a **scatterplot**. A scatterplot is a simple plot of data pairs (x_i, y_i), where the subscript i indicates an observed data pair, with $i = 1, 2, …, n$. In the scatterplot, each of the n pairs is indicated by a single dot. For example, suppose that data (income, housing expense) are as given as in Table 6.9.

Here, $i = 1, 2, 3, 4, 5$, with the pairs (x_i, y_i) given as $(x_1, y_1) = (78.1, 21.4)$, $(x_2, y_2) = (63.1, 15.8)$, $(x_3, y_3) = (112.0, 25.0)$, $(x_4, y_4) = (80.0, 18.5)$, and $(x_5, y_5) = (56.1, 12.3)$. Plotting these five points gives you the scatterplot shown in Figure 6.5.

Usually, the grid lines shown in Figure 6.5 are not included in a scatterplot. They are included here just so you can see clearly what the dots mean. For example, the point $(x_3, y_3) = (112.0, 25.0)$ is clearly located in the upper-right corner of Figure 6.5.

The scatterplot of Figure 6.5 shows a clear pattern of increasing housing expense with increasing income. It also shows that the relationship is not deterministic: While the trend is generally upward, one person has income = 78.1 and housing expense = 21.4, while another has higher income (80.0) and lower housing expense (18.5). There is an entire *distribution* of *possible* housing expenses for each income level, and these distributions continuously *morph*, as discussed in Chapter 5.

TABLE 6.9

Data on Income and Housing Expense

Income	Housing Expense
78.1	21.4
63.1	15.8
112.0	25.0
80.0	18.5
56.1	12.3

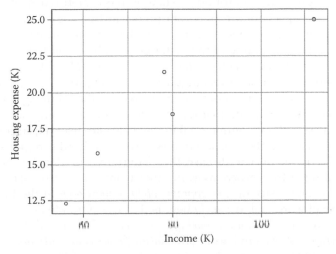

FIGURE 6.5
Scatterplot of $n = 5$ (income, housing expense) data points.

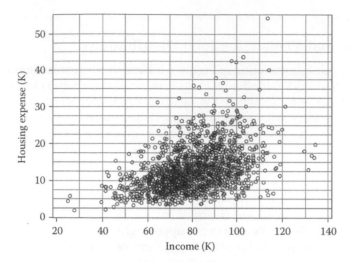

FIGURE 6.6
Scatterplot of (income, housing expense) pairs with $n = 1500$ households. Note the greater observed density in the central rectangular regions.

The scatterplot of Figure 6.5 also shows you how "pooling" method of Chapter 5 works for estimating conditional distributions. If you want to estimate the conditional distribution of housing expense when income = 80, for example, you could pool all the housing expense data where $70 \leq$ income < 90 and create a histogram of these values as a rough estimate of $p(y|X = 80)$. Here, there are only $n = 2$ such Y values, namely, 21.4 and 18.5, so the histogram will be a very poor estimate. However, if the n of Figure 6.5 were in the hundreds instead of just $n = 5$, then there would likely be plenty of observations in the 70–90 income range, making such a histogram adequate (see Figure 6.6).

The pooling method for estimating conditional distributions also shows the way to estimate joint pdfs $p(x, y)$ using a **bivariate histogram**. For example, consider using the data in Figure 6.5 to estimate the joint likelihood of the combination (income, housing expense) = (80, 20). Since there is no data at this particular point, you can instead use a rectangular region (i.e., a neighborhood) around it, such as the region ($70 \leq$ income <90, $17.5 \leq$ housing expense < 22.5). There are two data points in this region, and five data points total, so the probability is estimated as $2/5 = 0.40$. The density estimate would then be the height needed to make the volume of the rectangular cube equal to 0.40; using Volume = (Length × Width × Height) you get $(2/5) = (20 \times 5 \times \text{Height})$ or Height = 2/500. Other heights can be calculated similarly for other rectangular regions. Figure 6.7 shows a bivariate histogram from the more plentiful data shown in Figure 6.6, estimated using statistical software.

As in the case where univariate histograms are estimated using software, Figure 6.7 does not obey the area = 1.0 (or volume = 1.0 in the bivariate case) requirement of pdfs. The vertical axis is shown as counts of observations in the rectangular regions instead of density. Nevertheless, as in the univariate case, the histogram provides the correct appearance other than the scaling of the vertical axis. In other words, as in the case of the univariate histogram, the bivariate histogram is *approximately proportional to* the bivariate pdf. The reason for the weasel word *approximately* here is that the estimate is based on data and is therefore not the true distribution. (*Model produces data.*) The actual distribution is something more like what is shown in Figure 6.1. (*Nature favors continuity over discontinuity.*)

Also note that, as with univariate histograms, the choice of the dimensions of the rectangles is somewhat subjective. You should choose the dimensions so that you have enough

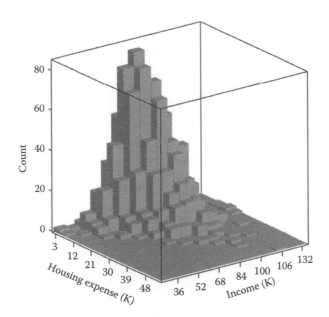

FIGURE 6.7
Bivariate histogram indicating relative joint density of the various (income, housing expense) combinations.

rectangles to visualize the joint distribution easily. At one extreme would be one rectangle covering all the data, which would give you a bivariate histogram that looks like a cardboard box. At the other extreme would be a rectangle for every (x, y) point, which would give you a bivariate histogram that looks like a bunch of needles.

You probably haven't seen many bivariate histograms before. Most statistics sources only discuss scatterplots for analyzing bivariate relationships. This misses the essential point that *Model produces data*. The scatterplot is just some data. The bivariate distribution is the producer of the DATA. You can visualize the data-producing distribution $p(x, y)$ from the scatterplot, though: Locations in the scatterplot where the data are more dense correspond to places in the bivariate histogram where the height of the bivariate distribution $p(x, y)$ is largest.

6.4 Conditional Distributions from Joint Distributions

Consider Table 6.3 again, given here as Table 6.10.

If your goal is to estimate the conditional distribution of age (X) given Y – red, the relevant data are shown in boldface in Table 6.10. The estimated conditional distribution is given in list form in Table 6.11.

To see the relationship between the conditional distribution and the joint distribution, look at the joint distribution of Table 6.7 again, repeated here as Table 6.12 with the relevant "slice" shown in boldface.

Notice that the numbers in the boldface column of Table 6.12 can't be a distribution, because they don't add to 1.0; they add to 0.227 instead. No problem, just make them add to 1! If you divide everything in the column by 0.227, you get estimated probabilities $0.045/0.227 = 0.20$ and $0.182/0.227 - 0.80$, just as shown in Table 6.11. But this isn't witchcraft;

TABLE 6.10

Contingency Table Showing Age and Car Color Selection

		Y			
		Red	Gray	Green	Total
X	Younger	50	20	30	100
	Older	200	400	400	1000
	Total	250	420	430	1100

TABLE 6.11

Estimated Conditional Distribution of Age, Given a Red Car Purchase

Age	Estimated Probability
Younger	50/250 = 0.20
Older	200/250 = 0.80
Total	1.00

TABLE 6.12

Estimated Joint and Marginal Distributions of (Age, Color), Using Data from Table 6.3

		Y			
		Red	Gray	Green	Total
X	Younger	0.045	0.018	0.027	0.091
	Older	0.182	0.364	0.364	0.909
	Total	0.227	0.382	0.391	1.000

there is a firm logic. Representing the estimated joint and marginal probabilities using the actual counts, you see that $0.045/0.227 = (50/1100)/(250/1100) = 50/250$ and similarly $0.182/0.227 = (200/1100)/(250/1100) = 200/250$. So the method of forcing the sum to be 1.0 makes logical sense when you look at the actual numbers.

This method of enforcing the sum to be 1.0 leads to a nice general formula: Notice that the formulas $0.045/0.227 = 0.20$ and $0.182/0.227 = 0.80$ are simply the joint probabilities divided by the marginal probabilities. This is the correct formula in general; here it is in the discrete case.

Obtaining the Conditional Distributions from the Joint and Marginal Distributions, Discrete Case

$$p(y|x) = \frac{p(x,y)}{p(x)}$$

$$p(x|y) = \frac{p(x,y)}{p(y)}$$

(6.3)

To understand formulas (6.3), you should first think of what follows the | as fixed and what precedes the | as varying. For example, use the second formula and fix Y = red; then X varies as (younger, older). The formula gives $p(\text{younger}|\text{red}) = p(\text{younger, red})/p(\text{red})$ and $p(\text{older}|\text{red}) = p(\text{older, red})/p(\text{red})$, estimated previously as $0.045/0.227$ and $0.182/0.227$, respectively.

Since you can get the marginal distributions from the joint distributions by summing them to the margins, you can rewrite the formulas. Recall the formula (6.1) for $p(y)$; it is just the sum of the joint probabilities $p(x, y)$ over all x values. The formula for $p(x)$ is obtained similarly by summing the probabilities $p(x, y)$ over all y values. Thus, you can rewrite the formulas (6.3) as follows:

Obtaining the Conditional Distributions from the Joint Distribution, Discrete Case

$$p(y|x) = \frac{p(x,y)}{\sum_{all\,y} p(x,y)}$$

$$p(x|y) = \frac{p(x,y)}{\sum_{all\,x} p(x,y)}$$

(6.4)

The way to think about conditional distributions in the discrete case is in terms of slices from the joint distribution; see the boldface columns in the distributions of (age, color selection) in Tables 6.10 and 6.12. The same slice idea applies when considering conditional continuous distributions. See Figures 6.2 and 6.3: They show the slice of the joint pdf of income and housing expense corresponding to income fixed at either 60 or 80. These slices look like pdfs, but they don't have area $= 1.0$. No problem, just make the areas of the slices equal 1.0! In the discrete case, Equation 6.4 shows that you can do this by dividing out the sum. In the continuous case, you just replace the Σ with an $\int$ hence, you divide out the area. Once you divide the slice function by the area of the slice, the resulting curve has area $= 1.0$ and is therefore a bona fide pdf.

Obtaining the Conditional Distributions from the Joint Distribution, Continuous Case

$$p(y|x) = \frac{p(x,y)}{\int_{all\,y} p(x,y)dy}$$

$$p(x|y) = \frac{p(x,y)}{\int_{all\,x} p(x,y)dx}$$

(6.5)

The denominators in Equations 6.5 are just the marginal pdfs, so the conditional distributions are obtained by dividing the joint probability distribution by the marginal probability distribution, with identical formulas as in the discrete case.

Obtaining the Conditional Distributions from the Joint and Marginal Distributions, Either Case

$$p(y|x) = \frac{p(x,y)}{p(x)}$$

$$p(x|y) = \frac{p(x,y)}{p(y)}$$

(6.6)

Simple algebra shows that you can also express the joint distribution in terms of the conditional and marginal distributions by multiplying both sides of either of Equations 6.6 by the denominator:

**Obtaining the Joint Distribution from the Conditional
and Marginal Distributions, Either Case**

$$p(x,y) = p(y|x)p(x)$$

$$p(x,y) = p(x|y)p(y)$$

(6.7)

6.5 Joint Distributions When Variables Are Independent

There is a simple but very powerful formula that applies when RVs X and Y are independent, and it applies for both the continuous and discrete cases.

The Joint Distribution *Under Independence*

$$p(x,y) = p(x)p(y)$$

(6.8)

A simple application: Flip a coin twice, let X be the outcome of the first toss, and let Y be the outcome of the second. Equation 6.8 tells you that $p(\text{heads, heads}) = p(\text{heads})p(\text{heads}) = 0.50 \times 0.50 = 0.25$. Thus, if you flip a coin twice, you have a 25% chance of seeing both tosses land on heads.

Equation 6.8 generalizes to any number of RVs; for example, $p(x, y, z) = p(x)\,p(y)\,p(z)$. Thus, the probability of 100 consecutive results of heads is

$$0.50 \times 0.50 \times \cdots \times 0.50 = (0.50)^{100} = 0.0000000000000000000000000000001$$

Or, in words, "extremely unlikely!"

The formula $p(x, y) = p(x)p(y)$ of Equation 6.8 is a consequence of the formula $p(x, y) = p(y|x)\,p(x)$ given in Equation 6.7. But why is $p(y|x)$ the same as $p(y)$? Glad you asked! Here's why.

As discussed in Chapter 5, if X and Y are independent, then $p(y|X = x_1) = p(y|X = x_2)$, for all x_1 and x_2. So, for example, if color choice (Y) were independent of age (X), then the conditional distributions $p(y|x)$ would have to be equal, as shown in Table 6.13.

TABLE 6.13

Examples of Conditional Distributions
$p(y|x)$ of Color Choice When Color Choice
Is Independent of Age

		Y			
		Red	Gray	Green	Total
X	Younger	0.40	0.30	0.30	1.00
	Older	0.40	0.30	0.30	1.00

What is the marginal distribution $p(y)$ of color choice (Y) in Table 6.13? It seems that it should be the same as the conditional distributions, and it is.

Relationship between Marginal and Conditional Distributions *Under Independence*

$$p(y) = p(y|x), \quad \text{for all } x$$

$$p(x) = p(x|y), \quad \text{for all } y$$

(6.9)

For example, let X = the first roll of a fair die, and let Y = the second roll of the same die. The marginal pdf of Y is the discrete uniform pdf on the numbers 1, 2, ..., 6: $p(y) = 1/6$, for $y = 1, 2, ..., 6$. If the rolls are independent, then all conditional distributions are the same as the marginal distribution: $p(y|X = 1)$, $p(y|X = 2)$, ..., $p(y|X = 6)$ are all the same as $p(y)$; namely, they are all discrete uniform pdf on the numbers 1, 2, ..., 6.

Be sure you don't take equations out of context. Often, there are assumptions attached to them. For example, if $x = 2$ and $y = 2$, then

$$xy = x + y$$

If you take this equation out of context, you will get very silly results, like $0 = 100$ when $x = 0$ and $y = 100$. So, when you see any equations, make sure you understand the *assumptions*. The assumption for Equations 6.9 is that X and Y are independent. Equations 6.9 are usually wrong, just like $xy = x + y$ is usually wrong.

You can see the logic for Equations 6.9 from the following statements. Assume the continuous case; the discrete case is similar except the integrals are replaced with summation signs.

Under independence, the conditional distributions $p(y|x)$ are the same, for every x, so you know that $p(y|x) = f(y)$, a function that doesn't depend on x. The goal is to show that this function $f(y)$ is in fact equal to the marginal distribution $p(y)$. The following sequence of equalities shows that this is true:

$p(y) = \int_{all\,x} p(x, y)dx$	(From Equation 6.2)
$= \int_{all\,x} p(y|x)p(x)dx$	(From Equation 6.7)
$= \int_{all\,x} f(y)p(x)dx$	(By independence, $p(y\|x) = f(y)$, where $f(y)$ does not depend on x)
$= f(y)\int_{all\,x} p(x)dx$	(Since $f(y)$ is constant with respect to x, it factors outside the integral; property I2 of integrals given in Section 2.6)
$= f(y)\,(1.0)$	(Since $p(x)$ is a pdf, its integral is 1.0)
$= f(y)$	(Because multiplying any number by 1.0 gives the same number)

This proves that the marginal distribution and the conditional distribution are identical when the variables X and Y are independent.

One application of the independence formula (6.8) is to check for dependence of discrete variables. Using the speaker preference rating data from Example 5.2, the contingency table of the observed data is shown in Table 6.14.

TABLE 6.14

Observed Joint Frequencies of (George H.W.
Bush, Barbara Bush) Preference Ratings

		Barbara Bush Rating					
		1	2	3	4	5	Total
George H.W. Bush Rating	1	5	1	0	0	0	6
	2	0	2	0	1	0	3
	3	1	1	3	1	1	7
	4	0	0	4	7	1	12
	5	0	1	0	1	3	5
	Total	6	5	7	10	5	33

TABLE 6.15

Estimated Marginal Distributions of Speaker Preference

G.H.W. Bush Rating	Estimated Probability	Barbara Bush Rating	Estimated Probability
1	6/33 = 0.182	1	6/33 = 0.182
2	3/33 = 0.091	2	5/33 = 0.152
3	7/33 = 0.212	3	7/33 = 0.212
4	12/33 = 0.364	4	10/33 = 0.303
5	5/33 = 0.152	5	5/33 = 0.152
Total	1.00	Total	1.00

The independence formula (6.8) tells you that the joint distribution is the product of the marginal distributions. The estimated marginal distributions for George H.W. Bush and Barbara Bush are given in Table 6.15.

Under independence, the estimated probability that a person rates George a "4" and Barbara a "1" is 0.364 × 0.182 = 0.066. Thus, out of 33 people, you would expect 6.6% of them, or 2.18, to be in this group. (Expected values are like averages; hence, 2.18 rather than 2 is correct.) Instead, there are none. Table 6.16 shows the entire table of similarly calculated expected frequencies under independence.

TABLE 6.16

Estimates of Expected Joint Frequencies
of (George H.W. Bush, Barbara Bush) Preference
Ratings, Assuming Ratings Are Independent

		Barbara Bush Rating					
		1	2	3	4	5	Total[a]
George H.W. Bush Rating	1	1.09	0.91	1.27	1.82	0.91	6
	2	0.55	0.45	0.64	0.91	0.45	3
	3	1.27	1.06	1.48	2.12	1.06	7
	4	2.18	1.82	2.55	3.64	1.82	12
	5	0.91	0.76	1.06	1.52	0.76	5
	Total[a]	6	5	7	10	5	33

[a] Numbers might not add to totals shown because of roundoff error.

Now, compare Table 6.14, the observed data frequencies, with Table 6.16, the expected data frequencies under independence. Under independence, you expect considerably fewer observations near the diagonal, where the responses for the two people are similar, than you see in the actual data. Similarly, under independence, you expect many more observations far from the diagonal, where the responses for the two people are dissimilar, than you see in the actual data.

The comparison of expected frequencies under independence and actual frequencies is the basis for the *chi-squared test of independence*, discussed in Chapter 17.

6.6 Bayes' Theorem

The question posed in the introduction, "What is the probability that the purchaser was younger, given that the car purchased was red?" is a classical application for **Bayes' theorem**. Many advanced statistical methods use this result, as you will see in Chapter 13.

Bayes' theorem is used to find conditional probabilities $Pr(A|B)$, when the reversed conditional information $Pr(B|A)$ is known. For example, you might know $Pr(Red|Younger)$, but you want to know $Pr(Younger|Red)$. (*Use what you know to predict what you don't know.*)

Bayes' theorem uses the conditional distribution formula. From Equation 6.6:

$$p(x|y) = \frac{p(x, y)}{p(y)} \tag{6.10}$$

Equation 6.10 applies in both the discrete case and the continuous case. By expressing the numerator with the reverse condition, using Equation 6.7, you get $p(x, y) = p(y|x)p(x)$. This gives the first expression of Bayes' theorem:

$$p(x|y) = \frac{p(y|x)p(x)}{p(y)} \tag{6.11}$$

Representing the denominator in terms of the joint probabilities using Equations 6.1 and 6.7 gives you an equivalent representation as follows:

Bayes' Theorem in the Discrete Case

$$p(x|y) = \frac{p(y|x)p(x)}{\sum_{all\,x} p(y|x)p(x)} \tag{6.12}$$

And similarly, using Equations 6.2 and 6.7, you get the following result.

Bayes' Theorem in the Continuous Case

$$p(x|y) = \frac{p(y|x)p(x)}{\int_{all\,x} p(y|x)p(x)dx} \tag{6.13}$$

The denominators are uglier and more mysterious than needed. Do you recall the idea of *slices* and making pdfs add (or integrate) to 1.0? That's all the denominators are doing here.

They make the area or total probability equal to 1.0. So an even simpler representation of Bayes' theorem, and the one you should memorize, is this form which applies to both the discrete and continuous cases:

Bayes' Theorem, Simplest Form

$$p(x|y) \propto p(y|x)p(x) \tag{6.14}$$

The symbol $\propto$ is read as "is proportional to," and means specifically that $p(x|y) = cp(y|x)$ $p(x)$, for some constant c. The constant c is whatever number is needed to make the total probability sum or integrate to 1.0. In the discrete case, it is the inverse of the sum of the numbers in the "slice," and in the continuous case, it is the inverse of the integral of the slice function.

Example 6.1: Probability of Death When Driving Drunk

In the introduction to this chapter, you saw a large difference between Pr(death|drunk) and Pr(drunk|death). How can you convert the statistic "40% of auto fatalities involve a drunk driver" to a probability that someone will die the next time you drive while intoxicated?

It's impossible to give a precise answer to this question, but it is possible to arrive at an answer that is at least in the same ball park. To do so, you need to make some assumptions. The reason that the answer we will give is not precisely true is that these assumptions are not precisely true.

The first step is to identify an X and a Y. One variable is the binary indicator of driving while drunk versus driving while sober, and the other is the indicator of whether the trip ends in fatality or no fatality. While it does not technically matter which of these binary variables you call X and which you call Y, Bayes' theorem, as stated previously, starts with knowledge of $p(y|x)$, then converts it to $p(x|y)$. In this framework, the given information, 40%, is part of $p(y|x)$. So let Y = driving method (drunk or sober), and let X = trip outcome (fatality or non-fatality). The distribution of $Y|X$ = fatality, as suggested by the roadside sign, is shown in Table 6.17.

But if you are planning to drive drunk—which, of course, we do not recommend—you will want the distribution of X, trip outcome, given Y = driving drunk, not vice versa. (*Use what you know to predict what you don't know.*) Bayes' theorem as given by Equation 6.14 states $p(x|y) \propto p(y|x) p(x)$, where you view x as variable (fatality or non-fatality) and y as fixed (drunk). So you also need the distribution of Y (drunk or sober) given X = non-fatality.

It is reasonable to assume that most driving excursions do not end in fatality, so the percentage of non-fatal car trips where the driver is drunk should be approximately the same as the percentage of drivers who are drunk. According to police checkpoint data, around 1% of drivers are drunk. Thus, a reasonable guess of $p(y|X$ = non-fatality) is as given in Table 6.18.

TABLE 6.17

Distribution of Drunk Drivers among Trips Ending in a Fatality

State of Driver, y	$p(y \mid X = \text{Fatality})$
Drunk	0.40
Sober	0.60
Total	1.00

TABLE 6.18

Distribution of Drunk Drivers among
Trips Not Ending in a Fatality

State of Driver, y	$p(y \mid X = \text{Non-Fatality})$
Drunk	0.01
Sober	0.99
Total	1.00

You now have all the information you need about $p(y|x)$ in the expression $p(x|y) \propto p(y|x)\,p(x)$. Once you know $p(x)$, you can plug everything in to get $p(x|y)$.

How many trips end in fatalities? Auto statistics show that there are around 1.5 fatalities per 100 million vehicle miles traveled. If a typical car excursion is 5 miles, then there are around 1.5 fatalities per 20 million excursions, or an approximate probability of death in an excursion of 1.5/20,000,000 = 0.000000075. This figure translates to 7 or 8 fatalities per 100 million trips. Table 6.19 gives the resulting estimated distribution of X.

You want to know Pr(fatality|drunk). Taking the relevant information from Tables 6.17 through 6.19, you get Table 6.20.

Notice that the numbers in the $p(\text{Drunk}|x)$ column of Table 6.20 do not add to 1.0, nor are they supposed to.

Now, in the expression $p(x|y) \propto p(y|x)\,p(x)$, the term $p(y|x)\,p(x)$ is the product of the last two columns in Table 6.20, as given by Table 6.21.

TABLE 6.19

Distribution of Outcomes
of Car Trips

Outcome of Car Trip, x	$p(x)$
Fatality	0.000000075
Non-fatality	0.999999925
Total	1.000000000

TABLE 6.20

Distribution of Outcomes of Car Trips along
with Conditional Probabilities of Drunken
Driving

Outcome of Car Trip, x	$p(x)$	$p(\text{Drunk}\,|\,x)$
Fatality	0.000000075	0.40
Non-fatality	0.999999925	0.01
Total	1.000000000	

TABLE 6.21

Distribution of Outcomes of Car Trips along with Conditional Probabilities
of Drunken Driving and Calculations for Bayes' Theorem

Outcome of Car Trip, x	$p(x)$	$p(\text{Drunk}\,|\,x)$	$p(\text{Drunk}\,|\,x)\,p(x)$
Fatality	0.000000075	0.40	$0.40 \times 0.000000075 = 0.00000003000$
Non-fatality	0.999999925	0.01	$0.01 \times 0.999999925 = 0.00999999925$
Total	1.000000000		0.01000002925

TABLE 6.22

Probability of Fatality, Given Drunk Driving, Calculated via Bayes' Theorem

Outcome of Car Trip, x	$p(\text{Drunk}\,|\,x)p(x)$	$p(x\,|\,\text{Drunk})$
Fatality	0.00000003000	0.00000003000/0.01000002925 = 0.000003
Non-fatality	0.00999999925	0.00999999925/0.01000002925 = 0.999997
Total	0.01000002925	1.000000

Again, the numbers in the last column of Table 6.21 do not add to 1.0. Those numbers are not the probabilities Pr(fatality|drunk), and Pr(non-fatality|drunk), but they are proportional to them. In other words, you have to multiply the numbers in the last column of Table 6.21 by a constant c to make them probabilities. Since the probabilities must add to 1.0, the number c is $c = 1/0.01000002925$. To complete the table, just divide the last column by 0.01000002925 to arrive at Table 6.22.

According to this estimate, your chances of dying or killing someone when you drink and drive are three in a million drunk driving excursions.

While this figure is very small, the chance of a death is *much smaller* when you don't drink and drive. Following the same process, you get Pr(Fatality|Sober) = 0.000000045, which translates to 4 or 5 fatalities out of *100 million* sober driving excursions.

Example 6.2: Age and Car Color Choice

To apply Bayes' theorem to the car color choice example from the introduction, consider the conditional distributions as given by Table 6.1, and suppose the marginal distribution of age is $p(\text{older}) = 1/3$ and $p(\text{younger}) = 2/3$. Let Y = color choice and X = age. If Y = red, what is the conditional distribution of X? The method shown in Example 6.1 repeats: Use Bayes' theorem $p(x\,|\,y) \propto p(y\,|\,x)\,p(x)$, and fill in the details. Table 6.23 shows the calculations.

What did you just find out? If the conditional distributions of car color purchasing are as given in Table 6.1, and if the age distribution is 2/3 younger and 1/3 older, then 83.3% of red car purchases are made by younger people, and 16.7% of red car purchases are made by older people. So if the car dealer told you she just sold a red car, and nothing else, it was much more likely sold to a younger person than an older one!

The previous examples used discrete distributions. Bayes' theorem also applies to continuous distributions; in fact, most of the applications of Bayesian statistics involve continuous distributions. The following example illustrates the famous formula $p(x\,|\,y) \propto p(y\,|\,x)\,p(x)$ in the continuous case.

TABLE 6.23

Illustrating Bayes' Theorem $p(x\,|\,y) \propto p(y\,|\,x)\,p(x)$ in the Car Color Choice Case

Age, x	$p(y\,|\,x) = p(\text{Red}\,|\,x)$	$p(x)$	$p(y\,|\,x)\,p(x)$	$p(x\,|\,y)$
Younger	0.50	0.667	$0.50 \times 0.667 = 0.333$	0.333/0.400 = 0.833
Older	0.20	0.333	$0.20 \times 0.333 = 0.067$	0.067/0.400 = 0.167
Total	—	1.000	0.400	1.000

Example 6.3: Income and Housing Expenses

For a continuous example, consider the case of income and housing expenses. Suppose that you work in a marketing company that provides targeted advertising to people in different income groups. Also, suppose that you have housing expense data from a regional database that has been segmented by income, in groups of low (less than 40K), medium (between 40K and 120K), and high (more than 120K) income categories.

Suppose also that, given income = x, your marketing research department tells you that housing expense Y is normally distributed with mean $0.30x$ and standard deviation $0.04x$. Then

$$p(y|x) = \frac{1}{\sqrt{2\pi}(0.04x)} \exp\left\{-0.5\frac{(y-0.3x)^2}{(0.04x)^2}\right\}, \quad \text{for } -\infty < y < \infty.$$

This is called a *heteroscedastic regression model*, meaning that the standard deviation of the distribution of Y depends on the value of X. Your market researchers' assumption of normal conditional distributions is not exactly right because it allows the possibility of negative housing expenses. But their model can be acceptable if the left tail of these normal distributions shows only minuscule probabilities below zero.

Given all this, suppose you happen to know someone who pays 60K per year on housing expenses and are in the middle-income group. What do you now know about their income, other than that it is some number between 40K and 120K?

Before seeing the 60K figure, you might have assumed, knowing that the person was in the middle-income group, that their income could have been any number between 40 and 120, all with equal likelihood. This is called your *prior distribution*. Thus, before seeing the housing expense data, you might have assumed $p(x) = 1/80$, for $40 < x < 120$.

But after seeing the 60K housing expense figure, you no longer think that income is uniformly distributed from 40 to 120. Think about it: The U(40, 120) distribution implies that you are just as likely to see an income of 45 as you are an income of 115. But if you know they are paying 60K per year on housing, then it is not likely that they are making 45K per year—it is much more likely that they are making 115K per year than it is that they are making 45K per year.

Bayes' theorem states that $p(x|y) \propto p(y|x)p(x)$. Here

$$p(y|x)p(x) = p(y|x) \times p(x) = \frac{1}{\sqrt{2\pi}(0.04x)} \exp\left\{-0.5\frac{(y-0.3x)^2}{(0.04x)^2}\right\} \times (1/80),$$

$$\text{for } -\infty < y < \infty \quad \text{and} \quad 40 < x < 120.$$

Substituting $y = 60$ gives you this equation:

$$p(x|60) \propto p(60|x)p(x) = \frac{1}{\sqrt{2\pi}(0.04x)} \exp\left\{-0.5\frac{(60-0.3x)^2}{(0.04x)^2}\right\} \times (1/80), \quad \text{for } 40 < x < 120.$$

Graphing this function of x over the range $40 < x < 120$ gives Figure 6.8.

Now this makes much more sense! If you know that someone's annual housing expense is 60K and that their income must have been in the 40K–120K range to begin with, then their income is probably very close to the upper limit of 120K.

Notice that the curve in Figure 6.8 is not quite the pdf, since the area is clearly too small. The entire area of the entire rectangular region of Figure 6.8 is $(120-40) \times (4 \times 10^{-9}) = 0.00000032$, and the area under the curve is much smaller than that. However, like the histograms whose vertical axes are not scaled to make the area = 1.0, the graph of the function

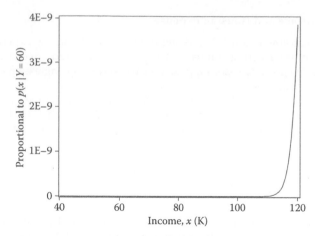

FIGURE 6.8
A graph that is proportional to $p(x|\text{housing expense} = 60)$ in the housing expense versus income example.

shown in Figure 6.8 is still useful because it shows the correct shape of the distribution, despite the fact that the vertical scale is wrong. In particular, the graph shows that the probability that income is greater than 110 is nearly 100%, given that housing expense = 60, since the area under the curve from 100 to 120 is nearly 100% of the total area.

You may think this conclusion is too strong, if you know of people who have little income but very expensive homes. If so, then you need to question the assumptions that gave you the curve of Figure 6.8. In particular, the market researchers' assumption of a normal distribution model could be flawed, in that the DATA* produced by their model don't quite match the real DATA. Always question the assumptions!

To get the precise values of the conditional distribution $p(x|\text{housing expense} = 60)$, you'll need to calculate the area under the curve (using software) in Figure 6.8 and divide the function in Figure 6.8 by that amount. The graph will look the same, except the numbers on the vertical axis will be much larger.

Example 6.4: Psychometric Evaluation of Employees

Some businesses practice the controversial policy of evaluating the traits of their employees using a survey. Employees typically answer the questions on the survey using the 1, 2, 3, 4, 5 scale. Each employee's responses then go into a complex formula—perhaps obtained using an advanced statistical method known as *discriminant analysis*—and voilà! The employer gets a number Y that purports to measure the employees' "fitness for the job." The controversial aspect is that the survey is generic, having nothing in particular to do with the employees' job tasks. Nevertheless, these data might be used in human resource decisions involving anything from reassignment to outright firing of employees. Yes, this should make you feel creepy. We are not endorsing this practice, but you have to admit the example is interesting.

Suppose, based on historical data, a company knows 5% of its employees are involved in stealing from the company, while 95% are not. Let X be the employees' stealing behavior; X is a binary variable with values (stealer, non-stealer). Also suppose that, based on historical data, the company believes that the employees' fitness—his or her Y value calculated from the survey—has the following conditional distributions:

$$Y|X = \text{Stealer} \sim N(60, \ 10^2)$$

$$Y|X = \text{Non-stealer} \sim N(75, \ 10^2)$$

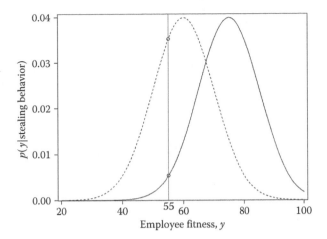

FIGURE 6.9
Conditional distributions of employee fitness (*Y*) for stealers (dashed curve) and non-stealers (solid curve).

Suppose also that you have just taken the test, and your *Y* value is *y* = 55. Are you a stealer? Now you should feel even more creeped out!

Figure 6.9 shows the conditional distributions of *Y*, as well as your *y* = 55 score.

Looking at the two circles in the conditional distribution graph of Figure 6.9, it appears that you are much more likely to be a stealer than a non-stealer, since the likelihood of your *y* = 55 observation is so much higher in the stealer group than in the non-stealer group. You can calculate the precise values of these likelihoods using the normal distribution formula:

$$p(55|\text{Stealer}) = \frac{1}{\sqrt{2\pi}10}\exp\left\{-0.5\frac{(55-60)^2}{10^2}\right\} = 0.035207$$

$$p(55|\text{Non-Stealer}) = \frac{1}{\sqrt{2\pi}10}\exp\left\{-0.5\frac{(55-75)^2}{10^2}\right\} = 0.005399$$

These numbers suggest that you are 0.0352/0.0054 = 6.5 times more likely to be a stealer than a non-stealer!

However, this calculation is wrong since it doesn't account for the marginal distribution of stealing behavior. To start with, there is only a 5% chance that you are a stealer in the company's eyes—from the standpoint of the cold, heartless human resources manager. The company must incorporate this information when calculating the probability that you are a stealer. Bayes' theorem $p(x|y) \propto p(y|x)\,p(x)$ appears again! The calculations are shown in Table 6.24. There is only a 25.6% chance that you are a stealer. From the cold, heartless human resources manager's perspective, 25.6% of employees who score 55 on the exam turn out to be stealers.

TABLE 6.24

Illustrating Bayes' Theorem $p(x|y) \propto p(y|x)\,p(x)$ in the Psychometric Evaluation Case

Stealing Behavior, *x*	$p(y\|x) = p(55\|x)$	$p(x)$	$p(y\|x)\,p(x)$	$p(x\|y)$
Stealer	0.035207	0.05	0.035207 × 0.05 = 0.00176	0.00176/0.00689 = 0.256
Non-stealer	0.005399	0.95	0.005399 × 0.95 = 0.00513	0.00513/0.00689 = 0.744
Total	—	1.00	0.00689	1.000

Still, it is a creepy example. The company now has its eye on you, based solely on this questionnaire. The management started out thinking you were only 5% likely to be a stealer, and now they have upped the probability to 25.6%.

Creepy or not, this is a good example to illustrate Bayes' theorem and also to introduce some terminology. The 5% figure in this example is called a **prior probability**. It's what you think before seeing the data. The 25.6% figure is called a **posterior probability**. It's what you think after seeing the data.

Bayesian methods are very attractive because they give you a formula showing how to update your knowledge as more data comes in. Your mind is changed by data. *Data reduce the uncertainty about the unknown parameters.* If the company actually caught you stealing, then their probability would be revised further, from 25.6% to 100%.

Vocabulary and Formula Summaries

Vocabulary

Marginal distribution	The ordinary distribution of a variable, specifically considered without constraining any other variable.
Joint distribution	The probability distribution of combinations of values of two variables X and Y.
Contingency table, cross-classification table	A table of counts, classified according to two distinct discrete variables such as sex (male or female) and purchase behavior (yes or no).
Scatterplot	A graph of data pairs (x_i, y_i), where each of the n pairs is indicated by a single dot.
Bivariate histogram	An estimate of the joint pdf $p(x, y)$.
Bayes' theorem	A theorem allowing you to find the conditional distribution $p(x|y)$ when you know $p(y|x)$ and $p(x)$; given specifically by $p(x|y) \propto p(y|x)\, p(x)$.
Prior probability	Your probability of an event, before you see the data.
Posterior probability	Your probability of the event, after you see the data.

Key Formulas and Descriptions

$p(y) = \sum_{all\, x} p(x, y)$	*Discrete case*: The sum of the joint probabilities over values of X gives the marginal probability distribution of Y.
$\sum_{all\, x} \sum_{all\, y} p(x, y) = 1.0$	*Discrete case*: The sum of all joint probabilities is 1.0.

$p(y) = \int_{all\,x} p(x,y)dx$ *Continuous case*: The integral of the joint distribution over values of X gives the marginal probability distribution of Y.

$\int_{all\,x} \int_{all\,y} p(x,y)dydx = 1.0$ *Continuous case*: The volume under the joint pdf is 1.0.

$p(y|x) = \dfrac{p(x,y)}{\sum_{all\,y} p(x,y)}$ *Discrete case*: The conditional distribution of Y given a particular $X = x$ is the slice of the joint distribution where $X = x$, but made to sum to 1.0.

$p(y|x) = \dfrac{p(x,y)}{\int_{all\,y} p(x,y)dy}$ *Continuous case*: The conditional distribution of Y given a particular $X = x$ is the slice of the joint distribution where $X = x$, but made to integrate to 1.0.

$p(y|x) = \dfrac{p(x,y)}{p(x)}$ *Discrete and continuous cases*: The conditional distribution of Y given a particular $X = x$ is the slice of the joint distribution where $X = x$, divided by the marginal distribution of X evaluated at $X = x$.

$p(x, y) = p(y|x)\, p(x)$ The joint distribution is the product of conditional and marginal distributions.

$p(x, y) = p(x)p(y)$ When RVs X and Y are independent, their joint pdf is equal to the product of their marginal pdfs.

$p(y) = p(y|x)$, for all x When RVs X and Y are independent, the marginal pdf of Y is equal to the conditional pdf of Y, given $X = x$, for all x.

$p(x|y) = \dfrac{p(y|x)p(x)}{\sum_{all\,x} p(y|x)p(x)}$ Bayes' theorem, discrete case.

$p(x|y) = \dfrac{p(y|x)p(x)}{\int_{all\,x} p(y|x)p(x)dx}$ Bayes' theorem, continuous case.

$p(x|y) \propto p(y|x)\, p(x)$ Bayes' theorem, either case.

Exercises

6.1   Suppose that 99.999% of Internet messages are not important for your business, Zia Technologies, and the remaining 0.001% are important. Among important messages, 90% contain the phrase *Zia Technologies*. Among non-important messages, 0.001% contain that phrase. Let Y be the binary variable indicating whether or not the message contains the phrase Zia Technologies. Let X be the binary variable indicating whether or not the message is important.

    A. Display the marginal distribution of X in list form.

    B. Find the conditional distributions $p(y|X=\text{important})$ and $p(y|X=\text{not important})$.

C. Suppose a message contains the phrase *Zia technologies*. Find the conditional distribution of X using Bayes' theorem, as shown in Table 6.23, for example.

D. Explain the relevance of the conditional distribution of X in Exercise 6.1C to your company. Why is it interesting?

6.2 A potato chip manufacturer samples 10 potatoes from a truckload and pays the farmer based on the results of the sample. If the truckload is from farm A, the distribution of bad potatoes in the sample looks like this:

Number Bad	Probability
0–3	0.1
4–7	0.3
8–10	0.6

If the truckload is from farm B, the distribution looks like this:

Number Bad	Probability
0–3	0.5
4–7	0.4
8–10	0.1

Ten percent of the truckloads come from farm A. Let Y be the variable *number bad* (treated as a three-level random discrete variable), and let X be the binary variable *farm* (A or B).

A. Display the marginal distribution of X in list form.

B. Suppose a crate has 8–10 bad. Find the conditional distribution of X using Bayes' theorem, as shown in Table 6.23, for example.

C. Explain the relevance of the conditional distribution of X in Exercise 6.2B to the chip manufacturer. Why is it interesting?

6.3 As part of a quality control initiative, a hospital hands out a satisfaction survey to $n = 7342$ patients as they leave the hospital. Patients rate their satisfaction with their care on a five-point scale, ranging from 1 = highly dissatisfied to 5 = highly satisfied. The patients are cross-classified by rating and insurance as shown in the following table.

	Satisfaction Rating				
	1	2	3	4	5
Insurance	587	1174	1179	1173	1762
No Insurance	440	290	296	147	294

A. Let X be the binary variable (insurance, no insurance). Show the estimated marginal distribution of X in list form, and explain why it is interesting.

B. Let Y be the discrete satisfaction measure. Show the estimated marginal distribution of Y in list form, and explain why it is interesting.

C. Display the two estimated conditional distributions of Y in list form, compare them, and explain why the comparison is interesting.

6.4 In Example 6.3 with Y = housing expense and X = income, the joint distribution was given as

$$p(y|x) \times p(x) = \frac{1}{\sqrt{2\pi}(0.04x)} \exp\left\{-0.5\frac{(y-0.3x)^2}{(0.04x)^2}\right\} \times \left(\frac{1}{80}\right),$$

for $-\infty < y < \infty$ and $40 < x < 120$

A. Use 3-D plotting software to display this joint distribution as shown in Figure 6.1.

B. Display the slice of this joint distribution where $y = 60$, similar to what is shown in Figure 6.2, but here using y instead of x. Compare the appearance of the slice to Figure 6.8.

6.5 Redo the analysis that produced Figure 6.8, but instead of assuming a U(40, 120) distribution for income, assuming a N(80, 15^2) distribution, so that

$$p(x) = \frac{1}{\sqrt{2\pi}15} \exp\left\{-0.5\frac{(x-80)^2}{15^2}\right\}, \quad \text{for } -\infty < x < \infty$$

Compare the result with Figure 6.8, and explain why they are different.

6.6 Simulate 1000 pairs (X^*, Y^*) from the joint distribution shown in Exercise 6.5. For each pair, first simulate X^* from N(80, 15^2) (the normal pdf with mean 80 and standard deviation 15), then simulate Y^* from N($0.3X^*$, $(0.04X^*)^2$) (the normal pdf with mean $0.3X^*$ and standard deviation $0.04X^*$).

A. Draw a scatterplot of the 1000 (X^*, Y^*) pairs and interpret it.

B. Draw a bivariate histogram of the 1000 pairs and explain how it provides similar information as the scatterplot in Exercise 6.6A.

6.7 We treat everyday observations as routine when in fact they are extraordinarily unlikely. As an example, flip a fair coin 30 times, and write down the sequence you get, such as H, H, T, ..., T, listing every single one of the 30 outcomes. Now, calculate the probability of seeing that precise sequence using the independence formula. What is the probability that you would have seen that particular sequence, before you saw it? Suppose it took you a minute to flip 30 times and record your sequence. Suppose you repeat the 30 flips, record the sequence, repeat the 30 flips, record the sequence, repeat the 30 flips, record the sequence, continuously. How many years do you expect it will take you to get the exact same sequence again?

6.8 Review the stealer versus non-stealer case (Example 6.4). Suppose the human resources manager is a pessimist, believing almost everyone steals from the company, or that Pr(Stealer) = 0.99. Suppose you take the test, and the human resources manager learns that your score was 95.

A. What is the probability that you are a non-stealer from the standpoint of the human resources manager?

B. Suppose instead that the company's prior probability is Pr(Stealer) = 1.0. Suppose you take the test and score a 95. What does the company think about you now? What if you scored 100? Would that make any difference?

C. A prior such as Pr(Stealer) = 1.0 is called a "dogmatic" prior. Using your answer to Exercise 6.8B, explain why it is dogmatic.

6.9   An Internet marketing company knows that only 1% of the visitors to a website ever click on the banner advertisement. Based on customer surveys, they have been able to estimate the income distributions for customers who click on their banner ad and for customers who do not click. The following graph shows these distributions: The solid line indicates non-clickers, and the dashed line indicates clickers.

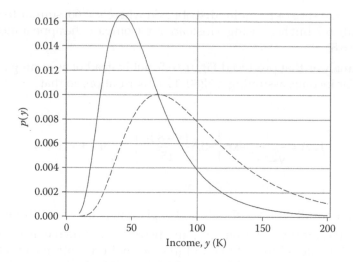

Based on this graph, about what percent of people with income = 100 will click?

6.10 In Example 6.1, you see the result Pr(Fatality|Sober) = 0.000000045. Show how that number is obtained.

6.11 Show how

A. Equation 6.11 follows from Equations 6.6 and 6.7

B. Equation 6.12 follows from Equations 6.1, 6.7, and 6.11

C. Equation 6.13 follows from Equations 6.2, 6.7, and 6.11

7

Sampling from Populations and Processes

7.1 Introduction

Recall the statistical science paradigm introduced in Figure 1.2 of Chapter 1, repeated in this chapter as Figure 7.1. Subsequent chapters discuss DATA that you can observe, called Y, or sometimes (X, Y) if the data are bivariate. Data values $Y_1, Y_2, ..., Y_n$ or pairs (X_1, Y_1), $(X_2, Y_2), ..., (X_n, Y_n)$ that might be produced are called a **sample**. Usually, the ultimate goal of using the sampled data is to learn about Nature, to reverse engineer its processes.

Your statistical model for how your DATA arise is shown in Figure 1.3, repeated here as Figure 7.2.

Many statistics sources use the term **population** in place of *Nature* and define it to be a finite collection of static, fixed data values $y_1, y_2, ..., y_N$. The population size, N, is denoted by a capital letter not because it is random—in the common usage of the term population, its size N is fixed. Instead, N is capitalized to emphasize that it is usually a much bigger number than your sample size, n.

In these statistics sources, you will see the assumption that the data values $Y_1, Y_2, ..., Y_n$ are a random sample of n values from the larger collection of the N values $y_1, y_2, ..., y_N$. They go on to suggest that the ultimate goal of using data $Y_1, Y_2, ..., Y_n$ is to learn about the population of N values $y_1, y_2, ..., y_N$. These sources refer to the distribution $p(y)$ as a *population distribution*, which is the distribution that puts probability $1/N$ on each of the values $y_1, y_2, ..., y_N$, and they define *population mean* and *population standard deviation* as numbers calculated from the N values $y_1, y_2, ..., y_N$.

By contrast, we have consistently defined $p(y)$ in *process* terms throughout this book: Before you see the data, there are natural processes, as well as design and measurement processes, that determine how your DATA will look. The part of the Mantra that states *model produces data* specifically means "$p(y)$ produces DATA." For example, if Y is income, then $p(y)$ tells you which income values you will see more often and which ones you will see less often. The processes at work that produce income include current macroeconomic conditions, especially those involving labor markets, regional effects, and government regulations such as minimum wage. There is no population of income values $y_1, y_2, ..., y_N$ that determine your process; to the contrary, it is the process that determines the income values for any population at any given time.

What *is* the specific population definition of $p(y)$? As given in Chapter 2, it is

$$\text{Pr(outcome)} = \frac{\text{No. of elements in population having the outcome}}{\text{Total no. of elements in population}}$$

FIGURE 7.1
The statistical science paradigm, from Chapter 1.

FIGURE 7.2
The statistical model, from Chapter 1.

Thus, the population definition of $p(y)$ is

$$p(y) = \Pr(Y = y) = \frac{\text{No. of elements in population having the outcome } (Y = y)}{\text{Total no. of elements in population}}$$

For example, if there are 1231 people in the population and you ask one of them, "What is your favorite drink?" the probability that Y = lemonade, using the population definition, is

$$p(\text{lemonade}) = \frac{\text{No. of people out of 1231 who say lemonade}}{1231}$$

To summarize, here is the population definition of $p(y)$.

The Population Definition of $p(y)$

$$p(y) = \frac{\#\{y_i = y\}}{N}, \quad \text{if } y = y_i \text{ for some } i = 1, 2, \ldots, N; \quad p(y) = 0 \text{ otherwise}$$

The process model is simply a quantification of the concept *model produces data* that you have seen many, many times already.

The Process Definition of $p(y)$

A model for how your DATA will appear, $p(y)$ governs their frequencies of occurrences.

In contrast to the population model, the process model is a mental model, one that does not have the concrete specificity of the population model. You *can* state precisely what the population model is, if you only had the entire population, but you usually *cannot* state precisely what the process model is. As you might have heard, *model has unknown parameters*.

Despite the lack of concreteness of the process model relative to the population model, we will argue strongly for the process model as being the more relevant and more scientific model. The main problems with the population definition of your model $p(y)$ are as follows. Section 7.3 provides further elaboration on these five points.

Problems with Population Definition of $p(y)$

1. The population data result from natural processes, so it is scientifically uninteresting to let the population data define the model, when the real goal is to understand the natural processes that lead to the population data.
2. The definition of *population* is ambiguous. The N values $y_1, y_2, ..., y_N$ are usually not static, unchanging; rather, they change continuously from one time point to the next. This makes it impossible to identify what the population really is.
3. The population conditional distributions $p(y|x)$ are discontinuous functions of x, unlike natural processes. *Nature favors continuity over discontinuity.*
4. The population distribution $p(y|x)$ does not exist at all when there is no x in the population taking a particular value such as $x = 28.92$.
5. Processes that produce your DATA include your design and measurement processes such as nonresponse and measurement error, which are mostly unrelated to the population of values $y_1, y_2, ..., y_N$ (assuming such values can be defined at all). To assume that the $p(y)$ producing your DATA is defined only by the population $y_1, y_2, ..., y_N$ is simply wrong because it ignores these processes.

Section 7.2 provides an excellent example of population sampling and proceeds to argue for a process definition of the model $p(y)$, even in that example. Later sections provide examples that are more directly process oriented and also define the concept of *independent and identically distributed* (iid) observations as the prototypical process model.

7.2 Sampling from Populations

Example 7.1: Estimating Inventory Valuation Using Sampling

Suppose you want to estimate the total value of all the electronic and furniture inventory items in your company. These include desktop and laptop computers, mobile devices, printers, fax machines, copiers, chairs, desks, couches, and other items, spread out company wide. Your records show inventory labels for a population having $N = 21,342$ such items. These items are in various conditions of depreciation, usually worth much less than their purchase prices. It would take a lot of time and money to ascertain the value of each of the 21,342 items, so instead, to save time, you decide to take a random sample of $n = 100$ of the $N = 21,342$ items and to assess each of their values carefully. You will then use the average value of the $n = 100$ randomly sampled items as a proxy for the average value of the $N = 21,342$ items. Assuming the average value from the random sample was \$350.21, you would then estimate the total value of the entire inventory to be $21,342 \times \$350.21 = \$7,474,181.82$. Pretty nice! You have saved time and money and arrived at a reasonably accurate estimate.

The set of $n = 100$ items sampled from the $N = 21{,}342$ items is an example of a **random sample from a population**.

Here you should have the question, "What n should I use?" In the inventory valuation case, we assumed $n = 100$, but you could sample more or less. Selecting n involves a trade-off between desired accuracy and cost: You get more accurate estimates with larger n, but it costs more.

We'll leave the discussion of selecting a sample size n for Chapter 18. Right now, let's talk about what it means to obtain a random sample of $n = 100$ observations from $N = 21{,}342$. It's easy to visualize: Just imagine 21,342 tennis balls in a large pit, each having an inventory label. Now mix those balls up thoroughly and choose 100 of them.

Of course, you don't need tennis balls and a pit; you just need a computer. First, you'll have to get a list of labels for all N items in the population. This is an absolute requirement of population sampling, no matter whether you are sampling from the $N = 21{,}342$ inventory items of Example 7.1 or sampling from the $N = 350{,}000{,}000$ people living in a particular country. In the case of sampling people, the labels include name, address, and/or other identifying information. The requirement that you obtain item labels for all the items in the population usually makes precise random sampling from a population infeasible.

Getting back to the inventory valuation example, assuming you have a list of item labels, you can enter the list in a spreadsheet or database, as shown in Figure 7.3. Here, the labels aren't names and addresses, but item code numbers.

The list shown in Figure 7.3 continues, with 21,342 rows of inventory item labels. To select a random sample of 100 of them, create a second column with 21,342 randomly generated $U(0, 1)$ values. It would be a good idea to specify the *random seed* such as 4323 in the input dialogue box (assuming you're using the "random number generation" of the data analysis add-in of Microsoft Excel) so that the numbers are not like others you have seen before. Figure 7.4 shows the resulting screenshot.

Sorting the data in the $U(0, 1)$ column and taking the top 100 inventory item labels give you the random sample you want, as shown in Figure 7.5.

So, the list of $n = 100$ items in the inventory to be sampled and assessed for their value consists of those labeled 45,192, 33,476, 12,656, (… 97 more). Now, have the assessor go look at those 100 items!

	A	B	C	D	E	F	G
1	Item						
2	1335						
3	13723						
4	43673						
5	29960						
6	30289						
7	19604						
8	1129						
9	16331						
10	9469						
11	927						
12	11295						

FIGURE 7.3
Inventory item labels consisting of item code numbers.

	A	B	C	D	E	F	G
1	Item	U(0,1)					
2	1335	0.43199					
3	13723	0.24586					
4	43673	0.42344					
5	29960	0.68697					
6	30289	0.04614					
7	19604	0.40364					
8	1129	0.87317					
9	16331	0.79727					
10	9469	0.6516					
11	927	0.83508					
12	11295	0.66576					

FIGURE 7.4
Inventory item labels and uniform random numbers.

	A	B	C	D	E	F	G
1	Item	U(0,1)					
2	45192	3.1E-05					
3	33476	0.00015					
4	12656	0.00018					
5	44103	0.00021					
6	6870	0.00027					
7	9203	0.00034					
8	17018	0.00037					
9	13731	0.00037					
10	29859	0.00052					
11	18477	0.00058					
12	17752	0.00061					

FIGURE 7.5
Inventory item labels, sorted from smallest to largest uniform random number.

In the Nature → design and measurement → DATA paradigm, Nature here is the 21,342 inventory items. Design refers to the obtaining of the list of labels, entering them in the computer, and the random sampling of the $n = 100$ items as shown in Figures 7.3 through 7.5 earlier. Measurement is the assessment of value, in dollars, for each item. The DATA are the numerical values of the $n = 100$ valuations that come from the assessments. The DATA are random because of the random sampling: A different stream of U(0, 1) numbers in Figure 7.4 gives a different sample of $n = 100$ items. Figure 7.6 illustrates this paradigm in the case of sampling from a population.

Your statistical model for the population sampling framework is still as in Figure 7.2, $p(y)$ → DATA. (*Model produces data.*) Here, the DATA produced by the model are assessments of the randomly sampled elements. Every element in the population has an equal chance of selection; hence, the distribution $p(y)$ that produces individual observations Y is as given in Table 7.1.

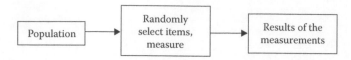

FIGURE 7.6
Nature, design and measurement, and DATA in the population sampling framework.

In Table 7.1, the numbers $y_1, y_2, \ldots, y_N$ are the numbers in the population (e.g., the assessments of all $N = 21{,}342$ inventory items). The distribution of Y, here a randomly sampled inventory item, is always discrete in the population sampling framework.

If there are repeats among the y values in the population, so that there are not N distinct y but instead only $k < N$ of them, then you should write the distribution showing only the distinct values, as in Table 7.2.

For example, suppose the inventory items are either electronic devices or furniture, with $N_1 = 17{,}980$ electronic devices and $N_2 = 3{,}362$ furniture items. Here there are $k = 2$ distinct y values in the population. Then the probability distribution of the inventory *type* variable for a randomly selected item is as given in Table 7.3.

The models in Tables 7.1 through 7.3 tell you the distribution of individual values sampled from a population, but they do not tell you the distribution of combinations of sampled values. The distributions of combinations depend upon whether the data are sampled **with replacement** or **without replacement**. *With replacement* samples are samples where an item is sampled, then put back into the population. With replacement samples are not usually used when sampling from populations, because,

TABLE 7.1

Probability Distribution
That Produces Data under
the Population Model

y	$p(y)$
y_1	$1/N$
y_2	$1/N$
...	...
y_N	$1/N$
Total	1.00

TABLE 7.2

Probability Distribution That
Produces Data under the Population
Model, When There Are Repeats

y	$p(y)$
y_1	N_1/N
y_2	N_2/N
...	...
y_k	N_k/N
Total	1.00

TABLE 7.3

Example of a Population Probability
Distribution When There Are Repeats

y	$p(y)$
Devices	17,980/21,342 = 0.842
Furniture	3,362/21,342 = 0.158
Total	1.000

for example, it means that you might sample the same inventory item twice. On the other hand, with replacement samples are used for *bootstrap sampling*, described in later chapters.

By far the more common sample is a *without replacement* sample, where you simply select n items from the N. All n sampled items are distinct, just as if you grabbed a handful of M&Ms from a jar and ate them. Your handful of M&Ms is a *without replacement* sample from the jar. You don't grab an M&M, pop it in your mouth, and then put it back in the jar! That would be called *with replacement* sampling and—needless to say—it also would be called *quite unsanitary*.

Random sampling from a population works very well to estimate the population average. Suppose the population has $N = 999$ numbers, the sequence 1, 2, …, 999. The population average is $(1 + 2 + \cdots + 999)/999 = 500.00$. If you randomly sample $n = 100$ items from the $N = 999$, your average from the $n = 100$ sampled items will be close to 500.00 because it is unlikely that a random sample will be all mostly low numbers or all mostly high numbers. Instead they will tend to be spread out roughly evenly above 500 and below 500, giving an average that is close to 500. Figure 7.7 shows the results of taking many random samples of size $n = 100$ and calculating the average of the 100 sampled observations. The sample average is random, but its value is relatively close to 500, usually within the range of 500 ± 50. This is remarkable when you consider that the original range of the data is from 1 to 999.

Figure 7.7 is the picture you should think of to understand the estimated total inventory valuation $21,342 \times \$350.21 = \$7,474,181.82$. The number \$350.21 is the average of the $n = 100$ sampled items which differs from the (unknown) average of the $N = 21,342$ items, in the same way that in Figure 7.7, the averages of the 100 sampled numbers differ from the (known) average value 500.00 calculated from the $N = 999$ numbers. In particular, there will not be any tendency to overestimate or underestimate, because, as shown in Figure 7.7, the estimates are as often too high as they are too low. If an estimation procedure does not systematically overestimate or underestimate, then the resulting estimates are **unbiased estimates**, discussed in more detail in Chapter 11.

Figure 7.7 is an estimate of a special kind of probability distribution called a **sampling distribution**, which is the probability distribution of a statistic calculated from a random sample. A result called *the central limit theorem* (CLT), discussed in Chapter 10, tells you that as the sample size n gets large, the distribution of the sample average gets closer to a normal distribution, regardless of the form of the sampled distribution. In the analysis of Figure 7.7, the sampled distribution is the discrete uniform on the numbers 1, 2, …, 999, yet you can see in Figure 7.7 that the distribution of the *average* is approximately a normal distribution.

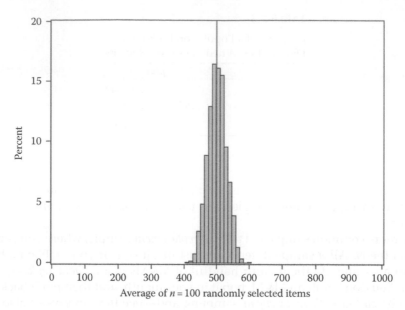

FIGURE 7.7
Histogram of sample averages calculated from many repeated random samples of $n = 100$ values from the population of numbers 1, 2, ..., 999.

7.3 Critique of the Population Interpretation of Probability Models

7.3.1 Even When Data Are Sampled from a Population

The inventory estimation example of Section 7.2 is a wonderful application of statistics—it shows that, through sampling, you can save time and money and still obtain acceptably accurate estimates. This is a very concrete application and relatively easy to understand. This is also the reason that other statistics sources use the term *population distribution* for $p(y)$ and define it as shown in Table 7.2.

Nevertheless, the interpretation of $p(y)$ as a population distribution is misleading at best and simply wrong at worst. The following discussion elaborates on the five problems with this interpretation that were given in the introduction to this chapter.

7.3.2 Point 1: Nature Defines the Population, Not Vice Versa

The population definition of the model $p(y)$ implicitly assumes that the population defines all of Nature. But the population is specific to a particular point in time and place, while Nature cuts across times and places. Thus, the population point of view allows no generalizability and can lead to extremely silly and sometimes harmful misunderstandings. The population data are the result of natural processes, but Nature is not the result of the population.

For example, suppose that, in the population of 21,342 inventory items, there were no items worth between $209.00 and $211.00. If you adopt the population interpretation of probability, then you would claim that there is zero chance of any item being valued between $209.00 and $211.00, since 0/21,342 = 0.00000000000000000000000000000000000000 00000000000000000000000000000....

Zero means zero; that's why we gave it so many zeros! The point is, while the number 0 may be correct for this company at this point in time, in the sense that there were no values between $209.00 and $211.00, the statement that there is zero probability does not generalize. For other companies, there could be items valued between $209.00 and $211.00, and even for this company at different times, or for different assessors, there could be items valued between $209.00 and $211.00. Zero probability implies absolute certainty, and such certainty is simply wrong outside the confines of this specific company at this specific instant in time, for this particular assessor.

A related point is that any population of numbers $y_1, y_2, \ldots, y_N$ always has a distinct maximum. If you adopt the population definition of $p(y)$, then you would state that anything larger than the maximum is impossible. For example, perhaps at this precise moment in time, the oldest human is 120 years old. Then, using the population definition of $p(y)$, you would say that it is absolutely impossible for a human to be older than 120 years, which is absolutely wrong! If you adopt the population definition of $p(y)$, then you have no ability whatsoever to generalize beyond the population values $y_1, y_2, \ldots, y_N$.

Strike one against the population definition of the model that produces your data.

7.3.3 Point 2: The Population Is Not Well Defined

A second problem with the population definition of $p(y)$ is that it applies only to instantaneous points in time. The population model defines $p(y)$ in terms of some fixed values $y_1, y_2, \ldots, y_N$. Which values are they? While the population model computed from the $N = 21{,}342$ inventory items is the right model assuming that all sampling is done precisely at time t (measured precisely, perhaps to the fraction of a second), the population model changes the instant a new purchase is made or inventory item scrapped, or when a new assessor is employed, because at that point in time, the list $y_1, y_2, \ldots, y_N$ changes. Thus, while the population interpretation may be comforting in its seeming concreteness, the population distribution is in fact not concretely defined, since the population values $y_1, y_2, \ldots, y_N$ continuously change. It is not clear *which* population distribution you are talking about when you employ the population interpretation. Do you mean the population as it exists at the beginning of the sampling period? At the end? Somewhere in the middle may be more logical, but when? Thus, despite its seeming concreteness, the population model is not clearly defined because of temporal effects.

Strike two against the population definition of the model that produces your data.

7.3.4 Point 3: Population Conditional Distributions Are Discontinuous

A third and even more troublesome problem with the population interpretation of the model is its implied definition of conditional probability. Suppose the boss wants a breakdown of electronics versus furniture at the $200 level, $300 level, and so on. There is a distribution of inventory type = (electronics or furniture) for every single assessed value in the population. Here are a few hypothetical (value, type) data points from the population, in ascending order of value: ($321.20, furniture), ($322.00, electronics), ($322.34, furniture), ... Now, in the population framework, you would conclude that there is 100% probability of furniture at the $321.20 valuation level, then 0% probability of furniture at the $322.00 valuation level, then back up to 100%, etc. These numbers are correct when thinking only about the population and about nothing more general than that, but from a process standpoint, they are illogical. The distribution of furniture and electronics should be a continuous function of valuation. The processes that produce the (furniture, electronics) DATA at

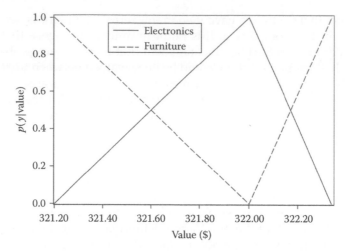

FIGURE 7.8
Conditional distributions of Y = inventory type (furniture, electronics) under the population interpretation of probability models, for value in the range \$321.20–\$322.34.

the \$321.20 valuation level should differ little, if at all, from the processes that produce the (furniture, electronics) DATA at the \$322.00 valuation level. In the population model, these distributions vary wildly. In the process model, they morph gradually and continuously, as shown in Figure 5.2 (Chapter 5), for example. *Nature favors continuity over discontinuity*, but population-based conditional distributions of Y are *always* discontinuous functions of $X = x$. Process-based conditional distributions of Y, on the other hand, are usually specified as continuous functions of $X = x$, but can be discontinuous if needed as well.

See Figure 7.8, which shows the conditional distributions of inventory type (furniture, electronics) for given value levels, when adopting the population interpretation of probability. As shown in Figure 7.8, the probabilities are sometimes 1.0, sometimes 0.0, and jump wildly back and forth.

Strike three against the population definition of the model that produces your data.

7.3.5 Point 4: The Conditional Population Distribution $p(y|x)$ Does Not Exist for Many x

See Figure 7.8 again: In reality, there is no population distribution whatsoever for inventory type when the value is between \$321.20 and \$322.00 or between \$322.00 and \$322.34. However, from the process standpoint, there is a distribution of inventory type at the \$321.50 valuation, and logically, it differs little, if at all, from the distribution of inventory type at the \$321.00 valuation level. On the other hand, the population distributions for such specific valuation levels are simply nonexistent.

A model that makes more sense than the population model is the process model that allows continuous morphing of distributions. *Nature favors continuity over discontinuity*; see Figure 7.9. As shown in Figure 7.9, the conditional distributions of inventory type differ very little across such small differences in value. In addition, unlike the population models shown in Figure 7.8, there is no problem discussing the distribution of inventory type when value is different from \$321.20, \$322.00, and \$322.34. Figure 7.9 shows a sensible way to define these distributions, but it requires the process model.

Strike four against the population definition of the model that produces your data.

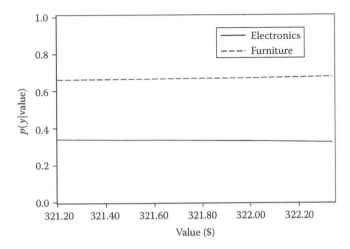

FIGURE 7.9
Plausible conditional distributions of inventory type (furniture, electronics) under the process interpretation of probability models, for value in the range $321.20–$322.34.

7.3.6 Point 5: The Population Model Ignores Design and Measurement Effects

There are no perfect design and measurement systems. When sampling from a population, the DATA you will see are affected not only by the collection of items $y_1, y_2, \ldots, y_N$ in the population but also by your design and measurement processes, which include nonresponse, interviewer/subject interaction, and measurement error. These processes make the DATA you will observe look different than DATA produced as a simple random sample from the values $y_1, y_2, \ldots, y_N$. The process model accommodates all such biases, while the population model does not.

Design and measurement processes are important to recognize when interpreting your observed data. In addition, if you understand that such processes can have bad effects on your DATA, it can help you to design studies to get better DATA.

Strike five against the population definition of the model that produces your data.

Examples 7.2 through 7.5 that follow provide more details in specific case settings.

There is nothing wrong with the application to inventory sampling in Section 7.2 to estimate the population average; that is a wonderful application of statistics. What we are saying is that the population-based *interpretation of the model* $p(y)$ is troublesome, awkward, and, in most cases, simply wrong. It leads to misunderstandings about the meanings of statistical models and parameters and makes it difficult to learn statistics as a logical subject.

Some use the term "population" in quotation marks to make it clear they don't really buy into the literal view of the population model. But there is no need for this ambiguity. You can use the more concrete alternative term, *process*, instead of the nebulous "population."

Rather than using the population framework of Figure 7.6, we suggest you use the more scientifically sound process framework shown in Figure 7.10.

Your statistical model for the framework shown in Figure 7.10 is still *model produces data* or $p(y) \to$ DATA. However, from the process point of view, your $p(y)$ refers ultimately to Nature as opposed to the population. This model posits that *the population itself* is a sample from Nature and is called the **superpopulation model**.

In the inventory sampling case of Example 7.1, you might argue that you really are interested in just the population average of the $N = 21,342$ inventory items and not in the

FIGURE 7.10
The revised Nature, design and measurement, and DATA diagram in the population sampling framework.

average of the process that produced the N = 21,342 items. That's okay—assuming a perfect population sampling system, there is little difference between the process average and the population average because of the law of large numbers, which is given in Chapter 8. Thus, it makes little difference whether you adopt the population model or the process model in cases where you think the population model really is valid. On the other hand, there is no perfect population sampling system, and adopting the population model can lead to problems as detailed by the five points earlier. Thus, it is generally safer to adopt the process model than the population model.

> **Example 7.2: Design and Measurement Process Elements
> in a Population Sampling Setting: Measurement Error**
>
> This example shows that design and measurement elements, which are completely outside the population, must be considered part of the definition of the model $p(y)$ that produces your data. Design and measurement processes affect your data dramatically.
>
> In the inventory example, the true values of each of the N = 21,342 items are really unknown. The numbers y are just guesses made by a person (the assessor); the true value is a **latent variable** v. Hopefully, the assessor's values y are close to the true values v, but they will differ. Further, the assessor's valuations are essentially random, depending upon how much research he did, on how many factors were taken into consideration, and how accurate the calculations were, and perhaps even on how much coffee he had. It is likely that the assessor will give different values to the same item, when presented with that same item in different circumstances.
>
> So the model that produces the data is *not* the population model—instead it is the population model mixed with the assessor's random errors. The valuation data produced by the model are the population values v plus a random assessor deviation D that is not part of the population at all:
>
> $$Y = v + D$$
>
> The assessor's deviations (the D) can have serious consequences. If the assessor is wildly erratic, missing the true values by a wide margin, then the estimate of the population average using the sample average will be less accurate than desired. For example, in Figure 7.7, the range of possible sample averages might go from $300 to $700, instead of from $400 to $600, had the assessor's D values covered a wide range.
>
> Even more sinister though is the possibility of bias in the D values. Suppose there is pressure from management to show more value to the shareholders. In this case, one might assume—cynically—that the assessor's deviations D are systematically larger than 0, indicating overvaluation of the inventory items. This causes bias in the estimated average. Figure 7.11 is a repeat of Figure 7.7, but using the model, $Y = v + D$. The v are again the population numbers 1, 2, ..., 999, and the D are generated from the $N(50, 50^2)$ distribution, which suggests that the assessor tends to overvalue items on an average by $50. Meanwhile, the standard deviation of $50 suggests (again, cynically) that the assessor is covering his or her tracks by occasionally *undervaluing* items.

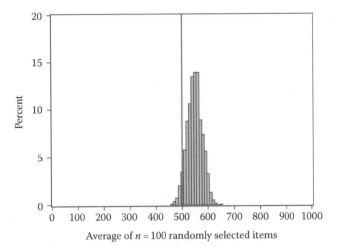

Average of $n = 100$ randomly selected items

FIGURE 7.11
Histogram of sample averages calculated from random samplings of $n = 100$ from the values 1, 2, ..., 999, when the values are assessed with bias.

Figure 7.11 shows that, unlike Figure 7.7, there is a systematic tendency to get estimates that are too large. A systematic tendency to get estimates that are either too large or too small is called **bias**. Figure 7.11 also shows that you need to consider random process effects that are not part of the population, even when you are sampling from a population. It also provides an example for why you shouldn't think that the population determines the $p(y)$ that produces your data: If the $p(y)$ that produced the data were only from the population, there would be no bias. Figure 7.11 also gives a practical take-home message: Understanding process elements helps you to understand that you need your measurements to be as accurate and as unbiased as possible. You should work on your design and measurement system so that the Ds in the measurements $Y = v + D$ are as close to zero as possible.

Example 7.3: E-mail Surveys and Nonresponse Processes

Here is another example showing how design and measurement elements, which are not part of the population, must be considered as part of the $p(y)$ that produces your data. You probably have received a request via e-mail to fill out an online survey. Maybe you filled it out; maybe you didn't—people do not always respond to e-mail requests for filling out surveys. Suppose the population of interest is a collection of $N = 50,000$ e-mail addresses are known, one of which is yours. The random selection process illustrated in Figures 7.3 through 7.5 is followed, and e-mails are sent to $n = 5,000$ of the $N = 50,000$ e-mail addresses. (Your e-mail address may or may not be one of the $n = 5,000$ selected addresses.) One question on the survey is "Do you live in an apartment?" Suppose for the sake of argument that in fact 20,000 out of the 50,000 people, or 40%, do live in an apartment.

Ideally, the random sampling procedure will give an unbiased estimate, in that the proportion of apartment renters found in the survey will not differ systematically from 40%, in repeated random samples from the population of 50,000. If there were no pesky process elements to deal with, this would be true. However, **nonresponse** is a process element that greatly affects the data Y that are produced. Nonresponse is a random element, not part of the population. A person typically makes a snap judgment as to whether to fill out a survey based on many process factors: Do they have time, is the survey interesting, will it take long, do they like taking surveys, and so on. If the act of responding to the survey is related to the measurement, then there will be bias.

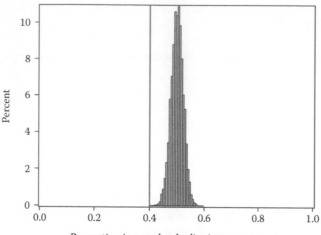

FIGURE 7.12
Histogram of sample proportions obtained by repeated random samplings of $n = 5{,}000$ values from a population of $N = 50{,}000$, 40% of whom live in an apartment, when there is 10% response and when response is related to apartment residency.

For instance, suppose that higher income people are less likely to find the time to fill out a survey. It is reasonable to expect that apartment residency is related to income. So, at the end of the survey collection, you will have a random number of surveys—somewhere around 500 if the response rate is around 10%—and this group will typically have lower income than in the population of $N = 50{,}000$ targeted people, leading to an estimate of apartment residency that is biased high (see Figure 7.12).

The survey example and Figure 7.12 provide another reason that the $p(y)$ that produces the data cannot be determined from the population. The model $p(y)$ that produces the data has process elements, which in this example includes the nonresponse process. Specifically, the distribution $p(y)$ that produces the data you ultimately see is affected by the process data R_i, where R_i is a 0/1 Bernoulli random variable (RV) indicating person i's decision to respond: $R_i = 1$ for response and $R_i = 0$ for nonresponse. These R_i data are process data, not part of the population. The actual data that you collect are the data y_i where $R_i = 1$. Thus, even if you intend to sample n people, the actual number you get will be a random number that is potentially much less than n; specifically

$$\text{Number of responders} = \sum_{i=1}^{n} R_i$$

Further, if the conditional distribution of R given Y depends on the value $Y = y$ (e.g., if the distribution of nonresponse depends on apartment residency), then the data resulting from the survey will provide biased estimates of the population distribution of the y's, as shown in Figure 7.12.

Some statistics sources make a distinction among types of populations to account for the problem of nonresponse. They note that the population of interest differs from the actual population sampled, because not everyone in the population will answer surveys. They use the term *sampled population* to refer to the subpopulation of the original N individuals defined by their willingness to answer surveys. But this subpopulation is completely random: People either fill out a survey or don't on a whim. One day they might fill it out; the next day they might delete it. In the population framework, individuals are continually popping in and out of the sampled population; hence, the sampled population simply does not exist as a definable set of values $y_1, y_2, \ldots, y_N$. The term *sampled population* is misleading and unnecessary once you adopt the process framework. What you really want to model is the process that leads to your actual observed data on apartment living.

Once again, the process framework carries an important take-home message: Nonresponse causes bias. Thus, you should aim to reduce nonresponse bias.

Another take-home message: No matter what you might have heard, there is no need to randomly sample from a population if you can get the entire population. For example, in the case of the e-mail survey, just send out all 50,000 surveys. The only concern is total cost—which includes not only the cost of tabulating survey responses but also the cost of potentially being called a spammer! But there are no requirements from statistical theory that you subsample from the 50,000 rather than sample them all.

A related take-home message is this: You may have heard that if you have the data for the entire population, then there is no need to make statistical inferences, since all the parameters—being population parameters—are completely known. This is also false. Even if you have the population data—and we hope by now that you are skeptical that you will ever really have *the* population, given temporal dependencies and design-based process elements—you still want to learn about the natural processes that produced the population data, and therefore, you still want to make inferences.

This is not to say that random sampling from populations is a bad idea. To the contrary, it is a great idea, with time- and money-saving potential. If you do decide to sample a subset of the population, you should do it via a random sample as shown in Section 7.2 to reduce the possibility of bias. Just don't think that your sampled population defines the known universe, and certainly don't think that it defines the model $p(y)$ that produces your DATA.

Example 7.4: Coffee Preferences of Students in a Classroom

Your classroom environment provides an ideal laboratory for data collection and analysis. Statistics becomes more understandable with real data, especially when *you* are part of the sample. One example is coffee consumption. How much coffee do you drink? Select none, 1 cup or less per week, 2–7 cups per week, 8–14 cups per week, or more than 14 cups per week. Each student gives a selection Y. What is the model $p(y)$ that you can assume to produce these data? Clearly it is *not* a population model such as shown in Table 7.4.

The reason that the model shown in Table 7.4 is wrong is that there is no population with N elements that you are all sampled from. You might try to imagine that you are

TABLE 7.4

An Incorrect Population Model for Coffee Consumption among Students in a Class

Weekly Coffee Consumption, y	$p(y)$
None	N_1/N
<1	N_2/N
2–7	N_3/N
8–14	N_4/N
>14	N_5/N
Total	1.00

TABLE 7.5

A Process Model for Coffee
Consumption among Students

Weekly Coffee Consumption, y	$p(y)$
None	π_1
<1	π_2
2–7	π_3
8–14	π_4
>14	π_5
Total	1.00

all a "random sample from all students," but that clearly is wrong. No one created a list of all students, selected 30 at random (or however many are in your class), and forced them to take this statistics class. Perhaps you might want to view the data in the class *as if* randomly selected? That would be wrong too, because if it were a random selection, then your classmates would have much more diverse class levels, majors, prerequisites, and international statuses. You can try to make the population more specific, for example, "the population of all graduate engineering students," and this is perhaps closer to reality, but then you end up with the same problems as with any population definition: generalizability, temporality, and conditionality—strikes one through four against the population interpretation discussed earlier. There is no possible way to interpret your classroom data as a random sample from a population nor is it useful to do so.

On the other hand, it is simple and useful to view your data as being sampled from a *process*. Year after year, students take classes like this one. The processes that put students in this class include major requirements, general interest, and a need to understand advanced statistical methods. The processes that produce the specific answers to the coffee consumption question include class time (perhaps students consume more coffee when the class is at 8:00 a.m.), culture (people from some cultures might prefer tea over coffee), and even attentiveness (a student might give a wrong answer based on misreading the question or not knowing how much coffee they drink). These processes all have variable elements, and they all work to produce the data that are similarly variable. Before seeing the data, you know that the selections Y will be either none, <1, 2–7, 8–14, or >14, but they are otherwise random. The distribution that you can assume to produce these random numbers is the generic discrete distribution given as shown in Table 7.5.

The parameters $\pi_1, \pi_2, \ldots, \pi_5$ are the unknown *process parameters*. (*Model has unknown parameters.*) They are *not* population parameters.

Example 7.5: Weight of Deer at Different Ages

Suppose you have identified an existing population of deer, and you want to find out their weights. Deer weight is an important indicator of health of the herd. For obvious logistical reasons, a sampling of the deer is preferable to a complete census. Suppose there are $N = 1000$ female deer ("Doe, a deer, a female deer…♪") in the population to be sampled. You collect data on weight (in kilograms) as well as age (in years). The values in the population are (age, weight) pairs. In ascending order of age, suppose the population data are …(1.22, 43.1), (1.22, 42.0), (1.24, 39.5), (1.27, 45.1), (1.27, 44.1), (1.31, 50.0), ….

Using the population-based definition of the conditional distribution $p(y \mid x)$, the conditional distribution of deer weight when age is 1.27 is given correctly as shown in Table 7.6.

TABLE 7.6

Population-Based Conditional Distribution
of Weight of 1.27-Year-Old Deer

y	$p(y \mid X = 1.27)$
44.1	$1/2 = 0.50$
45.1	$1/2 = 0.50$
Total	1.00

TABLE 7.7

Population-Based Conditional Distribution
of Weight of 1.24-Year-Old Deer

y	$p(y \mid X = 1.24)$
39.5	$1/1 = 1.00$
Total	1.00

When age is 1.24, the population-based definition of the conditional distribution $p(y \mid x)$ is given correctly as shown in Table 7.7.

These distributions are correct for the precise population of these 1000 deer at this particular instantaneous moment in time, but no wildlife scientist in his or her right mind would ever consider concluding that "50% of 1.27 year old deer weigh 44.1 kg and the rest of the 1.27 year old deer weigh 45.1 kg" or that "100% of 1.24 year old deer weigh 39.5 kg." The models $p(y \mid x)$ are best understood as models that *produce* the population data, not as models *constructed from* the population data.

From the standpoint of models that produce the data, the wildlife scientist would instead say, "There is a continuous distribution of deer weight among 1.27-year-old deer, and there is a continuous distribution of deer weight among 1.24-year-old deer. These distributions differ only a little because the age difference is so small, but the weights tend to be larger for older deer. The actual distributions are unknown but are estimated using the data as...," and at this point, the scientist would show the audience a couple of graphs of estimated continuous distributions, perhaps normal distributions. The scientist should go on to say "... while these distributions are not constructed from specific population data, the processes that these distributions represent are specific to the time and place of data collection, and these distributions are assumed to produce the population data in question."

The examples in this section show why you should use the process model instead of the population model. But statistics sources have been using the population model for decades, and most students of statistics worldwide have learned it this way. Can the population model really be so bad? Well, actually, the *data analysis* that you do will probably be the same either way. The average of the data values in your sample is the same, no matter whether you choose to view the sample as a subset of a finite population or as the result of a process. The error is in how you *interpret* your data analysis and how you *generalize* from your data analysis. *Data reduce your uncertainty about the unknown parameters.* As we have demonstrated in this section, it is safer to interpret those unknown parameters as characteristics of the data-generating process, rather than as characteristics of a population.

7.4 The Process Model versus the Population Model

To summarize the discussion so far, the population model $p(y)$ refers to a static, existing, never-changing set of observations $y_1, y_2, ..., y_N$, where N is the population size. All probabilities in the population-based $p(y)$ are determined in terms of the values $y_1, y_2, ..., y_N$. By contrast, the process model $p(y)$ refers to the model for the processes that produce DATA; that is, $p(y)$ refers to the model that produces your observable data values Y. These processes include natural processes, as well as design and measurement processes. The probabilities (likelihoods in the continuous case) $p(y)$ govern the frequencies of observable DATA that result from these processes.

Unlike the population definition of $p(y)$, which is always discrete, process distributions can be discrete or continuous. The most important distribution in our universe, the normal distribution, is continuous and is therefore a process distribution.

How can you define your process? To answer, you should envision the DATA, capital D, *before you see it*. Before your collect your data, there are many potential outcomes, which result in potentially many different data sets. Recall the model for the stock market given in Chapter 1, Section 1.11—there are many potential future trajectories the market can take. Recall the reason you wear a seat belt: In your potential future life trajectories, you will die of an auto crash more often if you do not wear a seat belt. To understand the point of view that Nature produces random DATA, whether it be a population or sample, all you need to do is to imagine going back in time: Before you observed the DATA, there were many potential outcomes.

In the absence of design and measurement biases, these potential outcomes arise as a result of the processes collectively called *Nature*. In the example of sampling inventory items, the processes include procurement decisions made at different times, human resource policies, warehousing practices and auctions, depreciation effects, categories of inventory items, and initial purchase prices, all of which themselves have variable future trajectories. These processes affect the potential future inventory DATA you will see. The sources of variability are everywhere. You should think of these processes as the DATA producers rather than the population. The real key to understanding empirical science is to understand the drivers of the variability—that is, the *processes* that produce variability.

Using the process definition of $p(y)$, you can define the probability $\Pr(Y \in A)$ as the percentage of *potential* outcomes where you will observe a Y in the set A. So, unlike the population definition of $p(y)$, you will not assign zero probability to plausible events, such as the event that income lies in the range \$1,002,122.12–\$1,002,122.89. Instead, using process definition of $p(y)$, there is a small, but nonzero, percentage of potential observable incomes whose values lie in the range \$1,002,122.12–\$1,002,122.89. This probability is given by the equation

$$\Pr\left(1,002,122.12 \leq Y \leq 1,002,122.89\right) = \int_{1,002,122.12}^{1,002,122.89} p(y)dy$$

This assumes your process model is continuous; otherwise, replace the $\int$ by Σ and remove the dy.

An easy way to understand the difference between *process* and *population* is to think of *potential* DATA versus *existing* data. The simplest example of a process is the coin flip—you can toss a quarter repeatedly, and all kinds of potential futures are possible, like H H T T H ... or T T T H H ... or T T H T T .... By contrast, an example of a population is the collection of all the coins in your change jar. There might be 43 quarters and 102 other coins, for a total of $N = 145$ coins, with values $y_1, y_2, ..., y_{145}$, each y either an H or a T, and that's it. In the population framework, the population of N items is static, frozen in time, never changing.

A second distinction between process and population is in the *infinite* versus the *finite*. In the case of processes, there are infinitely many potential sample sequences, whereas in the case of a population, there are only finitely many.

A third way to understand the process/population distinction is in terms of *prediction* or *generalization*. Process models are concerned with predicting or generalizing to outcomes that may come in the future or that could have come in the past. As discussed in Chapter 1, prediction is not necessarily about the future; it might concern what-if statements about what could have happened in the past, or even what might be happening in the present that you are otherwise unable to observe directly. Process models are for prediction, although not necessarily for the future. Population models, on the other hand, are always about an existing finite population at a precise point in time and provide no ability whatsoever to predict or generalize beyond those confines.

7.5 Independent and Identically Distributed Random Variables and Other Models

Simulations done so far in this book have used random number generators. The model for the data produced by random computer number generators, and for real process data such as the flips of coins, rolls of dice, and outcomes of a casino's roulette wheel, is that the DATA are **iid**. This is an important model since it forms the logical basis for most of the calculations performed by your statistical software. When the iid assumption is violated, the computer's calculations are simply wrong, to more or less of a degree, depending upon how badly violated is the assumption.

You need to verify two conditions before you can state that a sequence of RVs $Y_1, Y_2, ...$ is in fact an iid sequence.

Conditions Needed for a Sequence of RVs to be iid

1. All the RVs $Y_1, Y_2, ...$ must be independent.
2. All the RVs $Y_1, Y_2, ...$ must be produced by the same identical distribution, $p(y)$.

The coin flip is the simplest example. The coin tosses all come from the same distribution, called Bernoulli(0.5), with $p(y)$ as given in list form in Table 7.8 in the case of a fair coin.

TABLE 7.8

Distribution That Produces Fair Coin Outcomes

y	$p(y)$
H	0.5
T	0.5
Total	1.0

The coin tosses are also effectively independent, provided there is no trickery on the part of the coin flipper: The distribution of the second toss is the same Bernoulli distribution shown in Table 7.8, no matter whether the first toss is H or T.

The coin can be bent, so you don't know the probability of H, but that's okay. The tosses are still iid Bernoulli(π). The distribution that produces the data has an unknown parameter, π. By collecting data, you can reduce your uncertainty about this unknown parameter. *Data reduce the uncertainty about the unknown parameters.* Sounds familiar?

Successive rolls of a die are also iid, again assuming no mischief on the part of the die roller. In the case where the die is perfectly balanced, the distribution $p(y)$ that produces the data is the discrete uniform distribution, with probabilities 1/6 on each of the values 1, 2, 3, 4, 5, and 6.

RVs can be identically distributed but not independent. For example, suppose you roll a die once and call the outcome X_0. Then roll it three more times, calling the outcomes X_1, X_2, and X_3. Now let $Y_1 = X_0 + X_1$, let $Y_2 = X_0 + X_2$, and let $Y_3 = X_0 + X_3$. Then Y_1, Y_2, and Y_3 are identically distributed, all having the distribution $p(y)$ that is the distribution of the sum of two dice, shown in Figure 7.13. This distribution shows that 7 is the most likely total, while 2 and 12 are the least likely. Details of how you can get this distribution are shown in Chapter 9 (pp. 229–230).

But Y_1, Y_2, and Y_3 are not independent. For example, if you knew $Y_1 = 12$, then you could infer with certainty that $X_0 = 6$. Therefore, the conditional distribution of $Y_2 (= X_0 + X_2)$, given $Y_1 = 12$, is the same as the distribution of $6 + X_2$, since $Y_1 = 12$ implies $X_0 = 6$. This conditional distribution is graphed in Figure 7.14.

Since the conditional distribution of Y_2 given $Y_1 = 12$ shown in Figure 7.14 differs from the marginal distribution of Y_2 shown in Figure 7.13, the variables Y_1 and Y_2 are dependent.

You can apply the model of dependence suggested by the dice sums $Y_i = X_0 + X_i$ to **cluster sampling**. In cluster sampling, you first randomly sample clusters, and then you collect *all* the data within a cluster. For example, in an opinion poll, the clusters may be households, which you randomly sample and then in which you survey all the adult members of the household. People in the same household often share opinions and ideologies, and the dice model helps to explain why their data are dependent. Just like the sums $Y_2 = X_0 + X_2$ and $Y_1 = X_0 + X_1$ share the common value X_0, making Y_1 and Y_2 dependent, people within a household share common traits that make their data similarly dependent. You must use

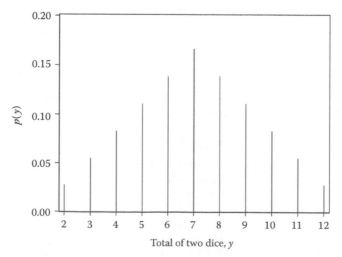

FIGURE 7.13
Distribution of the sum of two dice.

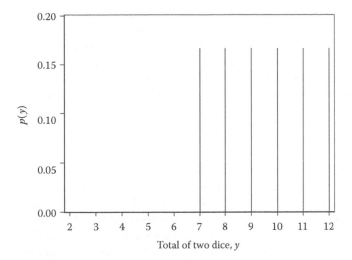

FIGURE 7.14
Conditional distribution of the sum $Y_2 = X_0 + X_2$ of two dice, given that $Y_1 = X_0 + X_1 = 12$.

special statistical models that account for such dependencies when analyzing clustered data; otherwise, the conclusions from your data analysis may be incorrect.

As it turns out, a *without replacement* random sample from a finite population also produces identically distributed but dependent data. Consider your coin jar having $N = 145$ coins, with 43 quarters and 102 other coins. Select a random sample of 10 coins from the jar and let $Y_1, Y_2, \ldots, Y_{10}$ be the binary measurements (quarter, other). Then the marginal distribution of each Y_i is as shown in Table 7.9.

The conditional distributions differ from what you see in Table 7.9. For example, given $Y_1 = $ quarter, then the conditional distribution of Y_2 is as shown in Table 7.10.

Thus, the data obtained from random sampling are not independent: Knowing the value of one of the data values changes the distributions of the other data values. However, notice

TABLE 7.9

Marginal Distribution of Y_i That Is Part of a Random Sample Selected from a Jar with 43 Quarters and 102 Other Types

y	$p(y)$
Quarter	$43/145 = 0.297$
Not a quarter	$102/145 = 0.703$
Total	1.000

TABLE 7.10

Conditional Distribution of Second Coin Selected from the Jar, Given the First Coin Selected Was a Quarter

y	$p(y)$
Quarter	$42/144 = 0.292$
Not a quarter	$102/144 = 0.708$
Total	1.000

that the dependence is slight because the distributions change very little. With larger population sizes (populations are usually much larger than $N = 145$), there is less dependence, and the observations are even closer to independent. This issue shouldn't concern you too much, because the dependence is slight and because the population definition of $p(y)$ is itself flawed. In the *process* definition of $p(y)$, you can usually ignore dependencies induced by without replacement sampling. Let's set that off so you can see it clearly:

> Despite potential (slight) dependencies, it is usually safe to model a random sample from a population as an iid sample from a process.

The examples earlier with the dice and with the quarter are cases of samples that are identically distributed but not independent. It is also possible to have samples that are independent but not identically distributed. For example, roll a die 100 times. Let $Y_1, Y_2, ..., Y_{50}$ be the first 50 rolls, and let $Y_{51}, Y_{52}, ..., Y_{100}$ be the results of adding the number 10 to each of the last 50 rolls. Then all 100 observations are independent, but the distributions are different: The first 50 come from the discrete uniform distribution on the numbers 1, 2,..., 6, and the second 50 come from the discrete uniform distribution on the numbers 11, 12, ..., 16.

This model of independent but not identically distributed data is applicable to the type of experiment known as a **two-sample comparison**. A common experiment is to randomly divide people (or other experimental units, like plots of land, production runs, stores) into two groups, apply different **treatments** (such as a pharmaceutical drug and a placebo) to the groups, and then compute the difference between the within-group averages. The totality of the data in the two samples can be modeled as independent but not identically distributed. If the treatment has an effect, then the distribution that produces the data in one group differs from the distribution that produces the data in the other group. The difference between these distributions is called a **treatment effect**, which is the number 10 in the dice example, but is an unknown process parameter in real life. The observed data from your experiment will reduce your uncertainty about this parameter. (*Data reduce the uncertainty* ... you know the rest.)

A common method for analyzing such two-sample data is via the *two-sample t test*, discussed in Chapter 16.

Example 7.6: Are Students' Coffee Preference Data iid?

As discussed in Example 7.4, the distribution in Table 7.11 is a reasonable model for producing students' coffee consumption.

TABLE 7.11

A Process Model for Coffee Consumption among Students

Weekly Coffee Consumption, y	$p(y)$
None	π_1
<1	π_2
2–7	π_3
8–14	π_4
>14	π_5
Total	1.00

Is it reasonable to assume that the student data are an iid sample from the distribution shown in Table 7.11? Without further information, the answer seems to be that, "yes, it is reasonable to assume that the students' coffee preferences are produced as an iid sample from this distribution." You can assume the observations are independent—why should knowledge of Hans' coffee consumption tell you anything about Mei's coffee consumption? There can be dependence if there are married couples in the class, where the coffee preference of one spouse provides information about the other's preference. There can also be dependence if a group of students all decide to answer the question with the same number. But barring such cases, independence is a reasonable assumption. Further, since the parameters $\{\pi_1, \pi_2, \pi_3, \pi_4, \pi_5\}$ can be any numbers whatsoever (provided that they are positive and add to 1.0), it is difficult to argue that the generic distribution $p(y)$ shown in Table 7.11 is somehow wrong. Hence, the assumption that the coffee data are iid from this $p(y)$ seems to be a reasonable assumption.

But wait a minute! Are the student observations really *identically distributed*? If so, then the same distribution $p(y)$ shown in Table 7.11 produces coffee consumption for *everyone in the sample*. What if half of the students are from the United States and half are from Great Britain? With a cultural tendency toward tea consumption, there might be lower coffee consumption among the Brits, and therefore, the distributions would not be the same. Instead there would be two distributions—one for Great Britain, say, $p_{GB}(y)$, and one for the United States, say, $p_{USA}(y)$. Carrying the logic further, if you knew the age of the student, that would tell you even more—older students perhaps have more of a liking for coffee, since the younger ones might prefer the convenient energy drinks. You might argue that the data can be assumed independent, but with a *different distribution for every student*, depending on the characteristics of the student!

If this is the argument you want to make, good for you! You are thinking very clearly and correctly. The resolution is simple: You can assume the coffee preference data are iid *marginally*. That is, without knowing any other information about the student, you can assume the coffee ratings all come from the same distribution. Once you know something about the student, however, the coffee rating distributions change. (*Use what you know to predict what you don't know.*) In other words, the conditional distribution of coffee preference, Y, depends on data about the student, X. The student's X data may include country of origin, age, sex, and a host of other variables. Given the data X on each student, you can assume the observations are independent, but not identically distributed: The distributions depend on the value of the students' X data. This model is the *regression model* that you have seen already and will continue to see in this book.

7.6 Checking the iid Assumption

The assumption that data, or at least some portions of the data, are from iid processes is fundamentally important for most advanced statistical methods. If the process that produces your data differs dramatically from iid, then such methods cannot be trusted. Thus, you should investigate whether the process that produced your data can be reasonably modeled as iid.

You can safely model a perfect random sample from a population as an iid sample. However, in the real world, samples from populations are not perfect random samples. For example, with nonresponse that is related to the measurement, as in the apartment housing example discussed earlier in this chapter, there can be a tendency for later response DATA to differ in distribution from earlier response DATA, making the distribution of the later data different from the distribution of earlier data, and hence, the data would not be iid.

Example 7.7: Non-iid Responses to an E-Mail Survey

Consider the e-mail survey that asks respondents whether they live in an apartment. If the nonresponders tend to be house residents rather than apartment residents, then on the initial round of e-mail requests, the data will be mostly apartment residents. But after a time, perhaps the initial nonresponders get bored and decide to fill out the survey. If this happens, then you will see a time trend in the responses Y to the question of apartment residency.

To assess conformance with the iid assumption, you can use the techniques discussed in Chapter 5 for estimating conditional distributions. In this example, you want to see whether the distribution of housing (apartment, home) changes over time. The simplest tool is a bar chart to represent housing as a function of time from initial e-mail solicitation, using time intervals that allow adequate data in each group. Figure 7.15 shows how these might look.

As shown in Figure 7.15, there is a tendency for later responders to be house owners. Thus, the survey response data $Y_1, Y_2, \ldots$ that are collected in time sequence appear non-iid. The consequence is that there is a strong potential for bias: If the remaining nonresponders follow the trend shown in Figure 7.15, then their distribution will show even more house ownership. Since they didn't respond, you would be missing their data, and the apartment residency estimate would be too high as a result, as shown in Figure 7.12.

If Figure 7.15 showed little difference between those three distributions, then the iid assumption would be more reasonable.

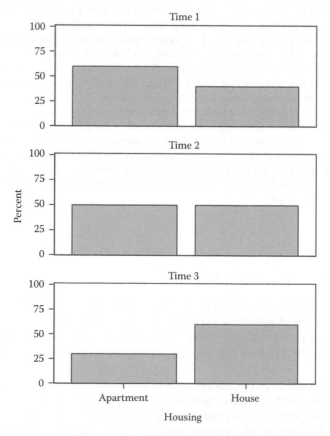

FIGURE 7.15
Distributions of housing (apartment or house) for early (time 1), middle (time 2), and late responders (time 3), showing different—not identical—distributions.

Other indications that the data are not produced as iid include clustering effects, where clusters of observations share a commonality. To see this type of non-iid character, you can plot the data using a scatterplot where the observation number is on the horizontal axis and the observed data are on the vertical axis. Assuming the observations in clusters have consecutive observation numbers, Figure 7.16 shows the appearance of cluster effects, indicating that the iid model is not a valid model for how the data were produced.

The examples described using Figures 7.15 and 7.16 concern **cross-sectional data**, where the data provide a snapshot of Nature at a particular point in time. With cross-sectional data, it makes some sense to use the population terminology, particularly when the items are indeed randomly sampled from some larger collection of items. However, as discussed previously in this chapter, the process terminology is still preferable, for a variety of reasons.

However, with **time series data**, where the data are collected sequentially in time, the process terminology is simply correct, and the population terminology is simply incorrect. For example, consider Example 5.1 concerning stock return data collected over successive trading days. It makes no sense to think of those returns as being randomly selected from some finite population of N returns. You might like to think that the population of returns would simply be all the historical returns, but this would be incorrect: Tomorrow's return is not randomly selected from the past historical returns; instead, tomorrow's return is produced by financial processes. The same financial processes that will produce tomorrow's return also produced yesterday's return. The population model makes no sense whatsoever for time series data.

Process models *are* appropriate for time series data. However, the iid process model might not be correct—you often need to use process models that allow for dependence. With time series data, you should always expect some degree of dependence due to **adjacency**: Observations that are closer to one another in time tend to be more similar than observations that are far apart in time. Think of your weight, for example. Your weight today is probably different only by a fraction of a kilogram from your weight yesterday. However, your weight today probably differs much more from your weight 5 years ago.

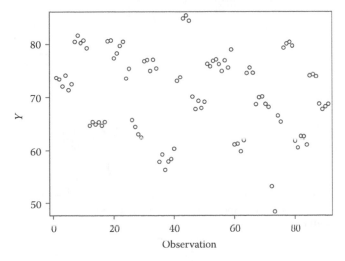

FIGURE 7.16
Scatterplot of data in observation sequence showing non-iid character due to cluster effects.

Adjacency applies to cluster sampling as well: Observations within a cluster are closer to one another (more adjacent) than observations in different clusters, causing dependence.

Adjacency also applies to physical space: Pairs of observations that are closer to one another in geographical coordinates tend to be more similar to one another than pairs of observations that are distant. Think of home sales prices, for example. The prices of homes in your neighborhood are more important for determining the price of your home than are prices of homes across town. When geographical information (e.g., longitude and latitude) is included in your data set, your data are called **spatial data**.

Indeed, dependencies induced by adjacency occur in many types of data.

Example 7.8: Detecting Non-iid Characteristics
of the Dow Jones Industrial Average (DJIA)

In Chapter 5, you saw that Dow Jones *returns* are approximately independent of previous returns. What about the index itself? Independence of the returns does not imply independence of the actual index values; some math shows why. Let R denote the return and I denote the actual index value. Then $R_t = (I_t - I_{t-1})/I_{t-1}$, by definition, implying that $I_t = I_{t-1} + R_t \times I_{t-1}$, by algebra. The latter equation shows that the index at time t (e.g., today) is related to the index at time $t - 1$ (e.g., yesterday) and suggests nonindependence. To verify this statistically, you can use a **time sequence plot** of the Dow Jones index as shown in Figure 7.17.

The adjacency effects are apparent in Figure 7.17 because the data values that are 1 day apart tend to be closer to one another than data values that are months apart. Look at the vertical distances between data points in Figure 7.17 to see the adjacency effects: The vertical distances tend to be smaller when the dates are close together, and the vertical distances tend to be larger when the dates are farther apart.

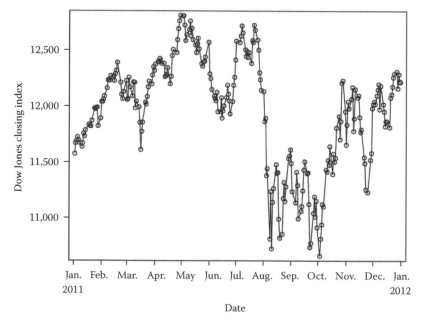

FIGURE 7.17
Time sequence plot of daily Dow Jones closing index values from January 3, 2011 through December 30, 2011.

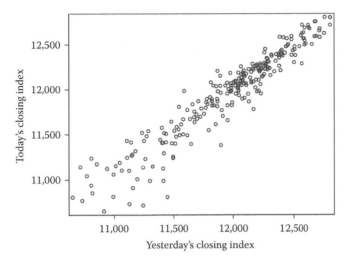

FIGURE 7.18
Lag scatterplot of today's closing index versus yesterday's closing index.

For another look at the adjacency effect, you can draw a scatterplot of today's value (on the vertical axis) versus yesterday's value, called the **lag** of the data value, on the horizontal axis. If today's data were independent of yesterday's data, then there should be no trend in the plot. Figure 7.18 shows that graph, called a **lag scatterplot**.

As seen in Figure 7.18, there is a strong dependence between today's and yesterday's values; hence, they are not independent. Figure 7.18 also shows clearly the relationship that is implied by the equation $I_t = I_{t-1} + R_t \times I_{t-1}$. Letting $D_t = R_t \times I_{t-1}$, the equation may be rewritten as $I_t = I_{t-1} + D_t$ and is exactly what Figure 7.18 shows: Today's index value is equal to yesterday's value plus a small deviation.

Recall the graphs of distributions of Dow Jones *returns* shown in Chapter 5 (Figures 5.5 and 5.6). We made the case there for *approximate independence* since the distributions of today's *return* differ little, if at all, depending on whether yesterday's return was low or high. It may be surprising that there is such a big difference between the actual Dow Jones index values and their returns: The index values are *strongly dependent*, while the returns are *nearly independent*.

Using the same method as displayed in Figure 5.5, you can look at the distribution of today's index depending on whether yesterday's index was either low or high. From Figure 7.18, you could say that values less than 11,750 are low and the rest high. Figure 7.19 shows the histograms of today's index, separated by cases where yesterday's index was either low or high.

Unlike the Dow Jones index *returns*, the distribution of *actual index* itself is strongly dependent on the value of the previous day's index.

In summary, we have shown the following methods.

Four Graphical Methods for Checking the iid Assumption

1. Construct separate histograms (continuous case) or bar charts (discrete case) for earlier observations and later observations. Gross dissimilarity suggests nonidentical distributions (see Figure 7.15).
2. Construct separate histograms (continuous case) or bar charts (discrete case) of the data for cases where the previous observation was high and where the previous observation was low. Gross dissimilarity suggests nonindependent observations (see Figure 7.19).

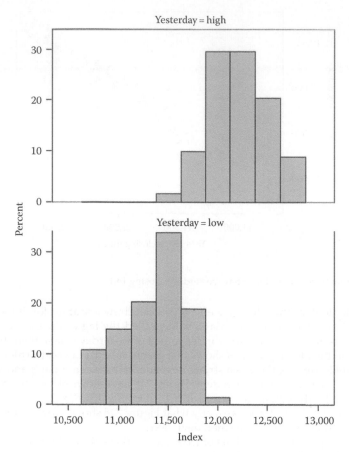

FIGURE 7.19
Histograms of closing 2011 Dow Jones index values, separated by cases where yesterday's index was low (<11,750) or high (≥11,750).

3. Construct a time sequence plot of the data. Look for evidence of adjacency effects, either from cluster or time sequence, to suggest nonindependence. Look for trends to suggest nonidentical distributions (see Figures 7.16 and 7.17). This plot is less useful when the data are highly discrete.
4. Construct a lag scatterplot. Look for evidence of trend suggesting noninde-pendence of current and previous data (see Figure 7.18). This plot is less useful when the data are highly discrete.

As stated in Chapter 5, you can't prove independence using data. Neither can you prove identical distributions. All you can hope for is reasonable conformance to expected appearances in these graphs, which would then suggest that the iid model is reasonable. In particular, when interpreting these four graphs, *do not expect perfection*. There will be deviations from the expected appearances due to randomness alone, even when data are in fact produced as an iid sample.

The best way to understand randomness is to simulate data. You can trust that random number generators produce iid data, and hence, you can use simulated data to understand the role randomness plays in the four graphs described above. Just remember that the sample size *n* plays a crucial role in determining how much variation is explainable by chance alone. To understand the role of chance variation in your data, you need to use the same n in your simulation study as the n of your data.

Example 7.9: The Appearance of the Diagnostic Graphs in the iid Case

Simulate a sample of $n = 100$ observations from the $N(70, 10^2)$ using the computer. (Yes, do it!) These data are indeed produced as an iid sample from a distribution; therefore, all deviations from the expected appearances are *explained by chance alone*. How do the four graphs look in this case? (see Figure 7.20). You should construct the same graphs with your data, and you will see that your graphs look slightly different from ours. Good! Now you understand the effects of chance variation even better.

Figure 7.20 from simulated data shows differences in the estimated distributions that are completely explained by chance alone. The corresponding Figure 7.15 from real data suggests that the distributions differ, but you should also consider the question, "Are the differences seen explainable by chance alone?" This question can be answered using simulations such as shown in Figure 7.20 but using the same n as the original data and choosing a model $p(y)$ that produces DATA* similar to the real DATA.

As in Figure 7.20, Figure 7.21 from simulated data shows differences in the estimated distributions that are completely explained by chance alone. The corresponding Figure 7.19 from real data suggests that the distributions differ, but you should also consider the

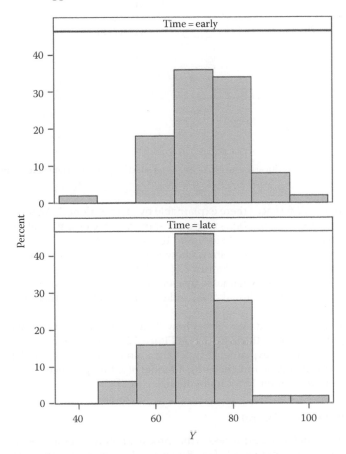

FIGURE 7.20
Histograms of 100 data values produced as iid $N(70, 10^2)$, separated by earlier observations (the first 50) or later observations (the second 50).

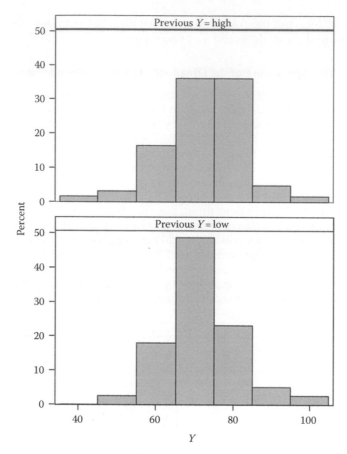

FIGURE 7.21
Histograms of 100 data values produced as iid N(70, 10²), separated by previous data value high (≥70) or low (<70).

question, "Are the differences there explainable by chance alone?" Again, this question can be answered using simulations using the same n as the original data and choosing a model $p(y)$ that produces DATA* similar to the real DATA.

Figure 7.22 from simulated data shows how data produced as iid look when graphed in order of appearance. Compare Figure 7.22 with the real data in Figures 7.16 and 7.17. Are the patterns seen there explainable by chance alone? You can answer using simulation: Use the same n as the original data, and choose a model $p(y)$ that produces DATA* similar to the real DATA.

Figure 7.23 from simulated data shows how data produced as iid look when you construct a lag scatterplot from them. Compare Figure 7.23 with the real data in Figure 7.18. Is the pattern seen there explainable by chance alone? (Answer: Simulate, using the same n as the original data and choosing a model $p(y)$ that produces DATA* similar to the real DATA.) See Chapter 15 for more details.

Example 7.10: Quality Control

In manufacturing and services processes that produce repetitive outcomes, consistency is desirable. A consistently high level of customer satisfaction is desirable in service industries. Manufactured products also need consistency, especially if the products require high precision as is the case, for example, with medical devices. Always, however, some

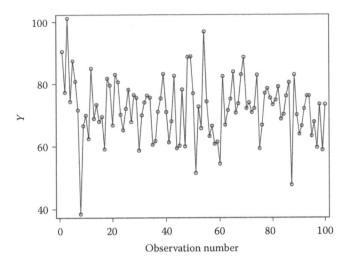

FIGURE 7.22

Time sequence plot of 100 data values produced as iid $N(70, 10^2)$.

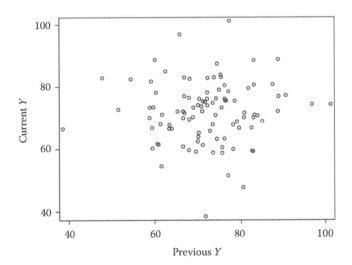

FIGURE 7.23

Lag scatterplot showing 100 data values produced as iid $N(70, 10^2)$ graphed against the previous value.

variation is unavoidable. Variation is everywhere; you can't get away from it. Even highly precise manufacturing machines produce items that differ slightly, from one to the next.

To the extent possible, it is desirable to control variation. Highly variable outcomes in manufacturing are a source of inefficiency and waste. Highly variable outcomes in service can cause loss of customer confidence.

One chart used to analyze quality control data is called an **individual chart**. It is a chart of the data observed in time sequence, the same as shown in Figures 7.16, 7.17, and 7.22. Non-iid characteristics are indications that the process is out of control and needs adjustment. For example, if there is a shift in the mean level of the observations for one shift of workers versus another, it suggests that perhaps the shift manager needs retraining.

The following data are measurements of computer chip width (in miniscule units) in time sequence: 311, 304, 316, 308, 312, 308, 314, 307, 302, 311, 308, 300, 316, 304, 316, 306,

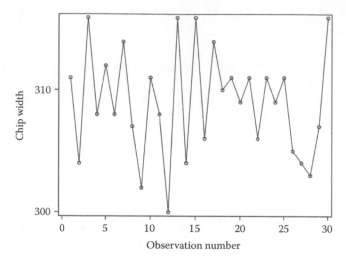

FIGURE 7.24
Individual chart of widths of manufactured computer chips.

314, 310, 311, 309, 311, 306, 311, 309, 311, 305, 304, 303, 307, and 316. The individual chart is shown in Figure 7.24.

There are no obvious indications that the data were produced by a non-iid process, and hence, no indication that the process is out of control. The other three graphs for checking the iid assumption can and should be examined; these are left for the exercises.

Vocabulary and Formula Summaries

Vocabulary

Sample	A data set, usually random because one data set differs from another.
Population	A finite, static, and fixed (nonrandom) set of numbers.
Random sample from a population	A set of items (the sample) that is a randomly selected subset of a larger set of items (the population).
With replacement sample	A sample where an item is sampled and then returned to the population, where it might be sampled again.
Without replacement sample	A sample where an item that is sampled cannot be sampled again later.
Unbiased estimate	An estimate that is neither systematically too large nor systematically too small.
Bias	A tendency for an estimate to be either systematically too large or systematically too small.
Biased estimate	An estimate that is either systematically too large or systematically too small.
Sampling distribution	The probability distribution of a statistic calculated from a DATA set.

Superpopulation model	The model which states that population data are themselves produced at random.
Latent variable	A variable whose values are not measured directly, a hidden outcome.
Nonresponse	The act of not responding to a survey or other data collection procedure; this can be a source of bias.
iid	A supposed property of a collection of RVs that states they are all (i) independent and (ii) produced by the same (identical) distribution, $p(y)$.
Cluster sampling	Sampling of clusters (e.g., households), where all observations within the cluster are obtained.
Two-sample comparison	A design where there are two groups (one example is drug/placebo, another example is male/female), and the goal is to compare the groups.
Treatment	An assignment of a particular condition in an experiment, e.g., in a clinical trial, one treatment is a drug, and the other is a placebo.
Treatment effect	The difference between the data in different treatment groups.
Cross-sectional data	Data used to provide a snapshot at a particular point in time.
Time series data	Data in consecutive time sequence.
Adjacency	The quality of being nearby, whether in time, space, or by other commonality such as familial tie.
Spatial data	Data that include geographical information.
Time sequence plot	A graph of data where the horizontal axis is time and the vertical axis is the data value. Sometimes the data values are connected with lines for easier visualization.
Lag	The previous value in time. If the current data are Y_t, then the lag of the current value is Y_{t-1}.
Lag scatterplot	A scatterplot of data at time t (vertical axis) against the previous data value at time $t-1$ (horizontal axis).
Individual chart	A graph of the data (on the vertical axis) versus time sequence (on the horizontal axis). Identical to a time sequence plot.

Key Formulas and Descriptions

n	The sample size.
N	The population size.
N_j/N	The population probability that an item of type j will be selected, when there are N items in the population, N_j of which are of type j.

$p(y) = \#\{y_i = y\}/N$, if $y = y_i$ for some $i = 1, 2, \ldots, N$; $p(y) = 0$ otherwise

The population-based definition of the probability model $p(y)$.

$\Pr(Y \in A) = \#\{y_i \in A\}/N$

The probability that the measurement Y lies in the set A when using the population-based probability model.

$Y = v + D$

An observed measurement Y is the true value v plus a deviation D from the true value.

Exercises

7.1 Use the data from Exercise 4.7.

   A. Describe, step by step, how you will obtain a *without replacement* sample of $n = 5$ observations from that data set.

   B. Obtain the sample and calculate the average of the five numbers you sampled.

   C. Compare the average of the five numbers you sampled with the average of the entire data set, and explain why those two numbers are different.

   D. Repeat Exercises 7.1A through C, but for a *with replacement* sample.

7.2 Students in a class are asked how often they read online newspapers. They answer using a 1, 2, 3, 4, 5 scale, where 1 = "never" and 5 = "every day." Their data are as follows: 4, 5, 1, 1, 5, 3, 1, 4, 1, 5, 3, 5.

   A. If this is a population, what is N?

   B. Give the population $p(y)$ in list form.

   C. Give the process $p(y)$ in list form. (Note: Process precedes the population.)

   D. Explain why the process interpretation of $p(y)$ in Exercise 7.2C is preferable to the population definition in Exercise 7.2B.

7.3 A quality control inspector wants to know the percentage of rotten potatoes in a truck shipment. She decides to take a sample of potatoes from the truck.

   A. What is N?

   B. What is n?

   C. Suppose the sample is taken from the top of the truck bed using a scoop. Describe process elements that can bias the quality control inspector's estimate.

   D. Explain how the lesson learned from Figure 7.12 applies to estimation of the percentage of bad potatoes using the method in Exercise 7.3C.

7.4 There is a population of $N = 1000$ deer that have been previously tagged. You randomly sample one of them and get its weight, Y, in kilograms. Explain the meaning of the following terms:

   A. $\Pr(79.2 < Y < 79.7)$, using the population interpretation of probability. Be sure that $N = 1000$ is part of your answer.

   B. $\Pr(79.2 < Y < 79.7)$, using the process interpretation of probability.

   C. Explain why the process interpretation of probability given in Exercise 7.4B is preferable to the population interpretation of probability in Exercise 7.4A.

7.5 It can be surprising how the sampling fraction has so little to do with the accuracy of estimates from samples of populations. The sample size n is mainly what matters. Redo the analysis that gives Figure 7.7 but using the list of 99,999 values 0.01, 0.02, ..., 999.99 for the population. The mean of the population is still 500, and the range is essentially the same. With 99,999 values in the population, and $n = 100$ in the sample, your sampling fraction is 100/99,999 = 0.001, 100 times less than the sampling fraction in the analysis leading to Figure 7.7. Are your estimates of the mean 100 times less accurate? Answer by comparing your histogram to that shown in Figure 7.7.

7.6 On many days, you will go from your home to the same place, be it work, school, the grocery store, or somewhere else. Let Y be the time it takes you to get there, measured precisely. Your number Y is variable—it is never exactly the same from 1 day to the next.

A. Describe all the process elements that contribute to the variability in your Y.

B. Explain why the population model for your Y is wrong. Start with a question about N.

7.7 Figure 7.20 is one of the four graphs that were suggested for checking the iid assumption with the quality control data. Three other graphs were suggested as well.

A. Construct and interpret the other three graphs.

B. Are the patterns in the graphs explainable by chance alone? Describe in detail how you can address that question by using simulations where the differences are *explained* by chance alone.

7.8 Use an example of your choosing where there is a Y and an X, where the X is reasonably continuous. You might search the internet for "effect of _____ on _____," filling in the blanks with terms that interest you. The first blank will be your X, and the second blank will be your Y. Choose an X that can be modeled as a continuous RV.

A. Identify the two specific x values of interest to the study, and explain why you chose those values. Draw (subjectively) graphs of two distributions $p(y|x)$ for this example, one for each of the two different values of x that you chose. Put numbers on all horizontal and vertical axes. Explain why you drew the graphs the way you did.

B. Explain the population interpretation of the models $p(y|x)$ in your example. Be sure that there is an N in each of your interpretations.

C. Explain the process interpretation of the models $p(y|x)$ in your example.

D. Explain why the process interpretation of your models $p(y|x)$ is preferred to the population interpretation of your models $p(y|x)$ in your example.

7.9 Bruce struggles with alcoholism. Every time Bruce goes see the doctor, his doctor asks how many alcoholic drinks he has had in the last week. Bruce reports the number Y.

A. Explain the meanings of the terms v and D in the latent variable model $Y = v + D$, in terms of Bruce's actual and reported drinking.

B. Do you think the D in the latent variable model is (1) always equal to zero, (2) always less than zero, (3) always more than zero, (4) usually less than zero, or (5) usually more than zero? Explain why you think that way. There isn't a single right answer; just be logical in your thought process and in your attempt to understand Bruce.

C. Explain why the model $Y = v + D$ is a process model and not a population model.

7.10 Suppose you roll a die successively, getting an iid sample $X_1, X_2, X_3, ..., X_n$ from the discrete uniform distribution on the numbers 1, 2, ..., 6. Then you calculate the successive averages $Ave_1 = X_1/1$, $Ave_2 = (X_1 + X_2)/2$, $Ave_3 = (X_1 + X_2 + X_3)/3$, ..., $Ave_n = (X_1 + X_2 + X_3 + ... + X_n)/n$. Without formal math calculation, give sensible arguments to justify your answers to Exercise 7.10A and B.

   A. $Ave_1, Ave_2, Ave_3, ..., Ave_n$ are not independent.

   B. $Ave_1, Ave_2, Ave_3, ..., Ave_n$ are not identically distributed.

   C. Assume $n = 5$. Use many (say 20 or more) repeated rolls of $n = 5$ dice, or use computer simulation to verify your answer to Exercise 7.10B.

7.11 Let X = a person's income and let Y = a person's housing expense. Explain why X, Y are (1) not independent and (2) not identically distributed. Don't refer to any external variables.

7.12 Suppose a random sample of three police department applicants gives Y_1, Y_2, Y_3, measures of their BMI (body mass index).

   A. Can you assume that Y_1, Y_2, Y_3 are independent? Explain.

   B. Can you assume that Y_1, Y_2, Y_3 are identically distributed? Explain.

8

Expected Value and the Law of Large Numbers

8.1 Introduction

If you flip a coin 10 times, you *expect* to get 5 heads. If you flip a coin 1000 times, you *expect* to get 500 heads. If you invest 100,000 in a stock that has had a consistent 5% annual return, you *expect* to earn 5,000 after a year. But you won't necessarily get 5 heads in 10 flips, it is very unlikely that you will get 500 heads out of 1000 flips, and it is extremely unlikely that your earnings will be precisely 5000 on the investment. *Expected value*, therefore, is not actual observed data; rather, it is kind of an average. An observed statistic may be higher than its expected value or it may be lower. Rarely—if ever—is the observed value equal to the expected value. Yet the expected value is useful for making decisions everywhere, no matter whether you are interested in how to invest your money, how to design an automobile tire, how to treat a hospital patient, or how many hunting licenses to issue. In this chapter, we clarify what precisely is meant by *expected value* and provide applications.

In case this concept slipped by, please recall that your statistical model is as follows:

A distribution $p(y)$ produces your data Y.

(Model produces data.)

Expected value is *never* from your data. Instead, expected value is a property of the model $p(y)$ that you assume to produce your data. The expected value is already there, whether or not you ever collect any data. In most cases, you don't know its numerical value because it's one of those unknown parameters (*model has unknown parameters*). But you can estimate the expected value by using the average of n independent and identically distributed (iid) data values that are produced by $p(y)$. The *law of large numbers* (LLN) states that as the sample size n increases, the averages of iid data get closer to the expected value (*data reduce the uncertainty about the unknown parameters*).

8.2 Discrete Case

Who pays for all those glittering lights in Las Vegas, Nevada? You do! Well, if you're a gambler, you do. The following example shows why.

Example 8.1: Roulette Winnings

In the game of American-style roulette, there are 38 slots. An attendant—called a *croupier*—spins the wheel in one direction and spins a white marble in the opposite direction. The marble will land in one of the slots at random. See Figure 8.1.

If you bet 10 (U.S. dollars, Euros, etc.) that the ball lands in a red slot, your earnings will be +10 if the ball lands in a red slot, and your earnings will be −10 if the ball lands in a black or green slot. There are 18 red slots, 18 black slots, and 2 green slots, so the probability distribution of your earnings Y, per play, is given as shown in Table 8.1.

What will your average earnings be? You might be tempted to say the average is $(-10 + 10)/2 = 0/2 = 0.00$, because that's how you compute the average of the numbers in the y column. This calculation suggests that, on average, you earn nothing and lose nothing. But if you lose nothing, the casino also earns nothing. So where do they get all that money for lights? Who pays for all those blackjack dealers? How do they afford to give cheap rooms and cheap buffets? Clearly, something is wrong with your calculation that the average earnings are zero.

Think of it another way. If you play 1000 times, then you *expect* (there's that word again) to lose 526 times, and you *expect* to win 474 times. Although your actual results will vary, in this hypothetical scenario, the results of your playing 1000 times might look like this:

> lose, lose, win, win, lose, win, win, …, lose (1000 outcomes with 526 losses and 474 wins).

This stream translates into earnings −10, −10, +10, +10, −10, +10, +10, …, −10, and your total earnings would then be

$$(-10) + (-10) + 10 + 10 + (-10) + 10 + 10 + \cdots + (-10) = 526(-10) + 474(10) = -520$$

Thus, your average earnings per play are not 0.00, they are $-520/1000 = -0.52$.

In other words, you lose 0.52 per play, on average. Now it makes sense! If you lose 0.52, then the casino wins 0.52 per play, on average. That explains the lights and buffets. It also explains why high-rollers are treated so well: They play 1000 every time and lose, on average, 52 per play. The casino therefore wins, on average, 52 per play. Just a few plays are enough to justify a complimentary luxury hotel room for the high roller.

Notice the formula for the average in the case of 10 per play: Average earnings = $\{526 \times (-10) + 474 \times (10)\}/1000 = (526/1000) \times (-10) + (474/1000) \times (10) = 0.526 \times (-10) + 0.474 \times (10)$. Your naïve calculation of average earnings, on the other hand, was the

FIGURE 8.1
A roulette wheel.

TABLE 8.1

Distribution of Earnings for One Play
of 10 on "Red" in Roulette

Earnings, y	$p(y)$
-10	$20/38 = 0.526$
10	$18/38 = 0.474$
Total	1.00

simple average $0.5 \times (-10) + 0.5 \times (10)$. Instead of a simple average, you need a **weighted average** to calculate the **mean** of the probability distribution function (pdf). The weighted average gives more weight to data values that will occur more frequently and less to data values that occur less frequently; this weighted average is the **expected value**, a term that is synonymous with the **mean**. The universal symbol used to denote the expected value, and the mean, is the Greek lowercase letter μ (pronounced "mew") and is defined as follows.

Expected Value of a Discrete Distribution

$$\mu = \sum_{\text{all } y} yp(y) \tag{8.1}$$

Sometimes this is written as

$$E(Y) = \sum_{\text{all } y} yp(y) \tag{8.2}$$

In Equation 8.2, the symbol E refers to "expected value," and the symbol Y refers to a random variable (RV) that is produced by $p(y)$. Thus, you can use the expressions μ and $E(Y)$ interchangeably, just like you can use the expressions $f'(x)$ and $\partial f(x)/\partial x$ interchangeably. Sometimes, it is more convenient to use one form than the other. For example, it is more convenient to write "$E(4X + 3)$" than to write "$\mu = \Sigma_{\text{all } y} yp(y)$, where $p(y)$ is the distribution of $Y = 4X + 3$."

Example 8.2: Difficulty of a Golf Hole

Historical data from Oakland Hills Country Club (South) show that, in professional tournaments, their hole number 9 (a par 3) is pretty tough, having estimated score distribution as given by Table 8.2.

The expected score for professionals using this estimated distribution is $\mu = \Sigma_{\text{all } y} yp(y)$, calculated conveniently in spreadsheet form as shown in Table 8.3.

This is a difficult hole because the pros average 3.396, much higher than the par of 3 where pros would ordinarily be.

While nominal variables are also discrete, there is no number that corresponds to $E(Y)$ when Y is nominal. Recall the car color choice example where Y is either red, green, or gray. It makes no sense, for example, to calculate red $\times$ 0.2 + green $\times$ 0.4 + gray $\times$ 0.4. On the other hand, you can recode nominal data using 0/1 binary variables, and the expected values then become probabilities as you will see in Section 8.5.

TABLE 8.2

Professional Golfers' Scores on
Hole Number 9 of Oakland Hills
Country Club (South)

Score, y	$p(y)$
2 (birdie)	0.070
3 (par)	0.525
4 (bogie)	0.348
5 (double bogie)	0.053
6 (triple bogie)	0.004
Total	1.00

TABLE 8.3

Calculation of the Expected Value
of a Discrete Distribution

Score, y	$p(y)$	$y\,p(y)$
2 (birdie)	0.070	$2 \times 0.070 = 0.140$
3 (par)	0.525	$3 \times 0.525 = 1.575$
4 (bogie)	0.348	$4 \times 0.348 = 1.392$
5 (double bogie)	0.053	$5 \times 0.053 = 0.265$
6 (triple bogie)	0.004	$6 \times 0.004 = 0.024$
Total	1.00	$\mu = 3.396$

8.3 Continuous Case

Again, the transition from discrete to continuous is simple. What was a summation Σ in the discrete case becomes an integral $\int$ in the continuous case.

Expected Value of a Continuous Distribution

$$\mu = \int_{\text{all } y} yp(y)dy \tag{8.3}$$

This formula provides the *center of gravity* of the function $p(y)$ or the point on the horizontal axis where the graph of $p(y)$ would balance if it were a cardboard cutout. See Figure 8.2.

In the discrete case, the formula is $\mathrm{E}(Y) = \Sigma_{\text{all } y}yp(y)$, a discretely weighted average of the observable Y. In the continuous case, the formula is $\mathrm{E}(Y) = \int_{\text{all } y}yp(y)dy$, a continuously weighted average of the observable Y.

You can view the continuous expected value formula just like the discrete formula using Riemann sums as shown in Equation 2.5 of Chapter 2:

$$\mu = \int_{\text{all } y} yp(y)dy \cong \sum_{y_i} y_i p(y_i)\Delta \tag{8.4}$$

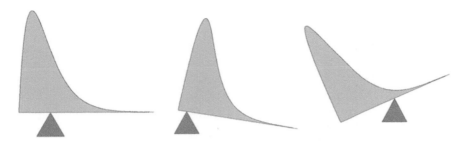

FIGURE 8.2
Expected value as the point of balance of the distribution $p(y)$. The first picture shows the location of the point of balance or the location of $E(Y)$. The middle picture shows a location to the left of the point of balance, so the curve falls to the right. The last picture shows a location to the right of the point of balance, so the curve falls to the left.

Since $p(y_i)\Delta$ is the approximate probability in a Δ-width range around y_i, the formula $\mu \cong \sum_{y_i} y_i p(y_i)\Delta$ can be viewed as the discrete formula (8.1) with the continuous data rounded off to the nearest Δ. Letting Δ tend to zero gives the integral formula (8.3).

Example 8.3: The Mean of the Exponential Distribution via Discrete Approximation

To help understand the calculus formula (8.3), it helps to walk through the discrete approximation (8.4) for a particular distribution. The waiting time distribution graphed in Figure 2.23 of Chapter 2 is given by $p(y) = 0.5e^{-0.5y}$, for $y > 0$. We claimed that this distribution produced waiting time data that were, on average, 2.0 minutes. Were we right? Table 8.4 shows how to apply Equation 8.4 using $\Delta = 2$.

Well, we were close! Using the discrete approximation, the waiting time average, where all numbers are rounded to the interval centers 1, 3, 5, 7, 9, and 11, is 2.039. Figure 8.3 shows the discrete approximation to the distribution that is used in Table 8.4. The approximate probabilities in the intervals are areas of rectangles.

If you make the interval width smaller than $\Delta = 2.0$ and extend the range farther to the right, you'll improve the approximation, and the discrete sums $\sum_{y_i} y_i(0.5e^{-0.5y_i})\Delta$ will converge to the integral $\int_0^\infty y(0.5e^{-0.5y})dy = 2.0$. You can solve for this integral exactly

TABLE 8.4

Discrete Approximation to the Expected Value Calculation for a Continuous Distribution

Interval, i	Wait Time Range (min)	Approximate y in Interval, y_i	Approximate Probability in Interval, $p(y_i)\Delta$	$y_i \times p(y_i)\Delta$
1	0–2	1.0	$0.5e^{-0.5(1)}(2) = 0.607$	$1.0 \times 0.607 = 0.607$
2	2–4	3.0	$0.5e^{-0.5(3)}(2) = 0.223$	$3.0 \times 0.223 = 0.669$
3	4–6	5.0	$0.5e^{-0.5(5)}(2) = 0.082$	$5.0 \times 0.082 = 0.410$
4	6–8	7.0	$0.5e^{-0.5(7)}(2) = 0.030$	$7.0 \times 0.030 = 0.210$
5	8–10	9.0	$0.5e^{-0.5(9)}(2) = 0.011$	$9.0 \times 0.011 = 0.099$
6	10–12	11.0	$0.5e^{-0.5(11)}(2) = 0.004$	$11.0 \times 0.004 = 0.044$
Total			0.957	$\mu \cong 2.039$

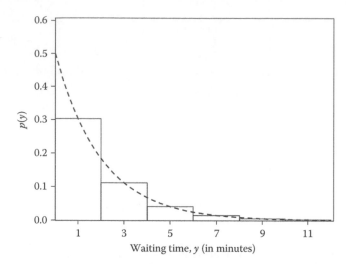

FIGURE 8.3
Discrete approximation (rectangular solid lines) to the waiting time distribution (dashed line).

using a technique called *integration by parts*, which is discussed in calculus texts but is not needed here.

In case you are not yet convinced that calculus is essential, note that you can also view the integral $\int_{all\,y} yp(y)dy$ in terms of the simple average of many data values, just as shown earlier with the discrete example and 1000 plays of roulette. Suppose there are 1000 observations from the exponential distribution shown in Figure 8.3. Then approximately 607 of them (from Table 8.1) will be in the range 0–2, approximately 223 will be in the range 2–4, and so on. The average of these 1000 observations will be, approximately, $(607 \times 1.0 + 223 \times 3.0 + 82 \times 5.0 + 30 \times 7.0 + 11 \times 9.0 + 4 \times 11.0)/1000 = 2.039$, exactly as in Table 8.1. Taking the interval widths to be shorter, the range to be wider, and taking a larger sample size, you can see that the resulting average of the data will converge to the integral $\int_0^\infty y(0.5e^{-0.5y})dy = 2.0$. Calculus really is essential if you want to understand statistics!

Example 8.4: The Triangular Distribution

In Chapter 2, Example 2.10, we introduced the triangular distribution $p(y) = 0.0002y$, for $0 \le y \le 100$, graphed it in Figure 2.22, and suggested that you could use it as a model for grades. What is the expected value of a Y produced by this distribution? According to expectation formula in Equation 8.3, it is given as follows:

$$\mu = \int_{all\,y} yp(y)dy = \int_0^{100} y(0.0002y)dy = \int_0^{100} 0.0002y^2dy = 0.0002 \int_0^{100} y^2dy = 0.0002 \left.\frac{y^3}{3}\right|_0^{100}$$

$$= 0.0002\left(\frac{100^3}{3} - \frac{0^3}{3}\right) = 66.67$$

Thus, the point of balance of this triangular pdf is 66.67; see Figures 8.2 and 2.22.

The median and mean are different numbers. As shown in Chapter 3, the inverse cumulative distribution function (or quantile) function for this distribution is

$P^{-1}(p) = (10{,}000p)^{1/2}$; hence, the median is $(10{,}000 \times 0.5)^{1/2} = 70.71$. If you tried to balance the cardboard cutout of the function $p(y) = 0.0002y$, for $0 \leq y \leq 100$, atop the median 70.71, it would fall down to the left. Further, more than 50% of the data values are larger than the mean 66.67.

The expected value (assuming the integral defining the mean exists and is finite) and the median coincide when the distribution is **symmetric**. A distribution $p(y)$ is symmetric if $p(\theta - c) = p(\theta + c)$, for any constant c, where θ is the median. In other words, a distribution is symmetric if its height $p(y)$ at the point c units to the left of the median is equal to its height c units to the right of the median. If so, then the point at which half of the distribution lies to the left and half to the right—the median—is precisely the point where the cardboard cutout would balance. Again, see Figure 8.2, but imagine the curve was symmetric.

The calculus formulas leading to calculations of $E(Y) = \int_{\text{all } y} yp(y)dy$ can be complicated, involving integration by parts or even fancier integration techniques. While you do not necessarily have to do the calculus, you should be able to recognize the answer from the form of the distribution, at least for the most common distribution forms. One famous formula involves the normal distribution: If Y is produced by the $N(\mu, \sigma^2)$ distribution, then

$$E(Y) = \int_{\text{all } y} yp(y)dy = \int_{-\infty}^{\infty} y \frac{1}{\sqrt{2\pi}\sigma} \exp\left\{-0.5\frac{(y-\mu)^2}{\sigma^2}\right\} dy = \mu$$

In other words, the expected value of a Y produced by the $N(\mu, \sigma^2)$ distribution is μ. Another famous formula involves the exponential distribution: If Y is produced by the exponential distribution with parameter λ, then

$$E(Y) = \int_{\text{all } y} yp(y)dy = \int_{0}^{\infty} y\lambda e^{-\lambda y} dy = \frac{1}{\lambda}$$

In other words, the expected value of a Y produced by the exponential distribution with parameter λ is $1/\lambda$.

Internet pages provide formulas for the means of various distributions in terms of their parameters; please have a look! Examples worth checking out, because you will see them later if not already, are the continuous distributions *beta, uniform, gamma, chi-squared*, and *Student's t* and the discrete distributions *Bernoulli, Poisson, binomial, geometric*, and *discrete uniform*.

8.4 Law of Large Numbers

You can estimate the probability of a bent coin landing heads by flipping it many times and then using the proportion (# of heads)/(# of flips) as your guess of Pr(heads). Your intuition should tell you that if you flip the bent coin 100 times instead of 10 times, then you should get a better estimate of Pr(heads). Your intuition should also tell you that your estimate will be even better with 1,000 flips, better still with 10,000 flips, and that your estimate should be extremely accurate with 1,000,000,000,000 flips.

This intuition is formalized in the **LLN**, which is stated as follows:

The Law of Large Numbers (LLN)

Let $Y_1, Y_2, \ldots$ be an iid sample from a pdf $p(y)$ whose expected value exists and is equal to μ. Let $\bar{Y}_n = (1/n)\sum_i Y_i$ be the sample average of the first n observations. Then

$$\lim_{n \to \infty} \bar{Y}_n = \mu$$

In other words, the LLN states that the average of data that are produced as an iid sample from a distribution $p(y)$ gets closer to the expected value of the distribution (either $\mu = \sum_{\text{all } y} yp(y)$ or $\mu = \int_{\text{all } y} yp(y)dy$, depending upon whether $p(y)$ is a discrete or continuous pdf, respectively), as the sample size n increases.

There is a *weak* LLN, and there is a *strong* LLN. For the purposes of this book, it does not matter which definition you adopt. Also, you will see in other sources the term *plim* instead of *lim* to emphasize that the convergence is probabilistic in nature; again, this makes no difference for the purposes of this book.

The LLN is a mathematical theorem, with a rigorous proof. It's not just a suggestion of something that might be true or one of those ugly rules of thumb. Under the conditions stated, it *is* true. You can't argue with it.

A beauty of the LLN is that it is true for any distribution with a finite mean. It doesn't matter whether the pdf is discrete, continuous, normal, Bernoulli, Poisson, or generic; it always works. Figures 8.4 through 8.8 show how $\bar{Y}_n$ gets closer to μ as n increases, for a variety of discrete and continuous distributions.

Figure 8.4 shows how your average of rolls of a die will converge to μ with more and more rolls. Here, $\mu = \Sigma yp(y) = 1 \times (1/6) + 2 \times (1/6) + 3 \times (1/6) + 4 \times (1/6) + 5 \times (1/6) + 6 \times (1/6) = 3.5$. Figures 8.5 through 8.8 show similar convergence for other distributions.

What you should notice in all of Figures 8.4 through 8.8 is the convergence. No matter what the distribution is, the average of iid observations gets closer to (i.e., converges to) the true mean, or expected value μ, of the distribution that produced the data, as the sample

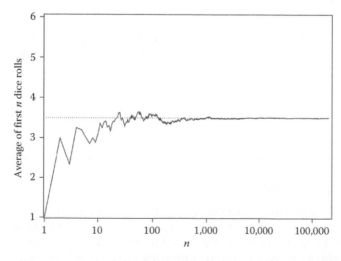

FIGURE 8.4
Convergence of a sequence of successive sample means of an iid sequence of dice data. Sample size, n, is shown in log scale.

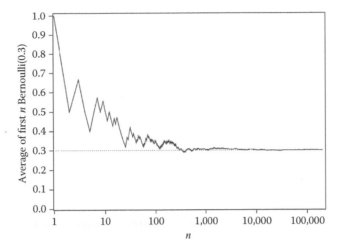

FIGURE 8.5
Convergence of a sequence of successive sample means calculated from an iid Bernoulli data sequence with $\pi = 0.3$. Sample size, n, is shown in log scale.

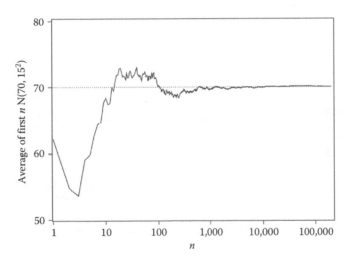

FIGURE 8.6
Convergence of a sequence of successive sample means of an iid sequence of N(70, 15²) data. Sample size, n, is shown in log scale.

size n increases. You should also notice that the convergence is *random*. The random sample averages can approach the fixed μ from the high side, or from the low side, or they can cross back and forth and back and forth before settling. But they always converge, meaning that they get ever-closer to the true mean.

Like many statistical results, the LLN requires assumptions for its validity. One is the iid assumption. If the observations are not identically distributed, then there is no single mean value μ; indeed, there may be many, one for each Y_i. So the LLN may not hold (or even make sense) in that case. But even when the observations are from the same distribution, with the same mean μ, the averages might not converge to μ. If the "independent" part of the iid assumption is false, then the average of the data, $\bar{Y}_n$, does not necessarily converge to the process mean, μ, as shown in the following example.

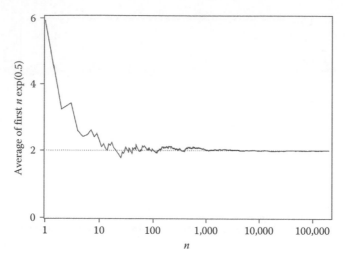

FIGURE 8.7
Convergence of a sequence of successive sample means of an iid sequence of exponential ($\lambda = 0.5$) data. Sample size, n, is shown in log scale.

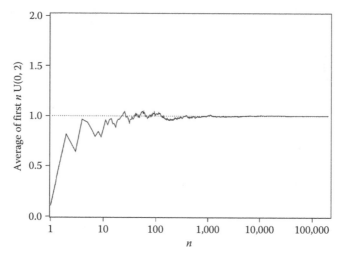

FIGURE 8.8
Convergence of a sequence of successive sample means of an iid sequence of U(0, 2) data. Sample size, n, is shown in log scale.

Example 8.5: Improper Convergence of the Sample Average When RVs Are Identically Distributed but Not Independent

In Chapter 7, we give an example where the data Y_i all come from the distribution graphed in Figure 7.13, the distribution of the sum of two dice, yet the observations are not independent because they share a common die. Again let X_0, X_1, X_2, ... be iid fair dice rolls having the discrete uniform distribution on the values {1, 2, 3, 4, 5, 6}, and let $Y_1 = X_0 + X_1$, $Y_2 = X_0 + X_2$, $Y_3 = X_0 + X_3$, ... The Y_i are dependent because they share the value X_0. In list form, but written sideways to save space, the pdf and the expected value calculation are as shown in Table 8.5.

So, the mean of the pdf $p(y)$ of the sum of two dice is 7.0. You should roll a couple of dice a few times to see that this makes sense—the sum is sometimes higher than 7.0, sometimes lower, but the long-run average is 7.0.

TABLE 8.5

Distribution and Expected Value of the Sum of Two Dice

					y							
	2	3	4	5	6	7	8	9	10	11	12	Total
$p(y)$	1/36	2/36	3/36	4/36	5/36	6/36	5/36	4/36	3/36	2/36	1/36	1.00
$y \times p(y)$	2/36	6/36	12/36	20/36	30/36	42/36	40/36	36/36	30/36	22/36	12/36	7.00

On the other hand, the average of the Y values calculated with the common value X_0 does not converge to 7.0. If you average the first 100 values, you get

$$\bar{Y}_{100} = \frac{1}{100}\sum_{i=1}^{100} Y_i = \frac{1}{100}(Y_1 + Y_2 + \cdots + Y_{100}) \qquad \text{(By definition of average and summation symbols)}$$

$$= \frac{1}{100}\{(X_0 + X_1) + (X_0 + X_2) + \cdots + (X_0 + X_{100})\} \qquad \text{(By substitution)}$$

$$= \frac{1}{100}(100X_0 + X_1 + X_2 + \cdots + X_{100}) \qquad \text{(By algebra)}$$

$$= X_0 + \frac{1}{100}(X_1 + X_2 + \cdots + X_{100}) \qquad \text{(By algebra)}$$

$$= X_0 + \bar{X}_{100} \qquad \text{(By definition of average)}$$

So the average of the Y values is equal to the value produced by your initial toss, X_0, plus the average of the 100 subsequent tosses. Since the X values are iid discrete uniform over the values {1, 2, 3, 4, 5, 6}, their expected value is 3.5, and by the LLN as applied to the iid Xs, the sample average $\bar{X}_n$ converges to 3.5. Hence, the sample average $\bar{Y}_n$ converges to $X_0 + 3.5$, which is the value of the first roll of the die plus 3.5. In this example, the average $\bar{Y}_n$ can't possibly converge to $E(Y) = 7.0$.

You could demonstrate this yourself via an experiment: Roll die once to get X_0. Then roll the same die repeatedly, adding the values to your initial X_0 to get Y_1, Y_2,.... Taking a running average of those Ys, you will see that they do not converge to 7.0. Figure 8.9 shows how your running averages would look, assuming your first roll is $X_0 = 2$.

Independence is a crucial concept in statistics. The intuitive notion that estimates get better with larger sample size (n) can be false, as shown in Figure 8.9, when the observations are dependent. Independence is also crucial in advanced statistical methods such as regression analysis, for the estimation of standard errors, and in latent variable structural equations models, for the estimation of interfactor correlations. When the independence assumption is violated, the computer software reports estimates that are grossly inaccurate, and the software usually does not even warn you that there is any problem!

Example 8.6: Non-Convergence of the Sample Average When the Mean Is Infinite

Another assumption needed for the validity of the LLN is that the mean is finite. How could you have an infinite mean? It is surprising how easily this can happen. Ratios are often used in statistics to measure percent changes; for example, a return is the ratio I_t/I_{t-1}, minus 1.0. If you made £50K salary last year and £60K this year, the ratio

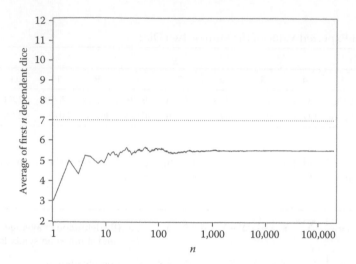

FIGURE 8.9
A sequence of successive sample means of dependent sums of two dice. The successive sample means do not converge to the true mean 7.0.

60K/50K = 1.20 tells you that you made 20% more this year than last year. And ratios can easily have an infinite mean when the denominator can be close to 0.

Consider the stoplight example, where the time X that the light stays green is distributed as $U(0, 2)$. For a given observation X, $Y = 2/X$ tells you how much longer the cycle time is than the observed time; for instance, if X is observed to be 1.25, then $2/1.25 = 1.60$ says that the cycle time was 60% longer than that observation.

What is the distribution of $Y = 2/X$? You can find its cdf as follows:

$P(y) = \Pr(Y \le y)$	(By definition of cdf)
$= \Pr(2/X \le y)$	(By substitution, since $Y = 2/X$)
$= \Pr(2 \le yX)$	(By algebra)
$= \Pr(X \ge 2/y)$	(By algebra)
$= (2 - 2/y) \times (1/2)$	(Since X has the $U(0, 2)$ distribution, and using the base × height formula for the area of a rectangle)
$= 1 - 1/y$	(By algebra)

The pdf of Y is $p(y) = P'(y)$—recall from Chapter 2 that the derivative of the cdf is equal to the pdf—and you get the pdf as follows:

$$p(y) = \frac{\partial P(y)}{\partial y} = \frac{\partial (1 - y^{-1})}{\partial y} = 0 - (-1)y^{-2} = y^{-2} = \frac{1}{y^2}, \quad \text{for } y > 1.$$

The mean value is infinity for this distribution; the calculation goes like this:

$$\mu = \int_{\text{all } y} yp(y)dy = \int_1^\infty y\left(\frac{1}{y^2}\right)dy = \int_1^\infty \left(\frac{1}{y}\right)dy = \ln(y)\Big|_1^\infty = \ln(\infty) - \ln(1) = \infty - 0 = \infty.$$

This means that, unlike the graphs shown earlier, the sample average of the ratios $2/X$ will not converge to any value. Figures 8.10 and 8.11 result from two different streams of iid samples from the distribution of $2/X$ and show what happens.

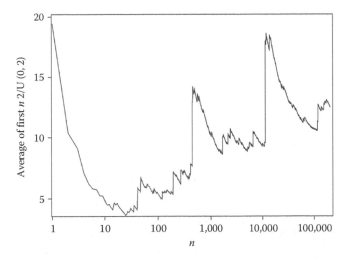

FIGURE 8.10
Non-convergence of a sequence of successive sample means calculated from an iid sample when $\int yp(y)dy = \infty$.

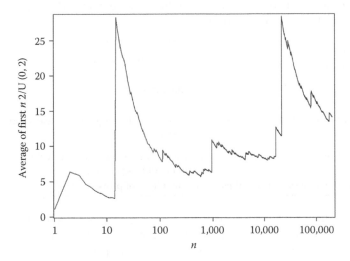

FIGURE 8.11
Non-convergence of a sequence of successive sample means calculated from an iid sample when $\int yp(y)\, dy = \infty$.

 The problems noted in Figures 8.10 and 8.11 regarding the sample mean are troubling. They show that the sample mean does not estimate any meaningful quantity, since the mean never settles to any particular value, even with very large sample sizes n.

In the examples graphed in Figures 8.10 and 8.11, with $Y = 2/X$ and $X \sim U(0, 2)$, it can easily happen that an extremely large Y is observed, on occasion. For example, if $X = 0.001$, then $Y = 2000$. If $X = 0.000001$, then $Y = 2,000,000$. Extreme values, observed only rarely, are called **outliers**. With outliers, the sample mean might not be estimating anything meaningful, as suggested by Figures 8.10 and 8.11.

TABLE 8.6

Bernoulli Distribution

y	$p(y)$
0	$1 - \pi$
1	π
Total	1.00

8.5 Law of Large Numbers for the Bernoulli Distribution

The Bernoulli distribution deserves special attention among the class of distributions. Recall that if Y is produced by a Bernoulli distribution with parameter π, then the probability distribution $p(y)$ that produces Y is given in list form as shown in Table 8.6.

The expected value of Y is given by $E(Y) = 0 \times (1 - \pi) + 1 \times \pi = \pi$; that is, the expected outcome is equal to the probability of observing a 1. Thus, the LLN tells you that the average of an iid sequence of Bernoulli outcomes $Y_1, Y_2, \ldots$ converges to π, the probability of observing a 1, as shown in Figure 8.5. This makes sense from the standpoint of the simple bent coin flip: Flip it 1000 times, and code the heads as one and the tails as zero. Then the average of the 0s and 1s is given as follows:

$\bar{Y}_{1000} = \dfrac{1}{1000} \displaystyle\sum_{i=1}^{1000} Y_i$	(By definition of average)
$= \dfrac{1}{1000}(Y_1 + Y_2 + \cdots + Y_{1000})$	(By definition of summation)
$= \dfrac{1}{1000}(0 + 1 + 0 + 0 + 1 + 0 + \cdots + 1)$	(Assuming such a sequence of heads and tails, heads coded as 1)
$= \dfrac{1}{1000}(\#\,\text{heads})$	(Since the sum of the 1s is simply the count of the number of heads you get)
$= \text{proportion of heads in 1000 flips}$	(By definition of proportion)

So, the average of the 0s and 1s just gives you the proportion, which is the estimated probability, and which intuitively should become more accurate with a larger sample size. This is precisely how you estimated probabilities in Chapter 3 via simulation, although it might not have seemed that way at the time. In Section 3.3, you estimated the probability $\Pr(Y \in A)$ via $\#\{Y_i^* \in A\}/NSIM$. This expression is an average of the Bernoulli(0/1) variables V_i, where $V_i = 1$ if $Y_i^* \in A$ and $V_i = 0$, otherwise. Hence, the LLN explains why the simulation-based probability estimates get better with larger $NSIM$.

8.6 Keeping the Terminology Straight: Mean, Average, Sample Mean, Sample Average, and Expected Value

The terms *mean, average, sample mean, sample average,* and *expected value* all refer to types of averages and are easily confused. Fortunately, there are only two concepts here. One is *model,* and the other is *data.* If you just remember *model produces data,* you'll get it quickly.

Expected value and *mean value* are the same thing. They are both properties of the model that produces the data. The mathematical expressions are the following:

In the discrete case

$$\text{Mean} = \text{expected value} = E(Y) = \mu = \sum_{\text{all } y} yp(y)$$

In the continuous case

$$\text{Mean} = \text{expected value} = E(Y) = \mu = \int_{\text{all } y} yp(y)dy$$

Recall not only that *model produces data* but also that *model has unknown parameters*. The term *mean*, or *expected value*, is one of those unknown parameters. In practice, you will never know μ, but you can reduce your uncertainty about μ by collecting data.

The terms *sample mean, average,* and *sample average* all refer to the data. Discrete or continuous, the formula is

$$\bar{y} = \frac{1}{n}\sum_{i=1}^{n} y_i$$

When viewed as a function of RVs (DATA), you should write the sample mean using capital letters like this:

$$\bar{Y} = \frac{1}{n}\sum_{i=1}^{n} Y_i$$

If you want to emphasize the dependence on n, as in the discussion earlier about the LLN, you can use an n subscript on the sample average:

$$\bar{Y}_n = \frac{1}{n}\sum_{i=1}^{n} Y_i$$

Sample averages all are from data (or DATA). You *will* get to observe the value of the sample average $\bar{y}$, unlike the mean μ, whose value you can't observe.

In some statistics sources, the term *population average* is used for μ, and it is defined as $\mu = (1/N)\sum_{i=1}^{N} y_i$ or the average of all elements in the population. This makes sense in the population sampling model of Table 7.1, repeated here as Table 8.7.

TABLE 8.7

Probability Distribution That Produces Data under the Population Model

y	$p(y)$
y_1	$1/N$
y_2	$1/N$
...	...
y_N	$1/N$
Total	1.00

(If some of the y values repeat, they should be collated and their probabilities accumulated as shown in Table 7.2).

Using the population definition of the probability distribution $p(y)$, the population mean is calculated as

$$\mu = E(Y) = \sum yp(y) = y_1 \left(\frac{1}{N} \right) + y_2 \left(\frac{1}{N} \right) + \cdots + y_N \left(\frac{1}{N} \right) = \frac{1}{N} \sum_{i=1}^{N} y_i$$

In this case, the expected value of Y is indeed the average of the population data. However, as discussed in Chapter 7, the process-based interpretation of the model $p(y)$ that produces the data is usually more relevant than the population-based interpretation. Hence, unless explicitly stated otherwise, we will *not* consider μ to be a *population mean*; instead, it is the *process mean*, with formulas either $\mu = \Sigma_{\text{all } y} \, yp(y)$ or $\mu = \int_{\text{all } y} yp(y)dy$ for discrete or continuous processes, respectively. Let's set this off to make sure you see it clearly.

How to Interpret μ

- *Do not* interpret the parameter μ as a *population mean*.
- *Do not* think that μ can be calculated by the formula $\mu = (1/N)\Sigma_{i=1}^{N} y_i$.
- *Do* interpret μ as a *process mean*.
- *Do* calculate μ by either $\mu = \Sigma_{\text{all } y} yp(y)$ or $\mu = \int_{\text{all } y} yp(y)dy$, where $p(y)$ is the process distribution.

Be aware that you will see the terms *average* and *mean* used sloppily elsewhere, not just on the Internet but even in software like structured query language (SQL), where the term *mean* is used to calculate a sample average.

In this book, and for understanding statistics in general, you should *never* confuse $\bar{y}$ with μ. A main point of statistical inference is to state how close $\bar{y}$ is to μ; in some cases, they are not close at all. To visualize the difference between $\bar{y}$ and μ, please have a look at Figures 8.4 through 8.8 again. The constant, flat line in all those graphs is μ, the true mean. The varying, erratic lines are the $\bar{y}$s for different sample sizes. There is a clear distinction between $\bar{y}$ and μ: The $\bar{y}$ values are outcomes from a random process, hence variable, while the parameter μ is a fixed, unchanging constant.

To make the distinction between $\bar{y}$ and μ even more stark, look at Figures 8.10 and 8.11 again. In those graphs, you see the $\bar{y}$s. The value of μ in both graphs is $\mu = \infty$, so the $\bar{y}$s you see in Figures 8.10 and 8.11 are nowhere close to $\mu = \infty$.

Do not confuse $\bar{y}$ with μ. They are quite different. This is a main lesson of the Mantra:

> Model produces data.
> Model has unknown parameters (like μ).
> Data (like $\bar{y}$) reduce the uncertainty about the unknown parameters.

8.7 Bootstrap Distribution and the Plug-In Principle

The model $p(y)$ produces the data Y, but the model $p(y)$ is unknown. One way to estimate $p(y)$ using your observed data $y_1, y_2, \ldots, y_n$ is to use the histogram. Another is to approximate it using the distribution shown in Table 8.8, sometimes called the **bootstrap population distribution** or simply the **bootstrap distribution** for short.

TABLE 8.8

Bootstrap Population Distribution

y	$\hat{p}(y)$
y_1	$1/n$
y_2	$1/n$
...	...
y_n	$1/n$
Total	1.00

If some of the ys repeat, they should be collated and their probabilities accumulated.

The term *bootstrap* comes from the famous statistician Brad Efron. His logic was that, since $p(y)$ is unknown, you can "pull yourself up by your bootstraps," using an estimate of $p(y)$ from the data. With a known distribution $\hat{p}(y)$, you can now do all the things you wanted to do with $p(y)$ but couldn't, because it is unknown. Bootstrap applications typically involve simulating data from $\hat{p}(y)$, as discussed in later chapters.

The bootstrap distribution $\hat{p}(y)$ looks like the population distribution presented in Table 8.7, except the y values are from the sample you selected, having only n observations instead of N. Since you have the sample in your hands, you know the bootstrap distribution, unlike the population distribution of Table 8.7, which is based on population data that you do not know. The bootstrap distribution $\hat{p}(y)$ is clearly incorrect as the producer of your data, because it assumes that the data produced by Nature's processes will be produced by the collection of data that you have already observed. Further, the bootstrap distribution is discrete even when the true distribution $p(y)$ is continuous, so again, it is clearly not the correct model, just an approximate one. Figure 8.12 compares the bootstrap distribution $\hat{p}(y)$ with a true $p(y)$, assuming $p(y)$ is the $N(70, 10^2)$ distribution and $n = 15$ observations are sampled.

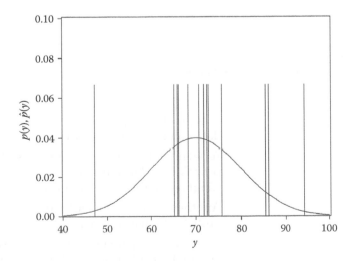

FIGURE 8.12
Comparing the bootstrap distribution $\hat{p}(y)$ (vertical lines) with the true distribution $p(y)$ (smooth curve).

Clearly, as shown in Figure 8.12, the bootstrap distribution is different from the true distribution. However, it does provide a useful way to estimate parameters. You know that μ is either $\int_{\text{all } y} y p(y) dy$ or $\Sigma_{\text{all } y} y p(y)$; either way, it depends on $p(y)$. The **bootstrap plug-in principle** is to estimate parameters that depend on a true distribution $p(y)$ by "plugging in" the estimate $\hat{p}(y)$ in the formula.

Bootstrap Plug-In Principle

If θ is a function of the distribution $p(y)$, that is, if $\theta = f\{p(y)\}$, you can estimate θ by plugging in an estimate of $p(y)$:

$$\hat{\theta} = f\{\hat{p}(y)\}$$

The bootstrap distribution $\hat{p}(y)$ is always discrete, so the application of the plug-in principle always involves summation rather than integration. Taking $\theta = \mu$, the plug-in principle gives

$$\hat{\mu} = \sum y\hat{p}(y) = y_1\left(\frac{1}{n}\right) + y_2\left(\frac{1}{n}\right) + \cdots + y_n\left(\frac{1}{n}\right) = \left(\frac{1}{n}\right)\sum y_i = \bar{y}$$

This is a perfectly sensible estimate of μ. In later chapters, we show how you can use the bootstrap plug-in principle to estimate many other parameters as well.

The bootstrap is often used to generate data. (*Model produces data!*) The distribution $p(y)$ is unknown, so you cannot generate samples from $p(y)$. But you know the bootstrap distribution $\hat{p}(y)$, and you can generate as many samples as you want from it; these are called *bootstrap samples*. For example, suppose your data set has $n = 7$ values, 45, 42, 12, 23, 25, 12, and 14. The bootstrap distribution $\hat{p}(y)$ puts 1/7 probability on each of the values except 12, which gets 2/7 probability. Using Excel, you can generate a sample using the random number generator, with the discrete distribution, using the bootstrap distribution $\hat{p}(y)$ as *value and probability input range*. If you want to read ahead and see how this works, have a look at Chapter 14, Section 14.3.

Vocabulary and Formula Summaries

Vocabulary

Weighted average	A type of average that gives more weight to some data values and less to others.
Mean	A weighted average of the possible values of a probability distribution, one that is continuously weighted in the case of a continuous distribution.
Expected value	Another name for the mean, or point of balance, of a probability distribution.
Symmetric	An object is symmetric if there exists at least one point where a line can be drawn through the object to produce two identical halves.

Law of large numbers	For a sample of n iid RVs from a process with a finite mean, the sample mean becomes closer and closer to the process mean as n gets larger.
Outlier	Extremely large or small values relative to the majority of the other values in a distribution, that occur infrequently.
Sample average	The sum of all of the observations in a sample divided by the number of observations in the sample.
Sample mean	Another name for the sample average.
Bootstrap population distribution	An estimated distribution created by assigning $1/n$ probability to every observation in a sample.
Bootstrap plug-in principle	A method of estimating parameters of a distribution by plugging in an estimate of the distribution.

Key Formulas and Descriptions

$\mu = E(Y) = \sum_{\text{all } y} yp(y)$	The definition of the mean of a discrete distribution $p(y)$.
$\mu = E(Y) = \int_{\text{all } y} yp(y)dy$	The definition of the mean of a continuous distribution $p(y)$.
$\mu \cong \sum_{y_i} y_i p(y_i)\Delta$	The discrete approximation to the mean of a continuous distribution.
$\int_{-\infty}^{\infty} y \frac{1}{\sqrt{2\pi}\sigma} \exp\left\{-0.5\frac{(y-\mu)^2}{\sigma^2}\right\} dy = \mu$	The mean of the normal distribution with parameters μ and σ^2 is μ.
$\int_0^{\infty} y\lambda e^{-\lambda y} dy = \frac{1}{\lambda}$	The mean of the exponential distribution with parameter λ is $1/\lambda$.
$\bar{Y}_n = \frac{1}{n}\sum_{i=1}^{n} Y_i$	The average of the first n observations sampled from a process is sometimes written as $\bar{Y}_n$. This is a RV.
$\mu = \frac{1}{N}\sum_{i=1}^{N} y_i$	The population mean, when the "population" definition of the probability model is used. Don't use this formula. Use the integral or summation formulas for μ instead.
$\bar{y} = \frac{1}{n}\sum_{i=1}^{n} y_i$	The average of the data in your sample, after you have seen the data, is called $\bar{y}$. This is a fixed constant, nonrandom.
$p(y) = 1/N$, when $y = y_i$	The population distribution, in the case of no repeats in the population data $y_1, y_2, \ldots, y_N$. Do not use this definition of $p(y)$. Use the process definition instead.

$\hat{p}(y) = 1/n$, when $y = y_i$

The bootstrap distribution, in the case of no repeats in the data $y_1, y_2, \ldots, y_n$.

$\theta = f\{p(y)\}$

A generic parameter that depends on the unknown process distribution $p(y)$.

$\hat{\theta} = f\{\hat{p}(y)\}$

The plug-in estimate of a generic parameter that depends on the unknown process distribution $p(y)$.

Exercises

8.1 Hans plans to play a lottery with a probability 1/1000 of winning. The game costs 1 (dollar, or euro, or franc, or …). He plans to play many, many times. How much would Hans have to receive every time he wins, if he wants to come out even, on average, in the long run?

8.2 A car dealership has been in business for 10 years, and it wants to determine how many luxury cars to order from the manufacturer for the next month. Based on historical information, it estimates the probabilities of the number demanded, as shown in the following table:

y	$p(y)$
0	0.18
1	0.16
2	0.14
3	0.34
4	0.10
5	0.05
6	0.03
Total	1.00

A. Find the expected number of cars sold for the next month.

B. Critique the dealership's model $p(y)$: In what situation(s) would their model be perfectly adequate for finding the expected number of cars sold next month? In what situation(s) would their model be terribly wrong for finding the expected number of cars sold next month?

8.3 Let Y be a RV with $p(y) = y^2/9$, for $0 \leq y \leq 3$.

A. Show that the area under the curve is 1.0 using calculus.

B. Graph the function $p(y)$ using software.

C. Find the mean of $p(y)$.

D. Find the median of $p(y)$.

E. Which is higher for this distribution, the mean or the median? Draw a graph (by hand) of another distribution $p(y)$ where this situation is reversed. Label and number both axes.

8.4 The LLN can be used to estimate probabilities that would be difficult (if not impossible) to find directly. Define the RV $Y = 203U^3 - 6U - 7$, with U a U(0, 1) RV. Generate 5000 U(0, 1) random numbers using software and compute the corresponding 5000 values of Y.

    A. Draw the histogram and normal distribution q–q plot of the resulting Y values using software. Does the distribution of Y appear to be approximately a normal distribution?

    B. Define B to be a Bernoulli RV that equals 1 when $Y > 9$ and equals 0, otherwise. Estimate $\Pr(Y > 9)$ using your simulated B data. What result from Section 8.5 are you using in your answer?

8.5 Use the data of Exercise 4.7, concerning cheese tasting.

    A. Construct the bootstrap distribution, $\hat{p}(y)$, in list form.

    B. Find the plug-in estimate of μ using this distribution. Show the formula and the result.

    C. Simulate 20,000 observations from $\hat{p}(y)$, and calculate the average of these 20,000 observations. Is this number close to the plug-in estimate from Exercise 8.5B? Why should it be?

8.6 Suppose you roll a die twice in succession, getting X_1 and X_2. Then divide them, getting $Y = X_1/X_2$. Thus, Y is discrete, ranging from a minimum of 1/6 to a maximum of 6.

    A. Use the fact that each of the 36 combinations (x_1, x_2) has probability 1/36 to find the distribution of Y, in list form. (Hint: Create a 6 × 6 table showing all combinations of (x_1, x_2) first.)

    B. Find the mean of the pdf you found in Exercise 8.6A.

    C. Simulate 10,000 (or more) iid observations Y_i (= X_{i1}/X_{i2}) having the distribution in Exercise 8.6A. Draw the graph of the successive averages of these Ys as shown in Figures 8.4 through 8.11, and comment.

8.7 The *Cauchy distribution* is an example of a distribution with no mean μ. It is a notoriously outlier-prone distribution and is sometimes used to model erratic behavior in financial markets. Its pdf is $p(y) = \{\pi(1 + y^2)\}^{-1}$, for $-\infty < y < \infty$. Simulate 10,000 (or more) iid observations Y_i from the Cauchy distribution, draw the graph of the successive averages as shown in Figures 8.4 through 8.11, and explain why this graph doesn't behave in the usual way.

8.8 Consider the function $f(y) = y^{-1}$, for $1 < y < e$, where e is Euler's constant (the classic $e = 2.718\ldots$).

    A. Show that $f(y)$ is a valid continuous probability distribution function.

    B. You showed that $f(y)$ is a valid continuous pdf in Exercise 8.8A, so now call it $p(y)$. Graph $p(y)$ and explain the "point of balance concept" without calculating E(Y).

    C. Calculate E(Y) using calculus and relate your answer to Exercise 8.8B.

    D. Find the cdf and inverse cdf of Y.

    E. Using the inverse cdf, find the median of the distribution of Y, and compare it to E(Y).

    F. Simulate 10,000 observations Y^* using the inverse cdf method and calculate their average. Why is the average of these 10,000 values different from E(Y)? What famous result tells you why the average of these 10,000 values is close to E(Y)?

8.9   Consider the function $f(y) = y^{-2}$, for $1 < y < \infty$. (This is an example of the famous *Pareto distribution*, a distribution that is often used as a model for income).

   A.   Show that $f(y)$ is a valid continuous probability distribution function.

   B.   You showed that $f(y)$ is a valid continuous pdf in Exercise 8.9A, so now call it $p(y)$. Graph $p(y)$.

   C.   Show that $E(Y) = \infty$ using calculus.

   D.   Find the cdf and inverse cdf of Y.

   E.   Using the inverse cdf, find the median of the distribution of Y, and compare it to $E(Y)$.

   F.   Simulate 100,000 observations Y^* using the inverse cdf method and graph the running average as shown in Figures 8.4 through 8.11. Explain why the graph doesn't behave in the usual way.

8.10 Consider the function $f(y) = y^3 - y^4$, for $0 < y < 1$. (This is the kernel of the famous *beta distribution*, a distribution that is often used as a model for proportions and probabilities).

   A.   Find the constant of proportionality c that makes $c \times f(y)$ a valid continuous probability distribution function.

   B.   You showed that $c \times f(y)$ is a valid continuous pdf in Exercise 8.10A, so now call it $p(y)$. Graph $p(y)$.

   C.   Find $E(Y)$ using calculus and explain it in terms of the "point of balance" of the graph in Exercise 8.10B.

   D.   This distribution is the *beta distribution* with parameters $\theta_1 = 4$ and $\theta_2 = 2$. Using software, simulate 100,000 observations Y^* and graph the running average as shown in Figures 8.4 through 8.11. Explain the appearance of graph in terms of a famous law. Which famous law is it?

9

Functions of Random Variables:
Their Distributions and Expected Values

9.1 Introduction

Functions of random variables (RVs) appear everywhere in statistics. Perhaps the most common statistic, the sample average $\overline{Y}$, is a function of the RVs: Specifically, $\overline{Y} = f(Y_1, Y_2, ..., Y_n) = (1/n)(Y_1 + Y_2 + \cdots + Y_n)$. But there are many other functions as well.

Applications of Functions of Random Variables

- The *variance*, discussed later in this chapter, is defined as the expected value of the function $f(Y) = (Y - \mu)^2$, a function of the RV Y.
- Your earnings on an investment, minus the transaction cost, are (initial principal) × (return) − (cost), a function of the RV *return*: Here, earnings = f(return) = (principal) × (return) − (cost).
- When there are outliers, it is common to log-transform the data to $f(Y) = \ln(Y)$.
- The *skewness* and *kurtosis* parameters, used to diagnose the degree of non-normality of a distribution, are related to cubic and quartic functions of Y.
- Linear regression models use a linear function of X to predict Y; specifically, $f(X) = \beta_0 + \beta_1 X$.
- The famous statistical distributions called *Student's t-distribution*, the *chi-squared distribution*, and the *F-distribution* are distributions of functions of normally distributed RVs $Y_1, Y_2, ..., Y_n$; these functions are called *test statistics*.

When you perform a statistical analysis, you instruct the computer to produce functions of your raw DATA such as percentages, averages, correlations, etc. Therefore, you can say that the *entire subject* of statistics concerns functions of RVs!

9.2 Distributions of Functions: The Discrete Case

Let Y be a discrete RV, and let $T = f(Y)$ be a function of Y. (T is for **transformation**, another word for a function.) You can easily construct the distribution of T in list form as follows.

Finding the Probability Distribution Function (pdf) of a
Transformation of a Discrete Random Variable

1. Write down the distribution of Y in list form with values y and probabilities $p(y)$.
2. Create new column $t = f(y)$.
3. If there are no repeats in the t column, you are done. The list of values $t = f(y)$ and the associated probabilities $p(y)$ are the distribution of T, in list form. If there are repeats among the values of t, collate them and add their associated probabilities, and then you are done.

Example 9.1: Finding the Distribution of $T = Y - 3$ When Y Is a Die Outcome

Consider a roll of a fair die. The outcome Y has the discrete uniform distribution on the values 1, 2, 3, 4, 5, and 6. Let $T = Y - 3$. What is the distribution of T? Table 9.1 shows how to get the distribution of T following the steps shown earlier.

Notice that $T = -2$ if (and only if) $Y = 1$; hence, $\Pr(T = -2) = \Pr(Y = 1) = 0.1667$. There are no repeats in the t column, so the distribution of T is given as shown in Table 9.2.

And that's all there is to it!

TABLE 9.1

Finding the Probability Distribution Function of a Transformation of a Discrete Random Variable When There Are No Repeats

y	$p(y)$	$t = y - 3$
1	0.1667	-2
2	0.1667	-1
3	0.1667	0
4	0.1667	1
5	0.1667	2
6	0.1667	3
Total	1.000	

TABLE 9.2

pdf of $T = Y - 3$ When Y Is a Die Outcome

t	$p(t)$
-2	0.1667
-1	0.1667
0	0.1667
1	0.1667
2	0.1667
3	0.1667
Total	1.000

TABLE 9.3

Finding the Probability Distribution Function
of a Transformation of a Discrete Random
Variable When There Are Repeats

y	$p(y)$	$t = (y - 3)^2$
1	0.1667	4
2	0.1667	1
3	0.1667	0
4	0.1667	1
5	0.1667	4
6	0.1667	9
Total	1.000	

TABLE 9.4

pdf of $T = (Y - 3)^2$ When
Y Is a Die Outcome

t	$p(t)$
0	0.1667
1	0.3333
4	0.3333
9	0.1667
Total	1.000

The function $t = f(y) = y - 3$ is an example of a one-to-one function, a function that associates each value of t with one, and only one, y value. There are no repeats for one-to-one functions. When there are repeats, it gets trickier.

Example 9.2: Finding the Distribution of $T = (Y - 3)^2$ When Y Is a Die Outcome

Suppose $T = (Y - 3)^2$ in the earlier example. Following the same steps, we have the data shown in Table 9.3.

Notice that there are repeats in the t column: You get $T = 4$ when $Y = 1$ or when $Y = 5$. So $\Pr(T = 4) = \Pr(Y \in \{1, 5\}) = \Pr(Y = 1) + \Pr(Y = 5) = 0.1667 + 0.1667 = 0.333$. Thus, you must collate the probabilities via summation, as shown in Table 9.4.

Figure 9.1 shows the difference between the distribution of Y and $T = (Y - 3)^2$ in the discrete uniform case.

9.3 Distributions of Functions: The Continuous Case

In the continuous case, you can't list the values of Y so you need to work with the functions. You saw the method in Chapter 8 to find the distribution of $T = 2/Y$ in Example 8.6—the one with the infinite mean. The method in general is as follows:

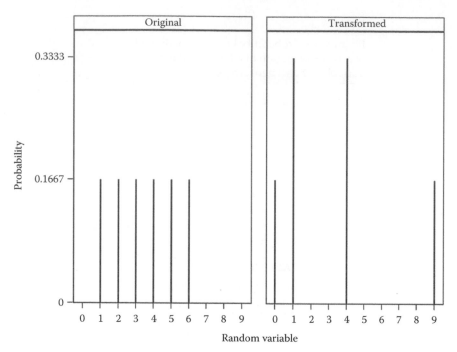

FIGURE 9.1
Distributions of a die roll Y (original) and of the transformed values $T = (Y - 3)^2$.

Finding the Probability Distribution Function $p(t)$ of a Transformation T of a Continuous Random Variable Y

1. Find the cumulative distribution function (cdf) of T, $P(t) = \Pr(T \leq t)$ as follows:
   A. Express $f(Y) \leq t$ as $Y \in A(t)$.
   B. Compute $P(t) = \Pr(T \leq t)$ as $\Pr\{Y \in A(t)\}$.
2. The pdf of T is $p(t) = P'(t)$.

This method works when the transformation is continuous and differentiable but not necessarily otherwise. For example, the transformation $T = 1$ if $Y > 10$, $T = 0$ otherwise, produces a discrete RV, and the identity $p(t) = P'(t)$ is true only for continuous RVs.

Example 9.3: The Distribution of $-\ln\{Y\}$ Where $Y \sim U(0, 1)$

Suppose $Y \sim U(0, 1)$ and $T = -\ln(Y)$. Following the earlier steps, the cdf of T is given as $P(t) = \Pr(T \leq t) = \Pr\{-\ln(Y) \leq t\}$. Now, $-\ln(Y) \leq t$ implies $\ln(Y) \geq -t$, which further implies $Y \geq e^{-t}$. The set $A(t)$ is thus the set $\{y; e^{-t} \leq y \leq 1\}$, since the $U(0, 1)$ RV Y can be no larger than 1.0. The probability of the set $A(t)$ can be calculated either using the rectangle formula for area or by calculus. Here is the calculus solution:

$$P(t) = \Pr(T \leq t) = \Pr\{Y \in A(t)\} = \Pr(e^{-t} \leq Y \leq 1) = \int_{e^{-t}}^{1} 1.0 \, dy = y \Big|_{e^{-t}}^{1} = 1 - e^{-t}$$

Taking the derivative of the cdf gives you the pdf:

$$p(t) = P'(t) = \frac{\partial(1 - e^{-t})}{\partial t} = 0 - \left\{ \frac{e^{-t}}{(-1)} \right\} = e^{-t}$$

This distribution is the exponential distribution with parameter $\lambda = 1$.

Thus, while the reciprocal transform of the uniform RV has an infinite mean as shown in Chapter 8, the natural logarithm transforms this RV from one where the mean of the distribution is infinite to one where the mean is finite.

There is a very practical moral to this story. The distribution of $2/Y$, where $Y \sim U(0, 2)$, presented in Chapter 8 is an example of a distribution that produces extreme outliers. The log transform reduces the outlier problem so dramatically that the distribution changes from one where the sample average is meaningless, since the true mean is infinity, to one where the sample average is in fact a meaningful estimate of a finite mean.

To elaborate, the log transformation makes high outliers more typical. For example, if most of the data fall in the range of 1–100, with an occasional value around 1000 (the odd outlier), then after the natural log transform, most of the data now fall in the range of 0–4.6 (ln(1) to ln(100)), with an occasional value around 6.9 (ln(1000)). The value 6.9 is not nearly such an outlier, when compared to the range 0–4.6, as is the value 1000 when compared to the range 0–100. You often need to log-transform your data before using standard statistics such as averages, particularly when the distribution that produced the data is extremely outlier-prone.

9.4 Expected Values of Functions and the Law of the Unconscious Statistician

Suppose T is a function of Y. What is the mean of the distribution of T? Simple enough: The mean of the distribution is the same as the expected value and is given, just as discussed in Chapter 8, by either $E(T) = \Sigma t p(t)$ or $E(T) = \int t p(t) dt$. For example, suppose as in Section 9.2 that Y is a discrete uniform dice outcome and $T = (Y - 3)^2$. Then, using the distribution of T shown in Table 9.4, $E(T) = 0 \times 0.1667 + 1 \times 0.3333 + 4 \times 0.3333 + 9 \times 0.1667 = 3.167$. You could also find $E(T)$ using Table 9.3 as $E(T) = 4 \times 0.1667 + 1 \times 0.1667 + 0 \times 0.1667 + 1 \times 0.1667 + 4 \times 0.1667 + 9 \times 0.1667 = 3.167$. This latter form is easier because you don't have to go through the trouble of finding the distribution of the transformation—instead you can use the distribution of the original untransformed variable. For continuous distributions, where it can be especially difficult to find the distribution of the transformed variable, this is a particularly handy trick. It is so handy that it has a special, amusing name.

The Law of the Unconscious Statistician

Let $T = f(Y)$. In the discrete case

$$E(T) = \sum f(y) p(y)$$

and in the continuous case

$$E(T) = \int f(y) p(y) dy$$

In either case, the beauty of the result is that there is no need to find the distribution $p(t)$ of T. The result is named for the "unconscious" statistician who is unaware that $E(T) = \int tp(t)dt$ in reality and is just plain lucky that $\int tp(t)dt$ and $\int f(y)p(y)dy$ turn out to be the same number. They are the same, as indicated in the discrete case earlier with the $T = (Y - 3)^2$ example. They are the same in the continuous case as well, although the mathematical theory needed to prove the result is somewhat advanced.

9.5 Linearity and Additivity Properties

Humans love linear functions! This is because linear functions are the easiest ones to understand. For example, the linear function $T = 2 + 3Y$ is much easier to understand than the nonlinear function $T = \exp[\cos^{-1}\{\ln(Y)\}]$. Because of its simplicity, the linear function is the most famous function of empirical research: In the vast majority of research where the effect of A on B is studied, researchers assume a linear relationship between A and B.

Nature, on the other hand, does not usually operate according to precise linear functions. Despite the human love and usage of linearity, linearity is usually wrong as a model for how Nature's data appear. Still, if DATA* produced by a linear model are, for some values of the model's parameters, similar to Nature's DATA, then the linear model is a reasonable model.

A **linear function** has the form $T = f(Y) = aY + b$. In this expression, Y is variable, while a and b are constants. When T is a linear function, its expectation is particularly simple to compute.

Linearity Property of Expected Value

For any RV Y with finite mean

$$E(aY + b) = aE(Y) + b$$

You can say this out loud—it has a pleasant rhythm: "The expected value of a linear function is equal to the linear function of the expected value."

Here is a proof of the linearity property in the continuous case; the discrete case is similar.

$E(T) = \int f(y)p(y)dy$	(By the law of the unconscious statistician)
$= \int (ay + b)p(y)dy$	(By substitution)
$= \int \{ayp(y) + bp(y)\}dy$	(By algebra)
$= \int ayp(y)dy + \int bp(y)dy$	(By the additivity property of integration; property I3 in Section 2.6)
$= a\int yp(y)dy + b\int p(y)dy$	(By property I2 in Section 2.6)
$= aE(Y) + b\int p(y)dy$	(By definition of $E(Y)$ in the continuous case)
$= aE(Y) + b\,(1.0)$	(Since $p(y)$ is a continuous pdf, its integral is 1.0)
$= aE(Y) + b$	(Since multiplying by 1.0 does not change anything)

Another property, the **additivity property of expectation**, is related to the linearity property in that it involves a sum. But unlike the linearity property, which concerns a function of one RV, the additivity property concerns a function of two RVs.

Additivity Property of Expected Value

For any RVs X and Y having finite means

$$E(X + Y) = E(X) + E(Y)$$

This property also has a pleasant rhythm when spoken: "The expected value of a sum is equal to the sum of the expected values."

The proof of the additivity property is relatively straightforward and somewhat similar to the proof of the linearity property. It involves slightly more advanced calculus, though, and is omitted here.

The additivity property can save you a lot of work, as shown in the following examples.

Example 9.4: The Expected Value of the Sum of Two Dice

If you roll a die twice, getting values X and Y, the sum is $T = X + Y$. (Here, the letter "T" could mean either "transformation" or "total"—take your pick!) You can find $E(T)$ by first finding the distribution $p(t)$ of T, then finding the weighted average $E(T) = \Sigma tp(t)$. Since the dice are independent, the probability of a 1 followed by another 1 is equal to $(1/6) \times (1/6) = 1/36$. Similarly, the probability of a 2 followed by a 4 is equal to $(1/6) \times (1/6) = 1/36$. In fact, every combination has probability 1/36, so the joint pdf is $p(x, y) = 1/36$, for $x = 1, 2, ..., 6$ and $y = 1, 2, ..., 6$. Figure 9.2 shows the joint distribution in table form, with combinations indicating where the sum is 2, 3, ..., 12.

Only one combination of die 1 and die 2 leads to $t = 2$, namely, the $(1, 1)$ combination. So $Pr(T = 2) = 1/36 = 0.02778$. By contrast, two combinations lead to $t = 3$, namely, $(1, 2)$ and $(2, 1)$, so $Pr(T = 3) = 1/36 + 1/36 = 0.0556$. Continuing this line of logic, the joint distribution of T is given as shown in Table 9.5.

The expected value calculation then gives $E(T) = \Sigma tp(t) = 2 \times 0.0278 + 3 \times 0.0556 + 4 \times 0.0833 + 5 \times 0.1111 + 6 \times 0.1389 + 7 \times 0.1667 + 8 \times 0.1389 + 9 \times 0.1111 + 10 \times 0.0833 + 11 \times 0.0556 + 12 \times 0.0278 = 7.00$. Whew!

The additivity property gives the same result but with much less work. You already know that the expected value of a single roll of a die is 3.5. So, $E(T) = E(X + Y) = E(X) + E(Y) = 3.5 + 3.5 = 7.0$. You don't have to go through the logic shown in Figure 9.2 and Table 9.5 to find the distribution of T after all. Wasn't that easier?

Die 2

		1	2	3	4	5	6
	1	0.02778	0.02778	0.02778	0.02778	0.02778	0.02778
	2	0.02778	0.02778	0.02778	0.02778	0.02778	0.02778
Die 1	3	0.02778	0.02778	0.02778	0.02778	0.02778	0.02778
	4	0.02778	0.02778	0.02778	0.02778	0.02778	0.02778
	5	0.02778	0.02778	0.02778	0.02778	0.02778	0.02778
	6	0.02778	0.02778	0.02778	0.02778	0.02778	0.02778

FIGURE 9.2

Joint distribution of two rolls of a die, indicating combinations leading to totals 2, 3, ..., 12. The combinations leading to a total of 9 are boxed. Darker shadings correspond to higher totals.

TABLE 9.5

Distribution of the Sum
of Two Dice

t	$p(t)$
2	$1/36 = 0.0278$
3	$2/36 = 0.0556$
4	$3/36 = 0.0833$
5	$4/36 = 0.1111$
6	$5/36 = 0.1389$
7	$6/36 = 0.1667$
8	$5/36 = 0.1389$
9	$4/36 = 0.1111$
10	$3/36 = 0.0833$
11	$2/36 = 0.0556$
12	$1/36 = 0.0278$
Total	1.000

The additivity property extends to any number of RVs. For example, if X, Y, and Z are all random, then the following are true:

$E(X + Y + Z)$	
$= E\{(X + Y) + Z\}$	(By algebra)
$= E\{(X + Y)\} + E(Z)$	(By the additivity property of expectation as applied to the two RVs $(X + Y)$ and Z)
$= E(X) + E(Y) + E(Z)$	(By the additivity property of expectation as applied to the two RVs X and Y)

Thus, there is no need to memorize the separate formula $E(X + Y + Z) = E(X) + E(Y) + E(Z)$. The simpler formula $E(X + Y) = E(X) + E(Y)$ is all you need to know.

Example 9.5: The Expected Value of the Sum of 1,000,000 Dice

The real beauty of the additivity property of expectation becomes apparent when a sum has many terms. If T is the sum of 1,000,000 dice values, so that $T = Y_1 + Y_2 + \cdots + Y_{1,000,000}$, you could calculate $E(T)$ by finding $p(t)$ and then using $E(T) = \Sigma tp(t)$. But just think for a second about finding $p(t)$. First, the list of possible values goes from 1,000,000 (in the unlikely event that all rolls are a 1) to 6,000,000 (in the equally unlikely event that all rolls are a 6). So there are 5,000,001 possible values of t in the list.

Now, the actual probabilities. Each combination of 1,000,000 rolls has probability $(1/6)^{1,000,000}$. And that's the easy part! The hard part is figuring out how many combinations lead to the sum t. There is only one way to get the sum $t = 1,000,000$ (a 1 every time), but imagine trying to figure out how many ways the sum can be $t = 3,453,121$. The distribution $p(t)$ can be calculated, but fortunately it's not needed. The additivity formula allows you to find the expected value of the sum much more easily: $E(T) = E(Y_1 + Y_2 + \cdots + Y_{1,000,000}) = E(Y_1) + E(Y_2) + \cdots + E(Y_{1,000,000}) = 3.5 + 3.5 + \cdots + 3.5 = 1,000,000 \times 3.5 = 3,500,000$. There is no need to find the distribution $p(t)$ at all. And that's the *real* beauty of the additivity property of expectation.

9.6 Nonlinear Functions and Jensen's Inequality

The linearity and the additivity properties, taken together, can be stated aloud in a simple, pleasing sentence: "When the function is linear and/or additive, the expected value of the function is equal to the function of the expected value."

What if the function is nonlinear or nonadditive? As the Urban Dictionary will tell you, "fuhgeddaboudit." You can't plug in expected values into functions and expect to get the right result. For example, in the dice function $T = (Y - 3)^2$ shown in Table 9.4, you can calculate the expected value of T correctly either as $E(T) = \Sigma t p(t) = 3.167$ or using the law of the unconscious statistician as $E(T) = \Sigma f(y)\, p(y) = 3.167$. What you can't do is to take the shortcut that the linearity and additivity properties suggest. In other words, you cannot say that $E\{(T - 3)^2\} = \{E(T) - 3\}^2$, because this would give you $\{3.5 - 3\}^2 = 0.25$, a far cry from the true value 3.167.

Always pay attention to the assumptions! The assumption leading to the conclusion that "the expected value of the function is equal to the function of the expected value" is that the function is linear and/or additive. When the assumptions are false, you can't trust the conclusions. A spreadsheet demonstration makes this crystal clear; see Figure 9.3.

In the spreadsheet screen shot shown in Figure 9.3, column B holds data values, column C is the linear function $f(y) = 2y + 1$, and column D is the nonlinear function $f(y) = y^2$. The average of the data in column B is 18.5. The average of the linear function in column C is 38, which is equal to the linear function of the average: $2(18.5) + 1 = 38$. The average of the nonlinear function in column D is 917.5, which is quite different from the nonlinear function of the average: $18.5^2 = 342.25 \neq 917.5$.

Nevertheless, people make the mistake of plugging averages into complex functions, expecting the result to be correct. The book, *The Flaw of Averages: Why We Underestimate Risk in the Face of Uncertainty*, by Sam L. Savage, documents this unfortunate practice.

While you can't plug averages into functions and get the right answers, you can anticipate the direction of the error using *Jensen's inequality*, when the function is either convex or concave. A **convex function** $f(y)$ is one whose slopes (i.e., derivatives) continuously increase as y increases; a **concave function** is one whose slopes continuously decrease as y increases. Since the slopes are the derivatives $f'(y)$, convexity or concavity depends on the changes in the slopes—that is, on the second derivatives.

A	B	C	D
	y	$2y+1$	y^2
	3	7	9
	5	11	25
	6	13	36
	60	121	3600
Average:	18.5	38	917.5
		$= 2 \times 18.5 + 1$	$\neq 18.5^2$

FIGURE 9.3
Illustration of the facts that the average of a linear function is equal to the function of the average and that the average of a nonlinear function is not equal to the nonlinear function of the average.

TABLE 9.6

Examples of Convex and Concave Functions

Function	Classification
y^2	Convex
$-y^2$	Concave
e^y	Convex
$y^{1/2}$, for $y > 0$	Concave
$\ln(y)$, for $y > 0$	Concave

Second Derivative Conditions for Convex and Concave Functions

If $f''(y) > 0$, for all y, then $f(y)$ is a convex function.
If $f''(y) < 0$, for all y, then $f(y)$ is a concave function.

Table 9.6 gives important examples of convex and concave functions, but does not contain an exhaustive list.

You can see convexity or concavity easily in the graph of the function. A convex function is one that lies above all of its tangent lines, and a concave function is one that lies below all of its tangent lines (see Figures 9.4 and 9.5).

Functions can be convex for some regions and concave for others, depending on the second derivative. For example, the function $f(y) = \exp(-y^2/2)$ is concave inside the region where $-1 < y < 1$ and convex outside; see Figure 9.6.

Jensen's inequality applies only to functions that are either entirely convex or entirely concave.

Jensen's Inequality

If $f(y)$ is a convex function, then $E\{f(Y)\} > f\{E(Y)\}$.
If $f(y)$ is a concave function, then $E\{f(Y)\} < f\{E(Y)\}$.

There are isolated cases where the inequality (> or <) can possibly be an equality (=), but those cases are so specialized they aren't worth worrying about. (Challenge: Find one!)

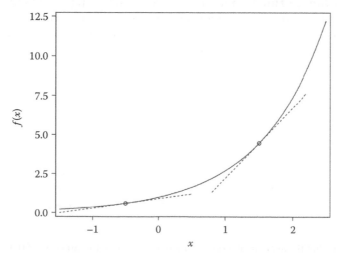

FIGURE 9.4
A convex function is one that lies above all of its tangent lines.

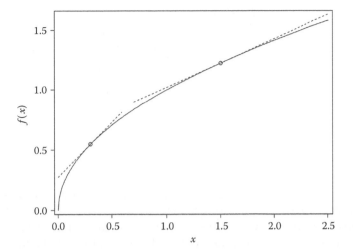

FIGURE 9.5
A concave function is one that lies below all of its tangent lines.

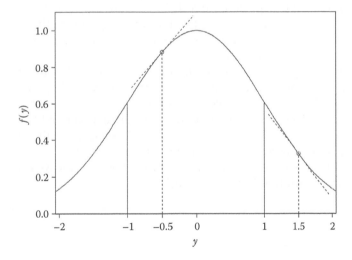

FIGURE 9.6
A function that is concave in the region $-1 < y < 1$, since the curve lies completely below its tangent lines in that region. In the region $y > 1$ and in the region $y < -1$, the curve lies above all of its tangent lines, and thus, the curve is convex in those regions.

  While we will not prove Jensen's inequality in general, you can see how it works from Figure 9.4. The two x values are -0.5 and 1.5, with an average of 0.5, and the function of the average is the corresponding value $f(0.5)$ on the curve. On the other hand, the average of $f(-0.5)$ and $f(1.5)$ is on the midpoint of a line segment joining the two data points, and this line segment lies entirely above the curve. Hence, the average of $f(-0.5)$ and $f(1.5)$ is higher than the value $f(0.5)$ on the curve. As predicted by Jensen's inequality for the convex function shown in Figure 9.4, the average of the function is more than the function of the average.
  The spreadsheet demonstration of Figure 9.3 used the function $f(y) = y^2$, a convex function. There you saw that the average of the squares was 917.5, much more than the square of the average, 342.25, also predicted by Jensen's inequality.

TABLE 9.7

Bootstrap Distribution for the
Data Set Shown in Figure 9.3

y	$\hat{p}(y)$
3	0.25
5	0.25
6	0.25
60	0.25
Total	1.00

But wait, the simple averages shown in Figure 9.3 are not expected values! Don't confuse *average* with *expected value*! Without further explanation, you can't assume that Jensen's inequality applies to simple averages, since Jensen's inequality refers to expected values, and the expected value is not the same as an average. However, the connection is simple: Recall the bootstrap distribution of Chapter 8, Section 8.7. You can create a bootstrap distribution from any observed data set by putting ($1/n$) probability on each of the observed values. In Figure 9.3, there are $n = 4$ observations, and bootstrap distribution is as given in Table 9.7.

For this distribution, the expected value E(Y) is identical to sample average $\bar{y} = 18.5$. Since Jensen's inequality applies for any distribution, it applies to the bootstrap distribution as well, and thus, you can be sure that Jensen's inequality applies to sample averages as well as to bona fide expected values.

For another example, consider the discrete uniform die distribution of Table 9.1 and the function $f(y) = (y - 3)^2$. You can see that $f(y)$ is convex by expanding $(y - 3)^2$ as $y^2 - 6y + 9$. Thus, $f'(y) = 2y - 6$. Taking the derivative again, you get $f''(y) = 2$. Since $f''(y) = 2 > 0$, the function is convex, and you know without doing any calculation that E$\{(Y - 3)^2\} > \{E(Y) - 3\}^2$. The calculations E$\{(Y - 3)^2\} = 3.167 > \{E(Y) - 3\}^2 = 0.25$ simply confirm what you already knew. Figure 9.7 provides a graph illustrating the result as well.

If the function $f(y)$ is concave, then E$\{f(Y)\} < f\{E(Y)\}$. For the concave function $f(y) = \ln(y)$, Figure 9.8 shows the result that E$\{\ln(Y)\} < \ln\{E(Y)\}$.

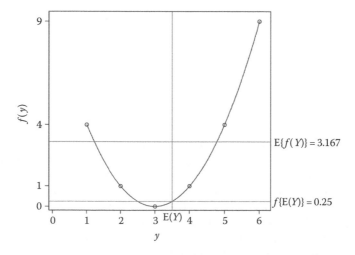

FIGURE 9.7
Graph of the convex function $f(y) = (y - 3)^2$, showing E$\{f(Y)\}$ versus $f\{E(Y)\}$ when Y has the discrete uniform die distribution.

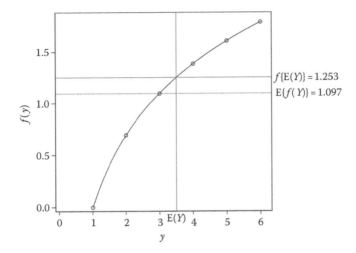

FIGURE 9.8
Graph of the concave function $f(y) = \ln(y)$, showing $E\{f(Y)\}$ versus $f\{E(Y)\}$ when Y has the discrete uniform die distribution.

If the function were a linear function, the graphs in Figures 9.7 and 9.8 would be straight lines, and the values $E\{f(Y)\}$ versus $f\{E(Y)\}$ would coincide, as the linearity property of expected value dictates.

In all cases, please be clear that $E\{f(Y)\}$ is the *correct result*. The value $f\{E(Y)\}$ is the *wrong result*. It's just easier to calculate $f\{E(Y)\}$; that's why people like to use it. That's also what the "flaw of averages" discussed by author Sam Savage refers to: It refers to the assumption that you get the right result when you plug the average into a function.

Example 9.6: Bank Profits and Housing Prices

In his book *The Flaw of Averages: Why We Underestimate Risk in the Face of Uncertainty*, Savage gives an example showing the harm of plugging averages into functions. Banks may try to predict profits using average home prices. Their profitability is actually a concave function of home prices, similar to that shown in Figure 9.8. When prices are high to start with and go up a little more, profits rise at a given rate. But when prices are low to start with and go down a little more, there is a more precipitous drop in profits because there is an attendant rise in mortgage defaults. Hence, the average profitably is not obtainable by plugging the average home price into the profitability function—that would give you too optimistic a forecast, as shown in Figure 9.8, since the profitability function is a concave function of housing prices. Savage argues that this problem was a contributing factor in the housing crisis of 2008.

9.7 Variance

Have a look at the two distributions shown in Figure 9.9. Both are distributions of $Y =$ future earnings, with means $\mu = 100.00$ for two different types of investments. Let's suppose the units are millions of U.S. dollars, although the story won't change if the units are millions of euros, British pounds, or other currency. We deliberately show the graphs as non-normal distributions.

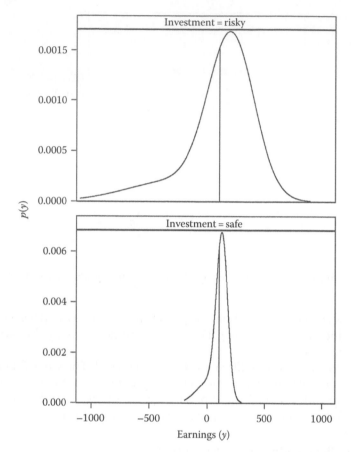

FIGURE 9.9

Distributions of potential future earnings on a risky investment and on a safe investment. In both cases, the mean of the distribution is $\mu = 100$ million dollars.

Clearly, the mean of earnings does not tell the whole story! In the case of the risky investment, it appears that you can easily lose 500. With the safe investment, such a loss appears unlikely if not impossible.

The **variance** is a measure of the spread of a distribution. If the number Y (earnings, for example) is far from the mean (either higher or lower), then the value $(Y - \mu)^2$ will be relatively larger. If the number Y (earnings, for example) is close to the mean (either higher or lower), then the value $(Y - \mu)^2$ will be relatively smaller. The function $(Y - \mu)^2$ measures *squared distance to the mean*. No matter whether Y is above or below the mean, the term $(Y - \mu)^2$ is always positive, with larger values indicating greater distance (either above or below) from the mean.

For example, in Figure 9.9, plausible future earnings values might be 600, −500, and 200 for the risky investment; for the safe investment, similar plausible future earnings values might be 225, −50, and 125. The squared deviations are $(600 - 100)^2 = 250,000$, $(-500 - 100)^2 = 360,000$, and $(200 - 100)^2 = 10,000$ for the risky investment and $(225 - 100)^2 = 15,625$, $(-50 - 100)^2 = 22,500$, and $(125 - 100)^2 = 625$ for the safe investment. It is clear that the risky investment has potential future earnings that can be much farther from the mean, as the squared deviations from the mean are much higher. The average of the three squared deviations from the mean is $(250,000 + 360,000 + 10,000)/3 = 206,667$ (millions of dollars)^2

for the risky investment and $(15{,}625 + 22{,}500 + 625)/3 = 12{,}917$ (millions of dollars)^2 for the safe investment.

Important note: These squared deviation numbers are not in units of millions of dollars. Since they are all squared measures, their units are (millions of dollars)^2.

The possible future earnings indicated earlier are just three values from each of the distributions. If you considered all the possible future earnings, and calculated the value $T = (Y - \mu)^2$ for each, you'd have a lot of numbers! These numbers would have a distribution $p(t)$, all on the positive side of zero. The mean of that distribution is called the *variance* of Y. Fortunately, by the law of the unconscious statistician, you don't need to know the distribution of $T = (Y - \mu)^2$ to know the variance. All you need is the distribution of Y itself.

Definition of Variance of an Random Variable Y

Assuming it exists, the variance of a RV Y is defined by

$$\mathrm{Var}(Y) = E\{(Y - \mu)^2\}$$

The variance is often given the symbol σ^2 (pronounced "sigma squared"). In other words, σ^2 and $\mathrm{Var}(Y)$ are different representations of the same numerical quantity:

$$\sigma^2 = \mathrm{Var}(Y) = E\{(Y - \mu)^2\}$$

The variance of an RV Y can be computed using the law of the unconscious statistician as either

$$\sigma^2 = \mathrm{Var}(Y) = \sum (y - \mu)^2 p(y), \text{ in the discrete case}$$

or

$$\sigma^2 = \mathrm{Var}(Y) = \int (y - \mu)^2 p(y) dy, \text{ in the continuous case}$$

Example 9.7: Variance of the Stoplight Green Signal Time

The distribution of the time the stoplight stays green (Y) is the continuous uniform distribution $p(y) = 0.5$, $0 < y < 2$. The mean time is $\mu = E(Y) = 1.0$, and the variance of Y is given by

$$\sigma^2 = \int_0^2 (y-1)^2 (0.5) dy = 0.5 \int_0^2 (y^2 - 2y + 1) dy = 0.5 \left(\frac{y^3}{3} - y^2 + y \right) \Big|_0^2 = 0.333$$

The units of $\sigma^2 = 0.333$ are not in minutes, but in (minutes)^2.

It is worth noting here that the "flaw of averages" applies especially well to the variance. If you were to apply crude, "flaw of averages"-type of thinking to the calculation of the variance, you might try to plug in $E(Y) = 1$ for Y in the expression $E\{(Y - \mu)^2\}$ for the stoplight example. But this would give you $(1 - 1)^2 = 0$ for the variance. Jensen's inequality explains it: The function $f(y) = (y - \mu)^2$ is a convex function, so $E\{f(Y)\} > f\{E(Y)\}$. In the case of the variance, $E\{(Y - \mu)^2\}$, which is the variance of Y, is greater than $\{E(Y) - \mu\}^2$, which is the same as $\{\mu - \mu\}^2$ or simply zero.

The formula $\text{Var}(Y) = E\{(Y - \mu)^2\}$ can be difficult to compute. The following formula is easier to compute but loses the clear interpretability in terms of squared deviations.

A Simple Computing Formula for Variance

$$\sigma^2 = \text{Var}(Y) = E(Y^2) - \mu^2$$

This formula follows easily from the linearity and additivity properties of expectation, which you can see as follows:

$\text{Var}(Y) = E\{(Y - \mu)^2\}$	(By definition)
$= E(Y^2 - 2\mu Y + \mu^2)$	(By algebra)
$= E(Y^2) - 2\mu E(Y) + \mu^2$	(By linearity and additivity properties of expectation, noting that μ is a constant)
$= E(Y^2) - 2\mu^2 + \mu^2$	(Since $E(Y) = \mu$ by definition)
$= E(Y^2) - \mu^2$	(By algebra)

In the case of the stoplight distribution in Example 9.7

$$E(Y^2) = \int_0^2 y^2 (0.5) dy = 0.5 \frac{2^3}{3} - 0.5 \frac{0^3}{3} = 1.333$$

So $\text{Var}(Y) = 1.333 - 1^2 = 0.333$, a simpler calculation than shown in Example 9.7.

The formula $\sigma^2 = E(Y^2) - \mu^2$ is great for remembering how Jensen's inequality works. Since variance is positive, $E(Y^2)$ must be greater than $\{E(Y)\}^2$. And since $f(y) = y^2$ is a convex function, you can now remember that $E\{f(Y)\} > f\{E(Y)\}$ when $f(y)$ is a convex function. You can also remember that it is the opposite for a concave function, or $E\{f(Y)\} < f\{E(Y)\}$.

Like the expected value μ, the variance σ^2 *precedes the data*. That is, the variance is part of the model that produces the data, and has the same value, no matter whether 10,000 observations are sampled, or 2, or 1 or none at all. For the uniform distribution earlier, $\sigma^2 = 0.333$ exists and is always $\sigma^2 = 0.333$, no matter whether $n = 10,000$ observations are sampled from the U(0, 2) distribution or whether $n = 2$ are sampled or whether none at all are sampled. *Model produces data.* The parameter σ^2 is part of that model, which is the U(0, 2) model is the stoplight example.

As with the term *population mean*, we will avoid using the term *population variance* for σ^2. For reasons discussed in Chapter 7, the *population* definition of $p(y)$ is simply wrong in most cases. The variance σ^2 is best thought of in terms of a *process* distribution $p(y)$ that you assume to have produced your data.

But of course, in practice, you don't know the distribution $p(y)$—all you have are the data. How can you estimate σ^2 using the data? The plug-in principle described in Chapter 8 is a great method. Simply use the bootstrap distribution as an estimate of $p(y)$, and calculate the estimated variance in terms of that estimated distribution. Recall the bootstrap distribution, shown here again as Table 9.8.

Table 9.8 shows a discrete distribution whose mean is $\sum y p(y) = y_1(1/n) + y_2(1/n) + \cdots + y_n(1/n) = \bar{y}$ and whose variance is $\sum (y - \mu)^2 p(y) = (y_1 - \bar{y})^2(1/n) + (y_2 - \bar{y})^2(1/n) + \cdots + (y_n - \bar{y})^2(1/n) = (1/n) \sum_i (y_i - \bar{y})^2$. This gives you an estimate of σ^2.

TABLE 9.8

Bootstrap Population
Distribution Based on a Sample
$y_1, y_2, \ldots, y_n$

y	$\hat{p}(y)$
y_1	$1/n$
y_2	$1/n$
...	...
y_n	$1/n$
Total	1.00

The Bootstrap Plug-In Estimate of Variance

$$\hat{\sigma}^2 = \frac{1}{n} \sum_i (y_i - \bar{y})^2$$

This estimate is slightly biased, and the divisor $n - 1$ is used instead of n by most computer software to correct this slight problem; we give more details about this in Chapter 11. However, dividing by n versus $n - 1$ makes little difference when n is large, and the plug-in estimate is a perfectly reasonable estimate in most cases. It is also intuitively appealing: Since variance is the expected squared difference from the true mean μ, it seems logical to estimate it as the average squared distance from the sample mean $\bar{y}$. A further point in favor of using n in the denominator is that it gives you the *maximum likelihood estimate* of σ^2; this is discussed in Chapter 12.

9.8 Standard Deviation, Mean Absolute Deviation, and Chebyshev's Inequality

Please have a look at the earnings distribution for the risky investment in Figure 9.9. Finding the variance σ^2 of that particular distribution requires complex calculations, but take our word for it that it is equal to 104,000 (millions of dollars)^2. The graph does not show the variance in any clear way, since the number $\sigma^2 = 104,000$ covers much more than the entire range of the horizontal axis.

But the variance is not supposed to be viewed in the graph. The units of measurement for earnings are millions of dollars, and the units of measurements for variance are (millions of dollars)^2. To make the units the same, just take the square root of the variance, and you get the **standard deviation**. In the case of the risky investment, the standard deviation is $\sigma = (104,000)^{1/2} = 322.5$ millions of dollars. Notice the units of the standard deviation: They are millions of dollars, not (millions of dollars)^2. This concept is so important that it deserves a shout out:

> The units of measurement for the standard deviation, σ, are identical to the units of measurement for the data, Y.

Since the variance is equal to the expected squared deviation from the mean, the standard deviation can also be thought of in terms of deviation from the mean. Figure 9.10

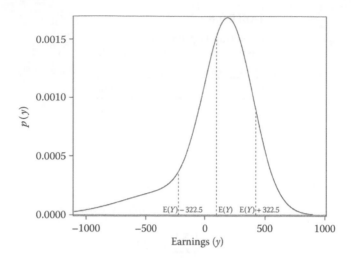

FIGURE 9.10
Distribution of future potential earnings for a risky investment, with vertical lines indicating the mean $(E(Y) = 100)$, one standard deviation above the mean $(E(Y) + 322.5)$, and one standard deviation below the mean $(E(Y) - 322.5)$.

shows the distribution of potential future earnings with the risky investment, with the mean $\mu = 100$ (millions of dollars), as well as the standard deviations ± 322.5 (millions of dollars) from the mean.

If the variance is the expected squared deviation from the mean, can you think of the standard deviation as the "expected absolute deviation from the mean"? The answer is no, and the reason is, once again, because this is "flaw of averages" thinking. The correct interpretation, through Jensen's inequality, is that the standard deviation is *larger* than the expected absolute deviation from the mean.

Sometimes abbreviated **MAD** for **"mean absolute deviation,"** the expected absolute deviation from the mean is defined as $MAD = E(|Y - \mu|)$. In words, it is the average—specifically, long-run average by the LLN—of the distance (whether high or low) from the data Y to the process mean μ.

Example 9.8: Expected Absolute Deviation and Standard Deviation for the Stoplight Green Signal Time

In the stoplight example, the MAD is computed as $E(|Y - \mu|) = \int_0^2 |y-1| p(y) dy = \int_0^2 |y-1|(0.5) dy$. Since $|y - 1|$ is not a smooth function, you need to break it into two smooth parts to integrate it: $|y - 1| = y - 1$ for $y > 1$, and $|y - 1| = 1 - y$ when $y < 1$. Hence, $MAD = \int_0^1 (1 - y)(0.5) dy + \int_1^2 (y - 1)(0.5) dy = 0.5$.

This makes sense from the graph shown in Figure 9.11: On average, the data are 0.5 away from the mean of 1.0. On the other hand, the standard deviation $\sigma = (0.3333)^{1/2} = 0.577$ does not seem to have such a clear-cut interpretation. The times are not, on average, 0.577 from the mean, they are on average 0.5 from the mean.

The relationship shown in the stoplight case with $0.577 > 0.500$, showing $\sigma > MAD$, holds in general because of Jensen's inequality. Let $V = |Y - \mu|$; then $MAD = E(V)$ by definition. Because $V^2 = |Y - \mu|^2 = (Y - \mu)^2$, it follows that $Var(Y) = E(V^2) > \{E(V)\}^2$, since $f(v) = v^2$ is a convex function. Hence, $\sigma^2 > \{E(V)\}^2$, implying that $\sigma > E(V)$ or $\sigma > MAD$.

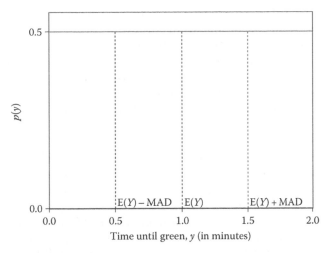

FIGURE 9.11
Uniform distribution in the stoplight example showing mean (E(Y) = 1), one MAD above the mean (E(Y) + 0.5) and one MAD below the mean (E(Y) − 0.5).

So if σ is not as interpretable as MAD, why on Earth should you use it? One reason is that σ is a natural parameter of the normal distribution, the most famous distribution in statistics. But still, that doesn't help you to interpret it, especially because real distributions usually differ from the normal distribution. If you are stuck with this σ beast, how *do* you interpret it? The answer lies in Chebyshev's inequality.

Chebyshev's Inequality

For any RV Y having finite variance:

$$\Pr\{Y \in (\mu - k\sigma, \mu + k\sigma)\} > 1 - \frac{1}{k^2}$$

Equivalently, at least $100(1 - 1/k^2)\%$ of the data will be within $\pm k$ standard deviations of the mean. You can plug in some particular values of k to make it more specific.

Famous Examples of Chebyshev's Inequality

- $k = 2$: *At least* 75% of the data will be within ±2 standard deviations of the mean.
- $k = 3$: *At least* 88.9% of the data will be within ±3 standard deviations of the mean.
- $k = 4$: *At least* 93.75% of the data will be within ±4 standard deviations of the mean.
- $k = 5$: *At least* 96% of the data will be within ±5 standard deviations of the mean.

These are not ugly rules of thumb, they are beautiful mathematical facts!

There are cases where the number can be precisely 75%, 88.9%, etc., that is, where the result is an equality instead of an inequality, but those cases are so unusual that you don't need to worry about them.

You can pick a smaller k, like $k = 0.5$, but the results won't be as interesting. With $k = 0.5$, you can conclude that the probability within ±0.5 standard deviations of the mean is at least $1 - 1/0.5^2 = -3.0$. This is certainly true, but not particularly useful, because you already knew the probability was more than −3.0!

This brings up a note about inequalities: Inequalities are more useful when they are more "sharp." A **sharp inequality** is one where the true number is close to the bound. A loose inequality is one where the number is not very close to the bound. For example, your statistics professor could say "My age is less than 200 years." That would be true, but uninformative. On the other hand, if he or she told you "My age is less than 30 years," the inequality is much sharper, and hence more informative, since it is now likely that your instructor is in his or her late twenties.

So, while the statement that the probability of observing Y to be within $\mu \pm 0.5\sigma$ is "> −3.0" is correct, it is not useful. Similarly, the $k = 2$ result, ">0.75," is not very sharp either, because it says the true probability could be anything between 0.75 and 1.0. On the other hand, the inequalities with $k = 3$ and higher are fairly sharp, because the true probability is now known to be between $1 - 1/k^2$, which is a number close to 1.0, and 1.0 itself.

The bottom line is that Chebyshev's inequality becomes more useful for larger k. But no matter what, *do not assume that* $1 - 1/k^2$ *is equal to* the true probability. The true probability is *more than* $1 - 1/k^2$. It is called Chebyshev's *inequality* specifically because the true probability is *not equal to* $1 - 1/k^2$.

On the other hand, Chebyshev's inequality is a beautiful and useful result because it is true for any distribution. As long as you know the mean and the standard deviation, you can make a factual statement about percentages of data values that will fall within ±2, ±3, and generally ±k standard deviations of the mean.

> **Example 9.9: Chebyshev's Inequality for the Stoplight Green Signal Time**
>
> For the stoplight green signal time scenario, you can apply Chebyshev's inequality with $k = 2$ to arrive at the statement, "The probability that the light will stay green between $1 - 2(0.577)$ and $1 + 2(0.577)$ minutes is at least 0.75. Or the probability that the light will stay green between −0.14 and 2.14 minutes is at least 0.75. This is true, but not informative: The actual probability is 1.0, since the light can stay green no longer than 2.0 minutes."

A smaller k gives a slightly more informative result in this example. With $k = 1.5$, the limits are $1.0 \pm 1.5(0.577)$ or between 0.134 and 1.866 minutes. Without knowing anything about the distribution, you can say that at least $100(1 - 1/1.5^2)\% = 55.5\%$ of the data will be in this range. Here, we know the distribution is uniform, so the true probability is $(1.866 - 0.1354)/2 = 86.6\%$. The inequality is correct since $86.6\% > 55.5\%$, although it is, again, not particularly useful. But the beauty of Chebyshev is that it is true for all distributions, not that it supplies the correct probability.

The following example shows a case where Chebyshev's inequality is more useful.

> **Example 9.10: Chebyshev's Inequality Applied to DJIA Return Data**
>
> The inequality applies to probability distributions and not to sample data, but you can make it apply to data by using the bootstrap trick, where you create a fake probability distribution $\hat{p}(y)$ out of the observed data, by putting $1/n$ probability on every observed value y_i. Then the mean of this fake distribution is $y_1(1/n) + \cdots + y_n(1/n) = \bar{y}$, and this is called the *plug-in* estimate of the mean. The variance of this fake distribution is $(y_1 - \bar{y})^2(1/n) + \cdots + (y_n - \bar{y})^2(1/n)$, and this is called the *plug-in* estimate of the variance. Taking the square root of the plug-in estimate of the variance yields the *plug-in* estimate of standard deviation, and now you can apply Chebyshev's inequality directly to the data.

For the n = 18,834 DJIA returns discussed in Chapter 4, the plug-in mean is 0.000200189, and the plug-in standard deviation is 0.0111868. Thus, using k = 5, Chebyshev's inequality tells you that there is at least a 96% probability that a return will lie inside the range 0.000200189 ± 5 × 0.0111868 or inside the range from −0.055733811 to 0.056134189. In the data set, there are 18,769 values in this range, so in fact 18,769/18,834 = 99.65% of the data are in that range. Chebyshev's inequality is validated here, since 99.65% > 96.0%.

The point is, you could state "At least 96% of the stock returns are between −0.055733811 and 0.056134189," if the only information you had was the mean and the standard deviation. You wouldn't need the data set. You wouldn't even have to know how many observations are in the data set! And that's the beauty of Chebyshev's inequality.

Example 9.11: The Normal Distribution, the 68–95–99.7 Rule, and Chebyshev's Inequality

Chebyshev's inequality works for any distribution. The normal distribution is so important, though, that you should memorize the specific probabilities for the $\mu \pm k\sigma$ ranges when data come from a normal distribution. Fortunately, it's easy: Just remember **68–95–99.7**.

The 68–95–99.7 Rule for a Normal Distribution

If Y is produced by the N(μ, σ^2) distribution, then

- 68% of the Y values will be between $\mu - \sigma$ and $\mu + \sigma$.
- 95% of the Y values will be between $\mu - 2\sigma$ and $\mu + 2\sigma$.
- 99.7% of the Y values will be between $\mu - 3\sigma$ and $\mu + 3\sigma$.

These are *long-run* statements about the observed data; recall from Section 8.5 that you can interpret probability as a log-run frequency.

Some statistics sources call the 68–95–99.7 rule the *empirical rule*, since empirical (or observed) data sets often follow these percentages. However, only data sets that look as if produced by a normal distribution will obey these percentages, so it's safer to call this rule the *68–95–99.7 rule for a normal distribution*.

Even though the percentages differ from those from Chebyshev's inequality, there is nothing contradictory here: Chebyshev simply tells you that these three probabilities are *at least* 0%, 75%, and 88.8%, respectively.

It doesn't matter what μ and σ are; you always get the same 68%, 95%, and 99.7% probabilities for normally distributed processes. For example, using μ = 70 and σ = 10 in Excel, you can calculate the three standard deviation range by entering into a cell of the spreadsheet the following expression:

= NORM.DIST(70+3*10, 70,10,TRUE) − NORM.DIST(70−3*10, 70,10,TRUE)

The software will return the result 0.9973. Change the mean and standard deviation from (70, 10) to any other numbers (provided the standard deviation is positive), and you will still get the same 0.9973 result. Change the range to ±2 standard deviations, and you will always get 0.9545, and change the range to ±1 standard deviation, and you will always get 0.6827, no matter what μ and σ you use.

Which raises the question, Why 68–95–99.7? Why not the more correct 68.27–95.45–99.73? The answer is simple: Keep it simple! You'll remember 68–95–99.7 more easily. You'll remember even more easily if you picture those numbers along with the Figure 9.12.

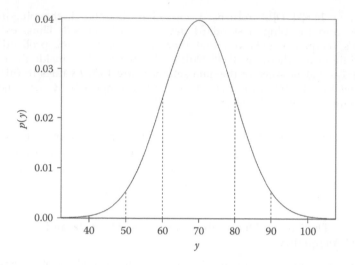

FIGURE 9.12
Graph of the N(70, 10^2) distribution showing the mean plus or minus one standard deviation ($\mu \pm \sigma$, or 70 ± 10), the mean plus or minus two standard deviations ($\mu \pm 2\sigma$, or 70 ± 20), and the mean ± 3 standard deviations ($\mu \pm 3\sigma$, or 70 ± 30).

In Figure 9.12, you can see that 68% of the area under the curve lies between 60 and 80, 95% of the area lies between 50 and 90, and 99.7% of the area lies between 40 and 100.

The 68%–95%–99.7% probabilities are correct for normal distributions but incorrect for other distributions. As shown with the stoplight example, Example 9.9, which involved the uniform distribution (a non-normal distribution), 100% of the times are within the range $\mu \pm 2\sigma$, not 95% as predicted by a normal distribution.

> **Example 9.12: The 68–95–99.7 Rule Applied to Dow Jones Industrial Average Daily Returns**
>
> The Dow Jones Industrial Average (DJIA) returns were shown to be clearly non-normal in Chapter 4, Figure 4.9. Using the bootstrap plug-in estimates, the $\hat{\mu} \pm 1\hat{\sigma}$, $\hat{\mu} \pm 2\hat{\sigma}$, and $\hat{\mu} \pm 3\hat{\sigma}$ ranges are −0.0109866 to 0.0113870, −0.0221734 to 0.0225738, and −0.0333602 to 0.0337606, respectively. By simple counting (using the computer!), there are 81.2%, 95.4%, and 98.3% of the actual Dow Jones returns, respectively, in these ranges, rather than 68%–95%–99.7% as predicted by the normal distribution model. While the actual percentage for the ±2 standard deviation range (95.4%) is close to the 95% predicted by the normal model, the other percentages (81.2% and 98.3%) differ markedly from the 68% and 99.7% that you would see if the distribution were normal.

9.9 Linearity Property of Variance

Recall the linearity property of expectation:

$$E(aY + b) = aE(Y) + b$$

What about variance? Naïvely, you might think the same equation works for variance or that $Var(aY + b) = aVar(Y) + b$. But this is wrong! Please forget you ever saw this formula! There are several ways to remember that this formula is wrong. The first is that it will give you a negative variance in the case where $a = -1$ and $b = 0$. But variances, being squared measures, can *never* be negative. The second way to understand that this formula is wrong is that variance only measures *deviation* from mean. If you shift the data upward by adding a positive constant b to all Y values, the variance of Y is not affected: Only the mean of Y is affected. Thus, there is no constant term b in the formula. The following example illustrates this concept.

Example 9.13: Gambler's Earnings versus Money in Pocket

Suppose you decide to play 10 (dollars or euros or ...) on red at the roulette wheel, 20 times. Your total earnings are $T = Y_1 + Y_2 + \cdots + Y_{20}$, where the Y_i are independent and identically distributed (iid) from the roulette distribution shown as Table 8.1, repeated here as Table 9.9.

The values of your total earnings, T, can range from -200 (a loss every time) to $+200$ (a win every time).

Suppose you start with 200 in your pocket. Then after 20 plays, the amount you have in your pocket is $M = T + 200$. (We use the letter M here because we're talking about moolah!) Figure 9.13 shows both the distributions of T and of M.

As you can see in Figure 9.13, the possible variation in your total earnings is exactly the same as the possible variation in your money in pocket. Adding a constant to a RV (like $b = 200$ in this case) does not affect its variance.

Here is the correct formula for the linearity property of variance.

The Linearity Property of Variance

If Y is a RV having finite variance, and a and b are constants, then:

$$\text{Var}(aY + b) = a^2 \text{Var}(Y)$$

You can see why the linearity formula is true using only algebra and the linearity property of expectation. Let $T = aY + b$. Then

$\text{Var}(T) = E\{T - E(T)\}^2$	(By definition)
$= E\{aY + b - E(aY + b)\}^2$	(By substitution)
$= E[aY + b - \{aE(Y) + b\}]^2$	(By linearity property of expectation)
$= E[a^2\{Y - E(Y)\}^2]$	(By algebra; note the disappearance of the constant b)
$= a^2 E\{Y - E(Y)\}^2$	(By linearity property of expectation)
$= a^2 \text{Var}(Y)$	(By definition of variance)

Taking the square root gives you the corresponding formula for the standard deviation.

TABLE 9.9

Distribution of Earnings for One Play of 10 on Red in Roulette

Earnings, y	$p(y)$
-10	$20/38 = 0.526$
10	$18/38 = 0.474$
Total	1.00

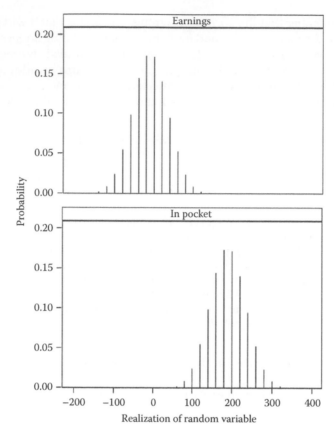

FIGURE 9.13
Distributions of total earnings and money in pocket (assuming 200 to start) after 20 plays of 10 on red in roulette.

Linearity Property of Standard Deviation

If Y is a RV having finite variance, and a and b are constants, then:

$$\text{StdDev}(aY + b) = a \times \text{StdDev}(Y)$$

Example 9.14: The Z-Score

Suppose you got a 75 on the midterm. Was your score good, bad, or typical? If the boundaries 90–100 = A, 80–89 = B, etc. are used, then 75 doesn't look very good. But what if it was a hard test? If the mean was 65, then your score seems okay, perhaps typical. The **z-score** is a statistic that measures distance from the mean in terms of number of standard deviations and is defined as follows:

Definition of Z-Score

If the distribution of Y has mean μ and standard deviation σ, the z-score for a generic y is

$$z = \frac{(y - \mu)}{\sigma}$$

If $z = 2$, then by solving for y you get $y = \mu + 2\sigma$; the data y are two standard deviations above the mean μ. If $z = -0.3$, then by solving for y, you get $y = \mu - 0.3\sigma$; that is, the data y are 0.3 standard deviations below the mean μ.

Because of Chebyshev's inequality, you should have an idea what z-scores mean, no matter what distribution may have produced your data. For example, if you have a value $z = 5.0$, then the corresponding y data value is five standard deviations above the mean. Chebyshev tells you that at least 96% of the observations are within ±5 standard deviations of the mean; hence, if you have a z-value of 5.0, then the corresponding y value is unusually high. The larger the z-score (in absolute value), the more unusual the y data value. Want to know whether an observation is outlier? Just compute the z-score. Time for another ugly rule of thumb!

Ugly Rule of Thumb 9.1

If the z-score is greater than +3.0 or less than −3.0, then the observation is an outlier.

With the raw data Y it is not so obvious, just by looking at the data values, which values are typical and which are outliers. To see that a z-score of 3.0 tells you that the data y is three standard deviations from the mean, just do some math: $z = 3.0$ implies $(y - \mu)/\sigma = 3.0$, which in turn implies that $y = \mu + 3\sigma$.

The z-score has the attractive property that it does not change when there is a constant shift in the distribution. For example, if the teacher added 10 to everyone's score, then your z-score wouldn't change. Your actual score would change, $75 + 10 = 85$, but the mean would also be 10 points higher, and the standard deviation would be unchanged (by the linearity property), so your z-score would also remain as it was.

The z-score also has the attractive property that it is does not change when the measurement units are changed. For example, the teacher decided to multiply everyone's score by 10, then the mean and standard deviation would also be multiplied by 10 (again by the linearity property), and the 10 would cancel from the numerator and denominator, again leaving your z-value unchanged.

Statistics that do not change in value when the units are changed are called **scale-free**. This is an important property—you shouldn't think that anything was fundamentally different if you decided to report data in millions of dollars rather than in actual dollars. A scale-free statistic does not change its value depending upon whether the data are in dollars or thousands of dollars. The z-score is one example of a scale-free statistic; other examples of scale-free statistics are the *skewness* and *kurtosis* statistics, discussed shortly, the *correlation coefficient*, discussed in the next chapter, and *test statistics* that researchers commonly use in hypothesis testing, discussed in later chapters.

By the linearity properties of expectation and variance, the mean of the RV z-score is zero, and its variance is one. This is shown as follows:

$$Z = (Y - \mu)/\sigma \quad \text{(By definition)}$$
$$= (1/\sigma)Y + \{-(\mu/\sigma)\} \quad \text{(By algebra)}$$

Letting $a = (1/\sigma)$ and $b = -(\mu/\sigma)$, the z-score is a linear function of Y, $Z = aY + b$. Hence, the expected value of Z is zero, which is shown as follows:

$E(Z) = E(aY + b)$	(By substitution)
$\quad = aE(Y) + b$	(By the linearity property of expectation)
$\quad = (1/\sigma)\mu + \{-(\mu/\sigma)\}$	(By substitution)
$\quad = 0$	(By algebra)

Further, the variance of Z is one:

$\text{Var}(Z) = \text{Var}(aY + b)$	(By substitution)
$\quad = a^2\,\text{Var}(Y)$	(By the linearity property of variance)
$\quad = (1/\sigma)^2\sigma^2$	(By substitution)
$\quad = 1$	(By algebra)

9.10 Skewness and Kurtosis

No real-world data-generating distribution $p(y)$ is a perfect normal distribution. Just like a perfectly symmetric circle is a mathematical ideal that does not exist in the real world, the normal distribution is also a perfect, symmetric mathematical ideal that does not exist in the real world. However, calculations based on perfect circles are used routinely by scientists and architects, and they are very useful. Calculations based on perfect normal distributions are also used routinely by scientists (although perhaps not so much by architects!), and they are also very useful.

The 68–95–99.7 rule is based on a perfect normal distribution, and it is *approximately correct* when the data-generating process distribution $p(y)$ is *approximately normal*. But remember, the word *approximately* is a weasel word that you should always question—a horse is *approximately* a cow, after all. You can judge how close the distribution is to normal using histograms and quantile–quantile plots, as described in Chapter 4. You can also use the skewness and kurtosis statistics to compare the distribution to a normal distribution.

Two important characteristics of the normal distribution are symmetry and lack of outliers. You can see the symmetry about the mean clearly in Figure 9.12. Lack of outliers is harder to visualize—after all, the distribution goes from $-\infty$ to $+\infty$, so aren't some extreme data values possible? The answer is yes, outliers are possible with the normal distribution, but they are extremely rare. The probabilities of the tails go to zero so quickly—the "$-(y - \mu)^2$" term in the exponent of the function form of the pdf explains it—that extreme outliers simply do not occur, for all intents and purposes.

The **skewness** and **kurtosis** parameters are measures of asymmetry and outlier-producing behavior, or **tail behavior**, of a distribution $p(y)$. Unlike the mean and variance, these measures reflect only the shape of the distribution and tell you nothing about the location (μ) or spread (σ). Like the variance, they are expected values of certain functions of the RV Y that is produced by $p(y)$.

Definitions of Skewness and Kurtosis

$$\text{Skewness} = E\left(\frac{Y - \mu}{\sigma}\right)^3$$

$$\text{Kurtosis} = E\left(\frac{Y-\mu}{\sigma}\right)^4 - 3$$

Notice, first of all, that these statistics are both functions of the z-score $Z = (Y - \mu)/\sigma$ and hence are location-free and scale-free. In other words, unlike mean and variance, skewness and kurtosis parameters remain unchanged following a linear transformation $T = aY + b$, provided $a > 0$. If $a < 0$, then the skewness changes in sign, but all else remains the same.

Both skewness and kurtosis measure deviation from normality, and both numbers are 0 for a normal distribution. The skewness can range from negative infinity to positive infinity, and kurtosis can range from –2 to positive infinity. How large do skewness and kurtosis have to be before you should be concerned? This calls for some more ugly rules of thumb!

Ugly Rule of Thumb 9.2

When the skewness is greater than or equal to +2 or less than or equal to –2, then the distribution is markedly different from a normal distribution in its asymmetry.

Ugly Rule of Thumb 9.3

When the kurtosis is greater than or equal to 3, then the distribution is markedly different from a normal distribution in its propensity to produce outliers.

NOTE: Ugly Rule of Thumb 9.3 for kurtosis applies to the kurtosis formula that includes the –3 term, as defined earlier, which is sometimes called *excess kurtosis*. The ugly rule would be "kurtosis greater than 6" if you use the kurtosis formula that does not include –3 term.

At this point, it may be a good idea to browse the Internet to find the skewness and kurtosis parameters for some common distributions. You'll find, for example, that the exponential distribution has skewness 2, enough to be called "markedly different" from normal by Ugly Rule of Thumb 9.2. Its kurtosis is 6, so the exponential distribution is much heavier tailed than the normal distribution by Ugly Rule of Thumb 9.3. The discrete Poisson distribution has skewness and kurtosis that depend upon its mean, $\lambda^{-1/2}$ and λ^{-1}, respectively. For λ near zero, the Poisson distribution differs markedly from the normal distribution, but for large λ, the Poisson distribution resembles a normal distribution.

Example 9.15: Calculating Mean, Variance, Standard Deviation, Skewness, and Kurtosis from a Discrete Distribution

While you'll have to use calculus to find skewness and kurtosis for continuous distributions, you can find them easily using simple spreadsheet operations in the discrete case. Table 9.10 depicts a discrete distribution that is both skewed and outlier-prone. Skewness and kurtosis require the mean and standard deviation, so these intermediate calculations are given in Table 9.10 as well.

From Table 9.10, $\sigma = (\sigma^2)^{1/2} = (185.61311)^{1/2} = 13.624$. Using the linearity property of expectation, you find

$$\text{Skewness} = E\left(\frac{Y-\mu}{\sigma}\right)^3 = \left(\frac{1}{\sigma^3}\right)E(Y-\mu)^3 = \left(\frac{1}{13.624^3}\right)9459.15 = 3.74$$

TABLE 9.10

Calculating the Mean, Variance, Skewness, and Kurtosis Parameters of a Discrete Distribution

y	$p(y)$	$y \times p(y)$	$(y - \mu)^2 \times p(y)$	$(y - \mu)^3 \times p(y)$	$(y - \mu)^4 \times p(y)$
1	0.02	0.02	3.06034	−37.856	468.3
3	0.15	0.45	16.13054	−167.274	1734.6
10	0.66	6.60	7.49555	−25.260	85.1
30	0.15	4.50	41.48354	689.871	11,472.6
90	0.02	1.80	117.44314	8999.668	689,644.5
Totals	1.00	$\mu = 13.37$	$\sigma^2 = 185.61311$	$E(Y - \mu)^3 = 9459.149$	$E(Y - \mu)^4 = 703{,}405.1$

According to Ugly Rule of Thumb 9.2, this distribution differs markedly from the normal distribution in its asymmetry, since $3.74 > 2$.

You also have:

$$\text{Kurtosis} = E\left(\frac{Y - \mu}{\sigma}\right)^4 - 3 = \left(\frac{1}{\sigma^4}\right)E(Y - \mu)^4 - 3 = \left(\frac{1}{13.624^4}\right)703{,}405.1 - 3 = 17.4$$

According to Ugly Rule of Thumb 9.3, this distribution differs markedly from the normal distribution in its propensity to produce outliers, since $17.4 > 3$.

The logic for why the skewness formula $E\{(Y - \mu)/\sigma\}^3$ measures asymmetry is as follows. First, note that the term $(Y - \mu)^3$ is either positive or negative, depending upon whether Y is above or below the mean μ. Now, since expectation (E) denotes a kind of average, the skewness will be either positive or negative, depending upon whether the positive terms or negative terms dominate in the average. Extreme values have an inordinately large effect on averages; hence, it is the extreme values that determine the sign. If the distribution of Y is asymmetric, with greater propensity for producing extreme values that are on the high side of μ, then these values will dominate the average, and you will have a positive skewness. Conversely, if the distribution of Y is asymmetric, with greater propensity for producing extreme values that are on the low side of μ, then these values will dominate the average and you will have a negative skewness. Finally, if the distribution is symmetric, then the positive and negatives all balance and your skewness will be 0.

See Figure 9.14 for examples of distributions with positive skewness, negative skewness, and zero skewness.

Meanwhile, the logic for the why the kurtosis formula $E\{(Y - \mu)/\sigma\}^4 - 3$ measures outlier propensity is as follows. First, note that the term $(Y - \mu)^4$ is always positive, with the occasional numbers that are far from μ (outliers) being greatly influential since they are taken to the 4th power. The net result is that kurtosis will be large when the distribution produces occasional outliers. The -3 term makes the kurtosis exactly 0 in the case of the normal distribution. Thus, if the kurtosis is greater than zero, then the distribution is more outlier-prone than the normal distribution; and if the kurtosis is less than zero,

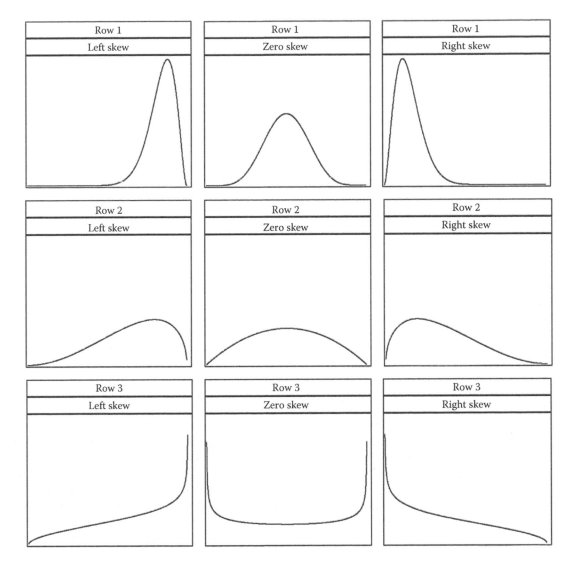

FIGURE 9.14
Distributions with negative skewness (left panels), zero skewness (middle panels), and positive skewness (right panels).

then the distribution is less outlier-prone than the normal distribution. See Figure 9.15 for distributions with varying degrees of kurtosis.

In Figure 9.15, the triangle distribution is bounded between 40 and 60 and hence is less outlier-prone than the normal distribution, explaining its negative kurtosis. The distribution that appears rectangular is actually a mixture of a uniform (30–70) distribution with a distribution that has tails extending to negative infinity and positive infinity; however, since the probability is so small on the part of the mixture that extends to infinity, it is barely visible. As in the case of histograms, the tails are hard to see in distribution plots; q–q plots are better ways to compare distributions when considering tail behavior.

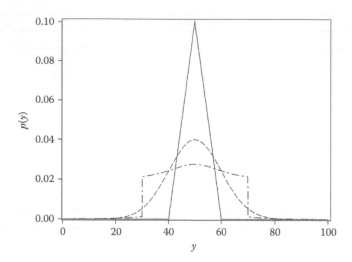

FIGURE 9.15
Distributions with negative kurtosis (solid), zero kurtosis (dashed), and positive kurtosis (dash-dot).

Kurtosis is one of the most useful measures of a distribution, but it is one of the most commonly misinterpreted measures as well. In ancient texts (some hewn into stone tablets by our evolutionary ancestors!), the terms *mesokurtic, platykurtic,* and *leptokurtic* appear, as descriptions of distributions with zero, negative, and positive kurtosis, respectively. These terms, while impressive sounding, are actually quite misleading. The problem with these terms is that the prefixes *platy-* and *lepto-* are descriptors of the *peak* of a distribution, rather than its *tails. Platykurtic* means "broad peaked" and *leptokurtic* means "thin peaked," and this is how kurtosis is often presented—as a descriptor of the peak of the distribution. The word *kurtosis* itself derives from a word meaning "curve."

But as you can see in Figure 9.15, the shape of the peak has nothing to do with positive or negative kurtosis. The thinnest peak in Figure 9.15 should be leptokurtic by the definition of "lepto," but it has negative kurtosis. The broadest peak in Figure 9.15 should be platykurtic by the definition of "platy," but it has positive kurtosis. Kurtosis has little to do with the peak of the distribution; it is determined mainly by the tails.

To lend some mathematical explanation to this point, note that the kurtosis $E\{(Y - \mu)/\sigma\}^4 - 3$ can be written as follows:

$$\text{Kurtosis} = \int_{\text{PeakSet}} \{(y-\mu)/\sigma\}^4 p(y)dy + \int_{\text{TailSet}} \{(y-\mu)/\sigma\}^4 p(y)dy - 3$$

Here, PeakSet $= \{y; |y - \mu|/\sigma \le 1\}$ and TailSet $= \{y; |y - \mu|/\sigma > 1\}$. Thus:

$$\text{Kurtosis} = \text{Peak} + \text{Tail} - 3$$

In this equation, Peak $= \int_{\text{PeakSet}} \{(y-\mu)/\sigma\}^4 p(y)dy$ and Tail $= \int_{\text{TailSet}} \{(y-\mu)/\sigma\}^4 p(y)dy$. Clearly, Peak ≥ 0. But since $\{(y - \mu)/\sigma\}^4 \le 1$ when y is in the PeakSet, it therefore follows that $\int_{\text{PeakSet}} \{(y - \mu)/\sigma\}^4 \, p(y)dy \leq \int_{\text{PeakSet}} (1)p(y)dy$. Further, since $\int_{\text{PeakSet}} (1)p(y)dy \leq \int_{\text{All } y} p(y)dy = 1$, it follows that $0 \leq \text{Peak} \leq 1$. Thus

$0 \leq \text{Peak} \leq 1$	(As just shown)
$\Rightarrow \text{Tail} - 3 \leq \text{Peak} + \text{Tail} - 3 \leq \text{Tail} - 3 + 1$	(By algebra, adding Tail − 3 to all)
$\Rightarrow \text{Tail} - 3 \leq \text{Kurtosis} \leq \text{Tail} - 2$	(Since Kurtosis = Peak + Tail − 3)

In other words, kurtosis is determined, up to ±0.5, by the *tail* behavior of the distribution $p(y)$. Figure 9.16 shows how much of the kurtosis is explained by the tail behavior of $p(y)$ according to this formula, along with the fact that kurtosis is mathematically ≥ -2.

So, any source you might read, whether textbook or online, is simply wrong if it states that "thin peaks" correspond to positive kurtosis and "broad peaks" correspond to negative kurtosis. The math illustrated by Figure 9.16, as well as the counterexamples shown in Figure 9.15, explains why these statements are wrong.

Do not promote confusion! Don't be a trained parrot! Avoid using the incorrect and misleading *platykurtic* and *leptokurtic* terms. If a distribution $p(y)$ has positive kurtosis, simply say that $p(y)$ is more outlier-prone than the normal distribution. Or, you can say that the tails of the distribution $p(y)$ are fatter than the tails of the normal distribution. If a distribution $p(y)$ has negative kurtosis, simply say that $p(y)$ is less outlier-prone than the normal distribution. Or, you can say that the tails of the distribution $p(y)$ are thinner than the tails of the normal distribution.

Using data, you can estimate skewness and kurtosis using the same plug-in principle shown earlier to estimate mean and standard deviation. However, as in the case of the plug-in estimate of the standard deviation, there will be slight discrepancies between the computer printout that shows skewness and kurtosis estimates and your plug-in estimates, again because the computer defaults employ minor bias-reducing corrections. But either estimate is fine. You can and should look at skewness and kurtosis estimates whenever you analyze a set of data.

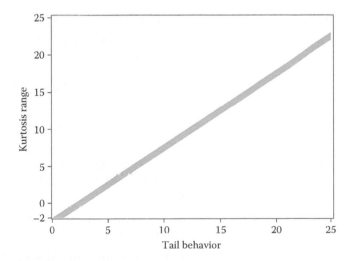

FIGURE 9.16
Kurtosis range, the shaded vertical range, determined by the portion of the distribution $p(y)$ that is more than 1 standard deviation from the mean.

Vocabulary and Formula Summaries

Vocabulary

Transformation	Another word for a function.
Law of the unconscious statistician	A formula that allows you to calculate the expected value of a transformed RV without having to find the pdf of the transformed RV.
Linear function	A function taking the form $aY + b$ for constants a and b.
Additivity and linearity properties of expectation	If a function is linear and/or additive, then the expected value of the function is equal to the function of the expected value.
Convex function	A function whose second derivative is greater than zero; one that lies above all of its tangent lines.
Concave function	A function whose second derivative is less than zero; one that lies below all of its tangent lines.
Jensen's inequality	A result that tells you how the expected value of a function differs from the function of the expected value, depending upon whether the function is concave or convex.
Variance (σ^2)	The average squared distance to the mean, defined as the expected value of the function $(Y - \mu)^2$.
Standard deviation (σ)	The (positive) square root of the variance.
Linearity property of variance	A result that tells you that constant shift terms do not affect variance and that constant multiplicative terms change the variance by the value of the constant, squared.
Mean absolute deviation (MAD)	The expected absolute distance to the mean, defined as the expected value of the function $|Y - \mu|$.
Chebyshev's inequality	A result that tells you how often data are within a certain number of standard deviations of the mean.
Sharp inequality	An inequality in which the true number is close to the bound.
68–95–99.7 Rule	How often normally distributed DATA are within ±1, ±2, and ±3 standard deviations of the mean.

Z-score	A statistic that measures distance from the mean in terms of standard deviations, defined as $z = (y - \mu)/\sigma$.
Scale-free statistics	Statistics that do not change in value when the units are changed.
Skewness	A measure of symmetry of a pdf.
Kurtosis	A measure of how outlier-prone a distribution is.
Tail behavior	The behavior of the tail of a probability distribution.

Key Formulas and Descriptions

$E(T) = \sum f(y)p(y)$	The expected value of a function of a discrete RV Y.
$E(T) = \int f(y)p(y)dy$	The expected value of a function of a continuous RV Y.
$E(aY + b) = aE(Y) + b$	The linearity property of expected value, for constants a and b.
$E(X + Y) = E(X) + E(Y)$	The additivity property of expected value, for RVs X and Y.
$E\{f(Y)\} > f\{E(Y)\}$	Jensen's inequality when the function f is convex.
$E\{f(Y)\} < f\{E(Y)\}$	Jensen's inequality when the function f is concave.
$\sigma^2 = \sum (y - \mu)^2 p(y)$	The variance of a discrete RV Y.
$\sigma^2 = \int (y - \mu)^2 p(y)\,dy$	The variance of a continuous RV Y.
$\hat{\sigma}^2 = (1/n) \sum_i (y_i - \bar{y})^2$	The bootstrap plug-in estimate of the variance of the RV Y.
$\text{M.A.D.} = E(|Y - \mu|)$	The MAD of an RV Y.
$\Pr\{Y \in (\mu - k\sigma, \mu + k\sigma)\} > 1 - 1/k^2$	Chebyshev's inequality (the proportion of observations that lie within k standard deviations of the mean is at least $1 - 1/k^2$).
$\text{Var}(aY + b) = a^2\text{Var}(Y)$	The linearity property of variance for an RV Y.
$\text{StdDev}(aY + b) = a \times \text{StdDev}(Y)$	The linearity property of the standard deviation for an RV Y.
$z = (y - \mu)/\sigma$	The z-score for a variable y.
$\text{Skewness} = E\left(\frac{Y - \mu}{\sigma}\right)^3$	The skewness of an RV Y.
$\text{Kurtosis} = E\left(\frac{Y - \mu}{\sigma}\right)^4 - 3$	The (excess) kurtosis of an RV Y.

Exercises

9.1 Consider the following data set:

x	y
1	2.2
3	3.2
2	1.7
3	4.3
5	72.1

A. Is the average of the *x* values plus the average of the *y* values equal to the average of the (*x* + *y*) values? Show how to do the calculation each way, so it is clear that you understand the concept. Also, give the name of the property that you just demonstrated, but be careful not to confuse $E(X)$ with the average of the *x* values.

B. Is the average of the *x* values, squared, equal to the average of the squared *x* values? Show how to do the calculation each way, so it is clear that you understand the concept. Relate your answer to Jensen's inequality. Concavity or convexity should be part of the answer; also, be careful not to confuse $E(X)$ with the average of the *x* values.

C. Is 100 minus two times the average of the *y* values equal to the average of the (100 − 2*y*) values? Show how to do the calculation each way so it is clear that you understand the concept. Also, give the name of the property that you just demonstrated, again being careful that you don't confuse *expected value* with *average* in your answer.

D. Is the average of the *x* values multiplied by the average of the *y* values equal to the average of the *xy* values? Show how to do the calculation each way so it is clear that you understand the concept.

9.2 Use the distribution shown in Example 9.15, presented again here:

y	p(y)
1	0.02
3	0.15
10	0.66
30	0.15
90	0.02
Total	1.00

A. Graph the pdf of *Y*, respecting the scale. In other words, the horizontal axis should show clearly that 90 is much farther from 30 than 3 is to 1. Describe the data that will be produced by the pdf that you have graphed. Don't use the terms *mean*, *standard deviation*, *skewness*, and *kurtosis* in your description of the data that are produced by this pdf.

B. Find the expected value, variance, standard deviation, skewness, and kurtosis of *Y*. Interpret each of these quantities.

C. Let $T = \ln(Y)$. Find and graph the pdf of T. Describe the distribution of T in comparison with the distribution of Y—is there anything remarkable or noteworthy?

D. Find $E(T)$ and verify that Jensen's inequality holds by comparing $E\{\ln(Y)\}$ with $\ln\{E(Y)\}$. Establish convexity or concavity as needed.

E. Find the skewness and kurtosis of T. Interpret each of these quantities, and compare them with those you found for Y. Was the logarithmic transformation "good"?

9.3 Consider* a simplified model of the insurance industry. A life insurance company sells a policy to a 21-year-old person. The policy pays 100K if the insured person dies within the next 5 years. The company collects 1K at the beginning of each year as payment for the insurance. Let the RV Y denote the company's earnings from the policy, 1K per year less the 100K it must pay if the insured dies in the next 5 years. The probability that a randomly chosen person will die each year at this age is approximately 0.002. Thus, the distribution of the company's earnings is given as shown in the following table. Note that 99% of the time the insurance company makes a small profit (the premium), but the other 1% of the time the company loses a lot of money.

Age at death:	21	22	23	24	25	26+
Earnings, y	−99K	−98K	−97K	−96K	−95K	5K
Probability	0.002	0.002	0.002	0.002	0.002	0.990

A. Find $E(Y)$, $Var(Y)$, $StdDev(Y)$, the skewness, and the kurtosis of Y. Interpret each of these quantities. Leave the y values in thousands of units, that is, as −99, −98, −97, etc., rather than in units −99,000, −98,000, −97,000, etc.

B. Find the ranges of (i) $E(Y) \pm StdDev(Y)$ (the range of earnings that are within one standard deviation of the mean), (ii) $E(Y) \pm 2StdDev(Y)$ (the range of earnings that are within two standard deviations of the mean), and (iii) $E(Y) \pm 3StdDev(Y)$ (the range of earnings that are within three standard deviation of the mean).

C. How often will the company's earnings on a single customer will be within one standard deviation of the mean? Within two standard deviations of the mean? Within three standard deviations of the mean? Use the probability distribution earlier to calculate these numbers, along with your answer to Exercise 9.3B.

D. Verify that your numbers in Exercise 9.3C satisfy Chebyshev's inequality.

E. The insurance company insures 10,000 people. Their earnings are $T = Y_1 + Y_2 + \cdots + Y_{10,000}$. Find $E(T)$, justifying all steps of the calculation carefully.

F. Simulate the future earnings if the company insures 10,000 people and report the value of $T = Y_1 + Y_2 + \cdots + Y_{10,000}$ for that simulation. Repeat 9 times, getting ten values of T in all. Explain how these simulated values of T, coupled with the law of large numbers as it applies to T, relate to the expected value you found in Exercise 9.3E.

9.4 Suppose $Y \sim U(-1, 1)$ so $p(y) = 0.5$ for $-1 < y < 1$, and $p(y) = 0$, otherwise. Let $T = Y^2$.

---

* Adapted from Moore, D.S. and McCabe, G.P. (1998), *Introduction to the Practice of Statistics*, 3rd Edition.

A. Graph $p(y)$.

B. Find and graph $p(t)$. (Hint: The set $A(t)$ is the set $\{y; -t^{1/2} < y < t^{1/2}\}$). Compare the result with the graph of Exercise 9.4A and comment.

C. Find the mean, variance, skewness, and kurtosis of $p(y)$ using calculus. (Hint: After you calculate the mean, you will see that all the other formulas simplify greatly.) Explain what these numbers tell you about the appearance of $p(y)$ in Exercise 9.4A.

D. Find the mean, variance, skewness, and kurtosis of $p(t)$ using calculus. (Hint: $(a - b)^3 = a^3 - 3a^2b + 3ab^2 - b^3$, and $(a - b)^4 = a^4 - 4a^3b + 6a^2b^2 - 4ab^3 + b^4$.) Explain what these numbers tell you about the appearance of $p(t)$ in Exercise 9.4B.

E. The LLN introduced in Chapter 8 also applies to functions of RVs. Verify the answers in Exercise 9.4D by simulating 20,000 values of Y, calculating T, and having the computer estimate the mean, variance, skewness, and kurtosis of the resulting T data.

9.5 The LLN introduced in Chapter 8 also applies to functions of RVs. What is the MAD for the normal distribution with mean 70 and standard deviation 10? To answer, simulate one million Y^* values from $N(70, 10^2)$ (or as many as your software will conveniently allow), and average the absolute values $|Y^* - 70|$. How different from the standard deviation, 10, is the result? Why, using Jensen's inequality, is the result larger or smaller than 10?

9.6 Hans wants your financial advice about a particular type of investment. So you collect thousands of returns for similar types of investments and find that the distribution is approximately bell shaped with a mean return of 5% and a standard deviation of 5%.

A. What is the most Hans could reasonably stand to lose in the investment? Use the 68–95–99.7 rule in your answer.

B. Why did you use the 68–95–99.7 rule, and not Chebyshev's inequality, in your answer to Exercise 9.6A?

C. Comment on the plausibility that the investment returns are produced by a normal distribution.

9.7 Show that $\Sigma(y_i - \bar{y}) = 0$ for any set of data $y_1, y_2, ..., y_n$. Do this first using summation algebra, also by finding $E(Y - \mu)$ when Y is sampled from the bootstrap distribution, and where μ is computed from the bootstrap distribution.

9.8 Show that $E(aX + bY) = aE(X) + bE(Y)$ carefully, justifying every step.

9.9 Use the data set from Exercise 9.1.

A. Give the bootstrap population distribution for x in list form.

B. Using the bootstrap plug-in principle, estimate $E(X)$, $Var(X)$, $StdDev(X)$, skewness(X), and kurtosis(X). Show details to illustrate how the bootstrap plug-in principle works. (Note: Computer software will usually give different answers for all but the estimate of $E(X)$ because there are built-in bias corrections in the default formulas used by most computers. For larger sample sizes n, there will be little difference between the computer defaults and the more intuitive plug-in estimates.)

9.10 Q: Which distributions have the smallest kurtosis in the universe? A: The Bernoulli(0.5) distribution is one. Show that this is true by calculating its kurtosis.

9.11 Consider the distribution function $p(y) \propto y^3 - y^4$, for $0 < y < 1$ of Exercise 8.10 in Chapter 8. (This is the *beta distribution* with parameters 4 and 2, abbreviated beta(4, 2).

A. Using an Internet search, find the general formula for the skewness and kurtosis of the beta distribution as a function of its parameters θ_1 and θ_2.

B. Using the formula in Exercise 9.11A, calculate the skewness and kurtosis for this $p(y)$ where $\theta_1 = 4$ and $\theta_2 = 2$.

C. Simulate 10,000 values Y^* from $p(y)$, and find (using software) the skewness and kurtosis of these 10,000 data values.

D. Why are the numbers in Exercise 9.11B and C different?

E. What famous law explains why the numbers in Exercise 9.11B and C close to one another?

10

Distributions of Totals

10.1 Introduction

The gamblers who play casino Roulette are bound to lose, since their expected earnings are less than zero. Why on Earth would anyone play? The answer is the subject of this book: statistics! More specifically, the answer lies in the variability (randomness) that all statistics exhibit. While the expected earnings are negative, the actual earnings can be positive. It is the positive outcomes that excite gamblers so much. Who doesn't enjoy a little free money? Some people attribute this free money to "good luck" or even worse, to "skill," but you should know better by now. Your "good luck" at casino Roulette is really just "randomness going in your favor"!

In this chapter, you will learn how variability in totals can be understood and predicted. While you *cannot* predict the *actual amount* of the gambler's earnings after n plays, you *can* predict the *possible range* of his or her earnings very precisely. Unfortunately for the gambler, but fortunately for the casino, the range of likely earnings values includes only negative numbers when n, the number of plays, is large.

The total is an interesting statistic, not just to gamblers, but to anyone who ever has analyzed or will analyze data. Hopefully this group includes you! The most common thing you will do with data is to compute an average. Remember from Section 8.5 that even a percentage is an average of binary (0 or 1) variables. But your average is just your total divided by n, so if you know the distribution of the total, then you also know the distribution of the average. Common statistical procedures depend on the assumption that this average has an approximately normal distribution. In this chapter, you will learn that this assumption is true, in many cases, because of the famous *central limit theorem* (CLT), which tells you when the total (and hence the average) has an approximately normal distribution. You will also see how the law of large numbers (LLN) works: It works because the variance of the (random) average tends toward zero with larger n, which explains how the (random) averages shown in Figures 8.4 through 8.8 eventually settle on the true mean with larger n.

10.2 Additivity Property of Variance

In Chapter 9, you saw that the linearity property of *variance* differs greatly from the linearity property of *expectation*: While $E(aY + b) = aE(Y) + b$, the variance property is completely different, with $Var(aY + b) = a^2Var(Y)$.

The additivity property of expectation described in Chapter 9 was $E(X + Y) = E(X) + E(Y)$. In contrast to the linearity property, this additivity property also works for variance, but with a caveat: It's true when the variables are independent, but not necessarily true otherwise.

Here's how the additivity property works. Let $T = X + Y$ denote the total (sum) of two random variables (RVs) X and Y. Also define $\mu_X = E(X)$ and $\mu_Y = E(Y)$. Then

$\text{Var}(T)$	
$= E\{T - E(T)\}^2$	(By definition)
$= E\{X + Y - E(X + Y)\}^2$	(By substitution)
$= E[X + Y - \{E(X) + E(Y)\}]^2$	(By the additivity property of expectation)
$= E\{X + Y - (\mu_X + \mu_Y)\}^2$	(By definition)
$= E\{(X - \mu_X) + (Y - \mu_Y)\}^2$	(By algebraic rearrangement)
$= E\{(X - \mu_X)^2 + (Y - \mu_Y)^2 + 2(X - \mu_X)(Y - \mu_Y)\}$	(By algebraic expansion $(a + b)^2 = a^2 + b^2 + 2ab$, with $a = X - \mu_X$ and $b = Y - \mu_Y$)
$= E\{(X - \mu_X)^2\} + E\{(Y - \mu_Y)^2\} + 2E\{(X - \mu_X)(Y - \mu_Y)\}$	(By the linearity and additivity properties of expectation)
$= \text{Var}(X) + \text{Var}(Y) + 2E\{(X - \mu_X)(Y - \mu_Y)\}$	(By definition of variance)

Notice that the variance does not have the same additivity property as expectation! The variance of the sum is the sum of the variances, plus twice the term $E\{(X - \mu_X)(Y - \mu_Y)\}$. This term is famous, having a special name, **covariance**.

Definition of the Covariance between RVs X and Y

If RVs X and Y have finite variance, then the covariance between X and Y is given by

$$\text{Cov}(X, Y) = E\{(X - \mu_X)(Y - \mu_Y)\}$$

The **additivity property of variance** requires this covariance term.

Additivity Property of Variance

For RVs X and Y having finite variance

$$\text{Var}(X + Y) = \text{Var}(X) + \text{Var}(Y) + 2\text{Cov}(X, Y)$$

Like the linearity property of variance differs from the linearity property expectation, the additivity property of variance also differs from the additivity property of expectation. But if the "2Cov(X, Y)" term were absent, the additivity property of variance would look just like the additivity property of expectation.

When is $\text{Cov}(X,Y) = 0$? It is true when X and Y are independent RVs, and this fact follows logically from the fact that $p(x,y) = p(x)p(y)$ when X and Y are independent. Before showing why $\text{Cov}(X,Y) = 0$ under independence, we first give another useful result about the expected value of the product of independent RVs.

Multiplicative Property of Expectation When X and Y Are Independent RVs

For *independent RVs* X and Y:

$$E(XY) = E(X)E(Y)$$

The mathematical logic for the multiplicative property of expectation under independence is shown in the continuous case as follows:

$$E(XY) = \iint xy\, p(x, y)\, dxdy \qquad \text{(By the law of the unconscious statistician)}$$

$$= \iint xy\, p(x)p(y)\, dxdy \qquad \text{(By independence of } X \text{ and } Y)$$

$$= \int yp(y) \left\{ \int xp(x)\, dx \right\} dy \qquad \text{(Since } y \text{ is constant with respect to integration over } x, \text{ it can be factored out using the linearity property of integrals, property I2 of Section 2.6)}$$

$$= \int yp(y)\, \{E(X)\}\, dy \qquad \text{(By the definition of } E(X))$$

$$= E(X) \int yp(y)\, dy \qquad \text{(By the linearity property of integrals, the constant } E(X) \text{ can be factored outside the integral)}$$

$$= E(X)\, E(Y) \qquad \text{(By the definition of } E(Y))$$

The multiplicative property of expectation is another unusual case where the expected value of the function is equal to the function of the expected values. Don't count on these cases! They only happen under very restrictive assumptions on the functions such as linearity or additivity or in isolated cases where the variables are independent.

The result $E(XY) = E(X)E(Y)$ under independence implies that $Cov(X, Y) = 0$ under independence, shown as follows:

$$Cov(X, Y) = E[\{X - E(X)\}\{Y - E(Y)\}] \qquad \text{(By definition of covariance)}$$

$$= E\{(X - \mu_x)(Y - \mu_y)\} \qquad \text{(By substituting } \mu_x \text{ for } E(X) \text{ and } \mu_y \text{ for } E(Y))$$

$$= E\{(X - \mu_x)\}E\{(Y - \mu_y)\} \qquad \text{(Because } X \text{ and } Y \text{ are independent and because } \mu_x \text{ and } \mu_y \text{ are constants, } (X - \mu_x) \text{ and } (Y - \mu_y) \text{ are also independent, and the multiplicative property of expectation can therefore be applied)}$$

$$= \{E(X) - \mu_x\}\{E(Y) - \mu_y\} \qquad \text{(Because } \mu_x \text{ and } \mu_y \text{ are constants, the linearity property of expectation can be applied)}$$

$$= (\mu_x - \mu_x)(\mu_y - \mu_y) \qquad \text{(By substituting } \mu_x \text{ for } E(X) \text{ and } \mu_y \text{ for } E(Y))$$

$$= 0 \qquad \text{(By algebra)}$$

This gives you a revised additivity formula for variance, one that looks just like the formula for expectation, but carries an essential caveat: There is an independence assumption.

Additivity Property of Variance When X and Y Are Independent RVs

For *independent RVs X* and *Y*:

$$Var(X + Y) = Var(X) + Var(Y)$$

Assumptions are important! The additivity property shown earlier can be grossly incorrect when X and Y are dependent. For example, suppose X and Y are actually the same number, and their variance is 1.0. Then $Var(X + Y) = Var(2X) = 2^2 Var(X) = 2^2 \times 1.0 = 4.0$, which is correct. But if you apply the additivity property assuming independence, you get $Var(X + Y) = Var(X) + Var(Y) = 1.0 + 1.0 = 2.0$, which is incorrect. This is not just a meaningless brainteaser either: Most statistical analyses you will see from computer software assume independent and identically distributed (iid) observations, by default. So if your data-generating process gives you dependent observations, then your software will calculate things incorrectly, unless you specify a model for the dependence.

TABLE 10.1

Earnings When Playing
10 on Red in Roulette

Earnings, y	p(y)
−10	20/38 = 0.526
+10	18/38 = 0.474
Total	1.000

Example 10.1: Predicting Your Gambling Losses

It's all random, right? You could win or lose, all depending on your luck, right? Wrong! While the word *random* itself is right, people commonly misinterpret what that means. Random doesn't mean "willy-nilly" or "anything can happen." While randomness does imply variable and therefore unpredictable individual outcomes, it also implies precise predictability concerning the *range* of possible outcomes. All you need to understand this precise predictability is the variance, along with Chebyshev's inequality.

Suppose you bet 10 on red in roulette, 10,000 times in succession. What is the tightly prescribed level of predictability concerning the range of your possible earnings? The question is easy to answer using the additivity properties of expectation and variance. First, your total earnings are $T = Y_1 + Y_2 + \cdots + Y_{10,000}$, where the Y_i are produced iid from the distribution $p(y)$ defined (hopefully very familiarly by now) as given in Table 10.1.

You know $E(T) = E(Y_1) + E(Y_2) + \cdots + E(Y_{10,000})$ by the additivity property of expectation. You also know that Y_1 comes from the distribution $p(y)$ shown in Table 10.1, so you know that

$$E(Y_1) = \sum yp(y) = -10\left(\frac{20}{38}\right) + 10\left(\frac{18}{38}\right) = -0.52632$$

But Y_2 comes from the same distribution $p(y)$, so you also know that

$$E(Y_2) = \sum yp(y) = -10\left(\frac{20}{38}\right) + 10\left(\frac{18}{38}\right) = -0.52632$$

Guess what? Your third earnings, Y_3, come from the same distribution $p(y)$ as well, so you also know that

$$E(Y_3) = \sum yp(y) = -10\left(\frac{20}{38}\right) + 10\left(\frac{18}{38}\right) = -0.52632$$

Since your earnings Y_i on the *i*th play come from the same roulette distribution $p(y)$, you know that $E(Y_i) = -0.52632$ for *every* i, where i ranges from 1 to 10,000. Hence, you can calculate the expected value of your total earnings T as follows:

$E(T) = E(Y_1 + Y_2 + \cdots + Y_{10,000})$	(By substitution)
$\quad = E(Y_1) + E(Y_2) + \cdots + E(Y_{10,000})$	(By the additivity property of expectation)
$\quad = (-0.52632) + (-0.52632) + \cdots + (-0.52632)$	(By substitution, using the fact that the Y_i are identically distributed)
$\quad = 10,000\,(-0.52632)$	(Since there are 10,000 terms in the sum)
$\quad = -5,263.2$	(By arithmetic)

The same arguments just made about expected values also apply to variances. You know that Y_1 comes from the distribution $p(y)$, so you know that

$$\text{Var}(Y_1) = \sum (y - \mu)^2 p(y) = \{-10 - (-0.52632)\}^2 \left(\frac{20}{38}\right) + \{10 - (-0.52632)\}^2 \left(\frac{18}{38}\right) = 99.723$$

But Y_2 comes from the same distribution $p(y)$, so you also know that

$$\text{Var}(Y_2) = \sum (y - \mu)^2 p(y) = \{-10 - (-0.52632)\}^2 \left(\frac{20}{38}\right) + \{10 - (-0.52632)\}^2 \left(\frac{18}{38}\right) = 99.723$$

Guess what? Your third earnings, Y_3, come from the same distribution $p(y)$ as well, so you also know that

$$\text{Var}(Y_3) = \sum (y - \mu)^2 p(y) = \{-10 - (-0.52632)\}^2 \left(\frac{20}{38}\right) + \{10 - (-0.52632)\}^2 \left(\frac{18}{38}\right) = 99.723$$

Since your earnings on the ith play Y_i come from the same roulette distribution $p(y)$, you know that $\text{Var}(Y_i) = 99.723$ for *every* i, where i ranges from 1 to 10,000.

Successive rolls of the roulette wheel are all independent; thus, you can apply the additivity property as follows:

$\text{Var}(T) = \text{Var}(Y_1 + Y_2 + \cdots + Y_{10,000})$	(By substitution)
$\quad = \text{Var}(Y_1) + \text{Var}(Y_2) + \cdots + \text{Var}(Y_{10,000})$	(By the additivity property of variance *when the RVs are independent*)
$\quad = 99.723 + 99.723 + \cdots + 99.723$	(By substitution, using the fact that the *Y*s are identically distributed)
$\quad = (10,000) \times (99.723)$	(Since there are 10,000 terms in the sum)
$\quad = 997,229.9$	(By arithmetic)

The standard deviation of your total earnings T is thus

$$\text{StdDev}(T) = \sqrt{997,229.9} = 998.614$$

Along with the mean and Chebyshev's inequality, this standard deviation tells the story about your total earnings, T. You saw earlier that $E(T) = -5263.2$. Therefore, using Chebyshev's inequality with $k = 3$, there is at least an 88.9% probability that your earnings will be within the range $-5263.2 \pm 3(998.614)$ or between -8259 and -2267 (dollars, euros, pounds, or whatever currency you are betting in). Good luck!

There are important general formulas for expected value, variance, and standard deviation of the sum of iid RVs that you can extract from this discussion of gambling. Suppose $Y_1, Y_2, \ldots, Y_n$ are produced as iid RVs from a probability distribution function (pdf) $p(y)$ having mean μ and variance σ^2. Define the total as $T = Y_1 + Y_2 + \cdots + Y_n$. Then you can find the expected value of the total as follows:

$E(T) = E(Y_1 + Y_2 + \cdots + Y_n)$	(By substitution)
$\quad = E(Y_1) + E(Y_2) + \cdots + E(Y_n)$	(By the additivity property of expectation)
$\quad = \mu + \mu + \cdots + \mu$	(Since the Y_i are all produced by the same identical pdf $p(y)$ and since μ is the mean of that pdf)
$\quad = n\mu$	(Since there are n terms in the sum)

This result that $E(T) = n\mu$ does not assume independence; it only assumes identical distributions. You can perform a similar derivation with variance, but here, independence is crucial.

$\text{Var}(T) = \text{Var}(Y_1 + Y_2 + \cdots + Y_n)$	(By substitution)
$= \text{Var}(Y_1) + \text{Var}(Y_2) + \cdots + \text{Var}(Y_n)$	(By the additivity property of variance *when the RVs are independent*)
$= \sigma^2 + \sigma^2 + \cdots + \sigma^2$	(Since the Y_i are all produced by the same identical pdf $p(y)$ and since σ^2 is the variance of that pdf)
$= n\sigma^2$	(Since there are n terms in the sum)

Now that you know that $\text{Var}(T) = n\sigma^2$, you can find the standard deviation of T easily:

$$\text{StdDev}(T) = \{\text{Var}(T)\}^{1/2} = (n\sigma^2)^{1/2} = n^{1/2}\sigma$$

This derivation makes it clear that there is no direct additivity property for standard deviation: You *cannot* say that $\text{StdDev}(Y_1 + Y_2 + \cdots + Y_n)$ is equal to $\text{StdDev}(Y_1) + \text{StdDev}(Y_2) + \cdots + \text{StdDev}(Y_n) = n\sigma$.

Instead, standard deviations "add like Pythagoras": Recall that the hypotenuse of a right triangle is equal to the square root of the sum of squares of its sides. Standard deviation works the same way: $\text{StdDev}(Y_1 + Y_2) = \{\text{Var}(Y_1) + \text{Var}(Y_2)\}^{1/2}$.

Using the linearity properties, these results for totals also translate to results for averages: Let $\bar{Y} = (1/n)T = (1/n)(Y_1 + Y_2 + \cdots + Y_n)$, where Y_i are produced by $p(y)$ whose mean is μ. Then $E(\bar{Y}) = \mu$, which can be shown as follows:

$E(\bar{Y}) = E\{(1/n)T\}$	(By substitution)
$= (1/n) E(T)$	(By the linearity property of expectation)
$= (1/n) (n\mu)$	(Since $E(T) = n\mu$ as shown earlier)
$= \mu$	(By algebra)

Notice that that assumption of independence is not needed for the result that $E(\bar{Y}) = \mu$; only identical distributions are assumed. There is a corresponding famous formula for variance, namely, $\text{Var}(\bar{Y}) = \sigma^2/n$, that assumes both iid observations:

$\text{Var}(\bar{Y}) = \text{Var}\{(1/n)T\}$	(By substitution)
$= (1/n)^2 \text{Var}(T)$	(By the linearity property of variance)
$= (1/n)^2 (n\sigma^2)$	(Since $\text{Var}(T) = n\sigma^2$ as shown earlier, when the T is a sum of iid RVs)
$= \sigma^2/n$	(By algebra)

A famous corollary is that $\text{StdDev}(\bar{Y}) = \sigma/n^{1/2}$, again assuming the data from which $\bar{Y}$ is calculated are an iid sample from a distribution $p(y)$ whose standard deviation is σ. This formula is seen in the output from any statistical software where the **standard error of the mean** is reported. The standard error of the mean is specifically the *estimated* standard deviation of $\bar{Y}$.

The Standard Error of the Mean

$$\text{StdErr}(\bar{y}) = \frac{\hat{\sigma}}{n^{1/2}}$$

The standard error matters because it tells you how close the sample average from your data, $\bar{y}$, is likely to be to the mean μ of the process $p(y)$ that produced your data.

Example 10.2: The Standard Error of the Mean Return for the Dow Jones Industrial Average (DJIA)

In Example 9.10, the bootstrap plug-in estimates of mean and standard deviation for the distribution of DJIA returns are given as $\bar{y} = 0.000200189$ and $\hat{\sigma} = 0.0111868$. The standard error of the mean is thus $\text{StdErr}(\bar{y}) = \hat{\sigma}/n^{1/2} = 0.0111868/(18,834)^{1/2} = 0.00008151$, which is an estimate of the standard deviation of $\bar{Y}$, the average of 18,834 returns produced by the DJIA model $p(y)$. By Chebyshev's inequality, it is at least 88.9% probable that $\bar{Y}$ will be no more than three standard deviations from μ; hence, it is highly likely that the mean μ is within the range $0.000200189 \pm 3(0.00008151)$ or in the range $-0.000044 < \mu < 0.000445$. For investors, this interval provides the uncomfortable suggestion that the mean return (μ) for the DJIA data-generating process could possibly be negative.

10.3 Covariance and Correlation

You saw in the previous section that the covariance between X and Y is zero when X and Y are independent. When X and Y are not independent, covariance provides you with a numerical measure of the dependence between X and Y.

Recall the definition of covariance: $\text{Cov}(X,Y) = E\{(X - \mu_x)(Y - \mu_y)\}$. It is handy to use shorthand symbols, so let the symbol $\hat{\sigma}_{xy}$ refer to the covariance between X and Y. Like the variance σ^2 and the mean μ, the covariance σ_{xy} is a model parameter, and since the model is there before you see any data, the covariance is there before you see any data as well. It's there and is always the same number, no matter whether you have sampled $n = 10,000$, $n = 10$, $n = 1$, or no observations at all from the process $p(x,y)$. It's part of the model ($p(x,y)$) that produces your data. The covariance σ_{xy} is not calculated from your data; it's calculated from your model. Specifically, using the law of the unconscious statistician, the computing formulas are as follows.

Definition of Covariance in Terms of Bivariate Distributions

In the discrete case:

$$\sigma_{xy} = \text{Cov}(X,Y) = E\{(X - \mu_x)(Y - \mu_y)\} = \sum\sum (x - \mu_x)(y - \mu_y)p(x,y)$$

In the continuous case:

$$\sigma_{xy} = \text{Cov}(X,Y) = E\{(X - \mu_x)(Y - \mu_y)\} = \iint (x - \mu_x)(y - \mu_y)p(x,y)dxdy$$

Example 10.3: Estimating Covariance Using (Income, Housing Expense) Data

To estimate σ_{xy} using the data, you can again use the bootstrap plug-in principle. Suppose you have sampled n observations, getting (income, housing expense) pairs (x_1, y_1), (x_2, y_2), ..., (x_n, y_n). Create the fictitious bootstrap joint probability distribution $\hat{p}(x,y)$ as given in Table 10.2.

TABLE 10.2

Bootstrap Joint Distribution $\hat{p}(x,y)$

Income (x)	Housing Expense (y)	$\hat{p}(x,y)$
x_1	y_1	$1/n$
x_2	y_2	$1/n$
...	...	...
x_n	y_n	$1/n$
	Total	1.00

TABLE 10.3

Bootstrap Joint Distribution $\hat{p}(x,y)$, in Cross-Classification Table Form, along with Marginal Bootstrap Distributions

		y_1	y_2	...	y_n	Total
				Y		
X	x_1	$1/n$	0	...	0	$1/n$
	x_2	0	$1/n$	...	0	$1/n$
	...	...	...	...	...	...
	x_n	0	0	...	$1/n$	$1/n$
	Total	$1/n$	$1/n$	...	$1/n$	1.000

This distribution puts $1/n$ probability on every observed combination (x_i, y_i) in the data set and zero probability elsewhere. It's clearly wrong, because it assumes that if your (income, housing expense) combination is not in that data set, you don't exist! Nevertheless, it provides an estimate of the true unknown joint distribution $p(x,y)$, and therefore, you can use it to estimate quantities like σ_{xy} that depend on the true unknown $p(x,y)$.

In terms of the two-way cross-classification tables you may remember from Chapter 6, the bootstrap joint distribution, assuming there are no repeats among the xs or ys, is shown in Table 10.3, along with the marginal bootstrap distributions.

Applying the law of the unconscious statistician with this distribution $\hat{p}(x,y)$, you get an estimate of σ_{xy}.

The Plug-In Estimate of Covariance

$$\hat{\sigma}_{xy} = \frac{1}{n}\sum_i (x_i - \bar{x})(y_i - \bar{y})$$

As in the cases of standard deviation, skewness, and kurtosis, the default estimates reported by computer software differ slightly from the plug-in estimates because of minor bias corrections. But again, the plug-in estimate is intuitively appealing: The true covariance σ_{xy} is the expected product of deviations of X and Y from their respective true means μ_x and μ_y, and the plug-in estimate of the covariance $\hat{\sigma}_{xy}$ is the sample average of the products of deviations of x_i and y_i from their respective sample means $\bar{x}$ and $\bar{y}$.

Figure 10.1 shows you how to understand covariance between X and Y in terms of these products of deviations from their means.

In Figure 10.1, there is a clear positive trend; this positive trend is reflected in the covariance as follows. Quadrant I shows observations i where both $(x_i - \bar{x}) > 0$ and $(y_i - \bar{y}) > 0$; hence, their product is also positive: $(x_i - \bar{x})(y_i - \bar{y}) > 0$. Quadrant II shows

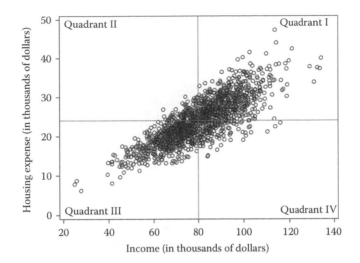

FIGURE 10.1

Scatterplot of (income, housing expense) data showing quadrants of positive and negative cross products. The mean of income is the vertical line at 80, and the mean of housing expense is the horizontal line at 24.

observations i where $(x_i - \bar{x}) < 0$ but $(y_i - \bar{y}) > 0$; hence, their product is negative: $(x_i - \bar{x})$ $(y_i - \bar{y}) < 0$. Quadrant III shows observations i where $(x_i - \bar{x}) < 0$ and $(y_i - \bar{y}) < 0$; hence, their product is positive: $(x_i - \bar{x})(y_i - \bar{y}) > 0$. Finally, quadrant IV shows observations i where $(x_i - \bar{x}) > 0$ but $(y_i - \bar{y}) < 0$; hence, their product is negative: $(x_i - \bar{x})(y_i - \bar{y}) < 0$.

In summary, the cross product $(x_i - \bar{x})(y_i - \bar{y})$ is positive in quadrants I and III, but negative in quadrants II and IV. The plug-in covariance estimate $\hat{\sigma}_{xy}$ is the simple average of the n cross products $(x_i - \bar{x})(y_i - \bar{y})$ and is therefore a positive number because there are far more large positive cross products than negative cross products. The positive covariance is therefore an indicator of the positive trend seen in Figure 10.1.

On the other hand, suppose you are analyzing the relationship between X = income and Y = weekly time spent cooking. Presumably, people with more income have more money to dine out and also perhaps less time with their hectic jobs to devote to preparing meals. The relationship might look as shown in Figure 10.2, although the strength of association is deliberately exaggerated to illustrate the point.

In Figure 10.2, you can see a clear negative trend reflecting the fact that the majority of the cross products $(x_i - \bar{x})(y_i - \bar{y})$ are less than zero, being in quadrants II and IV. The plug-in covariance estimate $\hat{\sigma}_{xy}$ is therefore negative in this case.

So the covariance σ_{xy} is positive when there is an increasing trend in the (X,Y) relationship and negative when there is a decreasing trend. What else does σ_{xy} tell you? Not much. For example, a "large" σ_{xy} does not necessarily tell you that the relationship is strong, because σ_{xy} is not scale-free. If you measure income in dollars, instead of thousands of dollars, then the covariance would be larger by a factor of 1000. However, the relationship fundamentally would not change: The graphs would look exactly the same, except for the numbers on the horizontal axis of Figure 10.2 would be 40,000, 60,000, ... instead of 40, 60, ...

The **correlation** coefficient removes the scale dependency by using z-scores. It is defined as the expected cross product of the z-scores for the X and Y measures and is denoted by the symbol ρ_{xy}.

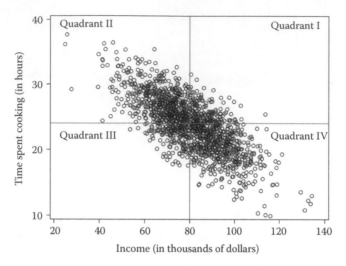

FIGURE 10.2
Scatterplot of (income, time spent cooking) data showing quadrants of positive and negative cross products. The mean of income is given by the vertical line at 80, and the mean of time spent cooking is given by the horizontal line at 24.

The Correlation Coefficient

$$\rho_{xy} = E\left[\left(\frac{X - \mu_x}{\sigma_x}\right)\left(\frac{Y - \mu_y}{\sigma_y}\right)\right]$$

By the linearity property of expectation, you can factor the denominator σ outside the expectation, getting

$$\rho_{xy} = \frac{1}{\sigma_x \sigma_y} E\{(X - \mu_x)(Y - \mu_y)\}$$

By the definition of covariance, $\sigma_{xy} = E\{(X - \mu_x)(Y - \mu_y)\}$, which gives you the following simple form:

$$\rho_{xy} = \frac{\sigma_{xy}}{\sigma_x \sigma_y}$$

In other words, the correlation is equal to the covariance divided by the product of the standard deviations.

Just as the variance is less useful than the standard deviation, the covariance is less useful than the correlation. On the other hand, both variance and covariance are necessary "stepping stones" to get to the more useful measures.

Properties of the Correlation Coefficient

- It is scale-free.
- $|\rho_{xy}| \leq 1$.
- $|\rho_{xy}| = 1$ if and only if Y is a deterministic linear function of X, that is, if $Y = aX + b$, with $a \neq 0$.

- Larger $|\rho_{xy}|$ (values closer to 1.0) indicates stronger relationship between Y and X, that is, a relationship where the observable (X,Y) data fall close to a line with $a \neq 0$.
- Smaller $|\rho_{xy}|$ (values closer to 0.0) indicates a weak or no linear relationship between Y and X, that is, a relationship where observable (X,Y) data are best represented by a horizontal line where $a = 0$.

The plug-in estimate of correlation uses the plug-in estimates of covariance and standard deviation.

The Plug-In Estimate of the Correlation Coefficient

$$\hat{\rho}_{xy} = \frac{\hat{\sigma}_{xy}}{\hat{\sigma}_x \hat{\sigma}_y} = \frac{(1/n)\sum_i (x_i - \bar{x})(y_i - \bar{y})}{\sqrt{(1/n)\sum_i (x_i - \bar{x})^2}\sqrt{(1/n)\sum_i (y_i - \bar{y})^2}}$$

This plug-in estimate also turns out to be the precise default estimate returned by software; ordinarily there is no bias correction in the reported sample correlation. Note also that you can cancel the $1/n$ terms from the numerator and denominator. In some sources, you will see the correlation formula presented without the $1/n$ term for this reason, but we choose to leave it in so that the connection between correlation, covariance, and standard deviations is clear.

As a point of reference, the estimated correlation using the data shown in Figure 10.2 is $\hat{\rho}_{xy} = -0.73$, which can be characterized as a strong inverse relationship. It's called *strong* because the correlation coefficient is near 1.0 in absolute value, and it's called *inverse* because the sign is less than zero, implying that larger X are associated with smaller Y.

Whether you call a particular correlation coefficient "strong" or "weak" depends on your subject matter. In chemistry, where relationships involving chemical reactions are essentially deterministic, any correlation less than 1.0 is a red flag indicating a poorly executed experiment. But in the social sciences, relationships involving the predictability of humans typically exhibit weak correlations because we humans are notoriously unpredictable. And that's a good thing! Otherwise, we'd be a bunch of robots. Hence, smaller correlations such as 0.3 can be quite interesting when predicting human behavior. In finance, the "efficient markets" theory states that you cannot predict future market movements using publicly available current data, which means that any nonzero correlation is interesting.

We will give the following rules of thumb, but they are very ugly because you should modify them depending on your subject area.

Ugly Rule of Thumb 10.1

If the correlation ρ_{xy} is ...	Then ...
Equal to 1.0	There is a perfect decreasing linear relationship between X and Y.
Between -1.0 and -0.7	There is a strong decreasing linear relationship between X and Y.
Between -0.7 and -0.3	There is a moderate decreasing linear relationship between X and Y.
Between -0.3 and 0.0	There is little to no linear relationship between X and Y.
Equal to 0.0	There is no linear relationship between X and Y.
Between 0.0 and 0.3	There is little to no linear relationship between X and Y.
Between 0.3 and 0.7	There is a moderate increasing linear relationship between X and Y.
Between 0.7 and 1.0	There is a strong increasing linear relationship between X and Y.
Equal to 1.0	There is a perfect increasing linear relationship between X and Y.

NOTE: There is nothing ugly about the cases where the correlation is either +1, −1, or 0; those interpretations are factual. It's the 0.3 and 0.7 thresholds that are ugly. Different thresholds make more sense in different contexts.

10.4 Central Limit Theorem

In his famous poem, "Ode on a Grecian Urn," John Keats wrote the lines

> Beauty is truth, truth beauty,—that is all
> Ye know on earth, and all ye need to know.

There are certain mathematical truths about our world that are beautiful, like the ratio $\pi = 3.14159 \ldots$ of the circumference to the diameter of a circle and the golden ratio $(1 + \sqrt{5})/2$ found in Greek art. The beauty in these numbers lies in their relationship to pleasing physical objects, such as a circle or a rectangle with artistic proportion. Their truth lies in their appearance in Nature: The numbers $3.14159 \ldots$ and $(1 + \sqrt{5})/2$ appear everywhere on our planet. They aren't just figments of the imaginations of a few deranged mathematicians.

The **Central Limit Theorem** (CLT) is one of those truths involving Nature that is also beautiful. Like π and the golden ratio, it too has a beautiful mathematical shape associated with it: The normal distribution is beautiful in its symmetry and pleasing bell-shaped appearance. Its truth comes from Nature. Natural processes everywhere are well modeled by the normal distribution. It's not just a figment of the imaginations of a few deranged statisticians.

The reason that natural processes so often can be modeled using the normal distribution is because of the famous CLT.

The Central Limit Theorem

Suppose $Y_1, Y_2, \ldots$ are produced as iid RVs from any probability distribution $p(y)$ having finite variance. Let $T_n = Y_1 + Y_2 + \cdots + Y_n$. Then the probability distribution $p(t_n)$ is approximately a normal distribution, and the approximation becomes better as n gets larger.

This is quite a remarkable statement. No matter what distribution produces the data—whether discrete, continuous, skewed, kurtotic, and anything—the *distribution* of the *sum* of DATA values is approximately a normal distribution. It's as if the normal distribution were a black hole, sucking up all distributions that come along.

The reason this applies to natural processes so often is that many natural processes themselves have several components that are essentially additive.

Example 10.4: The Central Limit Theorem and the Yield of a Plant

The yield of a crop plant (say, corn) can be thought of as an additive component of genetics and environment, and each of those components can be thought of as additive in their subcomponents. Genetic effects involve additive components from gene subtypes regulating growth and health of the plant. Environmental effects involve additive components of rain, irrigation, fertilization, and soil quality. The end result is an additive combination of many factors, which can result in an approximately normal distribution of plant yield.

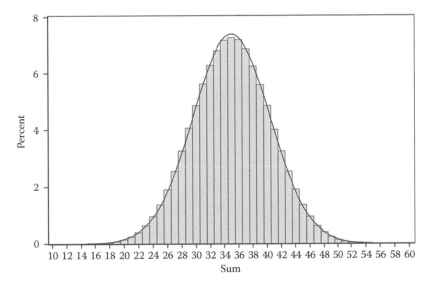

FIGURE 10.3
Distribution of the sum of $n = 10$ rolls of a six-sided die.

Illustrating the CLT via simulation is easy and instructive. First, pick a distribution $p(y)$. Then generate n observations Y_i^* and calculate the total T^*. Repeat many times (say thousands, this number is not the important one though—the n is the important number), and draw the histogram of the resulting totals T^*. According to the CLT, these histograms should have an approximate symmetric bell shape, with the approximation becoming better as you increase the n used in the calculation of the sum.

While the distribution of a single die is discrete uniform, the distribution of the sum of two dice becomes the discrete triangular distribution graphed in Figure 7.13. Thus, even with $n = 2$, the CLT effect begins to kick in: The distribution of the sum of two uniforms is symmetric, with a peak in the middle, similar to a normal distribution. With $n = 10$ in the sum, the distribution is even closer to a normal distribution, as shown in Figure 10.3.

From Figure 10.3, the sum of $n = 10$ dice looks to be very well modeled by a normal distribution. Of course, it is still discrete, but much less so than the original discrete uniform distribution of a single die, which only had six possible values. With a sum of 10 rolls, there are 51 possible values (10, 11, ..., 60). Apart from discreteness, the distribution is nearly perfectly described by the normal curve. The CLT in action!

For another example, consider the distribution of the sum of exponentially distributed RVs. Suppose the distribution producing the data is $p(y) = 0.5e^{-0.5t}$, for $t > 0$. This is the exponential distribution with mean 2.0 as shown in Figure 8.3. The exponential distribution looks nothing like a normal distribution, since it is so asymmetric and since its peak (or mode) is at the edge, not in the middle.

If you sample n iid observations $Y_1, Y_2, ..., Y_n$ from $p(y) = 0.5e^{-0.5t}$, and calculate the sum $T = Y_1 + Y_2 + \cdots + Y_n$, it turns out that T has a distribution known as the **gamma distribution**, and simulation is not needed to draw it. You can use any software that computes the gamma distribution. Figure 10.4 shows the distributions of sums of iid exponentials, for $n = 1, 3, 5,$ and 10.

Unlike the dice, notice in Figure 10.4 that the distributions are always continuous. Also, unlike the dice, the distributions of the sums of exponentials are always skewed—they

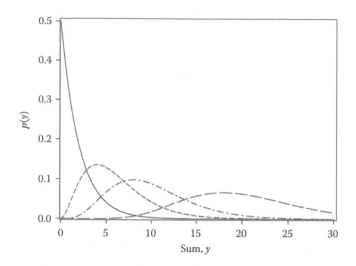

FIGURE 10.4
Distributions of an exponential RV with mean 2 (solid), the total of $n = 3$ iid exponential RVs (short dash), the total of $n = 5$ iid exponential RVs (dash dot), and the total of $n = 10$ iid exponential RVs (long dash).

inherit their skewness from the skewness of the parent exponential distribution. However, you can see the skewness lessening with sums of larger numbers of observations, with distributions looking more like the classic normal bell curve—again, the CLT in action.

How big does n have to be before you can assume that the normal approximation is adequate? Some ill-informed sources recommend $n > 30$ as a rule of thumb, but this rule is so laughably ridiculous that it is not even worthy of being called an "ugly rule of thumb." The rationale for the $n > 30$ rule harkens back to the ancient times, when tables for certain distributions were available only for $n \leq 30$ due to limited computing power.

The n needed to make the normal distribution a good approximation to the distribution of the total depends on the degree of non-normality in the parent distribution $p(y)$. If the $p(y)$ that produces the data is grossly skewed and/or outlier-prone, then you'll need a much bigger n before you can assume the normal distribution is a reasonable approximation.

In the case where $p(y)$ is normal to begin with, the distribution of the total, $p(t_n)$ is *exactly* a normal distribution *for all n*. Sums of normally distributed RVs are always normally distributed—this is the additivity property of the normal distribution. It's not a rule of thumb; it's a mathematical fact.

Figures 10.3 and 10.4 show that $n = 10$ is adequate when the parent distribution is either discrete uniform or exponential.

If the parent distribution is Bernoulli, there is an ugly rule of thumb that is worth stating.

Ugly Rule of Thumb 10.2

If $Y_1, Y_2, \ldots \sim_{iid}$ Bernoulli(π), then the distribution of $T = Y_1 + Y_2 + \cdots + Y_n$ is approximately normal provided $n\pi > 5$ and $n(1 - \pi) > 5$.

The logic is that you need a large enough sample size so that T, the total number of 1s, is usually far enough away from 0 (that's the $n\pi > 5$ stipulation) and also usually far enough away from n (that's the $n(1 - \pi) > 5$ stipulation) that its distribution has a chance to be reasonably symmetric. If T is usually 0 (in the case of very small π) or usually n (in the case of very large π), then its distribution will be strongly right-skewed (for small π) or strongly left-skewed (for large π).

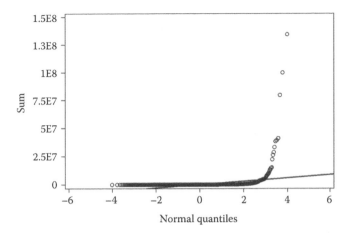

FIGURE 10.5
Quantile–quantile plot of 20,000 sums of $n = 10,000$ iid RVs $2/Y$, where $Y \sim U(0, 2)$.

For example, when playing a lottery where π is around 1/16,000,000, Ugly Rule of Thumb 10.2 implies that you will need $n > 80,000,000$ plays of the lottery before you can assume that the total number of wins is adequately modeled as a normal distribution!

And in some cases, the CLT won't work at all, for any n. There is an iid assumption, and if there is strong dependence, then the CLT will not apply. There is also an assumption of finite variance. If the variance is infinite, then $p(t_n)$ might never be close to a normal distribution, regardless of n. For example, the RV $2/Y$, where Y is the stoplight RV that is distributed as $U(0,2)$, has infinite variance. Figure 10.5 shows the q–q plot comparing the distribution of sums of $n = 10,000$ such RVs to a normal distribution. The distribution of the sums is nowhere close to normal, despite the huge sample size $n = 10,000$. It will not be close to normal for any n, no matter how large.

When the CLT is valid, it applies to any linear function of the sum. Thus, if the sum T is approximately normally distributed, then any transformation $aT + b$ has the same adequacy of approximation. The reason for this is that the adequacy of approximation depends on the discreteness, skewness, and kurtosis of the RV. Since linear transformations have exactly the same discreteness, skewness, and kurtosis characteristics as the original variable, they have the same adequacy of normal approximation.

One example comes from the estimate of total inventory value in the sampling case of Chapter 7: The estimated total was $N\bar{Y} = (N/n)T$. The CLT applies equally to this estimate as it does to the total T, since the estimate $(N/n)T$ is a linear transformation of the total T.

But the most famous example is the linear transformation $(1/n)T$ (for which $a = 1/n$ and $b = 0$), giving the sample average $\bar{Y} = (1/n)T$. Thus, the CLT applies equally to the sample average, $\bar{Y}$, as it does to the total, T. This fact, coupled with the variance and expectation calculations earlier showing that $E(\bar{Y}) = \mu$ and $\text{StdDev}(\bar{Y}) = \sigma/n^{1/2}$, gives you the following result.

The Central Limit Theorem Applied to Sample Averages

If $Y_1, Y_2, \ldots$ are produced as iid RVs from any probability distribution $p(y)$ having finite variance, then $\bar{Y} \sim N(\mu, \sigma^2/n)$. The approximation becomes better as n increases.

Figure 10.6 shows how this works for the exponential RV with $\lambda = 0.5$, that is, when $p(y) = 0.5e^{-0.5y}$, for $y > 0$.

Stare at Figure 10.6 for a while—at least long enough for it to burn an image on the back of your eyeballs. Many important lessons of statistics are wrapped up in that graph.

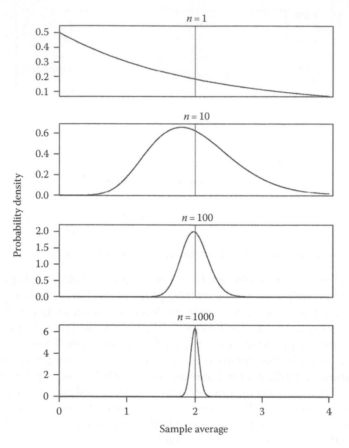

FIGURE 10.6

Distributions of the average of n iid RVs $Y_1, Y_2, ..., Y_n$ produced by the exponential distribution $p(y) = 0.5\exp(-0.5y)$, for n = 1, 10, 100, and 1000.

Important Lessons from Figure 10.6

1. The distribution of the average becomes closer to a normal distribution for larger n, illustrating the CLT.
2. The skewness of the distribution of the average is in the same direction as the skewness of the parent distribution.
3. The mean of the distribution of the sample average is equal to the mean of the parent distribution.
4. The standard deviation of the distribution of the sample average is equal to the standard deviation of the parent distribution, divided by $n^{1/2}$, or $\text{StdDev}(\bar{Y}) = \sigma/n^{1/2}$.
5. Points 3 and 4, taken together, explain how the LLN works. In Figures 8.4 through 8.8, you saw how the sample average converged, although in a random way, to the process mean μ. Figure 10.6 explains it all: The sample average is random, with mean μ, but with standard deviation that gets smaller as n gets larger.

Example 10.5: Predicting Your Gambling Losses, Revisited, Using the CLT

In Example 10.1, you saw that if you play 10 on red in roulette, 10,000 times in succession, then your expected total earnings are $E(T) = -5263.2$ and your standard deviation

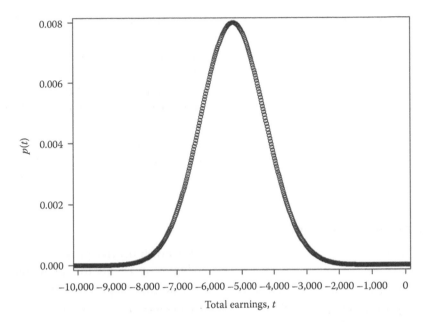

FIGURE 10.7
Distribution of the total earnings in $n = 10,000$ plays of 10 on red in roulette.

of earnings is $\text{StdDev}(T) = 998.6$. Now, the CLT tells you that your total earnings are approximately normally distributed, based on such a large n ($n = 10,000$) and based on the fact that the parent distribution (the Bernoulli distribution) easily satisfies Ugly Rule of Thumb 10.2, since $n\pi = 10,000(18/36) > 5$ and $n(1 - \pi) = 10,000(20/36) > 5$. Figure 10.7 shows this precise distribution; it's very closely approximated by a normal distribution.

Now you can tighten up your previous analysis that used the loose Chebyshev bound. While the statement that there is at least an 88.9% chance that your total earnings are between $-5263.2 \pm 3(998.614)$ or between -8259 and -2267 is correct, you can now say more. The actual chance that your earnings are between -8259 and -2267 is, by the 68–95–99.7 Rule, approximately 99.7%. Things are even worse than you feared. Good luck!

Vocabulary and Formula Summaries

Vocabulary

Covariance	A measure of dependence between RVs. It is not scale-free.
Additivity property of variance	A result that states that the variance of a sum is the sum of the variances plus twice the covariance.
Multiplicative property of expectation	A result that states that the expected value of the product is equal to the product of the expected values, when the RVs are independent.

Standard error of the mean

The estimated standard deviation of the RV $\bar{Y}$.

Correlation

A scale-free measure of dependence between RVs.

CLT

If T is the sum of iid RVs with finite variance, then T is approximately normally distributed.

Gamma distribution

The probability distribution that is the distribution of the sum of independent exponential RVs.

Key Formulas and Descriptions

$E(XY) = E(X)E(Y)$

The multiplicative property of expectation when X and Y are independent.

$Var(X + Y) = Var(X) + Var(Y) + 2Cov(X, Y)$

The additivity property of variance.

$Cov(X, Y) = E[\{X - E(X)\}\{Y - E(Y)\}]$

The covariance between RVs X and Y.

$Var(X + Y) = Var(X) + Var(Y)$

The additivity property of variance when X and Y are independent.

$E(T) = E(Y_1 + Y_2 + \cdots + Y_n) = n\mu$

The mean of the sum of n identically distributed RVs $Y_1, Y_2, ..., Y_n$ is $n\mu$.

$Var(T) = Var(Y_1 + Y_2 + \cdots + Y_n) = n\sigma^2$

The variance of the sum of n iid RVs $Y_1, Y_2, ..., Y_n$ is $n\sigma^2$.

$E(\bar{Y}) = \mu$

The expected value of the average of n identically distributed RVs is μ.

$Var(\bar{Y}) = \sigma^2/n$

The variance of the average of n iid RVs is σ^2/n.

$StdDev(\bar{Y}) = \sigma/\sqrt{n}$

The standard deviation of the average of n iid RVs is $\sigma/\sqrt{n}$.

$StdErr(\bar{y}) = \hat{\sigma}/\sqrt{n}$

The standard error of the average of n iid RVs is $\hat{\sigma}/\sqrt{n}$.

$\hat{\sigma}_{xy} = (1/n)\sum_i (x_i - \bar{x})(y_i - \bar{y})$

The bootstrap plug-in estimate of $Cov(X, Y)$.

$\rho_{xy} = E\left[\left(\dfrac{X - \mu_x}{\sigma_x}\right)\left(\dfrac{Y - \mu_y}{\sigma_y}\right)\right]$

The correlation between RVs X and Y.

$\hat{\rho}_{xy} = \dfrac{(1/n)\sum_i (x_i - \bar{x})(y_i - \bar{y})}{\sqrt{(1/n)\sum_i (x_i - \bar{x})^2}\sqrt{(1/n)\sum_i (y_i - \bar{y})^2}}$

The bootstrap plug-in estimate of correlation.

Exercises

10.1 Use the following distribution for observable customer satisfaction data:

y	$p(y)$
1	0.05
2	0.15
3	0.15
4	0.15
5	0.50

A. What do the numbers 0.05, 0.15, 0.15, 0.15, and 0.50 mean, specifically? Refer to customers and processes in your answer.

B. Find $E(Y)$ and $Var(Y)$.

C. Interpret $E(Y)$ in terms of the LLN and customers.

D. Suppose $Y_1, Y_2, \ldots, Y_{1000}$ are the outcomes of 1000 customers sampled as iid from the process $p(y)$. Let $\bar{Y} = (1/1000)(Y_1 + Y_2 + \cdots + Y_{1000})$. Find $E(\bar{Y})$. Carefully show all steps along the way indicating how linearity, additivity, independence, or identical distributions is used at each step, as appropriate.

E. Find $Var(\bar{Y})$. Carefully show all steps along the way indicating how linearity, additivity, independence, or identical distributions is used at each step, as appropriate.

F. Find $StdDev(\bar{Y})$, and apply Chebyshev's inequality using $E(\bar{Y})$ and $StdDev(\bar{Y})$, along with $k = 3$: What does it tell you about the values of $\bar{Y}$ that you will observe?

G. Using the earlier distribution, produce satisfaction data for $n = 1{,}000$ customers using software, calculate $\bar{Y}$, and repeat 10,000 times. Does Chebyshev's inequality with $k = 3$ work appropriately, based on the 10,000 $\bar{Y}$ values that you produced?

H. Have your software graph the histogram and q–q plot of the 10,000 $\bar{Y}$ values that you produced in Exercise 10.1G. Does the CLT seem to work here?

I. Does the 68–95–99.7 rule work reasonably well for the 10,000 $\bar{Y}$ values that you produced in Exercise 10.1G? Calculate the actual percentages from the 10,000 data values.

J. Which interpretation is best to use here: Chebyshev's inequality or the 68–95–99.7 Rule? Why?

10.2 Using the definition of variance of a RV V as $Var(V) = E\{V - E(V)\}^2$, as well as the linearity and additivity properties of expectation, show that $Var(aX + bY) = a^2Var(X) + b^2Var(Y) + 2abCov(X, Y)$. Follow the logic of the derivation in Section 10.2 that showed $Var(X + Y) = Var(X) + Var(Y) + 2Cov(X, Y)$, and provide similar step-by-step explanations.

10.3 You have 20 (millions of dollars, millions of euros, etc.) to invest, and there are two investment vehicles involving options such as mutual funds and hedge funds that you are considering. One possibility is that you could put all 20 in investment vehicle #1. This is a *non*diversified strategy. Alternatively, you could put 12 in investment vehicle #1 and 8 in investment vehicle #2. This is an example of a *diversified* strategy. The future returns on these investment vehicles are R_1 and R_2, respectively. Thus, if in the future you observe that $R_1 = 0.06$ and $R_2 = -0.02$, then you will have earned $20 \times 0.06 = 1.20$ under the nondiversified strategy, and you will have earned $12 \times 0.06 + 8 \times (-0.02) = 0.56$ under the diversified strategy. Suppose the RVs R_1 and R_2 have means of 0.05 and 0.05 and standard deviations of 0.04 and 0.04.

A. Let Y_1 be earnings under strategy #1 and Y_2 be your earnings under strategy #2. Give the equation that expresses Y_1 in terms of the RVs R_1 and/or R_2. Repeat for Y_2.

B. Find the expected earnings under the each of the two strategies. Be clear how the linearity and/or additivity properties are used in your calculations. Interpret the two numbers. Is either strategy preferred in terms of expected value?

C. Find the variance and standard deviation of your earnings under both strategies, assuming independence of returns. Be clear how the linearity and/or additivity properties are used in your calculations. Interpret the two standard deviations. Is either strategy preferred in terms of standard deviation? (Note: The answer to this question depends on your attitude toward risk.)

D. Repeat Exercise 10.3C for $\text{Corr}(R_1, R_2) = 0.9$. Use the result shown in Exercise 10.2, but note that correlation and covariance are different!

E. Repeat Exercise 10.3C for $\text{Corr}(R_1, R_2) = -0.9$.

F. Using your answers from Exercise 10.3C through E, what is the effect of correlation on diversification?

10.4 Correlation and regression are related. Regression models are often written as $Y = \beta_0 + \beta_1 X + D$, where β_0 and β_1 are constant (the model's slope and intercept, respectively), where X is the predictor variable, Y is the response variable, and D is a random error term that accounts for the fact that Y is not deterministically related to X. You may assume that X and D are independent and that $E(D) = 0$. Also, let $\sigma_x^2 = \text{Var}(X)$ and $\sigma^2 = \text{Var}(D)$.

A. Find $\text{Var}(Y)$, applying the linearity and/or additivity properties as appropriate.

B. In your answer to Exercise 10.4A, part of the variation in Y was explained by variation in X, and part explained by variation in D. What is the proportion of the variation in Y that is explained by the variation in X? (Comment: This is called *R-squared* in regression analysis.)

C. Using the definition of covariance and the linearity and additivity properties of expectation, show that $\text{Cov}(X, Y) = \beta_1 \text{Var}(X)$.

D. Show that the square of the correlation between Y and X is equal to the R-squared quantity you found in Exercise 10.4B.

10.5 Use the results of Exercise 10.4 to generate 1000 pairs X, Y that have mean 0 and variance 1 (for both variables)and correlation 0.9. Calculate $T = X + Y$, and compute the plug-in estimate of the variance of T for these 1000 observations. Is the estimate close to what the $\text{Var}(X + Y)$ formula dictates?

10.6 Show that $|\rho_{xy}| = 1$ if $Y = aX + b$, when a and b are constants with $a \neq 0$ and $\text{Var}(X) \neq 0$.

10.7 Using software, generate 10,000 observations (Z_1, Z_2) from an $N(0,1)$ distribution. You should have two columns when you're done.

A. Make a scatterplot of the variables. Explain its appearance.

B. Calculate the correlation between Z_1 and Z_2, again using software, and interpret its value.

C. Explain why this correlation is supposed to be 0.

D. Explain why the correlation is not exactly 0.

10.8 Using software, generate samples of $n = 3$ observations from an N(0,1) distribution. Then define a new variable *SumZsq* as the sum of squares of the 3 observations you generated or $\text{SumZsq} = Z_1^2 + Z_2^2 + Z_3^2$. Repeat 10,000 times, so you have 10,000 SumZsq values.

A. Make the histogram of these 10,000 values and describe its appearance.

B. Repeat Exercise 10.8A with $n = 8$ and then with $n = 25$.

C. The distribution known as the *chi-squared distribution with n degrees of freedom* is defined as the distribution of $\sum_i Z_i^2$, where $Z_1, ..., Z_n \sim_{\text{iid}}$ N(0,1). Apply the CLT to describe how the chi-squared distribution with n degrees of freedom looks when n gets larger, and relate this to your simulation and histogram results.

10.9 Roll a fair die; get X. Now, let $Y_1 = X$, $Y_2 = X$, ..., $Y_n = X$. In other words, the Y values are all the same number, X.

A. What is the distribution of $\bar{Y}$ when $n = 10$? When $n = 10,000$?

B. Why doesn't the CLT work here?

10.10 Show how the plug-in estimate of covariance, $\hat{\sigma}_{xy} = (1/n)\sum_i(x_i - \bar{x})(y_i - \bar{y})$, is derived from Tables 10.2 and 10.3 and the formula $\sigma_{xy} = \sum\sum (x - \mu_x)(y - \mu_y)\, p(x, y)$.

10.11 The total of n iid Bernoulli(π) RVs has the *binomial distribution* with parameters n and π. Thus, your number of wins at roulette in 10,000 plays has the binomial distribution with $n = 10,000$ and $\pi = 18/38$. If you bet 10 each time, then your earnings are (# of wins) $\times$ 10 $-$ (# of losses) $\times$ 10 $= T \times 10 - (10,000 - T) \times 10 = 20 \times T - 100,000$, where T is a binomial RV with parameters $n = 10,000$ and $\pi = 18/38$. Using the CLT, you saw that the probability that your earnings will be between $-\$8259$ and $-\$2267$ is *approximately* 99.7%. Find the *exact* probability using the binomial distribution, using whatever software you have handy. Explain your logic.

10.12 A very ugly rule of thumb you may see in some sources is that $n \geq 30$ is adequate to assure approximate normality of the distribution of the sample average $\bar{Y}$. Is $n \geq 30$ adequate for the insurance application shown in Exercise 9.3 of Chapter 9? To answer, take the following steps:

A. Generate a sample of $n = 30$ iid values from $p(y)$ given in Exercise 9.3 of Chapter 9 and calculate $\bar{Y}$ from the 30 values.

B. Repeat Exercise 10.12A many times, getting many different $\bar{Y}$ values, each comprised of $n = 30$ observations. (The actual number of $\bar{Y}$ you generate does not matter, so long as it is large enough to get a good estimate of the distribution. For example, you might generate 10,000 $\bar{Y}$ values. But don't confuse this number with n. The n is the crucial element of this analysis.) Draw the histogram and q–q plot of the resulting $\bar{Y}$ values. Is the distribution of $\bar{Y}$ approximately normal?

C. Larger n should help. Repeat Exercise 10.12A and B when $n = 100$. Is the distribution of $\bar{Y}$ approximately normal when $n = 100$?

D. Do some experimentation with different n. How large does n have to be before you would consider the distribution of $\bar{Y}$ to be approximately a normal distribution?

E. Does your conclusion of Exercise 10.12D change when you consider the distribution of the *total* of n observations, rather than the distribution of the *average* of n observations?

10.13 Using the definition of covariance and the linearity and additivity properties of expectation, show, step-by-step with justifications, why the computing formula $\text{Cov}(X,Y) = E(XY) - E(X)E(Y)$ is true. Use the mathematical demonstration showing $\text{Var}(Y) = E(Y^2) - \{E(Y)\}^2$ in Chapter 9 as your template.

10.14 A public opinion poll asks people whether they support or do not support the current president. Each person contributes a yes or a no. Code the yes responses as $Y = 1$ and the no responses as $Y = 0$.

A. What is the distribution of Y? Specify it in terms of an unknown parameter, π.

B. Find the expected value and variance of Y in terms of π.

C. Suppose $Y_1, Y_2, \ldots, Y_n$ are iid from the distribution in Exercise 10.14A. What is the importance and relevance of the statistic $\bar{Y}$ in this example?

D. Find the mean and variance of $\bar{Y}$ in terms of π and n.

E. Use the CLT, along with the answer to Exercise 10.14D, to identify the approximate distribution of $\bar{Y}$ in terms of π and n.

F. Suppose $\pi = 0.45$ and $n = 1000$. Draw a graph of the approximate distribution of $\bar{Y}$ using your answer to Exercise 10.14E.

G. Repeat Exercise 10.14F, but assuming $n = 4000$.

H. Apply the 68–95–99.7 Rule to the distribution of $\bar{Y}$, assuming $\pi = 0.45$, when $n = 1000$. Repeat for $n = 4000$.

I. State the practical interpretation of your answer to Exercise 10.14H in terms of public opinion polling.

10.15 We showed in this chapter that $\text{Var}(X + Y) \neq \text{Var}(X) + \text{Var}(Y)$ when $X = Y$. Show that the formula $\text{Var}(X + Y) = \text{Var}(X) + \text{Var}(Y) + 2\text{Cov}(X, Y)$ gives the correct answer when $X = Y$.

11

Estimation: Unbiasedness, Consistency, and Efficiency

11.1 Introduction

The plug-in estimation technique introduced in Chapter 9 showed one way to estimate parameters using data. This method produced sensible results, such as $\hat{\mu} = \bar{y}$ and $\hat{\sigma}^2 = (1/n)\sum_i(y_i - \bar{y})^2$. But remember, *estimate* is a weasel word! Anything can be an estimate of anything else. Ideally, you would like your estimated parameter to be *close* to the true parameter that it is estimating. The three concepts that relate to closeness, although in different ways, are *unbiasedness*, *consistency*, and *efficiency*.

To define these concepts precisely, we first must introduce some vocabulary terms having to do with estimation.

Vocabulary for Estimation Theory

- **Estimand:** The quantity being estimated, for example, μ, a process mean.
- **Estimator:** The random variable used to estimate the quantity, for example, $\bar{Y}$.
- **Estimate:** A particular, fixed observation on the random variable used to estimate the quantity, for example, $\bar{y} = 76.1$.

If you search the web for more information about these three terms, you'll probably run across someone using the word *estimate* when they really mean *estimator*. It's an unfortunate but common misuse of words. We're guilty ourselves of doing it occasionally, although we're going to be extra careful in this chapter to say it right! When you read about these topics elsewhere, the author's intended meaning should be clear from context even if the terminology is imprecise.

Now, back to the story. You want your estimate of the parameter to be close to the actual value of the parameter. In other words, using the new vocabulary, you want your *estimate* to be close to the *estimand*. Because your estimate is the result of a random process (*model produces data*), it can sometimes be close to the estimand, and sometimes far away. Thus, you need to understand how the random estimator behaves in relation to the fixed estimand in order to understand whether you can trust the estimate. The behavior of the random estimator is the topic of this chapter; the chapter topics unbiasedness, consistency, and efficiency all refer to the randomness of the estimator.

They all sound good, right? Unbiasedness sounds great—who wants to be biased? Ditto with consistency and efficiency—who wants to be inconsistent or inefficient? Many statistical formulas embedded into statistical software are mysterious-looking because they contain bias corrections; thus, the concept of bias is fundamental to understanding statistical theory and practice. Ironically, for all the effort at producing bias-corrected estimators, it turns out that unbiasedness is relatively unimportant compared to consistency and efficiency.

11.2 Biased and Unbiased Estimators

Terms with a hat $^\wedge$ on them will sometimes be *random* in this chapter; if so they are called *estimators*. As random variables, it makes sense to talk about their distributions—the distribution of possible values that the estimator might take on—as well as the expected values, or points of balance, of such distributions. Once an *estimator* becomes an *estimate*, there is no distribution, since there is only one number (like 76.1) and hence no expected value of interest either. (The expected value of 76.1 is, well, 76.1. It's not interesting.)

This notion of randomness of an estimator is essential to understanding the concept of unbiasedness. Recall that *model produces data*—specifically, the statistical model produces *random* data. The definition of an **unbiased estimator** refers to the distribution of possible values of the random estimator.

Definition of an Unbiased Estimator

An estimator $\hat{\theta}$ of an estimand θ is unbiased if $E(\hat{\theta}) = \theta$.

There is a lot of information packed into this simple definition. A longer, but equivalent statement is this:

> The estimator $\hat{\theta}$ is a function of random data and is therefore a random variable. As a random variable, $\hat{\theta}$ has a probability distribution. If the mean of the probability distribution of possible values of $\hat{\theta}$ is equal to θ, then $\hat{\theta}$ is an unbiased estimator of θ.

Example 11.1: Unbiasedness of the Sample Mean

Suppose $Y_1, Y_2, \ldots, Y_n$ are produced from the same distribution $p(y)$ (either independently or dependently, it doesn't matter). Let μ be the mean of the pdf $p(y)$; that is, $\mu = \int y p(y)\, dy$ or $\mu = \Sigma y p(y)$. Then $\bar{Y} = (1/n)\Sigma_i Y_i$ is an unbiased estimator of μ. Why is this true? To show unbiasedness, you need to show that $E(\bar{Y}) = \mu$. Here are the logical steps.

$E(\bar{Y}) = E\{(1/n)\Sigma_i\, Y_i\}$	(By substitution)
$\quad = (1/n)E(\Sigma_i\, Y_i)$	(By the linearity property of expectation)
$\quad = (1/n)E(Y_1 + Y_2 + \cdots + Y_n)$	(By the definition of the summation symbol Σ_i)
$\quad = (1/n)\{E(Y_1) + E(Y_2) + \cdots + E(Y_n)\}$	(By the additivity property of expectation)
$\quad = (1/n)(\mu + \mu + \cdots + \mu)$	(Since each Y_i is produced by the same distribution whose mean is μ)
$\quad = (1/n)(n\mu)$	(Since there are n terms in the sum)
$\quad = \mu$	(By algebra)

Hence, $\bar{Y}$ is an unbiased estimator of μ.

Example 11.2: Bias of the Sample Mean When There Is Systematic Measurement Error

In the example of estimating the value of inventory items, in Chapter 7, suppose the assessor's superiors charge him with showing that the inventory has a higher value than in it actually does. To comply, the assessor decides to add a random error that is normally distributed with mean $10 and standard deviation $15 to each of his valuations. Then his assessed valuation is $Y_i = T_i + D_i$, where the T_i is the true value of the inventory item i, and D_i is his $N(10, 15^2)$ "fudge" of the true value.

The process mean of the true values is μ, so $E(T_i) = \mu$, but the expected value of the assessor's average is not equal to μ. Instead, his average is biased, which you can see as follows:

$E(\bar{Y}) = E\{(1/n)\Sigma_i\, Y_i\}$	(By substitution)
$= (1/n)E(T_1 + D_1 + T_2 + D_2 + \cdots + T_n + D_n)$	(By linearity of expectation and substitution)
$= (1/n)\{E(T_1) + E(D_1) + E(T_2) + E(D_2) + \cdots$ $\quad + E(T_n) + E(D_n)\}$	(By the additivity property of expectation)
$= (1/n)(\mu + 10 + \mu + 10 + \cdots + \mu + 10)$	(Since each T_i is produced by the same distribution whose mean is μ and since each D_i is produced by the same distribution whose mean is 10)
$= \mu + 10$	(By algebra, noting that there are n μ terms and n 10s in the sum)

See Figure 7.11 for an illustration of bias. The value of $\bar{Y}$ is a random variable whose mean differs systematically on the high side from the target. That does not mean that $\bar{Y}$ is always too large: as shown in Figure 7.11, the average $\bar{Y}$ is lower than the target for some random samples, and higher than the target in other random samples. But on average, over all random samples, the value of $\bar{Y}$ is too large, so the estimate is biased high.

In summary, when there is systematic measurement error, the observable estimator of the mean, $\bar{Y}$, is a biased estimator of the process mean of the true values, μ, since $E(\bar{Y}) \neq \mu$.

The previous two examples might seem like a lot of math for some rather obvious conclusions. However, it can be surprising that seemingly sensible estimators are often biased, as the following example shows.

Example 11.3: Bias Induced by Nonlinearity: Estimating Percent Change in Averages

This year, the average age of students taking a class is 26.81 years. Ten years ago it was 24.64 years. The percent change formula tells you that this year's average is $100(26.81/24.64 - 1)\% = 8.8\%$ more than the average 10 years ago.

This is a perfectly sensible analysis—statistician approved, even. However, the procedure gives you a biased estimator of the percent change! To explain this seemingly counterintuitive result, suppose you can get an iid sample of positive numbers $Y_1, Y_2, \ldots, Y_n$ from a pdf whose mean is μ_Y—this year, for example. Suppose you can get another iid sample of positive numbers $X_1, X_2, \ldots, X_m$ from a pdf whose mean is μ_X—from 10 years ago, for example. You can reasonably assume that the X data from 10 years ago are independent of the current Y data.

The sample averages $\bar{Y}$ and $\bar{X}$ are unbiased estimators of the true (process) means μ_Y and μ_X, respectively, as shown in Example 11.1. Let $\hat{\theta} = 100(\bar{Y}/\bar{X} - 1)$. Is $\hat{\theta}$ an unbiased estimator of $\theta = 100(\mu_Y/\mu_X - 1)$? The answer is no; the reason lies in Jensen's inequality, and in the fact that $f(x) = 1/x$ is a convex function, since $f''(x) = 2x^{-3} > 0$ for $x > 0$.

Now, $\hat{\theta}$ is an unbiased estimator of θ if $E(\hat{\theta}) = \theta$. Is it true in this case? The answer is no; here's why:

$E(\hat{\theta}) = E\{100(\bar{Y}/\bar{X} - 1)\}$	(By substitution)
$\quad = 100\ \{E(\bar{Y}/\bar{X}) - 1\}$	(By the linearity property of expectation)
$\quad = 100\ [E\{\bar{Y}(1/\bar{X})\} - 1]$	(By algebra)
$\quad = 100\ [\{E(\bar{Y}) \times E(1/\bar{X})\} - 1]$	(By the multiplicative property of expectation for independent random variables given in Section 10.2)
$\quad > 100\ [\{E(\bar{Y}) \times 1/E(\bar{X})\} - 1]$	(By Jensen's inequality, since $1/x$ is a convex function)
$\quad = 100\ (\mu_Y/\mu_X - 1)$	(Since $\bar{Y}$ and $\bar{X}$ are unbiased estimators of μ_Y and μ_X)
$\quad = \theta$	(By definition of θ)

Because of the $>$ inequality in the chain of statements above, the obvious estimator $\hat{\theta} = 100(\bar{Y}/\bar{X} - 1)$ will tend to be larger than the estimand $\theta = 100(\mu_Y/\mu_X - 1)$, and $\hat{\theta}$ is therefore a biased estimator of θ.

If this seems mysterious, just do a simple simulation and you'll see for yourself. Roll five dice (like in the games Yahtzee and bar dice) and find the average, $\bar{Y}$. Then roll another five dice, getting $\bar{X}$. You would think that on average $\bar{Y}$ should be 0% higher than $\bar{X}$, right? After all, the means are $\mu_Y = 3.5$ and $\mu_X = 3.5$, and μ_Y is therefore 0% higher than μ_X. True enough, but on average, $\hat{\theta} = 100(\bar{Y}/\bar{X} - 1)$ is larger than 0.

Figure 11.1 shows the result of rolling the dice five times, computing $\hat{\theta}_1 = 100(\bar{Y}/\bar{X} - 1)$, and repeating, getting $\hat{\theta}_2, \hat{\theta}_3, \ldots$. The law of large numbers tells you that the long-run average of these $\hat{\theta}$s will converge to $E(\hat{\theta}) = E\{100(\bar{Y}/\bar{X} - 1)\}$. You can see that the running averages of the $\hat{\theta}$ values do not converge to 0; instead, they converge to a slightly larger number than zero (specifically, 5.555 in this case), as predicted by Jensen's inequality. This provides another illustration that $\hat{\theta}$ is a biased estimator.

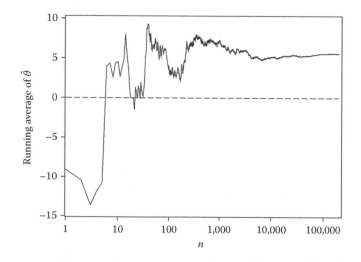

FIGURE 11.1
The law of large numbers applied to estimates $\hat{\theta} = 100(\bar{Y}/\bar{X} - 1)$, where the Y and X data are sampled from identical processes and where there are $n = 5$ values used to calculate each $\bar{Y}$ and $\bar{X}$. The running averages of values of $\hat{\theta}$ converge to $E(\hat{\theta})$, which is more than 0, the horizontal line, showing bias.

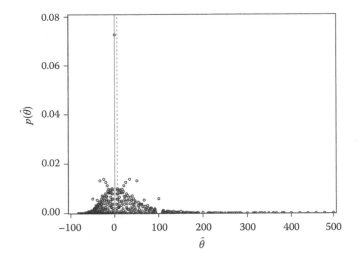

FIGURE 11.2
Distribution of $\hat{\theta} = 100(\bar{Y}/\bar{X} - 1)$, where there are $n = 5$ Xs and $n = 5$ Ys. The expected value is dashed (5.555) and the desired estimand is solid (0.000). The circles denote probabilities of observing the particular discrete $\hat{\theta}$ values. Since the expected value of the estimator $\hat{\theta} = 100(\bar{Y}/\bar{X} - 1)$ differs from the estimand 0, $\hat{\theta}$ is a biased estimator.

For another look at the bias in $\hat{\theta} = 100(\bar{Y}/\bar{X} - 1)$, see Figure 11.2. It shows the distribution of $\hat{\theta}$, calculated analytically by enumerating all combinations $(\bar{X}, \bar{Y})$. It also shows the true mean of $\hat{\theta}$, which turns out to be 5.555, and the desired estimand, which is 0.0.

Figure 11.2 shows that the statistician-approved estimator $\hat{\theta} = 100(\bar{Y}/\bar{X} - 1)$ is biased. Then why did the statisticians approve of it? The reason is apparent in Figure 11.2: While there is bias, it is very small compared to the variability. The estimator tends to be far from *both* its expected value *and* its estimand. The problem of bias is minor compared to the problem of variability. The estimator is logical, and the bias is small. Further, the statisticians approve of the fact that with larger sample sizes than five in each group, the bias of $\hat{\theta}$ will be even smaller; and more importantly, the variability of $\hat{\theta}$ will be smaller as well. Finally, the statisticians approve of common sense, and for all these reasons, the estimator $\hat{\theta} = 100(\bar{Y}/\bar{X} - 1)$ is viable. Unbiasedness isn't the only criterion for determining whether an estimator is good.

11.3 Bias of the Plug-In Estimator of Variance

The plug-in estimate of variance is given in Chapter 9 as

$$\hat{\sigma}^2 = \frac{1}{n}\sum_i (y_i - \bar{y})^2$$

This estimate, you may recall, is the variance of the bootstrap distribution $\hat{p}(y)$ that puts $1/n$ probability on each of your observed data values $y_1, y_2, \ldots, y_n$, with probabilities accumulated when there are repeats on some of the data values. Notice that this is an *estimate* since it is a function of the observed, fixed data values $y_1, y_2, \ldots, y_n$. As such it is fixed, non-random. It's a number, like 43.1, not a variable.

Viewed as a function of the potential data values $Y_1, Y_2, \ldots, Y_n$ that are random variables, the corresponding *estimator* is given by

$$\hat{\sigma}^2 = \frac{1}{n} \sum_i (Y_i - \bar{Y})^2$$

We apologize for using the same symbol $\hat{\sigma}^2$ for both estimator and estimate. But there are only so many symbols available! Please look for the context to tell whether $\hat{\sigma}^2$ is an estimator or an estimate. For now, think of it as an estimator, that is, as random.

Example 11.4: The Bias of the Plug-In Estimator of Variance When You Roll a Die Twice

The true variance of a roll of a die, calculated from the discrete uniform distribution, is

$$\sigma^2 = \sum (y - \mu)^2 p(y)$$

$$= (1 - 3.5)^2 \times \left(\frac{1}{6}\right) + (2 - 3.5)^2 \times \left(\frac{1}{6}\right) + (3 - 3.5)^2 \times \left(\frac{1}{6}\right)$$

$$+ (4 - 3.5)^2 \times \left(\frac{1}{6}\right) + (5 - 3.5)^2 \times \left(\frac{1}{6}\right) + (6 - 3.5)^2 \times \left(\frac{1}{6}\right) = 2.917$$

In practice you won't know σ^2; it is a process parameter. In fact, even for the die roll you really don't know that $\sigma^2 = 2.917$; this is true only if you can assume a perfectly symmetric die, which is not true in reality. For instance, the sides with more dots are slightly lighter for dice with recessed markings, since they have slightly less material in them than the other sides. Still, it is good to have an example where all the numbers are known in order to make the discussion concrete, so we will go ahead and assume a perfectly symmetric die with $\sigma^2 = 2.917$.

If you roll the die twice, getting Y_1 and Y_2, the plug-in *estimator* of variance is

$$\hat{\sigma}^2 = \frac{1}{n} \sum_i (Y_i - \bar{Y})^2 = \frac{1}{2} \{(Y_1 - \bar{Y})^2 + (Y_2 - \bar{Y})^2\}$$

where

$$\bar{Y} = \frac{1}{2}(Y_1 + Y_2)$$

For example, if you get $y_1 = 2$ and $y_2 = 6$, then your *estimate* of σ^2 is $\hat{\sigma}^2 = (1/2)\{(2 - 4)^2 + (6 - 4)^2\} = 4.0$. If you roll (3, 3), then your *estimate* is $\hat{\sigma}^2 = (1/2)\{(3 - 3)^2 + (3 - 3)^2\} = 0$.

Notice that combinations (1, 1), (2, 2), (3, 3), (4, 4), (5, 5), and (6, 6) *all* give $\hat{\sigma}^2 = 0$; no other combinations give $\hat{\sigma}^2 = 0$. Each combination has 1/36 probability, so the probability that your *estimator* will give the value 0 is 6/36 = 0.1667. There are only five other possible values for the estimator $\hat{\sigma}^2$ when $n = 2$; using Figure 9.2 you can identify these values and obtain the distribution shown in Table 11.1.

TABLE 11.1

pdf of the Plug-In Estimator of
Variance of Dice, When $n = 2$

$\hat{\sigma}^2$	$p(\hat{\sigma}^2)$
0.00	0.1667
0.25	0.2778
1.00	0.2222
2.25	0.1667
4.00	0.1111
6.25	0.0556
Total	1.000

The expected value of the *estimator* is thus

$$E(\hat{\sigma}^2) = \sum \hat{\sigma}^2 \times p(\hat{\sigma}^2)$$

$$= (0.00) \times (0.1667) + (0.25) \times (0.2778) + (1.00) \times (0.2222)$$

$$+ (2.25) \times (0.1667) + (4.00) \times (0.1111) + (6.25) \times (0.0556)$$

$$= 1.458$$

Egads! The plug-in estimator is seriously biased since $E(\hat{\sigma}^2) = 1.458$, much less than the true
value $\sigma^2 = 2.917$. Figure 11.3 illustrates the bias.

Since the expected value of the estimator differs from the estimand, the plug-in estimator
whose distribution is shown in Figure 11.3 is biased.

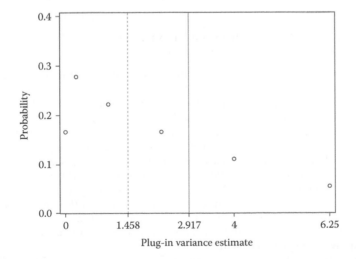

FIGURE 11.3
Probability distribution of the plug-in estimator of variance using $n = 2$ dice rolls, showing expected value of
the estimator, 1.458 (dashed vertical line), and the estimand, or the true variance, 2.917 (solid vertical line). The
circles denote points $(\hat{\sigma}^2, p(\hat{\sigma}^2))$.

What to do about this bias? Maybe try a different estimator. Here you know the mean is $\mu = 3.5$. Had you used the estimator $\hat{\theta} = (1/2)\{(Y_1 - 3.5)^2 + (Y_2 - 3.5)^2\}$, instead of the plug-in estimator $\hat{\sigma}^2 = (1/2)\{(Y_1 - \bar{Y})^2 + (Y_2 - \bar{Y})^2\}$, you would have had an unbiased estimator. Here's why:

$E(\hat{\theta}) = E[(1/2)\{(Y_1 - 3.5)^2 + (Y_2 - 3.5)^2\}]$	(By substitution)
$= (1/2)E\{(Y_1 - 3.5)^2\} + (1/2)E\{(Y_2 - 3.5)^2\}$	(By the linearity and additivity properties of expectation)
$= (1/2)\text{Var}(Y_1) + (1/2)\text{Var}(Y_2)$	(By the definition of variance)
$= (1/2)(2.917) + (1/2)(2.917)$	(Because each Y has the same discrete uniform die distribution)
$= 2.917$	(By arithmetic)

Apparently, then, the problem with the plug-in estimator is that it uses $\bar{Y}$ in the formula instead of μ. You get an estimate that is biased low because the average $\bar{Y}$ is *closer to the data values* Y_i than is μ. The reason is that $\bar{Y}$ *is calculated from the data*, whereas μ *is not calculated from the data*. Instead, μ is part of the model that *produces the data*; that is, μ *pre-exists* the data.

For instance, if the pair (1, 1) is observed, then $\bar{y} = 1$, but still, $\mu = 3.5$. The number $\bar{y} = 1$ is closer to the data values (1, 1) than is $\mu = 3.5$.

The fact that $\bar{Y}$ is *closer to the data values* Y_i than is μ implies that the estimated variance when using the squared deviations $(Y_i - \bar{Y})^2$ is smaller than the estimated variance when using the squared deviations $(Y_i - \mu)^2$. Hence, the plug-in estimator of variance is biased low, as shown in Figure 11.3.

So the solution to the bias problem is simple, right? Just use μ instead of $\bar{Y}$ in the plug-in formula for variance? Well, no, that won't work in practice. Unlike in the case of the dice distribution, you don't know the mean μ of the data-generating process for most data you might analyze in the real world, because *model has unknown parameters*. We'll solve that problem in a bit. For now, let's make these concepts more concrete by using a real example.

Example 11.5: Estimating the Mean and Variance of the Age Distribution of Students in a Graduate Class

A sample of self-reported ages, in years, will be collected from students in a class. Prior to data collection, you may assume the data values $Y_1, Y_2, ..., Y_n$ are generated as iid from a distribution $p(y)$ that reflects the processes that put students in the class. This distribution is a discrete distribution with possible values 0, 1, 2, 3, ... (years old), although values 0, 1, ..., 15 are very unlikely, as are values 110, 111, 112, ... The distribution has unknown mean $\mu = \Sigma y\, p(y)$ and unknown variance $\sigma^2 = \Sigma(y - \mu)^2\, p(y)$.

Assuming iid sampling, the *estimator* $\bar{Y} = (1/n)(Y_1 + Y_2 + \cdots + Y_n)$ has yet to be observed, and is therefore a random variable that is an *unbiased estimator* of the process mean μ. The plug-in estimator of σ^2 is $\hat{\sigma}^2 = (1/n)\Sigma_i(Y_i - \bar{Y})^2$, which also has yet to be observed, and is therefore also a random variable. But $\hat{\sigma}^2$ is a *biased estimator* of the process variance σ^2. The alternative estimator of variance that uses μ in place of $\bar{Y}$, $\hat{\sigma}^2 = (1/n)\Sigma_i(Y_i - \mu)^2$, also random, is an *unbiased estimator* of the process variance σ^2, but you can't use it because you don't know μ, the process mean.

Now, it so happens that some data come in. The numbers are $y_1 = 36$, $y_2 = 23$, $y_3 = 22$, $y_4 = 27$, $y_5 = 26$, $y_6 = 24$, $y_7 = 28$, $y_8 = 23$, $y_9 = 30$, $y_{10} = 25$, $y_{11} = 22$, $y_{12} = 26$, $y_{13} = 22$, $y_{14} = 35$, $y_{15} = 24$, and $y_{16} = 36$. The plug-in *estimate* of the mean (μ) of the distribution of age is $\bar{y} = (1/16)(36 + 23 + \cdots + 36) = 26.81$ years.

The plug-in *estimate* of the variance (σ^2) of the distribution of age is $\hat{\sigma}^2 = (1/16)$ $\{(36 - 26.81)^2 + (23 - 26.81)^2 + \cdots + (36 - 26.81)^2\} = 22.90$ years^2. This variance estimate is the result of a biased estimation procedure, so on average the estimates obtained using this estimation procedure are smaller than the true variance σ^2. However, the fact that there is bias does not imply that *this particular estimate*, 22.90 years^2, is too small (see Figure 11.3): despite being biased low, sometimes the plug-in estimate of variance is larger than the true variance. You have no way of knowing, in this particular sample, whether 22.90 is too large or too small. All you know is that, *on average*, the plug-in estimate tends to be too small.

Is the bias in the plug-in estimate of variance really that much of a problem? Figure 11.3 looks scary: the expected value of the estimator is one half the true value of the estimand. This looks like a lot of bias, and it is. On the other hand, you know that you can remove the bias by using μ instead of $\bar{Y}$ in the plug-in formula. You also know, by the law of large numbers, that $\bar{Y}$ is close to μ when n is large. Therefore, the extreme bias in Figure 11.3 diminishes rapidly with larger n, simply because $\bar{Y}$ becomes closer to μ with larger n.

Example 11.6: The Bias of the Plug-In Estimator of Variance with a Sample of $n = 16$ Observations

To illustrate the decrease in bias with larger n, suppose the true age distribution $p(y)$ of students in the class was a discrete uniform distribution on the numbers 23, 24, ..., 40. This is of course unrealistic in its restriction to 23 through 40 year olds, and also in its equal probabilities, but it will suffice to illustrate the point. Then the probability of observing any particular age in the $\{23, 24, ..., 40\}$ range is 1/18, and the true mean is

$$\mu = \sum yp(y) = 23\left(\frac{1}{18}\right) + 24\left(\frac{1}{18}\right) + \cdots + 40\left(\frac{1}{18}\right) = 31.5 \text{ years}$$

And the true variance is

$$\sigma^2 = \sum (y - \mu)^2 p(y) = (23 - 31.5)^2\left(\frac{1}{18}\right) + (24 - 31.5)^2\left(\frac{1}{18}\right) + \cdots + (40 - 31.5)^2\left(\frac{1}{18}\right)$$

$$= 26.92 \text{ years}^2$$

Suppose you sample $n = 16$ iid observations from this distribution and calculate the plug-in variance $\hat{\sigma}^2 = (1/n)\sum_i(Y_i - \bar{Y})^2$. For every sample of $n = 16$ observations, you get a different $\hat{\sigma}^2$, which is explained by randomness alone.

Figure 11.4 shows the distribution of the values of $\hat{\sigma}^2$ obtained in such repeated samples. You can relate Figure 11.4 to the actual in-class data collection described in Example 11.5. The number you got as an estimate, 22.9, might be larger than the true variance or smaller than the true variance. In Figure 11.4, it is smaller than the true variance, but Figure 11.4 is based on a purely hypothetical assumption of a discrete uniform distribution for which the true variance is 26.92. On average, the plug-in estimate

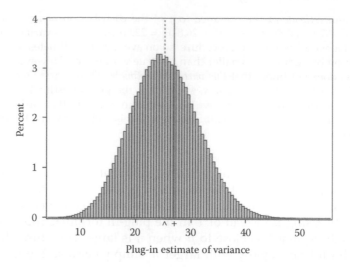

FIGURE 11.4

Estimated distribution of the plug-in estimator of variance based on a sample of $n = 16$ observations. The + symbol indicates the true variance, $\sigma^2 = 26.9$, while the $\wedge$ symbol indicates the mean of the distribution of the plug-in estimator, 25.2.

tends to be too small, but this does not tell you that your estimate, 22.9, is too small. As Figure 11.4 shows, despite being biased on the low side, the plug-in estimate of σ^2 is frequently *higher than* the estimand.

Compare Figure 11.4 with Figure 11.3. With a sample size of $n = 16$ rather than $n = 2$, the bias of the plug-in estimate is greatly lessened, again, simply because with a larger sample size, the estimator $\bar{Y}$ tends to be closer the μ, by the law of large numbers.

11.4 Removing the Bias of the Plug-In Estimator of Variance

The problem with the plug-in estimator is that the deviations $(Y_i - \bar{Y})^2$ tend to be smaller than the deviations $(Y_i - \mu)^2$. While $E(Y_i - \mu)^2 = \sigma^2$, by definition of variance, you can show that $E(Y_i - \bar{Y})^2$ is smaller than σ^2, specifically $E(Y_i - \bar{Y})^2 = \{(n-1)/n\}\sigma^2$, as follows:

$E(Y_i - \bar{Y})^2 = E\{(Y_i - \mu) - (\bar{Y} - \mu)\}^2$	(By algebra of subtracting and adding μ)
$\quad = E\{(Y_i - \mu)^2 + (\bar{Y} - \mu)^2 - 2(Y_i - \mu)(\bar{Y} - \mu)\}$	(By algebra of expanding the square: $(a-b)^2 = a^2 + b^2 - 2ab$)
$\quad = E\{(Y_i - \mu)^2\} + E\{(\bar{Y} - \mu)^2\} - 2E\{(Y_i - \mu)(\bar{Y} - \mu)\}$	(By the linearity and additivity properties of expectation)
$\quad = \text{Var}(Y_i) + \text{Var}(\bar{Y}) - 2\text{Cov}(Y_i, \bar{Y})$	(By definition of variance and covariance)

Now, $\text{Var}(Y_i) = \sigma^2$, by assumption that the data are sampled from a distribution whose variance is σ^2. Further, as shown in Chapter 10, $\text{Var}(\bar{Y}) = \sigma^2/n$.

So far, we have shown that $E(Y_i - \bar{Y})^2 = \text{Var}(Y_i) + \text{Var}(\bar{Y}) - 2\text{Cov}(Y_i, \bar{Y}) = \sigma^2 + \sigma^2/n - 2\text{Cov}(Y_i, \bar{Y})$. Now, for the covariance term:

$\text{Cov}(Y_i, \bar{Y})$	
$= E\{(Y_i - \mu)(\bar{Y} - \mu)\}$	(By the definition of covariance)
$= E[(Y_i - \mu)\{(1/n)(Y_1 - \mu) + (1/n)(Y_2 - \mu) + \cdots + (1/n)(Y_n - \mu)\}]$	(By substituting the definition of $\bar{Y}$ and by using properties of summation)
$= (1/n)E\{(Y_i - \mu)(Y_1 - \mu)\} + (1/n)E\{(Y_i - \mu)(Y_2 - \mu)\} + \cdots$ $+ (1/n)E\{(Y_i - \mu)(Y_n - \mu)\}$	(By the linearity and additivity properties of expectation)
$= (1/n)\text{Cov}(Y_i, Y_1) + (1/n)\text{Cov}(Y_i, Y_2) + \cdots + (1/n)\text{Cov}(Y_i, Y_n)$	(By the definition of covariance)

Now, since the Y_i are independently sampled, $\text{Cov}(Y_i, Y_j) = 0$, except when $i = j$, in which case $\text{Cov}(Y_i, Y_i) = \text{Var}(Y_i) = \sigma^2$. Hence, $\text{Cov}(Y_i, \bar{Y}) = (1/n)\hat{\sigma}^2$ (only one term is picked up in the last summation), implying that

$$E(Y_i - \bar{Y})^2 = \text{Var}(Y_i) + \text{Var}(\bar{Y}) - 2\text{Cov}(Y_i, \bar{Y}) = \sigma^2 + \sigma^2/n - 2(1/n)\sigma^2 = \sigma^2 - (1/n)\sigma^2 = \{(n-1)/n\}\sigma^2$$

That was the tedious part. The easy part is now to find the bias of the plug-in estimator:

$E(\hat{\sigma}^2) = E\{(1/n)\sum_i (Y_i - \bar{Y})^2\}$	(By substitution)
$= (1/n)\sum_i E(Y_i - \bar{Y})^2$	(By the linearity and additivity properties of expectation)
$= (1/n)\sum_i \{(n-1)/n\}\sigma^2$	(Because as shown earlier, $E(Y_i - \bar{Y})^2 = \{(n-1)/n\}\sigma^2$)
$= (1/n)[n\{(n-1)/n\}\sigma^2]$	(Because there are n identical terms in the summation)
$= \{(n-1)/n\}\sigma^2$	(By algebra)

So, the bias of the plug-in estimator gets smaller as n gets larger, since its expected value, $\{(n-1)/n\}\sigma^2$, gets closer to σ^2 as n increases. This explains why the bias seen in Figure 11.4, where $n = 16$, is so much smaller than the bias seen in Figure 11.3, where $n = 2$.

There is a simple correction for the bias of the plug-in estimator: just multiply it by $n/(n-1)$. The resulting estimator is $\{n/(n-1)\}(1/n)\sum_i(Y_i - \bar{Y})^2 = \{1/(n-1)\}\sum_i(Y_i - \bar{Y})^2$. This is the number that standard software reports for the variance estimate, rather than the plug-in estimate.

The Standard, Unbiased Estimator of σ^2

$$\hat{\sigma}^2 = \frac{\sum_i (Y_i - \bar{Y})^2}{n - 1}$$

While this formula is the standard, don't dismiss the plug-in estimator, which uses n in the denominator instead of $(n-1)$, too quickly. The standard formula is somewhat mysterious: dividing by $n - 1$ rather than n means that the estimator no longer has the direct interpretation as "the average squared deviation from the sample average." Further, the use of n rather than $n - 1$ makes very little difference when n is large—say, 100 or more.

11.5 The Joke Is on Us: The Standard Deviation Estimator Is Biased after All

All that trouble to show that by using $n - 1$ in the formula for the variance estimator you get an unbiased estimator of σ^2! But alas, when you take the square root of the unbiased estimator for variance, the resulting estimator is biased. Once again, Jensen's inequality explains it. Let $f(y) = y^{1/2}$; this is a concave function. Thus, for a non-negative random variable V, $E(V^{1/2}) < \{E(V)\}^{1/2}$. In particular, if $\hat\sigma^2$ is the unbiased estimator of σ^2, then $E\{(\hat\sigma^2)^{1/2}\} < \{E(\hat\sigma^2)\}^{1/2} = (\sigma^2)^{1/2} = \sigma$. More concisely, $E(\hat\sigma) < \sigma$.

The degree of bias in the estimated standard deviation is not at all troubling, though, as shown in Figure 11.5 for the case where $n = 16$ observations are sampled from the hypothetical discrete uniform age distribution of Example 11.6.

Despite the fact that the $(n - 1)$ standard deviation estimator is biased, it is still the default provided by software. Further, as shown in Figure 11.5, the bias of this estimator is small relative to its variability. Specifically, the bias is only $5.15 - 5.19 = -0.04$, whereas the variability shown Figure 11.5 is much higher. It appears likely that the estimated standard deviation can be 1.0 or more (years of age) from the true standard deviation (also in units of years of age) just by the variability inherent in a random sample of $n = 16$ observations. Returning to Example 11.5 with the real students' age data, the estimate $\hat\sigma = \sqrt{24.43} = 4.94$ (years) can easily be 1.0 (years) different from the true, unknown, process standard deviation σ.

While the bias is therefore not of great concern, the plug-in estimate of standard deviation, which uses n rather than $n - 1$, would have even greater downward bias, so it is typically not used. From now on, therefore, unless otherwise specified, the standard deviation estimate we discuss will be the one based on the $n - 1$ denominator, despite its bias.

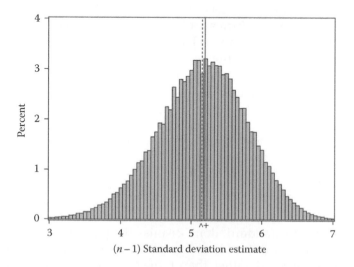

FIGURE 11.5
Estimated distribution of the standard deviation estimator that uses the $n - 1$ in the denominator. The true standard deviation is 5.19 (indicated by +). The expected value of the estimator is 5.15 (indicated by ^), showing slight downward bias.

The Usual Standard Deviation Estimator

$$\hat{\sigma} = \sqrt{\frac{1}{n-1}\sum_i (Y_i - \bar{Y})^2}$$

It might be desirable to try to remove the bias from this estimator, but it is difficult to do so in general, because the bias depends on the distribution $p(y)$ that produced the data. Since the distribution $p(y)$ is unknown (*model has unknown parameters*), you usually can't remove the bias. You can use the bootstrap, where you estimate the unknown distribution $p(y)$ via $\hat{p}(y)$, to approximately remove bias, but it may not be worth the effort. Again, the main problem is not bias: as shown in Figure 11.5, the main problem is variability. And to reduce the variability in the estimator, there is no simple statistical correction. Instead, you'll just have to take a larger sample than $n = 16$ students.

Example 11.7: Estimating the Mean, Variance, and Standard Deviation of Number of Children

How many brothers and sisters do you have? In general, the answer to this question depends heavily on nationality, religion, and culture. Suppose you want to know the average value for the United States. You set out a plan to collect data by randomly dialing a subset of a list of phone numbers; the interviewer will simply ask the question, "Are you an adult over the age of 21?" and if the answer is yes, the interviewer will ask, "How many children do you have?" While this design and measurement protocol is not necessarily the ideal one, it serves to make the point about processes and populations. The DATA that will arise in this study are a household's number of children, either 0, 1, 2, 3, ..., and can be viewed as an iid sample $Y_1, Y_2, ..., Y_n$ from a distribution $p(y)$, where $p(y)$ has the form shown in Table 11.2.

The mean of the distribution of each randomly generated Y is therefore $\mu = 0 \times \pi_0 + 1 \times \pi_1 + 2 \times \pi_2 + 3 \times \pi_3 + \cdots$ and its variance is $\sigma^2 = (0 - \mu)^2 \times \pi_0 + (1 - \mu)^2 \times \pi_1 + (2 - \mu)^2 \times \pi_2 + (3 - \mu)^2 \times \pi_3 + \cdots$. If you had the entire collection of phone numbers, and if we could find out the number of children from everyone on the list, this mean μ and variance σ^2 might be interpretable as the average and variance calculated from such a population of numbers. But that is population thinking. Think instead of the process: the process that produces the data is not pristine. The values μ and σ^2 instead refer to the actual process of data collection, which includes biases such as nonresponse, interviewer effects, untruthful responses, etc. You can view the parameters μ and σ^2 as population values that are attenuated by such biasing influences.

TABLE 11.2

Distribution of Number of Children per Household

y	$p(y)$
0	π_0
1	π_1
2	π_2
3	π_3
...	...
Total	1.00

But even without such biasing influences, a better way to interpret the values μ and σ^2 is as values of the process that actually produced the population itself, as discussed in Chapter 7. You can interpret the parameters μ and σ^2 as the mean and variance of the number of children, resulting from social and demographic processes in effect at this particular time in the United States, attenuated by the biases of the design and measurement processes. Or, in short, μ and σ^2 are parameters of the data-generating process. Even shorter: *Model produces data.*

Suppose the observed data ($n = 50$) are 0, 2, 2, 2, 6, 1, 0, 2, 1, 1, 3, 3, 4, 2, 1, 3, 2, 3, 0, 2, 3, 1, 3, 1, 1, 2, 0, 1, 0, 0, 3, 2, 1, 1, 1, 1, 2, 2, 1, 2, 2, 1, 2, 2, 0, 4, 2, 1, 2, and 1. Then the estimate of the process mean, μ, is 1.7 children, and the (usual) estimate of the process standard deviation is 1.2 (children). The estimate of the mean is from an unbiased procedure as regards the *process*: That is, in different samples using the same process, the estimates obtained—you'll get a different estimate from 1.7 in almost every other sample of $n = 50$ people—average out to the process mean μ. Since all estimates are obtained from the same sampling procedure, this μ necessarily inherits whatever attenuation biases are present from the sampling procedure. In other words, the estimator of μ, $\bar{Y}$, is unbiased for the process mean μ, but it is necessarily a biased estimator of the population mean. The parameter μ differs from the population mean because of sampling procedure-based attenuation biases.

Even though $\bar{Y}$ is an unbiased estimator of μ, there is a lot of uncertainty about the value of μ, even after data collection. The value of μ is either larger or smaller than 1.7 children, and you don't know how much larger or smaller it is.

The estimate 1.2 of the process standard deviation σ results from the estimator $\hat{\sigma} = \sqrt{\{1/(n-1)\}\sum_i (Y_i - \bar{Y})^2}$ which is, by Jensen's inequality, a slightly biased estimator. However, as shown in Figure 11.5, the bias is only a minor concern relative to the variability. Variability is caused by randomness in the observed sample of $n = 50$ observations and is the main source of our uncertainty about the process standard deviation σ. The value of σ is either larger or smaller than 1.2 (children), and you don't know how much larger or smaller it is.

All you have at this point are estimates of the mean and the standard deviation of the process, and you do not know how accurate they are. You can use confidence intervals and credible intervals, discussed in later chapters, to quantify the accuracy of these and other estimates of the process parameters.

11.6 Consistency of Estimators

The discussion about unbiased estimators hopefully left you feeling that bias is not all that important, so long as the bias is not too great. If so, good! There are other, more important properties that you should look for in an estimator, including *consistency* and *efficiency*.

Definition of a Consistent Estimator

Suppose an estimator $\hat{\theta}$ is a function of n observations $Y_1, Y_2, ..., Y_n$. The estimator $\hat{\theta}$ is a consistent estimator of θ if $\lim_{n \to \infty} \hat{\theta} = \theta$.

As in the definition of unbiasedness there is a lot of information packed into the definition of consistency. A longer, but equivalent statement defining consistency is the following:

> The estimator $\hat{\theta}$ is a function of random data, and is therefore a random variable. As a random variable, $\hat{\theta}$ has a probability distribution. If the distribution of $\hat{\theta}$ narrows for larger n, collapsing around the true value of θ, then $\hat{\theta}$ is a consistent estimator of θ.

In more advanced probability texts, you will find multiple definitions of consistency, involving strong convergence, weak convergence, and convergence in rth mean. For our purposes, a type of convergence called *weak convergence in probability* is sufficient and corresponds to our definition. Econometricians sometimes call such convergence the *probability limit* and denote it by $\text{plim}_{n \to \infty} \hat{\theta} = \theta$.

A simple example of a consistent estimator is the sample mean. If $Y_1, Y_2, \ldots, Y_n \sim_{\text{iid}} p(y)$, where $E(Y) = \mu$, then by the law of large numbers, $\lim_{n \to \infty} \overline{Y}_n = \mu$. Thus, if $\theta = \mu$ and $\hat{\theta} = \overline{Y}$, then $\hat{\theta}$ is a consistent estimator of θ.

The standard estimator of standard deviation is $\hat{\sigma} = \sqrt{\{1/(n-1)\} \Sigma_i (Y_i - \overline{Y})^2}$. Although $\hat{\sigma}$ is biased, it is consistent. Figure 11.6 shows what happens to the sample standard deviation when the data Y_i are sampled as iid from a process distribution $p(y)$ whose true standard deviation is $\sigma = 1.4$.

While the sample standard deviation estimator is biased for every n, the estimator converges to the right quantity for larger n. As long as the sample size is large, the estimator will be close to the true value, despite bias. Clearly, for this to happen, the bias has to get smaller for large n. Even more importantly, the variability caused by randomness gets smaller for large n as well, leading to the convergence shown in Figure 11.6.

So, the sample standard deviation estimator is biased but consistent. Can an estimator be unbiased but inconsistent? The answer is yes.

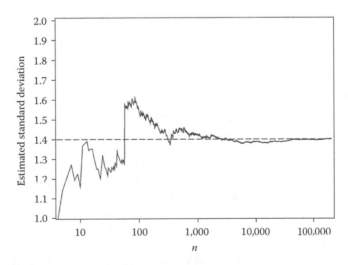

FIGURE 11.6
Convergence of a sequence of successive sample standard deviations calculated from an iid sequence of n observations from a pdf $p(y)$ for which $\sigma = 1.4$. Sample size (n) shown in log scale.

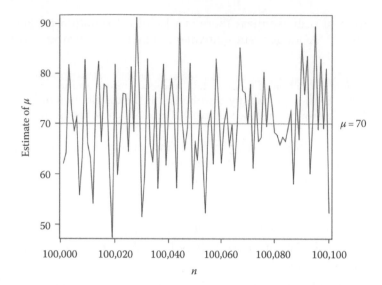

FIGURE 11.7
Non-convergence of the unbiased estimator Y_n sampled from the N(70, 10²) distribution. Even after 100,000 observations are sampled, the estimator Y_n is not close to $\mu = 70$ (horizontal dashed line).

Example 11.8: An Estimator That Is Unbiased but Inconsistent

Suppose you take a sample of n iid observations from $p(y)$, whose mean is μ. Suppose you decide to look only at the last observation, Y_n, and you call this your estimator of μ; that is, you take $\hat{\theta} = Y_n$. This is an unbiased estimator: since Y_n is sampled from $p(y)$, you know that $E(Y_n) = \mu$. But as n increases, the number Y_n doesn't get any closer to μ; it's just random, bouncing all over the place. Figure 11.7 shows how this estimator looks when the sample is iid from N(70, 10²). Even after $n = 100,000$ samples, the numbers are not converging to 70, they are just random. So an estimator can indeed be unbiased but inconsistent!

11.7 Efficiency of Estimators

It may be surprising how many different estimates there are for a single parameter θ. You have seen plug-in estimates and biased-corrected estimates of the variance. You can use either the sample mean or the sample median to estimate the center of a distribution. You will find likelihood-based estimates in Chapter 12. There are literally *infinitely many* different estimates for the same parameter, all based on the same data set, and all giving different numbers. Which one should you pick?

While unbiasedness and consistency are both nice properties, what you really want is an estimate $\hat{\theta}$ that is *close* to θ. If you have a choice between two estimates $\hat{\theta}_1$ and $\hat{\theta}_2$, both estimates of the same parameter θ, your choice is simple: pick the estimate that is closer to θ. But wait, you don't know θ, so how can you know which is better? For example, if one public opinion polling agency said $\hat{\theta}_1 = 45.1\%$ of the population approve of a candidate, and another polling agency said $\hat{\theta}_2 = 52.4\%$, how can you know which estimate is better? You can't tell because you don't know the true approval rating, θ.

The choice of an estimator is based on *distributions*—like everything else in statistics! When faced with a choice between *estimates*, look at the *distributions* of the *estimators*. If an estimator $\hat{\theta}_1$ *tends to be closer* to the estimand θ than another estimator $\hat{\theta}_2$, then you should prefer to use $\hat{\theta}_1$.

One way to measure closeness of an estimator to an estimand is by using the **expected squared difference** (or ESD). If the squared difference $(\hat{\theta}_1 - \theta)^2$ is, *on average*, smaller than the squared difference $(\hat{\theta}_2 - \theta)^2$, then you can say that $\hat{\theta}_1$ *tends to be closer* to θ than $\hat{\theta}_2$.

The Expected Squared Difference between an Estimator $\hat{\theta}$ and an Estimand θ

$$\text{ESD}(\hat{\theta}) = \text{E}\left\{(\hat{\theta} - \theta)^2\right\}$$

In other sources, you may find the ESD called the *mean squared error* (MSE). But in regression analysis MSE refers to an unbiased estimator of conditional variance, which is completely different than ESD. So we choose the expression ESD to avoid this confusion.

The ESD is related to both the variance and the bias of the distribution of the estimator $\hat{\theta}$. To simplify the notation, define $\theta_B = \text{E}(\hat{\theta})$. The subscript B reminds you that the estimator $\hat{\theta}$ may be biased.

$\text{ESD} = \text{E}\{(\hat{\theta} - \theta)^2\}$	(By definition)
$\quad = \text{E}\{(\hat{\theta} - \theta_B) + (\theta_B - \theta)\}^2$	(By algebra: subtracting and adding the term θ_B leaves the result unchanged)
$\quad = \text{E}(\hat{\theta} - \theta_B)^2 + (\theta_B - \theta)^2 + 2(\theta_B - \theta)\text{E}(\hat{\theta} - \theta_B)$	(By algebra: $(a + b)^2 = a^2 + b^2 + 2ab$; and by linearity and additivity properties of expectation)
$\quad = \text{Var}(\hat{\theta}) + (\theta_B - \theta)^2$	(By the definition of variance, and since $\text{E}(\hat{\theta} - \theta_B) = 0$ by the linearity property of expectation and by the definition of θ_B)

The term $\text{E}(\hat{\theta}) - \theta = \theta_B - \theta$ is the bias of the estimator: bias is negative if the estimator $\hat{\theta}$ is smaller, on average, than θ; bias is positive if the estimator $\hat{\theta}$ is larger, on average, than θ; and bias is zero for an unbiased estimator. Summarizing, the ESD is related to variance and bias as follows:

$$\text{ESD}(\hat{\theta}) = \text{Var}(\hat{\theta}) + \{\text{Bias}(\hat{\theta})\}^2$$

Thus, you may choose a biased estimator over an unbiased estimator provided that the variance of the biased estimator is much smaller than the variance of the unbiased estimator. In this case, the biased estimator will tend to be closer to the parameter than the unbiased estimator. See Figure 11.8 for an illustration.

In Figure 11.8, the variance of the biased estimator is 1^2 while the variance of the unbiased estimator is 10^2. The bias of the biased estimator is 1.0, and the bias of the unbiased estimator is 0. Thus, the ESD of the biased estimator is $1^2 + 1^2 = 2.0$, while the ESD of the unbiased estimator is $10^2 + 0^2 = 100.0$. On average, the unbiased estimator of θ is farther from the actual parameter θ, as seen in Figure 11.8, and as confirmed by the ESD comparison: $100.0 > 2.0$.

This leads us to the notion of **efficient estimators**.

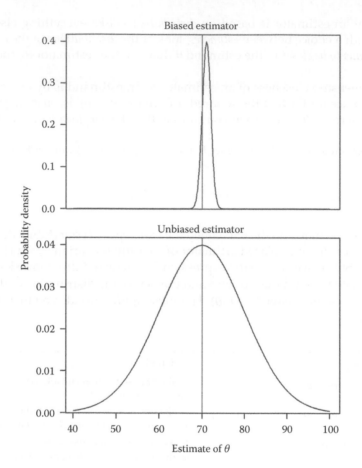

FIGURE 11.8
An example showing when a biased estimator might be preferred to an unbiased estimator. The parameter is $\theta = 70$. The distribution of the biased estimator is $N(71, 1^2)$, while the distribution of the unbiased estimator is $N(70, 10^2)$.

Comparing Estimators Based on Efficiency

If $\hat\theta_1$ and $\hat\theta_2$ are estimators of θ, then the one with smaller ESD is more efficient.

The word *efficient* here refers to the way the estimator uses data. The more efficient estimator uses the same data to produce an estimate that tends to be closer to the estimand. The less efficient estimator uses the same data but produces an estimate that tends to be farther away from the estimand.

Example 11.9: Mean or Median?

If the distribution is symmetric, the mean and median of the distribution are identical. So you could use either the sample mean, $\hat\mu_1 = \bar Y$, or the sample median, where $\hat\mu_2 = Y_{((n+1)/2)}$ if n is odd, and $\hat\mu_2 = (Y_{(n/2)} + Y_{(n/2+1)})/2$ if n is even, to estimate the center of the distribution. Which is better? You already know that $\bar Y$ is unbiased and that its variance is σ^2/n; therefore, $\text{ESD}(\bar Y) = \text{Var}(\bar Y) + \{\text{Bias}(\bar Y)\}^2 = \sigma^2/n + (0)^2 = \sigma^2/n$. The corresponding formula for $\text{ESD}(\hat\mu_2)$ is trickier, but you can estimate it easily via simulation.

Algorithm for Comparing the Efficiencies of the
Sample Mean and the Sample Median

1. Generate a sample of n observations from a known symmetric distribution $p(y)$ (e.g., a normal distribution or some other symmetric distribution) with known mean μ.
2. Calculate $\hat{\mu}_1$ and $\hat{\mu}_2$ from the sample of n observations.
3. Repeat steps 1 and 2. NSIM times: Choose NSIM to be very large, in the thousands or millions.
4. Estimate the ESD of $\hat{\mu}_1$ using the law of large numbers as the average of the NSIM values of $(\hat{\mu}_1 - \mu)^2$. Estimate the ESD of $\hat{\mu}_2$ using the law of large numbers as the average of the NSIM values of $(\hat{\mu}_2 - \mu)^2$.

In a simulation study with the normal distribution $N(70, 10^2)$, with $n = 10$ observations sampled, the average of $(\bar{Y} - 70)^2$ was found (using NSIM = 1,000,000 simulated data sets) to be 10.001, agreeing well with the theory, since $\sigma^2/n = 10^2/10 = 10.000$. The average of (Median $- 70)^2$ (using the same NSIM = 1,000,000 simulated data sets) was found to be 13.84. Thus, the sample mean tends to be closer to the process mean than is the sample median when sampling from the $N(70, 10^2)$ distribution, so the sample mean is a better estimate than the sample median with the $N(70, 10^2)$ distribution. (It's also true for any other normal distribution.)

The normal distribution does not produce extreme outliers. What about distributions that do produce extreme outliers? One way to model outlier-producing distributions is to use *mixture* distributions, where most of the time the data come from one distribution, but occasionally they come from another distribution that has much larger variance. Suppose the data are produced by the $N(70, 10^2)$ distribution 95% of the time, but 5% of the time the data come from the $N(70, 100^2)$ distribution. Here, the average of $(\bar{Y} - 70)^2$ was found (using NSIM = 1,000,000 simulated data sets) to be 59.7, while the average of (Median $- 70)^2$ (using the same NSIM = 1,000,000 simulated data sets) was found to be 15.44. Thus, the sample median tends to be closer to the process mean of a symmetric distribution when the distribution is outlier-prone. The logic is clear: the sample median is relatively unaffected by the occasional extreme outlier, whereas the sample mean is strongly affected by such outliers.

Example 11.10: The "C Chart" in Quality Control

If you know the following technique, then you'll know much of what you need to know about quality control. The technique is simply this: measure the process, look at the resulting data, and make operational decisions using these measurements. Without measurement, you can't tell how well or poorly you are doing, and you therefore can't understand how to improve your process.

A measure of poor quality is the number of defects (or mistakes) made in day-to-day operations, whether 0, 1, 2, … . Ideally there would be no defects at all, but in the real world this doesn't happen! Defects are unavoidable. The system has glitches, the input material is sometimes ill-prepared, someone dozed off … things happen. It's all part of the process, and you must plan for it. You need to keep track of defects and hopefully improve the process so that they happen less frequently.

If the level of defects spikes in a particular day, on a particular shift, or for a particular plant, then something should be done, and quickly. To identify whether there has been such a spike, a **C chart** is used; see Figure 11.9.

The outer boundary lines in the graph show the typical limits of the process, and values outside the boundaries show exceptional cases. If the defects exceed the upper limit, then there is a problem that needs immediate resolution. If the defects fall

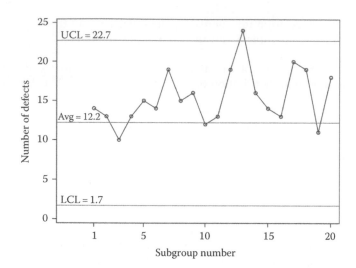

FIGURE 11.9
A C chart.

below the lower limit, then someone is doing something very well; they need accolades and emulation.

The limits are calculated as $\bar{y} \pm 3\sqrt{\bar{y}}$, where $\bar{y}$ is the average number of defects from historical data when the process is in control. Hmmm … this seems odd. Wouldn't it make more sense to use $\bar{y} \pm 3\hat{\sigma}$, where $\hat{\sigma}$ is the standard deviation of the values? Why use $\sqrt{\bar{y}}$ instead of $\hat{\sigma}$? The answer is that the C chart assumes that processes that are "in control" produce counts of defects that look like counts coming from a Poisson distribution $p(y) = e^{-\lambda}\lambda^y/y!$. And if Y is produced by the Poisson(λ) distribution, then $E(Y) = \lambda$ and $Var(Y) = \lambda$; that is, the mean of the Poisson distribution is identical to its variance. That explains the logic for using the square root of the sample mean as an estimator of the standard deviation, but it does not answer the following question: Is the square root of the sample mean really a better estimator of the standard deviation than the sample standard deviation? It still seems odd. To answer the question, you can evaluate and compare their ESDs.

We performed a simulation study with the Poisson ($\lambda = 2$) distribution, whose standard deviation is $2^{1/2} = 1.4142$. Using $n = 20$ observations sampled, the results for the simulation are shown in Table 11.3.

Notice the layers of nuance in the third column of Table 11.3: These are simulation-based estimates of the standard deviation of two different estimators of a standard deviation!

Compare the means of the estimators, 1.4097 and 1.3923, with the true value 1.4142: You can see that both estimators are slightly biased low, as expected by Jensen's inequality with the concave square root function. Similar to what is shown in Figure 11.5, the bias is very small compared to the variability, which, as shown in the third column of Table 11.3, is relatively much more than the bias. Further, the variability of the estimator $\sqrt{\bar{Y}}$ is much smaller than that of the estimator $\sqrt{\{1/(n-1)\}\sum_i(Y_i - \bar{Y})^2}$: the estimated standard deviations are 0.1123 and 0.2497, respectively. This gives the estimator $\sqrt{\bar{Y}}$ a much smaller ESD than that of the usual estimator $\sqrt{\{1/(n-1)\}\sum_i(Y_i - \bar{Y})^2}$. And that's why you should use $\sqrt{\bar{Y}}$ to estimate standard deviation in your C chart.

TABLE 11.3

Simulation-Based Estimation of $E\left(\sqrt{\bar{Y}}\right)$, StdDev$\left(\sqrt{\bar{Y}}\right)$, $E\left(\sqrt{\{1/(n-1)\}\sum_i(Y_i-\bar{Y})^2}\right)$, and

StdDev$\left(\sqrt{\{1/(n-1)\}\sum_i(Y_i-\bar{Y})^2}\right)$, When $n = 20$ Observations Are Sampled from the Poisson(2)

Distribution

Estimator of σ	Estimated Mean (Based on 1,000,000 Simulations)	Estimated Standard Deviation (Based on 1,000,000 Simulations)	Estimate of ESD = Variance + Bias²
$\sqrt{\bar{Y}}$	1.4097	0.1123	$(0.1123)^2 + (1.4097 - 1.4142)^2$ $= 0.0126$
$\sqrt{\{1/(n-1)\}\sum_i(Y_i-\bar{Y})^2}$	1.3923	0.2497	$(0.2497)^2 + (1.3923 - 1.4142)^2$ $= 0.0628$

Note: All estimates are based on 1,000,000 samples, each of size $n = 20$.

The take-home message of the preceding analysis is as follows: while the usual estimator of the standard deviation will work, if you have reason to believe that your process is well modeled by a specific probability distribution for which you know the form of the variance (and thus, the standard deviation), you can potentially gain accuracy when you use a model-based version of the estimator instead of the usual estimator. If you use *maximum likelihood*, discussed in Chapter 12, you will get these improved model-based estimators automatically.

Vocabulary and Formula Summaries

Vocabulary

Estimand	The quantity being estimated, e.g., the process mean μ.
Estimator	The random variable used to estimate a quantity, e.g., the sample average $\bar{Y}$ is an estimator of the process mean μ.
Estimate	A particular fixed observation of the random variable used to estimate a quantity, for example, $\bar{y} = 76.1$.
Unbiased estimator	An estimator having a pdf whose mean is equal to the estimand.
Consistent estimator	An estimator that converges to the estimand as the sample size increases.
Expected squared difference	A measure of how far it is from an estimator to an estimand, on average.
Efficiency	An estimator $\hat{\theta}_1$ is efficient relative to another estimator $\hat{\theta}_2$ if it uses the same data to produce estimates that are generally closer to the estimand than those produced by $\hat{\theta}_2$. The closeness is measured by ESD.
C chart	A graph used in quality control to track defects.

Key Formulas and Descriptions

$\hat{\theta} = 100(\bar{Y}/\bar{X} - 1)$ The estimator of percent change of averages.

$E(Y_i - \mu)^2 = \sigma^2$ The expected squared difference from a data value to the process mean is equal to the variance.

$E(Y_i - \bar{Y})^2 = \{(n-1)/n\}\sigma^2$ The expected squared difference from a data value to the sample average is less than the variance.

$\hat{\sigma}^2 = \dfrac{\sum_i (Y_i - \bar{Y})^2}{n-1}$ The usual, unbiased estimator of the process variance σ^2.

$\hat{\sigma} = \sqrt{\dfrac{1}{n-1}\sum_i (Y_i - \bar{Y})^2}$ The usual (but biased) estimator of the process standard deviation σ.

$E(\hat{\sigma}) < \sigma$ The standard deviation estimator is biased low.

$\lim_{n\to\infty}\hat{\theta} = \theta$ The definition of a consistent estimator.

$E\{(\hat{\theta} - \theta)^2\} = \mathrm{Var}(\hat{\theta}) + \{E(\hat{\theta}) - \theta\}^2$ The expected squared difference between an estimator and its estimand.

$\bar{y} \pm 3\sqrt{\bar{y}}$ The C chart for quality control.

Exercises

11.1   You will put a batch of material through a stress test. If the material passes, you will score it as $Y = 1$; if it fails, you score it as $Y = 0$.

   A.   Show that Y is an unbiased estimator of the probability that the material passes the stress test.

   B.   What assumption(s) are you making in Exercise 11.1A?

11.2   Figure 11.1 shows that $\hat{\theta} = 100(\bar{Y}/\bar{X} - 1)$ is biased when $n = 5$ is used for both averages. Show that it is consistent by simulating data and drawing a graph where increasing n are used in the calculations of $\bar{Y}$ and $\bar{X}$.

11.3   The exponential distribution has pdf $p(y) = \lambda e^{-\lambda y}$, for $y > 0$. Its mean is $1/\lambda$ and its standard deviation is also $1/\lambda$. If you have reason to believe an exponential model would be appropriate for your process, should you use the usual estimator of the standard deviation (the one that uses $n - 1$ in the denominator), or should you use the sample average $\bar{Y}$ to estimate the standard deviation? Perform a simulation study for a sample of $n = 20$ observations from an exponential distribution with mean 4. To accomplish this, perform the following steps, similar to the analysis shown in Table 11.3 for the Poisson distribution.

   A.   Generate a set of $n = 20$ observations from an exponential distribution with mean 4.0, and hence with standard deviation that is also 4.0. Use either built-in random number generators from your software or use the inverse cdf method. Calculate both the usual estimator of the standard deviation and the estimator $\bar{Y}$ using your sample of $n = 20$ observations. Compare these estimates to the true standard deviation: for this sample, which is the better estimate of standard deviation, the usual estimate or $\bar{Y}$?

B.   Repeat Exercise 11.3A NSIM times, where NSIM is large (say, 10,000 or more) and calculate the average and standard deviation of the resulting 10,000 (or however many you simulate) estimates that you computed in Exercise 11.3A.

C.   Using the results in Exercise 11.3B, which appears to be the better estimator in terms of bias? In terms of variance? In terms of ESD?

11.4   Use the sample average $\bar{Y}$ calculated from an iid sample of $n = 10$ observations to estimate the *median* of the exponential distribution $p(y) = 0.5\exp(-0.5y)$, a distribution with mean 2.0 and standard deviation 2.0.

A.   Find the true median of this exponential distribution using the inverse of the cdf.

B.   Show that $\bar{Y}$ is a biased estimator of the median. Do not use simulation.

C.   Find $\text{Var}(\bar{Y})$ and $\text{ESD}(\bar{Y})$ as an estimator of the median. Do not use simulation.

D.   The sample median of $n = 10$ observations is $\{Y_{(5)} + Y_{(6)}\}/2$. Estimate the bias, variance, and ESD of the sample median of the $n = 10$ observations by simulating 10,000 or more samples of size $n = 10$ each.

E.   Based on ESD, which estimator is preferred, the sample mean or the sample median?

11.5   Show how the distribution in Table 11.1 is derived. Use Figure 9.2.

11.6   Calculate the standard deviation of the $n = 999$ numbers 1, 2, ..., 999 using the bootstrap plug-in estimate and the usual estimate. Is there a big difference between the two? Comment on the effect of n on the difference between these two estimates.

11.7   Suppose $\hat{\theta}_1$ and $\hat{\theta}_2$ are unbiased estimators of θ.

A.   Is $(\hat{\theta}_1 + \hat{\theta}_2)/2$ an unbiased estimator of θ? Apply the definition.

B.   Is $0.3\hat{\theta}_1 + 0.7\hat{\theta}_2$ an unbiased estimator of θ? Apply the definition.

C.   Is $2\hat{\theta}_1 - \hat{\theta}_2$ an unbiased estimator of θ? Apply the definition.

D.   Is $0.8\hat{\theta}_1 + 0.4\,\hat{\theta}_2$ an unbiased estimator of θ? Apply the definition.

E.   Is $(\hat{\theta}_1\,\hat{\theta}_2)^{1/2}$ an unbiased estimator of θ? Assume independence, apply the definition, and apply Jensen's inequality.

F.   Is $\hat{\theta}_1^2$ an unbiased estimator of θ_1^2? Apply the definition, and apply Jensen's inequality.

11.8   Opinion polls typically estimate percentage approval. Suppose $\hat{\theta}_1$ and $\hat{\theta}_2$ are independent unbiased estimators of θ, with $\text{Var}(\hat{\theta}_1) = 100$ and $\text{Var}(\hat{\theta}_2) = 1$. These variances correspond to an unreliable poll giving an estimate that is easily wrong by $10 = (100)^{1/2}$ percentage points, and a more reliable poll whose estimate typically differs from the truth only by one percentage point.

A.   If $c_1\hat{\theta}_1 + c_2\hat{\theta}_2$ is an unbiased estimator of θ, what must be true about the constants c_1 and c_2?

B.   Find c_1 and c_2 that provide the unbiased estimator $c_1\hat{\theta}_1 + c_2\hat{\theta}_2$ having minimum variance. (Hint: use Exercise 11.8A to solve for c_2 in terms of c_1. Then find the variance as a function of c_1. Then use calculus to find the c_1 that minimizes the variance.)

11.9   Using simulation and an appropriate graph show that the sample median of an iid sample of n N(70, 10^2) random variables appears to be a consistent estimator. Note that the sample median is $Y_{(n+1)/2}$ if n is odd, and it is $\{Y_{(n/2)} + Y_{(n/2+1)}\}/2$ if n is even.

11.10 Suppose $X_i \sim_{iid} U(0, 2)$ and $Y_i = 2/X_i$. Is $\bar{Y}$ a consistent estimator? Explain.

11.11 Suppose Y_i is the cholesterol level of a sampled person and that the data are iid from $p(y)$. Let $B_i = 1$ if $Y_i < 200$, and $B_i = 0$ otherwise. Explain why the average of the B values is a consistent estimator of $\Pr(Y < 200)$.

11.12 Suppose $Y_1, Y_2, \ldots$ are iid with finite mean μ. Let $\hat{\theta} = (1/n)(Y_1 + Y_2 + \cdots + Y_n) + 1/n$.

    A. Show that $\hat{\theta}$ is a biased estimator of μ by calculating $E(\hat{\theta})$.

    B. Using the law of large numbers provide a logical argument that $\hat{\theta}$ is a consistent estimator of μ.

    C. Perform a simulation analysis that illustrates bias of $\hat{\theta}$ when $n = 4$.

    D. Perform a simulation analysis that illustrates consistency of $\hat{\theta}$.

11.13 Give the mathematical explanation for the result $Cov(Y_i, Y_i) = Var(Y_i)$, which was used in the demonstration that the plug-in estimator of variance is biased.

11.14 Suppose $Y_1, Y_2, \ldots, Y_n \sim_{iid}$ Bernoulli(π). Show that

    A. $\bar{Y}$ is an unbiased estimator of π.

    B. Y_1 is an unbiased estimator of π.

    C. $\bar{Y}$ is more efficient than Y_1.

    D. $\bar{Y}(1 - \bar{Y})$ is a biased estimator of $\pi(1 - \pi)$. You can do this using Jensen's inequality, but don't. Instead, show that $\sigma^2 = \pi(1 - \pi)$ and that $\bar{Y}(1 - \bar{Y}) = (1/n) \Sigma (Y_i - \bar{Y})^2$. Then apply what you have learned in this chapter.

    E. Using Exercise 11.14D identify an unbiased estimator of $\pi(1 - \pi)$ and show that it is unbiased.

11.15 Suppose $(X_1, Y_1), (X_2, Y_2), \ldots, (X_n, Y_n)$ are an iid sample from $p(x, y)$.

    A. Show that $(1/n)\Sigma (X_i - \mu_X)(Y_i - \mu_Y)$ is an unbiased estimator of σ_{xy}.

    B. Suppose you observe a sample (x_i, y_i), $i = 1, 2, \ldots, n$, and replace the (X_i, Y_i) in Exercise 11.15A with the observed (x_i, y_i) data. Is the result still an unbiased estimator?

    C. Unbiased or not, why can't you use the method in Exercise 11.15B in practice?

12

Likelihood Function and Maximum Likelihood Estimates

12.1 Introduction

So, you now have to consider unbiased estimators, consistent estimators, efficient estimators, $n - 1$, n, mean versus median, using the square root of the sample average to estimate a standard deviation, and plug-in estimators. Which estimator should you use? The possibilities are endless, and the choice is daunting.

Fortunately, there is a guiding principle that leads you to an excellent answer. It produces estimates that are usually very efficient and consistent, although not necessarily unbiased. It also provides solutions to complex estimation problems, such as in logistic regression models, when there is no other obvious way to proceed. In cases where there is an obvious way to proceed—$\bar{Y}$ is an obvious estimator for μ, for example—it usually gives you the obvious estimator. The principle is the **likelihood principle**, which states that all information in your sample is contained in your **likelihood function**. This principle leads to an estimation method called **maximum likelihood**, a standard method used to analyze data for many advanced statistical techniques such as regression analysis, logistic regression analysis, time series analysis, categorical data analysis, survival analysis, and structural equation models.

In addition to providing estimates of parameters, the likelihood function also provides a set of values for your model's parameters that are consistent with your observed data—showing how your data reduce your uncertainty about your model's parameters. The reduction in uncertainty can be seen in the *range* of values of the parameter that is supported by the likelihood function. In this chapter, you will see many examples that show how to use likelihood functions, both to estimate parameters and to quantify the uncertainty of the estimates.

12.2 Likelihood Function

Somewhere, sometime, somebody might have mentioned to you that *model produces data*, that the *model has unknown parameters*, and that *data reduce the uncertainty about the unknown parameters*. Whoever said that didn't mention it at the time, but they were talking about the *likelihood function*.

Recall your statistical model for how your DATA are produced:

$$p(y|\theta) \rightarrow \text{DATA}$$

Usually DATA are samples, often assumed independent and identically distributed (iid), called $Y_1, Y_2, \ldots, Y_n$, in which case your model looks like this:

$$p(y|\theta) \rightarrow Y_1, Y_2, \ldots, Y_n$$

If the data $Y_1, Y_2, \ldots, Y_n$ are iid, then the joint distribution of the entire sample is, by independence, given as follows:

$$p(y_1, y_2, \ldots, y_n|\theta) = p(y_1|\theta) \times p(y_2|\theta) \times \cdots \times p(y_n|\theta)$$

The function $p(y_1, y_2, \ldots, y_n|\theta)$ gives you the relative likelihood of all possible configurations of observable data $(y_1, y_2, \ldots, y_n)$, for a given value of the parameter θ. The larger the value of the function $p(y_1, y_2, \ldots, y_n|\theta)$, the more likely is the configuration $(y_1, y_2, \ldots, y_n)$ for that given θ.

The discussion earlier is *prior to* data collection. It describes possible configurations of data values $(y_1, y_2, \ldots, y_n)$ that *can be* observed.

Suppose now that you collect some data and observe the configuration $(Y_1, Y_2, \ldots, Y_n) = (y_1, y_2, \ldots, y_n)$. For instance, you might observe the data values $(y_1, y_2, \ldots, y_n) = (21.4, 43.7, \ldots, 32.0)$. The $(y_1, y_2, \ldots, y_n)$ are actual, *fixed, and known* numbers, rather than random variables (RVs) Y_i that can assume different values. Still, θ is a fixed and unknown parameter, and the function $p(y_1, y_2, \ldots, y_n|\theta)$ (e.g., $p(21.4, 43.7, \ldots, 32.0|\theta)$) still describes the likelihood of the configuration of your observed data $(y_1, y_2, \ldots, y_n)$ for different θ. But now the only variable in the likelihood function $p(y_1, y_2, \ldots, y_n|\theta)$ is θ because your data $(y_1, y_2, \ldots, y_n) = (21.4, 43.7, \ldots, 32.0)$ are fixed.

While the true parameter θ is constant, it is unknown. (*Model has unknown parameters.*) In the likelihood function, θ is a variable, reflecting the fact that you do not know the value of θ. Let the symbol θ_T denote "true value of the parameter," which is not to be confused with the symbol θ used in the likelihood function. By plugging in different potential values for θ into $p(y_1, y_2, \ldots, y_n|\theta)$, you can see that the values of θ are more consistent with the observed data, but keep in mind that you are never going to find the true value, θ_T.

For some values of θ, the likelihood $p(y_1, y_2, \ldots, y_n|\theta)$ of your observed data is higher, and for other values of θ, the likelihood is lower. The likelihood function, then, is the joint probability distribution function (pdf) $p(y_1, y_2, \ldots, y_n|\theta)$, but viewed as a function of θ, with the data $(y_1, y_2, \ldots, y_n)$ fixed. The likelihood function is identical to the joint pdf; however, the joint pdf is a function of the observable data values $(y_1, y_2, \ldots, y_n)$, for fixed θ, while the likelihood function is a function of the parameter θ, for fixed data values $(y_1, y_2, \ldots, y_n)$.

Definition of the Likelihood Function for θ

$$L(\theta|y_1, y_2, \ldots, y_n) = p(y_1, y_2, \ldots, y_n|\theta)$$

The set of values that θ might take on is called the **parameter space**, sometimes abbreviated Θ, the Greek capital letter theta. To be more specific, you can identify the likelihood function as

$$L(\theta|y_1, y_2, \ldots, y_n) = p(y_1, y_2, \ldots, y_n|\theta), \quad \text{for } \theta \in \Theta.$$

When the data are sampled independently, the joint distribution is the product of the marginal distributions, giving the most commonly used form of the likelihood function:

The Likelihood Function for θ Resulting from an Independent Sample

$$L(\theta|y_1, y_2, \ldots, y_n) = p(y_1|\theta) \times p(y_2|\theta) \times \cdots \times p(y_n|\theta)$$

It's a simple concept, as the following examples show.

Example 12.1: Likelihood Function for the Parameter of an Exponential Distribution Based on a Sample of $n = 1$

It's usually a good idea to sample more than one observation! But one is better than none. In the customer service call center waiting time scenario from previous chapters, we assumed that the exponential model $p(y|\lambda) = \lambda e^{-\lambda y}$, for $y > 0$, produced the waiting time DATA. The parameter λ is the θ here; recall that θ is just a generic symbol for a parameter. Also recall that $1/\lambda$ is the mean of this process; that is, $E(Y) = 1/\lambda$. Since the mean of the exponential distribution, $1/\lambda$, can be any positive number, the parameter space is $\Theta = \{\lambda; \lambda > 0\}$. In this example, the sample space $S = \{y; y > 0\}$ and the parameter space $\Theta = \{\lambda; \lambda > 0\}$ are exactly the same sets, but usually they are different, as shown in later examples. Now, the true λ_T is unknown; how to estimate it?

Suppose you sample a single observation (again, it's better to sample more than just one!), $y_1 = 2.0$. Since there is only one observation, we'll drop the 1 subscript for now to make the notation cleaner and state simply that $y = 2.0$. What values of λ are plausible given this observation? Since there is only one data value, and it is observed to be 2.0, you should guess that the mean of the distribution should be somewhere around 2.0; hence, you should also guess that the value of λ is somewhere around $1/2.0 = 0.5$. Figure 12.1 shows the exponential pdf of Y, $p(y|\lambda) = \lambda e^{-\lambda y}$, for different values of λ in a neighborhood of 0.5, and also indicates the likelihood of observing the single data value $y = 2.0$ for each of those distributions.

For the values $\lambda = 0.2, 0.5, 1.0$, and 2.0 shown in Figure 12.1, the likelihoods of observing $y = 2.0$ are, respectively, $(0.2)e^{-(0.2)(2)} = 0.134$, $(0.5)e^{-(0.5)(2)} = 0.184$, $(1.0)e^{-(1.0)(2)} = 0.135$, and $(2.0)e^{-(2.0)(2)} = 0.037$. Thus, among the values of λ shown, $\lambda = 0.5$ provides the highest likelihood and is most consistent with the observed data.

The likelihood function provides the likelihoods of observing $y = 2.0$ for all possible values of λ, not just for those four values. It is identical to the pdf with the observed value $y = 2.0$ fixed and the parameter λ variable and is given by $L(\lambda|y = 2.0) = \lambda e^{-\lambda(2)}$, as graphed in Figure 12.2.

Figure 12.2 shows that the most likely value of λ, in the sense of providing the highest likelihood given the observed data, is $\lambda = 0.5$, corresponding with the intuition described earlier. But more important than the maximizing value 0.5 is the entire likelihood function shown in Figure 12.2. You don't expect to know everything about the process based on a sample of $n = 1$ from the process, so you certainly do not know λ. As the graph shows, there is an infinite range of possible values of λ. In particular, λ_T is *not* 0.5. It could be more or it could be less than 0.5. Your $n = 1$ data point *reduces* your uncertainty about the unknown parameter, but does not *eliminate* your uncertainty.

What is interesting, though, is how much your uncertainty *is* reduced just from your $n = 1$ observation. Large values such as $\lambda = 4.0$ and higher are *effectively ruled out* just by $y = 2.0$. The reason? If the true λ_T were 4 or higher, then it would be *extremely unlikely* to see a value $y = 2.0$. To understand this, look at Figure 12.1 again. The lower right

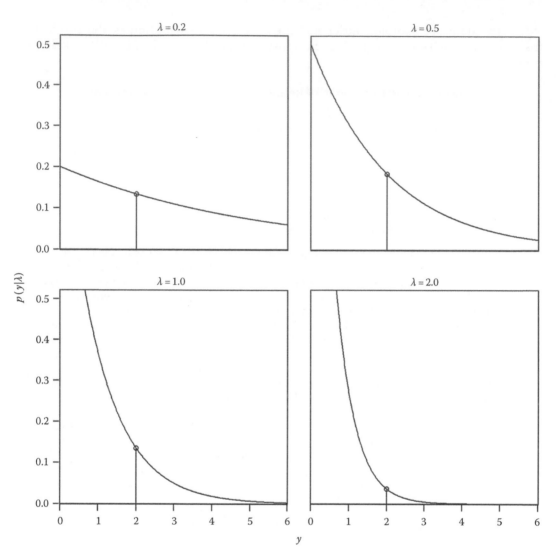

FIGURE 12.1
Exponential pdfs for different values of λ (shown as "lambda" in the graphs). The observed value $y = 2.0$ is indicated in all graphs.

panel has $\lambda = 2$. Already, the observation $y = 2.0$ is out in the tail of the distribution and therefore unlikely. For larger values of λ, the distribution would be pushed even closer toward zero (remember, the mean is $1/\lambda$), making the observation $y = 2.0$ even further out in the tail and therefore even less likely. *Data reduce the uncertainty about the unknown parameters.*

We can now make the Mantra more explicit.

- *Model produces data.* This statement refers to your statistical modeling assumption that your DATA, $Y_1, Y_2, \ldots, Y_n$, are produced by the model $p(y_1, y_2, \ldots, y_n | \theta)$.

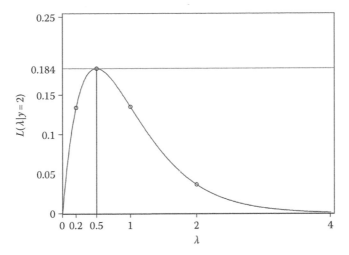

FIGURE 12.2
Likelihood function for λ based on a single observation of $y = 2.0$ from an exponential distribution. Circles correspond to the specific four likelihoods shown in Figure 12.1.

- *Model has unknown parameters.* This statement refers to the fact that you don't know the true parameter(s) θ_T of your statistical model $p(y_1, y_2, ..., y_n | \theta)$.
- *Data reduce the uncertainty about the unknown parameters.* This statement refers to the range(s) of values of the parameter(s) θ wherein θ_T might lie, given your observed data $(y_1, y_2, ..., y_n)$, as shown by the likelihood function $L(\theta | y_1, y_2, ..., y_n)$.

It's true: The data do reduce your uncertainty about the unknown parameters. And with more data, there is a greater reduction in uncertainty, as the following example shows.

Example 12.2: Likelihood Function for the Parameter of an Exponential Distribution Based on a Sample of $n = 10$

Consider the waiting times as in Example 12.1 but with a sample $n = 10$ potential observations $Y_1, Y_2, ..., Y_{10}$, assumed iid, instead of just $n = 1$ observation Y_1. By independence, the joint distribution of the sample of actual observations $y_1, y_2, ..., y_{10}$ is given as follows:

$p(y_1, y_2, ..., y_{10} \| \lambda)$	
$= p(y_1\|\lambda) \times p(y_2\|\lambda) \times \cdots \times p(y_{10}\|\lambda)$	(By independence)
$= \lambda e^{-\lambda y_1} \times \lambda e^{-\lambda y_2} \times \cdots \times \lambda e^{-\lambda y_{10}}$	(Substituting the exponential pdf for $p(y_i\|\lambda)$)
$= \lambda^{10} e^{-\lambda y_1 - \lambda y_2 - \cdots - \lambda y_{10}}$	(By algebra of exponents: $x^a x^b = x^{a+b}$)
$= \lambda^{10} e^{-\lambda \sum_i y_i}$	(By definition of Σ and by algebra)
$= \lambda^{10} e^{-10\lambda \bar{y}}$	(By algebra and by noting that $(1/10)\Sigma y_i = \bar{y}$ implies $\Sigma y_i = 10\bar{y}$)

By definition, the likelihood function is identical to the pdf, but viewed as a function of the parameter rather than as a function of the data:

$$L(\lambda | y_1, y_2, ..., y_{10}) = \lambda^{10} e^{-10\lambda \bar{y}}$$

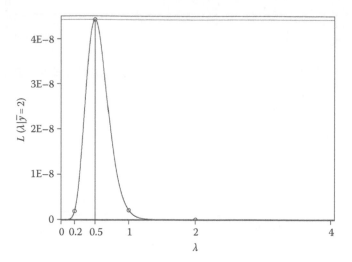

FIGURE 12.3
Likelihood function for λ based on an iid sample $n = 10$ observations from an exponential distribution for which $\bar{y} = 2.0$.

Note that this likelihood function depends on the data only through the average. That means all other statistics, such as median and standard deviation, are irrelevant. When the likelihood function depends on the data $(y_1, y_2, ..., y_n)$ only through a function $t = f(y_1, y_2, ..., y_n)$, then the function t is called a **sufficient statistic**. With the exponential distribution, the average, $\bar{y}$, is a sufficient statistic.

Suppose you observed the data values $y_1 = 2.0$ (as before), $y_2 = 1.2$, $y_3 = 4.8$, $y_4 = 1.0$, $y_5 = 3.8$, $y_6 = 0.7$, $y_7 = 0.3$, $y_8 = 0.2$, $y_9 = 4.5$, and $y_{10} = 1.5$ (all in minutes). The average is still 2.0, as $\bar{y} = (2.0 + 1.2 + 4.8 + 1.0 + 3.8 + 0.7 + 0.3 + 0.2 + 4.5 + 1.5)/10 = 2.0$, but the likelihood function is different because the sample size $n = 10$ appears in it:

$$L(\lambda|y_1, y_2, ..., y_{10}) = \lambda^{10} e^{-10\lambda(2.0)}$$

Figure 12.3 shows how the likelihood function looks now.

The most likely value of λ remains 0.5, but, more importantly, your uncertainty has been greatly reduced. In Figure 12.2, when there was just $n = 1$ observation, the value $\lambda = 2$ seemed relatively plausible, since its likelihood was not too close to zero. However, with $n = 10$ observations, values $\lambda = 2$ and higher are effectively ruled out. Further, values $\lambda = 0.1$ and lower are also effectively ruled out. The range of plausible values of λ, as shown by the likelihood function in Figure 12.3, is much narrower than the range of plausible values of λ shown in Figure 12.2.

Data reduce the uncertainty about the unknown parameters. A wider range of plausible values of the unknown parameter equates to greater uncertainty about the value of the unknown parameter. A narrower range of plausible values of the unknown parameter equates to less uncertainty about the value of the unknown parameter.

Data reduce uncertainty about the unknown parameters. With more data, you get more reduction in uncertainty. Figure 12.4 shows how this works when sampling from the exponential distribution. With larger sample sizes, the likelihood function $\lambda^n e^{-n\lambda y}$ collapses on the true parameter value, in this case $\lambda_T = 0.667$. While different random samples will show different graphs than in Figure 12.4, all will show the same collapsing behavior.

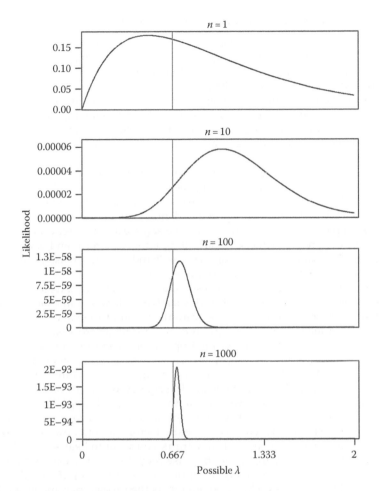

FIGURE 12.4
Likelihood functions for λ using different sample sizes of iid exponential data where $\lambda_T = 0.667$ (vertical line). Bottom panel shows likelihood $\times 10^{500}$.

Notice the tiny values on the vertical axes of Figures 12.3 and 12.4. Clearly, the likelihood function is not a pdf, since the area under the likelihood curve is not equal to 1.0. Let's set that off, so you can find it easily later:

> The likelihood function is not a pdf, since the area under the likelihood curve is not equal to 1.0.

You can solve this problem easily. As shown in Chapter 5, you can turn "slices" into pdfs, simply by multiplying the slice by an appropriate constant. To turn the likelihood function $L(\theta \,|\, y_1, y_2, \ldots, y_n)$ into a pdf, just multiply it by $c = 1/A$, where:

$$A = \int L(\theta|y_1, y_2, \ldots, y_n)d\theta$$

Note that A is just the area under the likelihood function. The result, $c \times L(\theta \,|\, y_1, y_2, \ldots, y_n)$, is a probability distribution for the parameter θ. In a nutshell, this is what Bayesian statistics is all about: It's all about converting likelihood functions into pdfs. We'll talk more about this in Chapter 13.

Example 12.3: The Likelihood Function for Public Opinion Percentage

Scotland became a part of the United Kingdom in 1707 and has remained that way ever since. But not without controversy! A survey of Scots found that 392 out of 1002 agreed with the statement "Scotland should be an independent country." How reliable are these data? A model for the 1002 Scots is $Y_1, Y_2, ..., Y_{1002} \sim_{iid}$ Bernoulli(π), where $Y = 1$ refers to a Scot who supports independence and $Y = 0$ refers to a Scot who either does not support independence or has no opinion. With this model, you view the Scots exactly like you would view tosses of a bent coin. Flip the coin and get heads—the Scot supports independence. Flip and get tails—the Scot either does not support independence or has no opinion.

The Scots can rightly complain about being compared to a bent coin! It's a good model, in that the DATA* produced by this model, for some values of π, really do look like the Scots' data. But in fairness, bent coins have nothing to do with the Scots, per se: The same model would apply equally well to denizens of any country.

In this example, the parameter space and the sample space differ dramatically. The sample space is discrete: It is the set of all configurations of zeros and ones among the 1002 Y_i values. The parameter space, on the other hand, is continuous: It is the set of all possible values of π in the 0 to 1 range, or $\{\pi; 0 \leq \pi \leq 1\}$.

The Bernoulli parameter π is a process parameter. You can interpret the true π_T to be the unobserved measure of political, social, and demographic factors influencing Scottish opinion at the time of the administration of the survey, as attenuated by the biasing influences of the design and measurement processes that produced these 1002 values. You *cannot* call π_T the population proportion of Scots who favor independence, because that is simply incorrect. It may be considered close to the population proportion when there is a good design and measurement process, as performed, for example, by a reputable polling agency, but π_T is not exactly the population proportion. Even if you were to ask the entire adult population of Scotland this question, it still makes more sense to think of π_T as a process parameter. The true π_T remains just out of reach because many factors go into how people respond. They may lie to you or tell you what they think you want to hear so you'll go away, or they might just mentally flip a coin because they don't really care but don't want to give a "no opinion" answer. Thus, as before, think of the population data as random, produced by a process. The parameter π_T is the process parameter. By the law of large numbers (LLN), the population proportion will be very close to this π_T but only with a perfect design and measurement system—which is impossible to achieve in public opinion polling.

The Bernoulli distribution is $p(y|\pi) = \pi$, if $y = 1$, and $p(y|\pi) = 1-\pi$, if $y = 0$. If the data are modeled as iid, then the probability of observing $(Y_1, Y_2, ..., Y_{1002})$ to have the configuration $(y_1, y_2, ..., y_{1002})$ is the product of the marginal probabilities, by independence:

$$p(y_1, y_2, ..., y_{1002}|\pi) = p(y_1|\pi) \times p(y_2|\pi) \times \cdots \times p(y_{1002}|\pi)$$

The likelihood function is identical, but viewed as a function of the Bernoulli parameter π rather than as a function of the observable data $(y_1, y_2, ..., y_{1002})$:

$$L(\pi|y_1, y_2, ..., y_{1002}) = p(y_1|\pi) \times p(y_2|\pi) \times \cdots \times p(y_{1002}|\pi)$$

In the Scottish data, there are 392 cases where y_i is 1 and 610 cases where y_i is 0. When y_i is 1, $p(y_i|\pi) = \pi$. When y_i is 0, $p(y_i|\pi) = 1 - \pi$. Hence, the likelihood function for π is

$$L(\pi|392 \text{ ones and } 610 \text{ zeros}) = \pi^{392}(1-\pi)^{610}$$

Another way you can deduce this likelihood function is by expressing the Bernoulli distribution in function form $p(y) = \pi^y (1 - \pi)^{1-y}$. Here

$L(\pi \mid y_1, y_2, \ldots, y_{1002})$ $\quad = p(y_1 \mid \pi) \times p(y_2 \mid \pi) \times \cdots \times p(y_{1002} \mid \pi)$	(By definition of likelihood function for an independent sample)
$\quad = \pi^{y_1}(1-\pi)^{1-y_1} \times \pi^{y_2}(1-\pi)^{1-y_2} \times \cdots \times \pi^{y_{1002}}(1-\pi)^{1-y_{1002}}$	(By substituting the function form of the Bernoulli pdf)
$\quad = \pi^{y_1+y_2+\cdots+y_{1002}}(1-\pi)^{1-y_1+1-y_2+\cdots+1-y_{1002}}$	(By properties of exponents)
$\quad = \pi^{392}(1-\pi)^{610}$	(Because 392 of the y_i's are 1s and the remaining 610 are 0s)

Figure 12.5 is a graph of this likelihood function.

As shown in Figure 12.5 there is quite a reduction in uncertainty: You now know that π_T is almost certainly less than 0.5. You have to be careful how you interpret this, however. While it is tempting to conclude that "Most likely, fewer than half of the Scots support independence," based on Figure 12.5, you have to remember that π_T is not a population parameter. Rather, it is attenuated by any biases of the sampling process, so if the sampling process is improper or simply wrong (e.g., sampling urban dwellers disproportionately, undersampling young voters who don't have traditional telephones, or pushing respondents to answer the question a certain way), then any conclusion about Figure 12.5 in terms of the general population of Scots is similarly inaccurate.

What you *can* say is this: The process parameter π_T lies almost certainly in the range 0.30 to 0.50—probably in a more narrow range like 0.35 to 0.45 as well—we'll narrow it down more precisely later. Appealing to the law of large numbers as applied the Bernoulli observations (Section 8.5), you can say that in a many hypothetical replications of the sampling—many, many more than the $n = 1002$ sampled—using an identical sampling process, the resulting proportion of Scots favoring independence will almost certainly be in the range from 0.35 to 0.45.

Thus, the sampling procedure is reasonably *reliable*: If the study is replicated, the results will not differ too much. Whether the results are *valid* in the sense of representing true Scottish opinion is not clear—you'd have to know more about how the polling agency operates.

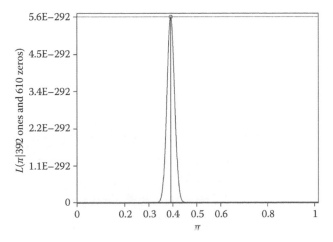

FIGURE 12.5
Likelihood function for the Bernoulli proportion based on an iid sampling of $n = 1002$ that produced 392 ones and 610 zeros.

From here on, we'll dispense with the T subscript on π_T, λ_T, θ_T, etc. that indicates *true value* of the parameter and ask you to recognize from context whether we are talking about the true value of the generic parameter θ or potential values of θ in a likelihood function.

Usually, θ contains more than one value so that $\theta = (\theta_1, \theta_2, \ldots, \theta_k)$; the most famous example is $\theta = (\mu, \sigma^2)$. When there are many parameters in θ, you call θ a **parameter vector**, which simply means a *list* containing more than one parameter value.

Example 12.4: The Likelihood Function for Public Opinion Percentage: Really, There Is More than One Parameter

Some of the Scots answered that they support independence, some answered that they do not support independence, and some answered with no opinion. So there are really two parameters: $\pi_1 = \Pr(\text{Support Independence})$ and $\pi_2 = \Pr(\text{Do Not Support Independence})$. The probability of no opinion is a function of the first two probabilities: $\Pr(\text{No Opinion}) = 1 - \pi_1 - \pi_2$. The data y_i are of the nominal type. When $y_i = $ support independence, $p(y_i | \pi_1, \pi_2) = \pi_1$. When $y_i = $ do not support independence, $p(y_i | \pi_1, \pi_2) = \pi_2$. When $y_i = $ no opinion, $p(y_i | \pi_1, \pi_2) = 1 - \pi_1 - \pi_2$. The parameter vector here is $\theta = (\pi_1, \pi_2)$, and the parameter space is $\{\pi_1, \pi_2; 0 \leq \pi_1, \pi_2, \pi_1 + \pi_2 \leq 1\}$.

Suppose the 610 respondents who were not explicitly in favor of independence break down into 401 respondents who do not support independence and 209 who have no opinion. Following the logic shown earlier for the Bernoulli case, the likelihood function for (π_1, π_2) is

$$L(\pi_1, \pi_2 \,|\, 392 \text{ support, } 401 \text{ do not support, } 209 \text{ no opinion}) = \pi_1^{392} \pi_2^{401} (1 - \pi_1 - \pi_2)^{209}$$

Figure 12.6 shows a 3-D rendering of this likelihood function.

Figure 12.7 is a contour plot corresponding to Figure 12.6. Think about looking down from straight above Figure 12.6 and finding "rings" of equal likelihood, where the rings radiate from the center in order of decreasing likelihood. This plot is exactly the same

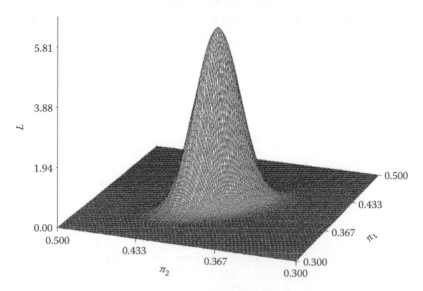

FIGURE 12.6
Likelihood function ($\times 10^{28}$) for (π_1, π_2), where $\pi_1 = \Pr(\text{support})$ and $\pi_2 = \Pr(\text{do not support})$, based on 392 support, 401 do not support, and 209 no opinion.

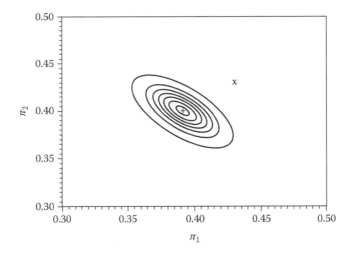

FIGURE 12.7
Figure 12.6 viewed as a contour plot, with peak at $(\pi_1, \pi_2) = (0.391, 0.400)$ indicated by the + symbol and another combination $(\pi_1, \pi_2) = (0.430, 0.430)$ indicated by the × symbol.

kind of plot that you use when you go hiking except that instead of elevation above sea level, the contours measure likelihood.

In Figure 12.7 you can see that the peak of the likelihood function occurs in the obvious place, namely, at the combination $(\hat{\pi}_1, \hat{\pi}_2)$, where $\hat{\pi}_1 = 392/1002 = 0.391$ and $\hat{\pi}_2 = 401/1002 = 0.400$. Looking at the vertical range of the contour plot in Figure 12.6, you see the same conclusion about π_1 that you saw in Figure 12.4: π_1 is roughly between 0.35 and 0.45. However, the joint likelihood tells you more: Looking at the horizontal range, π_2 is also (very roughly) between 0.35 and 0.45. But even more interesting is what appears to be a negative correlation: If π_2 is higher, then π_1 is lower. Thus, the combination $(\pi_1, \pi_2) = (0.43, 0.43)$ has low likelihood. This makes sense because this combination would imply Pr(no opinion) $= 1 - 0.43 - 0.43 = 0.14$, quite far from the observed proportion $209/1002 = 0.21$ of no opinion responses.

Multiparameter likelihood functions are the usual case. In cases of complex advanced statistical models such as regressions, structural equation models, and neural networks, there are often dozens or perhaps even hundreds of parameters in the likelihood function. In such cases, it is impossible to graph the likelihood function, but it is still possible to find values that maximize the likelihood, and it is possible to exploit the likelihood function to assess parameter uncertainty via Bayesian analysis; see Chapter 13. Let's consider for now another classic example with just two parameters, the case of the normal distribution with unknown mean and standard deviation.

Example 12.5: The Joint Likelihood Function for the Parameters (μ, σ) of a Normal Distribution

College students' ages are not normally distributed; rather, the distribution is positively skewed (right-skewed) since there is a virtual lower bound on age (around 17) and occasional large values of age for nontraditional students. Still, you may find it instructive to put data into the normality-assuming likelihood machine and see what happens. The normal distribution is particularly appealing because its two parameters, the mean

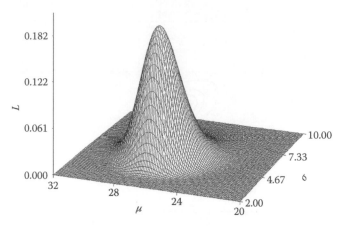

FIGURE 12.8
The joint likelihood ($\times10^{20}$) of (μ, σ) assuming the age data came from a normal distribution.

and the standard deviation, are the main two statistical summaries that you will use—whether or not the distribution happens to be normal.

The age data presented in Example 11.5 are $y_1 = 36$, $y_2 = 23$, $y_3 = 22$, $y_4 = 27$, $y_5 = 26$, $y_6 = 24$, $y_7 = 28$, $y_8 = 23$, $y_9 = 30$, $y_{10} = 25$, $y_{11} = 22$, $y_{12} = 26$, $y_{13} = 22$, $y_{14} = 35$, $y_{15} = 24$, and $y_{16} = 36$, all in years. Assuming—incorrectly, but just for the sake of investigation—that these data are produced by a normal distribution, the likelihood function for μ and σ is as follows:

$$L(\mu,\sigma\,|\,y_1,y_2,y_3,\ldots,y_{16})$$

$$= L(\mu,\sigma\,|\,y_1)\times L(\mu,\sigma\,|\,y_2)\times L(\mu,\sigma\,|\,y_3)\times\cdots\times L(\mu,\sigma\,|\,y_{16})$$

$$= \frac{1}{\sqrt{2\pi}\sigma}\exp\frac{-(36-\mu)^2}{2\sigma^2}\times\frac{1}{\sqrt{2\pi}\sigma}\exp\frac{-(23-\mu)^2}{2\sigma^2}\times\frac{1}{\sqrt{2\pi}\sigma}\exp\frac{-(22-\mu)^2}{2\sigma^2}\times\cdots\times\frac{1}{\sqrt{2\pi}\sigma}\exp\frac{-(36-\mu)^2}{2\sigma^2}$$

When you graph this function, you get Figure 12.8.

Figure 12.8 describes the uncertainty in both μ and σ after a sample of $n=16$ observations. The process mean μ appears to be in the range from 23 to 29, and the process standard deviation σ appears to be in the range from 3 to 8. The actual peak occurs where $(\hat{\mu}, \hat{\sigma}) = (26.81, 4.79)$, both in units of years. The peak is found where the mean takes on the ordinary sample average, as is intuitive; but interestingly the value of the standard deviation at the peak is the plug-in estimate (the n version) rather than the square root of the unbiased estimate of variance (the $n - 1$ version).

12.3   Maximum Likelihood Estimates

The likelihood function shows you the uncertainty you have about your parameter(s) θ, following your collection of data. Another use of the likelihood function is to provide a specific estimate $\hat{\theta}$ of θ. The value $\hat{\theta}$ that *maximizes* the likelihood function $L(\theta\,|\,y_1, y_2, \ldots, y_n)$ is called the MLE of θ.

Definition of the MLE

If $L(\hat{\theta} \mid y_1, y_2, ..., y_n) > L(\theta \mid y_1, y_2, ..., y_n)$ for all permissible θ, then $\hat{\theta}$ is the MLE of θ.

You can find the MLE in various ways. One is by **inspection**: Just look at the graph! That is essentially what we did in Section 12.2—we looked at the graphs and located their peaks. You can make this a little more precise by calculating the likelihoods for a list of values θ, then sorting the likelihoods from largest to smallest, and then picking the θ that gives you the largest. For example, using the waiting time likelihood where $L(\lambda \mid y_1, y_2, ..., y_{10}) = \lambda^{10}e^{-10\lambda(2.0)}$, consider the Excel screenshots shown in Figures 12.9 and 12.10.

After sorting, Figure 12.9 becomes Figure 12.10, which agrees with Figure 12.3, where $\hat{\lambda} = 0.5$ was identified as the value that maximized the likelihood function.

While inspection is simple and intuitively appealing, it does not necessarily find the precise value. What if the real maximum were between 0.50 and 0.51? You can choose finer increments such as 0.001 and get an answer that is closer, but the question would remain, what if the real maximum is between 0.500 and 0.501? Further, inspection becomes less useful when there are multiple parameters, as the list of possible combinations of values increases multiplicatively with each additional parameter.

A better solution for finding the MLE is to use calculus. If the likelihood function is continuous and differentiable, and if the maximum occurs in the interior of the set of

	A	B	C	D
1	lambda	L(lambda | data)		
2	0	=A2^10*EXP(-10*A2*2)		
3	0.01	8.18731E-21		
4	0.02	6.86408E-18		
5	0.03	3.24068E-16		
6	0.04	4.71156E-15		
7	0.05	3.59257E-14		
8	0.06	1.82121E-13		

FIGURE 12.9
Likelihood for different values of λ, based on $L(\lambda \mid y_1, y_2, ..., y_{10}) = \lambda^{10}e^{-10\lambda(2.0)}$.

	A	B	C	D
1	lambda	L(lambda | data)		
2	0.5	4.43359E-08		
3	0.51	4.42484E-08		
4	0.49	4.42461E-08		
5	0.52	4.39917E-08		
6	0.48	4.39729E-08		
7	0.53	4.3575E-08		
8	0.47	4.35121E-08		

FIGURE 12.10
Likelihood for different values of λ, based on $L(\lambda \mid y_1, y_2, ..., y_{10}) = \lambda^{10}e^{-10\lambda(2.0)}$, sorted from largest to smallest value of likelihood.

permissible θ values—that's a lot of ifs but often they are all true—then the derivative of the likelihood function is zero at the value $\theta = \hat{\theta}$. More briefly:

$$\left.\frac{\partial L(\theta \mid y_1, y_2, ..., y_n)}{\partial \theta}\right|_{\theta=\hat{\theta}} = 0 \tag{12.1}$$

The derivative can be zero at values of θ other than the MLE. For example, when the likelihood function has **a local minimum** or a **local maximum** in addition to a **global maximum**, then the derivative will be zero at each of these places. Hence, when Equation 12.1 has multiple solutions, you need to check them all to find the one that gives the largest likelihood—that is, the one that gives the global maximum. Fortunately, (12.1) often has only one solution, and that solution provides the global (and only) maximum.

While Equation 12.1 can be used to solve for the MLE, it is easier and better to work with the **log-likelihood function** instead. The log-likelihood function is just the natural logarithm of the likelihood function.

The Log-Likelihood Function

$$LL(\theta \mid y_1, y_2, ..., y_n) = \ln\{L(\theta \mid y_1, y_2, ..., y_n)\}$$

Reasons for Using the Log-Likelihood Function Instead of the Likelihood Function

- Many pdfs have the "e" term in them; taking the natural log removes the "e" and leaves the exponent since $\ln(e^y) = y$.
- The likelihood of an iid sample is the product of the likelihoods for the individual observations; taking the logarithm gives you the sum since $\ln(xy) = \ln(x) + \ln(y)$. Sums are easier to work with than products.
- Notice how small the likelihoods can be—see Figures 12.3 and 12.4, for example. It can easily happen that the likelihood values become so infinitesimally small that the computer can't work with them and just calls them 0. The logarithm makes small values easier to work with: For example, the computer may call $10^{-10,000}$ simply 0 and therefore could not tell the difference between $10^{-10,000}$ and $10^{-20,000}$. But $\ln(10^{-10,000}) = -10,000 \ln(10) = -23,025.9$, and $\ln(10^{-20,000}) = -20,000 \ln(10) = -46,051.7$, which the computer knows are different numbers.
- Statistical software that performs maximum likelihood analysis often reports the log likelihood in measures of model fit, with a higher log likelihood implying a better-fitting model. Two such measures are the *likelihood ratio chi-square statistic* and *Akaike's information criterion*, which are discussed in Chapter 17.

Also, the log-likelihood function provides the same MLE. The reason is that the function $\ln(x)$ is a **monotonically increasing** function, implying that $x_1 > x_2$ if and only if $\ln(x_1) > \ln(x_2)$. For example, $x_1 = 100$ is more than $x_2 = 10$, and $\ln(100) = 4.605$ is more than $\ln(10) = 2.303$.

Thus, $L(\hat{\theta} \mid y_1, y_2, ..., y_n) > L(\theta \mid y_1, y_2, ..., y_n)$ is equivalent to $\ln\{L(\hat{\theta} \mid y_1, y_2, ..., y_n)\} > \ln\{L(\theta \mid y_1, y_2, ..., y_n)\}$ by monotonicity of $\ln(x)$. So you can maximize either the log-likelihood function or the likelihood function; either way you get the same MLE $\hat{\theta}$. See Figure 12.11 for an illustration that the same parameter value maximizes both the likelihood function L and

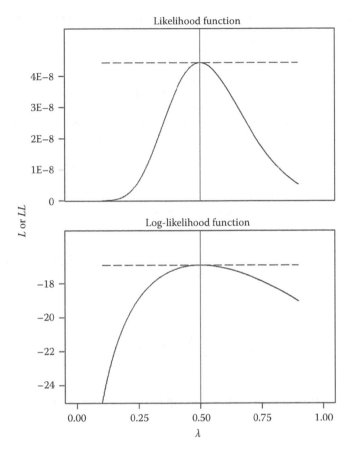

FIGURE 12.11
The likelihood function and log-likelihood function, both maximized at the same parameter value (0.5). The derivative of the function is zero at the maximum, in either case, as indicated by the flat tangent line.

the log-likelihood function LL from Example 12.2 where $L(\lambda \,|\, y_1, y_2, ..., y_{10}) = \lambda^{10} e^{-10\lambda(2.0)}$ and $LL = \ln(\lambda^{10} e^{-10\lambda(2.0)}) = 10 \ln(\lambda) - 20\lambda$.

The Derivative Condition for Identifying the MLE

If the likelihood function is continuous and differentiable, and if its maximum occurs in the interior of the set of permissible θ values, then the derivative of the log-likelihood function is zero at the value $\theta = \hat{\theta}$ or

$$\left. \frac{\partial LL(\theta \,|\, y_1, y_2, ..., y_n)}{\partial \theta} \right|_{\theta = \hat{\theta}} = 0 \qquad (12.2)$$

The same comments concerning local minima, local maxima, and the global maximum of the likelihood function, as given in the text following Equation 12.1, apply equally here for the log-likelihood function: If there are multiple solutions to Equation 12.2, you simply have to check them all to see which one provides the largest value of the log likelihood LL.

Example 12.6: Finding the MLE by Differentiating the Log-Likelihood Function

In Example 12.2 with $L(\lambda \mid y_1, y_2, \ldots, y_{10}) = \lambda^{10}e^{-10\lambda(2.0)}$, you can find the MLE for λ using the log-likelihood function as follows:

$LL(\lambda \mid y_1, y_2, \ldots, y_{10})$	(By definition of LL and by substituting the specific
$= \ln(\lambda^{10}e^{-10\lambda(2.0)})$	likelihood from Example 12.2)
$= \ln(\lambda^{10}) + \ln(e^{-10\lambda(2.0)})$	(By property of logarithms that $\ln(xy) = \ln(x) + \ln(y)$)
$= 10\ln(\lambda) - 10\lambda(2.0)$	(By property of logarithms that $\ln(e^x) = x$)
$= 10\ln(\lambda) - 20\lambda$	(By algebra)

To unpack Equation 12.2 in this instance, first note that $\theta = \lambda$ in this example and that $n = 10$. So, by substitution and properties of derivatives:

$$\frac{\partial LL(\theta \mid y_1, y_2, \ldots, y_n)}{\partial \theta} = \frac{\partial LL(\lambda \mid y_1, y_2, \ldots, y_{10})}{\partial \lambda} = \frac{\partial(10\ln(\lambda) - 20\lambda)}{\partial \lambda} = \frac{10}{\lambda} - 20$$

Further:

$$\left.\frac{\partial LL(\theta \mid y_1, y_2, \ldots, y_n)}{\partial \theta}\right|_{\theta = \hat{\theta}} = \left.\frac{\partial LL(\lambda \mid y_1, y_2, \ldots, y_n)}{\partial \lambda}\right|_{\lambda = \hat{\lambda}} = \frac{10}{\hat{\lambda}} - 20$$

Setting the derivative to zero as shown in Equation 12.2 gives $10/\hat{\lambda} - 20 = 0$, or $10/\hat{\lambda} = 20$, or $\hat{\lambda} = 0.5$.

With a multiparameter likelihood function, you still want to locate the peak; see Figure 12.8, for example, where the peak of $L(\mu, \sigma \mid \text{data})$ occurs at $(\hat{\mu}, \hat{\sigma}) = (26.81, 4.79)$. The slice of the function where $\hat{\sigma} = 4.79$ is $L(\mu, 4.79 \mid \text{data})$ and is shown in Figure 12.12. For this slice

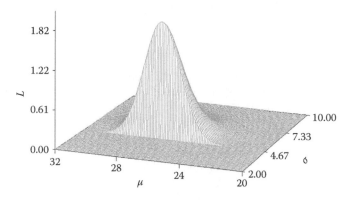

FIGURE 12.12
The slice of the normal likelihood function ($\times 10^{21}$) where the standard deviation σ is fixed at the maximum likelihood value $\hat{\sigma} = 4.79$.

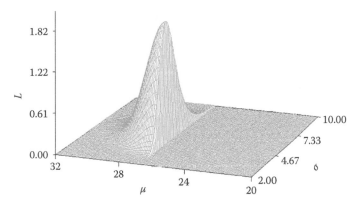

FIGURE 12.13
The slice of the normal likelihood function ($\times 10^{21}$) where the mean, μ, is fixed at the maximum likelihood value $\hat{\mu} = 26.81$.

function, the maximizing value of μ is the value at which the curve is flat, that is, at the μ for which $(\partial/\partial \mu)L(\mu, 4.79 | \text{data}) = 0$.

Applying the same logic to the standard deviation, the slice of the function where $\hat{\mu} = 26.81$ is $L(26.81, \sigma | \text{data})$ and is shown in Figure 12.13. For this slice function, the maximizing value of σ is again the value at which the curve is flat, that is, at the σ for which $(\partial/\partial \sigma)L(26.81, \sigma | \text{data}) = 0$.

Summarizing, you want the derivatives of both of the slice functions to be zero to locate the combination of parameter values that maximizes the likelihood function. And, again, you can (and should) use the log-likelihood function.

Derivative Conditions to Locate the Multiparameter MLE

When the parameter vector is $\theta = (\theta_1, \theta_2, ..., \theta_k)$, and if the likelihood function is differentiable with maximum occurring in the interior of the parameter space, then the MLE $\hat{\theta} = (\hat{\theta}_1, \hat{\theta}_2, ..., \hat{\theta}_k)$ satisfies:

$$\frac{\partial LL(\theta | y_1, y_2, ..., y_n)}{\partial \theta_j} \Bigg|_{\theta = \hat{\theta}} = 0 \tag{12.3}$$

This is true for all $j = 1, 2, ..., k$.

Again there can be multiple solutions for (12.3), if so, you must check all solutions, one by one, to find the one that gives the global maximum—the largest value of LL.

Example 12.7: The MLEs of μ and σ for a Normal Distribution

The likelihood functions shown in Figures 12.8, 12.12, and 12.13 all come from the data $y_1 = 36$, $y_2 = 23$, $y_3 = 22$, $y_4 = 27$, $y_5 = 26$, $y_6 = 24$, $y_7 = 28$, $y_8 = 23$, $y_9 = 30$, $y_{10} = 25$, $y_{11} = 22$, $y_{12} = 26$, $y_{13} = 22$, $y_{14} = 35$, $y_{15} = 24$, and $y_{16} = 36$. We stated earlier that

$(\hat{\mu}, \hat{\sigma}) = (26.81, 4.79)$ maximized this function. How do we know this? The following derivation shows how:

$L(\mu, \sigma \mid y_1, y_2, y_3, \ldots, y_{16})$

(By definition of likelihood function for an iid sample and by using the normal distribution for the data values)

$$= \frac{1}{\sqrt{2\pi}\sigma} \exp\frac{-(36-\mu)^2}{2\sigma^2} \times \frac{1}{\sqrt{2\pi}\sigma} \exp\frac{-(23-\mu)^2}{2\sigma^2}$$

$$\times \frac{1}{\sqrt{2\pi}\sigma} \exp\frac{-(22-\mu)^2}{2\sigma^2} \times \cdots \times \frac{1}{\sqrt{2\pi}\sigma} \exp\frac{-(36-\mu)^2}{2\sigma^2}$$

$$= \left(\frac{1}{\sqrt{2\pi}\sigma}\right)^{16} \times \exp\left(-\frac{(36-\mu)^2}{2\sigma^2} - \frac{(23-\mu)}{2\sigma^2}\right.$$

(By algebra of exponents)

$$\left. -\frac{(22-\mu)^2}{2\sigma^2} - \cdots - \frac{(36-\mu)^2}{2\sigma^2}\right)$$

$$= \left(\frac{1}{\sqrt{2\pi}\sigma}\right)^{16} \times \exp\left[-\left(\frac{1}{2\sigma^2}\right)\{(36-\mu)^2 + (23-\mu)^2\right.$$

(By algebra)

$$\left. + (22-\mu)^2 + \cdots + (36-\mu)^2\}\right]$$

The log-likelihood function is simpler, with the exponential terms ("exp") removed, and with products becoming sums:

$LL(\mu, \sigma \mid y_1, y_2, y_3, \ldots, y_{16}) = \ln\{L(\mu, \sigma \mid y_1, y_2, y_3, \ldots, y_{16})\}$ (By definition)

$= -16 \ln(2\pi)^{1/2} - 16 \ln(\sigma) - (1/2\sigma^2)\{(36-\mu)^2$ (By properties of logarithms)
$+ (23-\mu)^2 + (22-\mu)^2 + \cdots + (36-\mu)^2\}$

Taking the derivative of LL with respect to μ using properties D2 and D8 of derivatives in Chapter 2, you get:

$$\left(\frac{\partial}{\partial\mu}\right) LL(\mu, \sigma \mid y_1, y_2, y_3, \ldots, y_{16})$$

$$= \left\{\frac{-1}{(2\sigma^2)}\right\}\{2(\mu - 36) + 2(\mu - 23) + 2(\mu - 22) + \cdots + 2(\mu - 36)\}$$

Taking the derivative with respect to σ, you get:

$(\partial/\partial\sigma) LL(\mu, \sigma \mid y_1, y_2, y_3, \ldots, y_{16})$

$= \partial/\partial\sigma\left[-16 \ln(2\pi)^{1/2} - 16 \ln(\sigma) - (1/2\sigma^2)\{(36-\mu)^2 + (23-\mu)^2\right.$ (By substitution)
$\left. + (22-\mu)^2 + \cdots + (36-\mu)^2\}\right]$

$= -16\sigma^{-1} - (-2)\frac{1}{2\sigma^3}\{(36-\mu)^2 + (23-\mu)^2 + (22-\mu)^2 + \cdots + (36-\mu)^2\}$ (By properties of derivatives)

$= -16\frac{1}{\sigma} + \frac{1}{\sigma^3}\{(36-\mu)^2 + (23-\mu)^2 + (22-\mu)^2 + \cdots + (36-\mu)^2\}$ (By algebra)

Equation 12.3 translates, in this example, to the requirement that the derivative with respect to μ and the derivative with respect to σ both be zero at the MLE $(\hat{\mu}, \hat{\sigma})$. In other words, you need:

$$\left.\frac{\partial LL(\mu, \sigma \mid y_1, y_2, \ldots, y_n)}{\partial\mu}\right|_{(\mu, \sigma) = (\hat{\mu}, \hat{\sigma})}$$

$$= \left(\frac{-1}{2\hat{\sigma}^2}\right)\{2(\hat{\mu} - 36) + 2(\hat{\mu} - 23) + 2(\hat{\mu} - 22) + \cdots + 2(\hat{\mu} - 36)\} = 0$$

and

$$\frac{\partial LL(\mu,\sigma \mid y_1, y_2, \ldots, y_n)}{\partial \sigma}\Bigg|_{(\mu,\sigma)=(\hat{\mu},\hat{\sigma})}$$

$$= -16\frac{1}{\hat{\sigma}} + \frac{1}{\hat{\sigma}^3}\left\{(36-\hat{\mu})^2 + (23-\hat{\mu})^2 + (22-\hat{\mu})^2 + \cdots + (36-\hat{\mu})^2\right\} = 0$$

You can solve the first equation without knowing what $\hat{\sigma}$ is, getting

---

$(-1/2\hat{\sigma}^2)\{2(\hat{\mu}-36) + 2(\hat{\mu}-23) + 2(\hat{\mu}-22) + \cdots + 2(\hat{\mu}-36)\} = 0$ (By derivative requirement)

$\Rightarrow (\hat{\mu}-36) + (\hat{\mu}-23) + (\hat{\mu}-22) + \cdots + (\hat{\mu}-36) = 0$ (By multiplying both sides by $-\hat{\sigma}^2$)

$\Rightarrow 16\hat{\mu} = 36 + 23 + 22 + \cdots + 36$ (By algebra)

Solving, you get the MLE $\hat{\mu} = (1/16)(36 + 23 + 22 + \cdots + 36) = 26.81 = \bar{y}$, the ordinary sample average.

Incorporating $\hat{\mu} = 26.81$ into the second equation, you get

---

$-16\frac{1}{\hat{\sigma}} + \frac{1}{\hat{\sigma}^3}(366.44) = 0$ (By substitution)

$\Rightarrow \hat{\sigma}^2 = 366.44/16 = 22.902$ (By multiplying both sides by $-\hat{\sigma}^3/16$)

$\Rightarrow \hat{\sigma} = \sqrt{22.902} = 4.79$ (By arithmetic)

It is noteworthy that the MLE of the standard deviation is the square root of the plug-in estimate of the variance presented in Chapter 11. In other words, the MLE of the standard deviation of the normal distribution uses the n formula rather than the $n-1$ formula.

In practice, likelihood functions tend to be much more complicated, and you won't be able to solve the calculus problem even if you excel at math. Instead you'll have to use **numerical methods**, a fancy term for "letting the computer do the calculus for you." But it is still important to understand the calculus solution because you'll see output that refers to the computer's attempt to do the calculus. Understanding how the calculus works will help you if the computer gets stuck and can't figure out how to bail itself out.

Numerical methods for finding MLEs work by **iterative approximation**. They start with an initial guess at θ, say θ_0, then update the guess to some value θ_1 by climbing up the likelihood function. Then the algorithm replaces θ_0 with θ_1 and iterates—that is, it applies the same mathematical logic as applied to θ_0, to obtain a new value, θ_2, even higher up the likelihood function. The iteration continues until the successive values θ_i and θ_{i+1} are so close to one another that the computer is willing to assume that the peak has been achieved. When this happens, the algorithm is said to **converge**.

Iterative methods for optimizing functions have been around for centuries; many algorithms still bear the name of Isaac Newton, one of the original inventors of calculus (Gottfried Leibniz was the other). There are many methods available; you can find some of

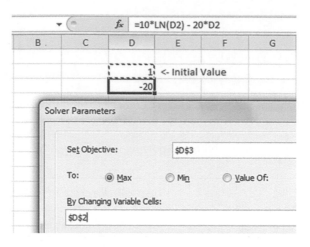

FIGURE 12.14
Using Microsoft Excel's Solver to find MLEs.

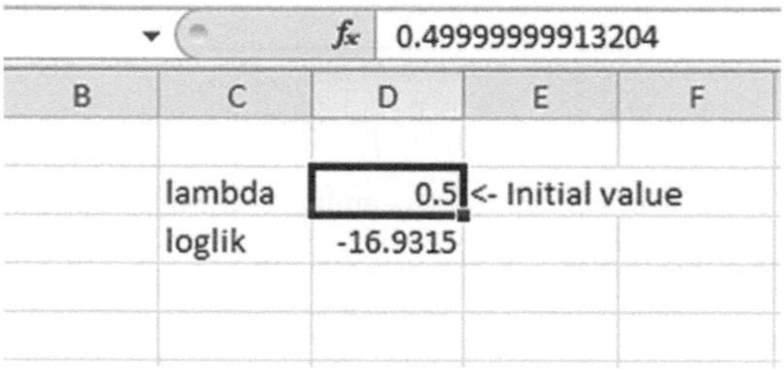

FIGURE 12.15
The result of using Microsoft Excel's Solver to find an MLE.

them in Microsoft Excel's Solver add-in. The following screenshots show how to use it to maximize the log-likelihood function $10\ln(\lambda) - 20\lambda$, starting with Figure 12.14.

In Cell D3 of Figure 12.14 is the log-likelihood formula = 10*LN(D2) – 20*D2, a function of the initial value in Cell D2. The initial value θ_0 (or here, for λ_0) is your choice, though some software selects a default for you. The algorithm then optimizes the function internally, for example, after you click the "Solve" button in Excel's Solver, giving the screenshot shown in Figure 12.15.

Notice that the solution value for the MLE of λ shown in Figure 12.15 is not precisely 0.5 (the actual MLE), but the converged value 0.49999999913204. The algorithm stopped at this step because the difference between λ_i and $\lambda_i + 1$ at successive iterations was so small that the computer deemed it close enough.

While the computer generally works well, there are serious potential problems that you should watch out for when you use numerical methods for finding MLEs.

Potential Pitfalls When Using Numerical Methods to Find MLEs

- If the likelihood function has more than one peak, the numerical method might converge to the wrong peak, depending on the initial value.

- If the data and/or the model is inadequate, or if the likelihood function is very complicated, the method might not converge at all.
- If there are parameter constraints (e.g., variances must be positive), the usual methods can have trouble locating cases where the solution is on the boundary, where the derivative is not zero.

Right now, you might be thinking, "So what? Maximum likelihood just gives me obvious estimates, such as the average, the proportion, and the plug-in standard deviation. Why don't I just use the obvious estimates and not bother with all this calculus and iterative methods?" Great question! There are two reasons why you should bother with likelihood, the first of which we already discussed: The likelihood function shows you the specific *range of uncertainty* about the value of the parameter, given your observed data. It's not just a way to estimate the parameter, which you can often do easily using the obvious methods.

The second reason you should care about likelihood-based methods is that they provide a way to estimate parameters in advanced statistical models where there is no obvious estimator such as a mean, proportion, or standard deviation. The following example is a beautiful application where likelihood-based methods are needed. The application is called *logistic regression analysis,* and it is used to predict Bernoulli ($Y = 0$ or $Y = 1$) outcomes as a function of a predictor X. Here are some uses of this methodology.

Applications of Logistic Regression

- **Medicine**. Predict patient survival (Y = lived or died) as a function of therapy and patient characteristics.
- **Biology**. Predict animal survival (Y = lived or died) as a function of human and environmental pressures and animal characteristics.
- **Engineering**. Predict the reliability of a material (Y = break or no break) as a function of material composition and environmental factors.
- **Marketing**. Predict customer purchasing behavior (Y = purchased the company's product or purchased competitor's product) as a function of customer demographics.
- **Finance**. Predict customer default on loan (Y = default or paid) as a function of customer demographics.

The following example is an application of logistic regression in the field of human resource management.

Example 12.8: Predicting Success as a Function of Experience: Estimating the Logistic Regression Model

Prospective employers often want to see experience on your résumé. Is experience an important predictor of success in job? One might guess that the answer would be "yes," but it's not necessarily true. Less experienced people might be more motivated, and they also might have the ability to adapt more quickly to newer technology. If so, it is possible that there is a greater success rate for the less experienced employees.

Data can shed light on the relationship between experience and success. Suppose that a company measures the success at completing a particular work task for a collection of employees during a specified time frame. The data from $n = 32$ employees are as follows:

$$(Y,10), (Y,3), (N,3), (N,4), (N,6), (Y,2), (Y,20), (Y,12), (N,5),$$

$$(Y,12), (Y,15), (N,0), (Y,16),$$

$(Y,15)$, $(Y,10)$, $(N,2)$, $(Y,10)$, $(Y,3)$, $(N,3)$, $(N,4)$, $(N,6)$, $(N,2)$, $(N,8)$,

$(N,2)$, $(Y,20)$, $(Y,0)$, $(N,6)$, $(N,2)$, $(N,2)$, $(Y,20)$, $(N,2)$, $(N,5)$

Each pair of observations refers to an employee, with the first value in the pair either Y for "successful in task" or N for "not successful in task." The second value in the pair number refers to experience, ranging from 0 to 20 years in this sample.

You should expect that the probability of success for someone with 1.0 years of experience should not differ much from a person with 1.1 years of experience. The probabilities should morph, as described in Chapter 5. *Nature favors continuity over discontinuity.*

One way to estimate the probabilities of success for each level of experience is to simply count successes and failures. For the two employees with experience = 0 in the sample, one was successful and one failed, leading to an estimated probability $1/2 = 0.50$ of success. There is no employee in the sample with experience = 1, so no such estimate of success probability is available using this method. There are 7 employees having experience = 2, with only 1 success, leading to an estimated $1/7 = 0.143$ probability of success. Among employees with experience = 3, the estimated probability is $2/4 = 0.50$, and so on. Figure 12.16 shows how these estimates of probability of success vary as a function of experience.

Notice that the estimates shown in Figure 12.16 are not sensible. There are estimates of 0.0 probability of success and estimates of 1.0 probability of success. But 0.0 probability of success means absolute impossibility, and 1.0 probability means absolute certainty. Obviously, years of experience cannot confer such absolute certainty about success and failure, because countless other factors also affect an employee's success, all the way down to whether they had a bad headache that day. The problem is simply that the sample sizes in the different experience groups are too small to achieve reliable estimates. One solution is to group the data in terms of ranges of experience, as done in Chapter 5 where we estimated conditional distributions. But this is not an ideal solution, because there is doubt as to where to draw the boundaries. In addition, the idea of using experience ranges suggests that everyone within the range, for example, $0.0 \leq$ experience ≤ 5.0 years, has the same probability of success and that the probability somehow instantaneously changes as you go from 5.00 to 5.01 years of experience.

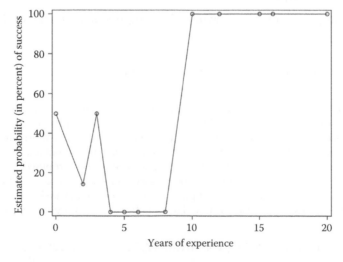

FIGURE 12.16
Estimated probability of success as a function of years experience using simple frequency tabulations (in dots), connected by lines.

TABLE 12.1

Logistic Regression Model

y	$p(y \mid x, \theta)$
Failure	$1/\{1 + \exp(\beta_0 + \beta_1 x)\}$
Success	$\exp(\beta_0 + \beta_1 x)/\{1 + \exp(\beta_0 + \beta_1 x)\}$
Total	1.00

Nature favors continuity over discontinuity. A more realistic model is that the probability of success continuously morphs. A commonly used model for such morphing is the **logistic regression model**, which supposes that the success data (Y) are produced independently according to a distribution $p(y \mid x, \theta)$, which morphs continuously as a function of X; see Table 12.1.

This model may seem mysterious. The following notes help to explain it.

Properties of the Logistic Regression Model

- The probabilities of failure and success add to 1.0, as required.
- The probabilities of success and failure continuously morph as a function of x; hence, the model is more realistic than the one shown in Figure 12.16 where the probabilities bounce all over.
- The fact that $\exp(\beta_0 + \beta_1 x) > 0$ implies that $0 < 1/\{1 + \exp(\beta_0 + \beta_1 x)\} < 1$ and that $0 < \exp(\beta_0 + \beta_1 x)/\{1 + \exp(\beta_0 + \beta_1 x)\} < 1$. Hence, this function ensures that the probabilities of success and failure are always between 0 and 1—this would not happen if the success probability were a simple linear function like $\beta_0 + \beta_1 x$.

Figure 12.17 shows how the success probability function looks for different settings of the parameter vector $\theta = (\beta_0, \beta_1)$.

Figure 12.17 shows that the logistic regression model provides a rich set of probability functions that continuously morph and that these functions depend on the values of the parameters $\theta = (\beta_0, \beta_1)$, which can lie anywhere in the parameter space $\Theta = \{\beta_0, \beta_1; -\infty < \beta_0, \beta_1 < \infty\}$.

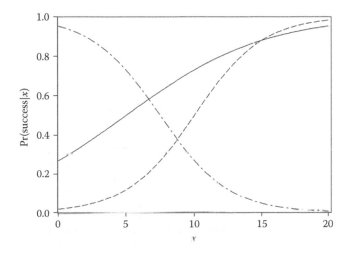

FIGURE 12.17

Logistic regression models for $(\beta_0, \beta_1) = (-1.0, 0.2)$ (solid), $(\beta_0, \beta_1) = (-4.0, 0.4)$ (dash), and $(\beta_0, \beta_1) = (3.0, -0.4)$ (dot-dash).

How can you estimate the parameters $\theta = (\beta_0, \beta_1)$ using the data? There is no intuitively obvious way to do it, such as taking simple averages or percentages, yet maximum likelihood provides a simple answer. Just like in Example 12.3, when there is a success, the contribution to the likelihood function is the probability of a success, and when there is a failure, the contribution to the likelihood function is the probability of a failure. The only difference here is that the probabilities of success and failure depend on x as shown in Table 12.1: The success probability is $\exp(\beta_0 + \beta_1 x)/\{1 + \exp(\beta_0 + \beta_1 x)\}$, and the failure probability is $1/\{1 + \exp(\beta_0 + \beta_1 x)\}$.

Hence, by independence, the likelihood of the sample $(Y, 10)$, $(Y, 3)$, $(N, 3)$, …, $(N, 5)$ is

$$L(\beta_0, \beta_1 | \text{data})$$

$$= \exp\frac{\{\beta_0 + \beta_1(10)\}}{\left[1 + \exp\{\beta_0 + \beta_1(10)\}\right]}$$

$$\times \exp\frac{\{\beta_0 + \beta_1(3)\}}{\left[1 + \exp\{\beta_0 + \beta_1(3)\}\right]}$$

$$\times \frac{1}{\left[1 + \exp\{\beta_0 + \beta_1(3)\}\right]}$$

$$\times \cdots$$

$$\times \frac{1}{[1 + \exp\{\beta_0 + \beta_1(5)\}]} \tag{12.4}$$

You can estimate the parameters via maximum likelihood using a variety of software, including Microsoft Excel. Figure 12.18 shows how to set up the calculations.

Column C of Figure 12.18 contains the formula for likelihood for success, expressed as EXP(\$B\$1+\$B\$2*B5)/(1+EXP(\$B\$1+\$B\$2*B5)). Column D contains the likelihood for failure, expressed as 1/(1+EXP(\$B\$1+\$B\$2*B5)). Column E contains the choice between either column C or D, depending upon outcome, with the formula: IF(A5="Y",C5,D5). Column F contains the logarithm of the likelihood, or LN(E5). Finally, the log likelihood in Cell G5 for the sample shows the sum of the logarithms of the likelihoods for the individual observations, or SUM(F5:F36), because the logarithmic transformation turns products into sums.

	A	B	C	D	E	F	G
1	$\beta_0 =$	0	<- Initial values				
2	$\beta_1 =$	1					
3							
4	Success	Years	Pr(Success)	Pr(Failure)	Likelihood	Log Likelihood	LL(β_0, β_1 | data)
5	Y	10	0.9999546	4.54E-05	0.9999546	-4.53989E-05	-64.52678207
6	Y	3	0.9525741	0.047426	0.9525741	-0.048587352	
7	N	3	0.9525741	0.047426	0.0474259	-3.048587352	
8	N	4	0.9820138	0.017986	0.0179862	-4.018149928	
9	N	6	0.9975274	0.002473	0.0024726	-6.002475685	
10	Y	2	0.8807971	0.119203	0.8807971	-0.126928011	
11	Y	20	1	2.06E-09	1	-2.06115E-09	
12	Y	12	0.9999939	6.14E-06	0.9999939	-6.14419E-06	
13	N	5	0.9933071	0.006693	0.0066929	-5.006715348	
14	Y	12	0.9999939	6.14E-06	0.9999939	-6.14419E-06	
15	Y	15	0.9999997	3.06E-07	0.9999997	-3.05902E-07	

FIGURE 12.18
Setting up the likelihood calculations for logistic regression using Microsoft Excel.

▲	A	B	C	D	E	F	G
1	$\beta_0 =$	0	<- Initial values				
2	$\beta_1 =$	1					
3							
4	Success	Years	Pr(Success)	Pr(Failure)	Likelihood	Log Likelihood	LL(β_0, β_1 | data)
5	Y	10	0.9999546	4.54E-05	0.9999546	-4.53989E-05	-64.52678207
6	Y	3	0.95257413	0.0474259	0.9525741	-0.048587352	
7	N	3	0.95257413	0.0474259	0.0474259	-3.048587352	
8	N						
9	N						
10	Y						
11	Y						
12	Y						
13	N						
14	Y						
15	Y						

Solver Parameters

Set Objective: $G$5

To: ⦿ Max   ○ Min   ○ Value Of: 0

By Changing Variable Cells:

$B$1:$B$2

FIGURE 12.19
Maximizing the likelihood calculations for logistic regression using Microsoft Excel.

To optimize the likelihood function, you can use Excel's Solver to change the initial values in *both* Cells B1 and B2 to maximize the log-likelihood function in Cell G5, as shown in Figure 12.19.

Applying "Solve" in Excel's Solver as shown in Figure 12.19 yields the estimates $\hat{\beta}_0 = -2.296$ and $\hat{\beta}_1 = 0.333$.

Figure 12.20 shows the resulting estimated model for the probability of success, whose equation is

$$\hat{Pr}(\text{success}|\text{experience}) = \frac{\exp\{-2.296 + 0.333(\text{experience})\}}{1 + \exp\{-2.296 + 0.333(\text{experience})\}}$$

The form of the estimated model shown in Figure 12.20 is much more sensible and believable than the model estimated in Figure 12.16; in particular, there is a continuous morphing of estimated probabilities as experience continuously rises. Further, the

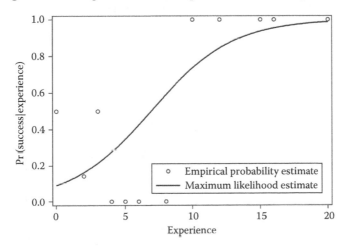

FIGURE 12.20
Estimated probabilities of success using the logistic regression model (the smooth curve), as estimated via maximum likelihood, compared to the empirical probability estimates shown in Figure 12.16 (circles).

model provides estimates of success probability for all experience levels, not just for those in the observed data set.

You can construct the likelihood function that leads to estimates $\hat{\beta}_0 = -2.296$ and $\hat{\beta}_1 = 0.333$ by plugging values (β_0, β_1) into the likelihood function $L(\beta_0, \beta_1 | \text{data})$ shown in Equation 12.4 and having software graph the results. Figures 12.21 and 12.22 show the 3-D rendering of the likelihood function and the associated contour plot.

Figures 12.21 and 12.22 show values of the combination (β_0, β_1) that are more likely (those in the center of the contours) and less likely (those on the outer contours), given the data. The MLE $(\hat{\beta}_0, \hat{\beta}_1) = (-2.296, 0.333)$ has the highest likelihood. But many other

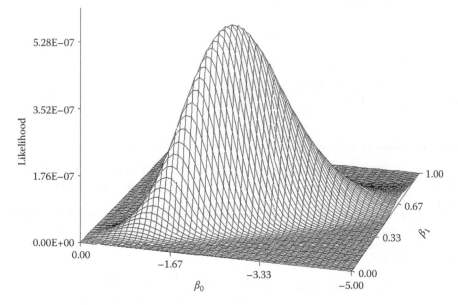

FIGURE 12.21
Likelihood function for (β_0, β_1) in the success in task example.

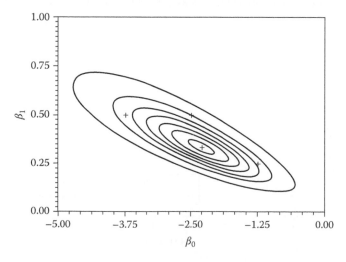

FIGURE 12.22
Contour plot corresponding to the likelihood function shown in Figure 12.20. Plausible combinations (β_0, β_1) are shown by + symbols.

FIGURE 12.23
Probability functions that are consistent with the observed data: $(\beta_0, \beta_1) = (-2.5, 0.5)$ (solid line), $(\beta_0, \beta_1) = (-1.25, 0.25)$ (short-dashed line), $(\beta_0, \beta_1) = (-3.75, 0.5)$ (dot-dash line), and $(\beta_0, \beta_1) = (-2.296, 0.333)$ (the MLE, long-dashed line).

combinations are nearly as well supported by the data; that is, there is uncertainty about the values of (β_0, β_1). Your uncertainty about (β_0, β_1) translates into uncertainty about the logistic probability curve shown in Figure 12.20, since that curve is a graph of the function $\text{Pr}(\text{success}|\text{experience} = x) = \exp(\beta_0 + \beta_1 x)/\{1 + \exp(\beta_0 + \beta_1 x)\}$, which depends upon these uncertain values (β_0, β_1).

Figure 12.22 shows that the combinations $(-2.5, 0.5)$, $(-1.25, 0.25)$, and $(-3.75, 0.50)$, shown with + symbols, are all reasonably plausible values for (β_0, β_1). Figure 12.23 shows how the probability function $\exp(\beta_0 + \beta_1 x)/\{1 + \exp(\beta_0 + \beta_1 x)\}$ looks for these values and for the MLE $(\hat\beta_0, \hat\beta_1) = (-2.296, 0.333)$.

The selection of the four combinations of (β_0, β_1) shown in Figure 12.23 is limited; many other combinations (β_0, β_1) are consistent with the data. A better method would be to sample many more reasonable combinations. In Chapter 13, you will see how to do so using Bayesian methods.

Most statistical software will perform logistic regression analysis. The output shown in Figure 12.24 is from SAS.

```
                    The LOGISTIC Procedure

              Analysis of Maximum Likelihood Estimates

                              Standard        Wald
     Parameter    DF    Estimate    Error    Chi-Square    Pr > ChiSq

     Intercept     1     -2.2959    0.8181      7.8759       0.0050
     experience    1      0.3330    0.1198      7.7319       0.0054


                      Odds Ratio Estimates

                          Point           95% Wald
            Effect      Estimate      Confidence Limits

            experience    1.395       1.103       1.764
```

FIGURE 12.24
Output of logistic regression analysis from the SAS software showing maximum likelihood parameter estimates and Wald standard errors.

Notice in Figure 12.24 that the MLEs of (β_0, β_1) are shown under the column titled *Estimate*. Notice also that there are two references to (Abraham) Wald. The **Wald standard error** is a measure of the uncertainty surrounding an estimated parameter. So far, we have suggested that you gauge your uncertainty about a parameter simply by looking at the likelihood function. The Wald standard error gives you a more precise quantification.

12.4   Wald Standard Error

When you perform likelihood-based analysis using statistical software, the software will tell you how accurate the MLEs are by reporting their Wald standard errors. If you take the MLE plus or minus two Wald standard errors, then you get an estimate of the range of plausible values of the parameter that is determined by the range of the likelihood function. In other words, the Wald standard error tells you how much reduction in uncertainty you have about the true parameter value, given the data that you have observed.

To understand the Wald standard error, first have a look at Figure 12.2, which shows the likelihood function $L(\lambda \mid y = 2) = \lambda e^{-\lambda(2)}$ for a single observation $y = 2$ from an exponential distribution. Also look at the likelihood function $L(\lambda \mid y_1, y_2, ..., y_n) = \lambda^{10} e^{-10\lambda(2.0)}$ shown in Figure 12.3. Have you done that? Good! You should notice that these graphs have an approximately bell-shaped appearance, with the approximation looking better for the $n = 10$ example graphed in Figure 12.3 than for the $n = 1$ example graphed in Figure 12.2. Further, in Figure 12.4 you see that they become even closer to bell-shaped with larger n. Notice also that the peaks of these approximate bell-shaped distributions occur at $\theta = \hat{\theta}$; that is, the peaks occur at the MLEs.

The Wald standard error is calculated by approximating the likelihood function using a normal distribution with mean $\hat{\theta}$. The standard deviation of that approximating normal distribution is the Wald standard error. For specific details, assume that

$$L(\theta \mid y) \cong c \times \exp\left\{-\frac{1}{2}\frac{(\theta - \hat{\theta})^2}{\hat{\sigma}^2}\right\}$$

In other words, assume that the likelihood function is approximately proportional to a normal distribution with mean $\hat{\theta}$ and standard deviation $\hat{\sigma}$. If this approximation is good, then a reasonable range of values for θ is $\hat{\theta} \pm 2\hat{\sigma}$, since this range would capture approximately 95% of the total area under the likelihood function. The value $\hat{\sigma}$ of the normal distribution that approximates the likelihood function is the famous Wald standard error of the parameter estimate $\hat{\theta}$.

How to find the value of $\hat{\sigma}$? Suppose for a minute that the likelihood is exactly proportional to a normal pdf, so the logic begins with the assumption that:

$$L(\theta \mid y) = c \times \exp\left\{-\frac{1}{2}\frac{(\theta - \hat{\theta})^2}{\hat{\sigma}^2}\right\}$$

The log-likelihood function would then be

$$LL(\theta \mid y) = \ln(c) - \frac{1}{2}\frac{(\theta - \hat{\theta})^2}{\hat{\sigma}^2}$$

To find the value of $\hat{\sigma}$, use the following steps:

$$\frac{\partial}{\partial\theta}LL(\theta\,|\,y)$$

$$= \frac{\partial}{\partial\theta}\ln(c) - \frac{\partial}{\partial\theta}\left\{\frac{1}{2}\frac{(\theta-\hat{\theta})^2}{\hat{\sigma}^2}\right\}$$ (By linearity and additivity properties of derivatives, D2 and D3 of Section 2.5)

$$= 0 - \frac{1}{2}\left\{\frac{2(\theta-\hat{\theta})}{\hat{\sigma}^2}\right\}$$ (By derivative properties of a constant is zero, property D1, D2, and D8)

$$= \frac{(\hat{\theta}-\theta)}{\hat{\sigma}^2}$$ (By algebra)

Taking the derivative with respect to θ again, you get

$$\frac{\partial^2}{\partial\theta^2}LL(\theta\,|\,y) = -\frac{1}{\hat{\sigma}^2}$$

This implies that

$$\hat{\sigma}^2 = \left[-\frac{\partial^2}{\partial\theta^2}LL(\theta\,|\,y)\right]^{-1}$$ (12.5)

Equation 12.5 says the following: If the likelihood function is proportional to the normal distribution, then the inverse of the negative of the second derivative of the log-likelihood function is equal to the variance of that normal distribution. Further, this second derivative is constant for all θ.

However, the likelihood function is not exactly proportional to a normal distribution; typically it is skewed as shown in Figures 12.2 and 12.3. This means that the second derivative is not constant for all θ. So, to find a value of $\hat{\sigma}$, you need to pick a value of θ in the equation for $\hat{\sigma}^2$ in Equation 12.5. If you pick $\theta = \hat{\theta}$, then the curvature of the approximating normal function will match the curvature of the likelihood function at the MLE.

The Square of the Wald Standard Error

$$\hat{\sigma}^2 = \left[-\frac{\partial^2}{\partial\theta^2}LL(\theta\,|\,y)\bigg|_{\theta=\hat{\theta}}\right]^{-1}$$

The second derivative of a function is called the *Hessian* of a function. If the function has many variables (parameters in our case), then the matrix of mixed partial derivatives is called the *Hessian matrix*. In the multiparameter case, there is a *multivariate normal* approximation to the likelihood function, and the Wald standard errors are obtained from the inverse of the Hessian matrix. If you use statistical software to estimate parameters via maximum likelihood, it is likely that you will see a reference to this Hessian matrix in the output.

Example 12.9: Calculating the Wald Standard Error

Consider the waiting time data of Example 12.2, a sample of $n=10$ observations from the exponential distribution where the average of the y values was 2.0. Recall that the likelihood function is $L(\lambda\,|\,y_1, y_2, \ldots, y_{10}) = \lambda^{10}e^{-10\lambda(2.0)}$, and the log-likelihood function for λ is $LL(\lambda\,|\,y_1, y_2, \ldots, y_{10}) = 10\ln(\lambda) - 20\lambda$. Hence:

$$\frac{\partial^2}{\partial\lambda^2}LL(\lambda\,|\,y_1,\ldots,y_{10}) = -\frac{10}{\lambda^2}$$

Since the MLE of λ is $\hat{\lambda} = 0.5$, the estimated variance of the approximating normal distribution is

$$\hat{\sigma}^2 = -\left[-\frac{10}{\hat{\lambda}^2}\right]^{-1} = -\left[-\frac{10}{0.5^2}\right]^{-1} = \frac{1}{40}$$

And the Wald standard error is

$$\hat{\sigma} = \sqrt{\frac{1}{40}} = 0.158$$

Thus, the range of plausible values of λ, according to the 95% **Wald confidence interval**, is

$$0.50 - 2(0.158) < \lambda < 0.50 + 2(0.158)$$

or

$$0.184 < \lambda < 0.816$$

Figure 12.25 shows the actual likelihood function, the normal approximation, and the 95% Wald limits from the normal approximation.

Note that the Wald intervals are of limited use when the likelihood function is not symmetric like a normal distribution. It is clear from Figure 12.25 that the lower limit of the interval range from the actual likelihood function should not be as low as 0.184 and that the upper limit of the interval range from the actual likelihood function should be higher than 0.816. Bayesian methods (coming very soon to a statistics book near you!) provide a simple way to construct asymmetric interval ranges for the true parameter value when the likelihood function is skewed.

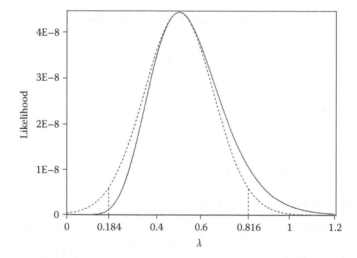

FIGURE 12.25
The likelihood function for λ with $n = 10$ observations sampled from an exponential distribution (solid line), along with the Wald approximation using the normal distribution (dashed line). 95% of the area of the approximating normal distribution lies between 0.184 and 0.816.

Vocabulary and Formula Summaries

Vocabulary

Likelihood principle	A principle that states that all information in your sample is contained in your *likelihood function*.
Likelihood function	The joint pdf of a sample viewed as a function of the parameters.
Maximum likelihood	A generally efficient and flexible method of obtaining good estimates of parameters; the method is to maximize the likelihood function.
Parameter space	The set of possible values of the parameter (vector) θ.
Sufficient statistic	A function of the data $f(y_1, y_2, ..., y_n)$ such that the likelihood function depends on the data $y_1, y_2, ..., y_n$ only through f.
Parameter vector	A list containing more than one parameter.
MLE	The value of θ, denoted $\hat{\theta}$, that maximizes the likelihood function.
Inspection	A method of determining MLEs by looking at the graph of the likelihood function.
Local minimum	A value $f(x_0)$ for which $f(x_0) < f(x)$ for all x near x_0, even though $f(x) < f(x_0)$ for some x that are far from x_0.
Local maximum	A value $f(x_0)$ for which $f(x_0) > f(x)$ for all x near x_0, even though $f(x) > f(x_0)$ for some x that are far from x_0.
Global maximum	The value $f(x_0)$ for which $f(x_0) > f(x)$ for all $x \neq x_0$.
Log-likelihood function	The natural logarithm of the likelihood function.
Monotonically increasing	A function $f(x)$ for which $f(x_1) > f(x_0)$ whenever $x_1 > x_0$; examples are $2x$, $\ln(x)$, and $\exp(x)$.
Numerical methods	Computer-based approaches to solving mathematical problems that arrive at a numerical solution.
Iterative approximation	A method used to solve equations that starts with an initial guess, updates it using an algorithm to obtain a new guess, and then applies the algorithm again to the new guess, repeating until convergence.
Convergence	When a numerical method arrives at a solution that satisfies some stopping criterion, such as the change between the current estimate and the next estimate becoming very small.
Logistic regression model	A model that relates the probability of category membership of a Y variable to a set of X variables, allowing for continuous morphing of probabilities as the X variables change.
Wald standard error	The value $\hat{\sigma}$ of the normal distribution that approximates the likelihood function for θ.
Wald confidence interval	An interval for θ that is based on the Wald standard error (typically $\hat{\theta} \pm 2\hat{\sigma}$).

Key Formulas and Descriptions

θ_T	The true value of the parameter θ.
$L(\theta \mid y_1, y_2, \ldots, y_n) = p(y_1, y_2, \ldots, y_n \mid \theta)$	The likelihood function for θ is the joint pdf of $y_1, y_2, \ldots, y_n$ viewed as a function of θ.
$L(\theta \mid y_1, y_2, \ldots, y_n) = p(y_1 \mid \theta) \times p(y_2 \mid \theta) \times \cdots \times p(y_n \mid \theta)$	The likelihood function when the data values are sampled independently.
$L(\theta \mid y_1, y_2, \ldots, y_n) = L(\theta \mid y_1) \times L(\theta \mid y_2) \times \cdots \times L(\theta \mid y_n)$	The likelihood function when the data values are sampled independently.
$LL(\theta \mid y_1, y_2, \ldots, y_n) = \ln\{L(\theta \mid y_1, y_2, \ldots, y_n)\}$	The log-likelihood function.
$LL(\theta \mid y_1, y_2, \ldots, y_n) = LL(\theta \mid y_1) + LL(\theta \mid y_2) + \cdots + LL(\theta \mid y_n)$	The log-likelihood function when the data values are sampled independently.
$\left. \dfrac{\partial L(\theta \mid y_1, y_2, \ldots, y_n)}{\partial \theta} \right\|_{\theta = \hat{\theta}} = 0$	Under commonly observed conditions, the derivative of the likelihood function is 0 when the parameter is equal to the MLE.
$\left. \dfrac{\partial LL(\theta \mid y_1, y_2, \ldots, y_n)}{\partial \theta} \right\|_{\theta = \hat{\theta}} = 0$	Under commonly observed conditions, the derivative of the log-likelihood function is 0 when the parameter is equal to the MLE.
$\hat{\sigma}^2 = \left[-\left. \dfrac{\partial^2}{\partial \theta^2} LL(\theta \mid y) \right\|_{\theta = \hat{\theta}} \right]^{-1}$	The square of the Wald standard error.
$\Pr(\text{Success} \mid X = x) = \exp(\beta_0 + \beta_1 x)/\{1 + \exp(\beta_0 + \beta_1 x)\}$	The logistic regression model.

Exercises

12.1 These data are from an iid sampling from the Bernoulli(π) distribution: 1, 1, 0, 0, 1, 1, 0, 0, 0.

    A.   State the likelihood function, then graph it.

    B.   Find the MLE using the inspection method.

    C.   State the log-likelihood function and find its maximum using calculus.

    D.   Find the maximum of the log-likelihood function using an iterative method.

    E.   Find the Wald standard error, the 95% Wald interval, and interpret them with respect to the graph in Exercise 12.1A.

12.2 Example 12.4 gives the parameter space $\Theta = \{\pi_1, \pi_2; 0 \le \pi_1, \pi_2, \pi_1 + \pi_2 \le 1\}$. Draw a graph of this space, putting π_1 on the horizontal axis and π_2 on the vertical axis.

12.3 The parameter space of the logistic regression model is $\Theta = \{\beta_0, \beta_1; -\infty < \beta_0, \beta_1 < \infty\}$. Draw a graph of this space, putting β_0 on the horizontal axis and β_1 on the vertical axis.

12.4 Using the function form of the multinomial distribution shown in Example 2.4 of Chapter 2, show that the likelihood function for Example 12.4 is $\pi_1^{392}\pi_2^{401}(1-\pi_1-\pi_2)^{209}$.

12.5 These data are from an iid sampling from the Poisson(λ) distribution: 0, 0, 2, 0, 0, 4, 1, 0, 0, 0, 0.

    A. State the likelihood function, then graph it.

    B. Find the MLE using the inspection method.

    C. State the log-likelihood function and find its maximum using calculus.

    D. Find the maximum of the log-likelihood function using an iterative method.

    E. Find the Wald standard error, the 95% Wald interval, and interpret them with respect to the graph in Exercise 12.5A.

12.6 If the likelihood that $\lambda = 0.5$ is 0.45, then the likelihood that $1/\lambda$ is $1/0.5 = 2.0$ is also 0.45. Redraw the graphs of Figures 12.2 and 12.3 so that the horizontal axis is $1/\lambda$ instead of λ. Interpret the resulting likelihood functions in terms of expected waiting time.

12.7 How do the parameters of the logistic regression model affect $\Pr(Y = 1 | X = x)$? Answer by drawing the following graphs of the logistic regression function, drawn over the range $0 \le x \le 10$.

    A. Use $(\beta_0, \beta_1) = (-1.0, 0.5)$, and then $(\beta_0, \beta_1) = (-1.0, 0.0)$. What happens when $\beta_1 = 0$?

    B. Use $(\beta_0, \beta_1) = (-1.0, 0.3)$, and then $(\beta_0, \beta_1) = (-1.0, 0.7)$. What happens when β_1 increases?

    C. Use $(\beta_0, \beta_1) = (-1.0, 0.3)$, and then $(\beta_0, \beta_1) = (1.0, 0.3)$. What happens when β_0 increases?

    D. Use $(\beta_0, \beta_1) = (-1.0, 0.3)$, and then $(\beta_0, \beta_1) = (0.0, 0.3)$. What happens when β_0 is zero?

    E. Use $(\beta_0, \beta_1) = (-0.3, 0.6)$, and then $(\beta_0, \beta_1) = (-0.3, -0.6)$. What happens when β_1 is negative?

12.8 Suppose data (y, x) are as follows: (1, 2.0), (0, 0.5), (1, 4.0), (1, 0.7), (0, 2.0), (1, 2.8), (1, 2.2), (1, 2.3), (0, 1.5), (1, 4.4), (0, 0.8), (0, 2.1), (1, 2.9), (1, 3.2), (1, 2.2), (1, 2.3), (1, 4.5), (1, 4.4), (1, 5.8), (0, 2.1), (1, 2.9), (1, 3.2). The variable y is binary, so logistic regression is appropriate.

    A. Write down the log-likelihood function for the parameters (β_0, β_1) of the logistic regression model.

    B. Find the MLEs using software and an iterative method.

    C. Report the Wald standard error using software. Using the 95% range of plausible values for β_1, answer the question, "Does $\Pr(Y = 1 | X = x)$ get higher with larger x, or does it get lower for larger x?" Justify your answer. Refer to Exercise 12.7 to see how β_1 affects $\Pr(Y = 1 | X = x)$.

12.9 Sometimes data modeled using regression exhibit nonconstant variance, called *heteroscedasticity*. For example, when X is larger, the variance of Y may be larger. In the classic, normally distributed regression model, you assume that Y is normally distributed with mean $\beta_0 + \beta_1 x$ and standard deviation σ. On the other hand, a heteroscedastic model assumes that the standard deviation is related to x; for example, you might assume that Y is normally distributed with mean $\beta_0 + \beta_1 x$ and standard deviation $x\sigma$ (x times σ). (Example 6.3 uses a heteroscedastic model with $\beta_0 = 0$, $\beta_1 = 0.3$, and $\sigma = 0.04$.) Use the following data set containing (x, y) values: (6.11, 8.99), (1.80, 5.50), (2.32, 7.14), (1.17, 5.59), (5.28, 1.58), (0.62, 2.93), (0.68, 0.81), (0.43, 0.47), (1.18, 3.73), (2.20, 9.64), (1.24, 0.62), (1.92, 6.03), (0.63, 0.93), (1.18, 4.97).

A. Estimate the unknown parameters β_0, β_1, and σ of the classic model via maximum likelihood using an iterative method and the computer. Explain your method and give the results.

B. Estimate the unknown parameters β_0, β_1, and σ of the given heteroscedastic model via maximum likelihood using an iterative method and the computer. Explain the method and give the results.

C. Graph the estimated lines from Exercise 12.9A and B, as well as the scatterplot, all on the same graph. Comment on the difference between the estimated line based on the constant variance assumption and the estimated line under heteroscedasticity.

12.10 In the course of clinical trials, the data are *blinded*; that is, the researchers don't know which data came from the treatment and which came from the control. This blinding is necessary to prevent experimenter bias. Still, you can sometimes glean information from the blinded data. The following data are blinded, but come from a clinical trial where roughly half of the observations are from the treatment group and roughly half are from the control group:

12 19 22 16 21 14 19 13 17 21 20 17 14 18 20 21 20 11 14 19 21 19 22 15 11 15 19 16 19
14 17 12 20 17 21 17 23 19 15 15 20 15 12 23 20 21 19 21 21 14 13 18 21 12 22 19 17
21 22 10 21 12 14 22 14 16 16 23 13 20 12 16 16 13 16 20 10 16 23 18 18 15 12 23 21
15 18 21 22 17 18 20 15 16 21 19 21 24 13 20

An observation comes from one group with probability 0.5 and the other with probability 0.5, but you don't know which group. Assume an $N(\mu, \exp(\theta))$ distribution for one group and an $N(\mu + \delta, \exp(\theta))$ distribution for the other; then δ is the difference due to treatment.

A. Using the pdf for each observation $p(y \mid \mu, \delta, \theta) = 0.5N(y \mid \mu, \exp(\theta)) + 0.5N(y \mid \mu + \delta, \exp(\theta))$, estimate the parameters (μ, δ, θ) via maximum likelihood using an iterative method and the computer. Can you tell, just from the estimate of δ, whether the treatment made the data values generally higher or generally lower?

B. Using statistical software that reports the Wald standard error, report the 95% range of plausible values of δ, given the data. Can you tell whether there is a difference between the treatment and control? Can you tell whether the treatment made the data values generally higher or generally lower?

12.11 Here is a case where the derivative of the likelihood function is not zero at the MLE. Consider the stoplight example from Chapter 4, where the green signal duration time is unknown (call it θ). The likelihood for an individual observation y is $1/\theta$, for $y < \theta$, and 0 otherwise. Suppose you observe a sample of 30 days, with time the light stays green (in minutes) being as follows:

1.26 2.22 2.24 1.65 0.86 0.58 0.86 2.11 1.53 0.14 0.67 0.05 1.41 0.01 1.07 0.93 0.46 0.19 1.02 0.25 0.40 1.09 0.79 1.30 1.99 1.03 2.39 0.86 0.15 0.36

A. State the likelihood function for θ. In other words, write down its mathematical form of the likelihood function in this particular case.

B. Graph the likelihood function for θ and identify its MLE.

C. Using the graph of Exercise 12.11B, give an approximate range of plausible values for θ.

D. Can you tell from the graph of Exercise 12.11B whether or not the derivative is equal to zero at the MLE?

12.12 A common measure of statistical significance is the p-value (pv), discussed more in Chapter 15. Smaller p-values indicate that the results are unlikely to be explained by chance alone and hence indicate real effects rather than randomness. The thresholds $pv < 0.05$ and $pv < 0.10$ are commonly used to indicate significance. A meta-analysis of studies on a particular subject finds p-values of 0.012, 0.001, 0.043, 0.008, and 0.059. A model that describes p-values from similar studies is $p(y) = \theta\, y^{\theta-1}$, for $0 < y < 1$, where $Y = 1 - PV$.

A. Show that $p(y)$ is a valid pdf. (The data are irrelevant here.)

B. Using the log-likelihood function for θ and calculus, find its MLE.

C. Using the MLE of θ, draw a graph of the estimated pdf $p(y) = \theta\, y^{\theta-1}$. Using the resulting graph and calculus, estimate the proportion of studies that will result in a p-value greater than 0.10.

12.13 In logistic regression, it can happen that the binary y data are perfectly separated by the x data; this is sometimes called *quasi-complete separation*. Here are some (y, x) data: (1, 2.0), (0, 0.5), (1, 4.0), (1, 2.8), (1, 2.2), (1, 2.3), (0, 1.5), (1, 4.4), and (0, 0.8).

A. Draw a (x, y) scatterplot and comment on how the y data are perfectly separated by the x data.

B. State the likelihood function for (β_0, β_1) when using the logistic regression model.

C. Allow your software to try to maximize the function. Choose the best guess provided by your software and graph the estimated logistic regression function. Do you see any problems?

12.14 The following data (x, y) were collected from a supermarket at random times. The variables are $X =$ number of open checkout lines at a given time and $Y =$ number of shoppers in line for the shortest line at that time: (1 4), (1 0), (4 0), (2 1), (2 4), (1 8), (1 4), (3 1), (4 1), (1 6), (2 4), and (4 0). The Poisson regression model is appropriate for such data. It assumes that $Y|X = x$ has a Poisson distribution with mean $\mu(x)$, where $\mu(x) = \exp(\beta_0 + \beta_1 x)$.

A. Find the log-likelihood values when using the following parameter settings for (β_0, β_1): (0, 0), (0, −0.5), (1, 1), and (2, −0.5). (You should report four log likelihoods, one for each parameter setting.) Which of these settings is most consistent with the observed data?

B. Find the MLE of (β_0, β_1) and compare the log likelihood at the MLE with the log likelihoods you found in Exercise 12.14A. Why are the log likelihoods in Exercise 12.14A all smaller?

C. Draw graphs of the estimated functions $\mu(x)$ corresponding to the four parameter settings in Exercise 12.14A and to the MLE setting. Which of these five graphs is best supported by the data?

13

Bayesian Statistics

13.1 Introduction: Play a Game with Hans!

Hans has a thumbtack. He won't tell you its shape. It might have a long point and a narrow circular base, or it might have a wide circular base and a short point.

If Hans tosses the thumbtack in the air, what is the chance that it will land point up? Well, it depends on the shape. If it has a long point and a narrow circular base, then the probability is low—well below 50%. If it has a short point and a very rounded base, then the probability is high—well above 50%.

Suppose Hans tosses the thumbtack 10 times in a random way—that is, he throws it way up in the air and lets it fall, with no tricks. Suppose it lands 2 out of 10 times with the point up. What is your best guess of the probability (π) of the thumbtack landing point up?

Using what you learned in Chapter 12, you would calculate the likelihood function for π and pick the value that maximizes the likelihood. If the data are 0, 1, 0, 1, 0, 0, 0, 0, 0, and 0, where 1 denotes the thumbtack lands point up, then the likelihood function for π is the product of Bernoulli probabilities, with each 1 contributing π and each 0 contributing $1 - \pi$ the likelihood function. Hence, $L(\pi \,|\mathrm{data}) = (1 - \pi) \times \pi \times (1 - \pi) \times \pi \times (1 - \pi) \times (1 - \pi) \times (1 - \pi) \times (1 - \pi) \times (1 - \pi) \times (1 - \pi) = \pi^2(1 - \pi)^8$.

Figure 13.1 shows a graph of the likelihood function $L(\pi \,|\mathrm{data}) = \pi^2(1 - \pi)^8$.

As expected, the MLE is $\hat{\pi} = 0.2$. This is intuitive: Based on the data, the most likely value of the probability of the thumbtack landing point up is 2/10.

But wait! Now Hans has a new game! He has a coin—a U.S. quarter. It is an ordinary quarter, regulation U.S. mint, and you think Hans is not a magician or trickster.

If Hans flips the coin way up in the air, and lets it bounce around on the ground and settle, what is the probability π that it will land with heads facing up? Again, it depends on the coin's shape. The U.S. quarter is not perfectly symmetric; it has very mild irregularities. So the chance may be slightly above 50% or slightly below 50%, but it should be darn close to 50%.

Suppose Hans tosses it 10 times in a random way—that is, he flips it way up in the air and lets it bounce on the ground and settle, with no tricks. Suppose it lands with heads facing up on two out of ten flips. What is your best guess of the probability π that the coin lands on heads?

Again, using what you learned in Chapter 12, you would calculate the likelihood function for π and pick the value that maximizes the likelihood. If the data are identical to the thumbtack toss, with values 0, 1, 0, 1, 0, 0, 0, 0, 0, and 0, where 1 denotes coin lands heads up, then your likelihood function for π is again $L(\pi \,|\mathrm{data}) = (1-\pi) \times \pi \times (1-\pi) \times \pi \times (1-\pi) \times (1-\pi) \times (1-\pi) \times (1-\pi) \times (1-\pi) = \pi^2(1-\pi)^8$.

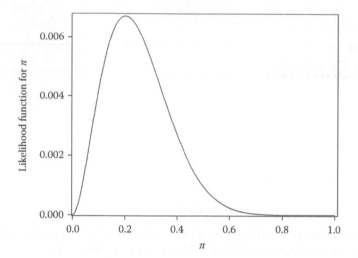

FIGURE 13.1
Likelihood function $L(\pi \mid \text{data}) = \pi^2(1-\pi)^8$ for $\pi = \text{Pr}(\text{Thumbtack lands point up})$, based on two observations of point up and eight observations of point down.

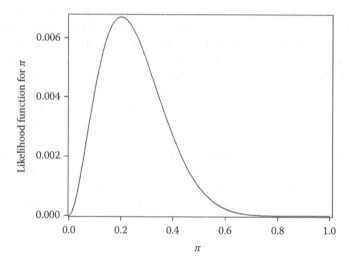

FIGURE 13.2
Likelihood function $L(\pi \mid \text{data}) = \pi^2(1-\pi)^8$ for $\pi = \text{Pr}(\text{Coin lands heads})$, based on two observations of heads and eight observations of tails.

Figure 13.2 shows the likelihood function for π when there are two heads and eight tails. It is identical to the likelihood function for the thumbtack data having two results of point up and eight of point down as shown in Figure 13.1.

Again, the MLE is $\hat{\pi} = 0.2$. But this answer is no longer intuitive: Because you know something about coins, you know that the most likely value of the probability of landing on heads is nowhere near 20%—it is somewhere around 50%.

What's going on here? This example shows clearly that the observed data are not completely adequate to make inferences about your parameters. You also need to consider your *prior information*.

This chapter shows you how to incorporate your prior information into the analysis of your data. The topic is called *Bayesian statistics*, which, after decades on the sidelines, is now a standard method for analyzing statistical data due to advances in statistical software. It is based on—no surprise—the same Bayes' theorem you already learned in Chapter 6. One reason for its popularity is that it offers a very natural way to understand statistics: The transition from probability, to likelihood, to Bayesian methods is seamless. Another reason for its popularity is that you simply cannot do many kinds of statistical analyses in any other way; this chapter provides several such examples.

13.2 Prior Information and Posterior Knowledge

Before any empirical study, you always have some prior information. You are not ignorant! **Prior information** is what you know *prior to collecting the data.*

Your Sources of Prior Information

- Other similar studies. For example, in presidential elections, you have seen that the percentage of votes for the winner is typically between 40% and 55%.
- Life experience and common sense. For example, in the coin toss experiment, you know that, in the absence of trickery, the true probability of heads should be around 50%. This is common sense, based on the fact that the coin is symmetric and balanced. You have also seen that coins come up heads about 50% of the time in your life experience.
- Parameter constraints. Examples: (1) variances are always positive, (2) probabilities are always between 0 and 1, and (3) the mean of the Poisson distribution is never negative. When there are parameter constraints, you are 100% certain that the parameters must lie in the permissible region.

It would be hard to think of a study where you *wouldn't* have any prior information.

In Chapter 12, you learned that the part of the Mantra that says *data reduce uncertainty about the unknown parameters* can be interpreted in terms of what the likelihood function $L(\theta \,|\, \text{data})$ tells you about the plausible range of values of θ given the data. You saw that with more data, the likelihood function becomes narrower. In this chapter, we are telling you that this is not quite enough. As Hans' coin toss example in this chapter shows (see Figure 13.2), the likelihood alone does not always provide a commonsense solution. You need *both* the likelihood function *and* your prior information.

How does this work specifically? Bayes' theorem presented in Chapter 6 explains it all.

First, remember this: *Model has unknown parameters.* This is the beginning of the Bayesian statistics story. Yes, the parameters are unknown, but you are not completely ignorant! If it's a coin toss, you know something about π. So you start by stating your uncertainty about the parameter(s) θ using a probability distribution $p(\theta)$, which is called your **prior distribution** because it comes first before you collect data for your current study. After you collect the data, your knowledge about θ increases—*Data reduce uncertainty about the unknown parameters.* Your state of uncertainty about θ after seeing the data is given by your **posterior distribution** for θ, and this distribution is obtained directly via Bayes' theorem.

The following sections give specific details on finding and interpreting these distributions.

13.3 Case of the Unknown Survey

Suppose that you have a pile of employee satisfaction surveys coming from five different companies. Each employee answered one of the numbers 1, 2, 3, 4, or 5 to indicate their job satisfaction. Table 13.1 shows the tabulated distributions. The numbers in the table are not real; they are just made up to make the points clearer. The company names are also fictitious.

Now suppose that a survey got separated from the pile, so that you do not know what company it was from. You look at the survey and see that the person answered $Y = 4$. Call the unknown company θ; this is a discrete parameter that may take on any one of the five values in the parameter space Θ = {BankTen, DoggyTreats, CraftyCrafts, AgBus, InternetSavvy}. Logic would suggest that it is least likely that the survey was from DoggyTreats, since relatively few (2.5%) of their employees selected 4 on the survey. On the other hand, you recall from Chapter 6 that there is a big difference between Pr(death|drunk driving) and Pr(drunk driving|death). So there is also a difference between Pr(Y = 4|DoggyTreats), whose value is 2.5%, and Pr(DoggyTreats|Y = 4), whose value you don't know at this point. You need to use Bayes' theorem to find this probability.

What is the probability distribution of θ given that you see that the person answered $Y = 4$ on the survey? It depends on your prior information—on your state of uncertainty about the unknown parameter θ.

Example 13.1: Uniform Prior (Prior Ignorance) Case

Suppose you really have no prior idea which group the survey came from. In that case, your prior distribution on the companies is as shown in Table 13.2.

Now, your posterior distribution of θ given that $Y = 4$ is given by Bayes' theorem for Chapter 6—recall that $p(\theta\,|y) \propto p(y\,|\,\theta)\,p(\theta)$. Table 13.3 shows the details, which are essentially identical to what you saw in Chapter 6.

The posterior distribution for θ shown in Table 13.3 is intuitive: All you have done is take the likelihood values 0.196, 0.025, 0.085, 0.238, and 0.045 and make them a probability distribution function (pdf) by enforcing the condition that they sum to 1. The prior probabilities did not play any essential role here because they are the same for every company.

You can interpret these results as with any pdf. For example, given that you have observed a 4, the probability that the survey came from an employee who works for DoggyTreats is 0.042. It's not 0.025, but still pretty unlikely! It's much more likely (73.7% likely, in fact) that the survey was from either BankTen or AgBus.

TABLE 13.1

Distributions of Employee Job Satisfaction for Five Different Companies, with Observed Value $y = 4$ in Bold

Satisfaction Rating, y	Company Name, θ				
	BankTen (%)	DoggyTreats (%)	CraftyCrafts (%)	AgBus (%)	InternetSavvy (%)
1	24.0	33.6	16.3	21.6	16.2
2	20.0	23.9	32.7	20.9	24.5
3	27.6	0.9	9.1	27.6	3.1
4	**19.6**	**2.5**	**8.5**	**23.8**	**4.5**
5	8.8	39.1	33.4	6.1	51.7
Total	100.0	100.0	100.0	100.0	100.0

TABLE 13.2
Uniform Prior Distribution
on the Discrete Parameter θ

θ	$p(\theta)$
BankTen	$1/5 = 0.20$
DoggyTreats	$1/5 = 0.20$
CraftyCrafts	$1/5 = 0.20$
AgBus	$1/5 = 0.20$
InternetSavvy	$1/5 = 0.20$
Total	1.00

TABLE 13.3
Posterior Distribution of θ, Given $Y = 4$ and a Uniform Prior

θ	$p(\theta)$	$p(4 \mid \theta)$	$p(4 \mid \theta)\, p(\theta)$	$p(\theta \mid 4)$
BankTen	0.20	0.196	$0.196 \times 0.20 = 0.0392$	$0.0392/0.1178 = 0.333$
DoggyTreats	0.20	0.025	$0.025 \times 0.20 = 0.0050$	$0.0050/0.1178 = 0.042$
CraftyCrafts	0.20	0.085	$0.085 \times 0.20 = 0.0170$	$0.0170/0.1178 = 0.144$
AgBus	0.20	0.238	$0.238 \times 0.20 = 0.0476$	$0.0476/0.1178 = 0.404$
InternetSavvy	0.20	0.045	$0.045 \times 0.20 = 0.0090$	$0.0090/0.1178 = 0.076$
Total	1.00	—	0.1178	1.00

In Example 13.1, the likelihood function $p(4 \mid \theta)$ is a maximum when $\theta =$ AgBus, and the value of θ having the largest posterior probability is also AgBus. Thus, the uniform prior lets the data guide the definition of your posterior distribution. In other words, the uniform prior distribution "let the data do the talking" (more on this later in the chapter). With nonuniform priors, however, your specification of prior probabilities makes a big difference, as the next example shows.

Example 13.2: Partial Information (Informative Prior) Case

Suppose the surveys had been presorted into two piles, one with the surveys from DoggyTreats and AgBus combined and the other pile containing the surveys from BankTen, CraftyCrafts, and InternetSavvy. Suppose that you know that the survey response you dropped comes from either DoggyTreats or AgBus because the response came from that pile, but you know nothing else. In this case, your prior distribution is as shown in Table 13.4.

The solution proceeds exactly as shown in Table 13.3, albeit with a different prior; see Table 13.5.

The result shown in Table 13.5 is again intuitive: If DoggyTreats and AgBus are the only possible companies, and if the likelihood is much higher for AgBus, then it is only sensible that the probability of AgBus should be much higher than that of DoggyTreats (0.905 vs. 0.095). Notice also that even though the survey is much less likely to have come from DoggyTreats than from AgBus, the probability that it *did* come from DoggyTreats is much higher in this case than in the previous case with the non-informative prior: 9.5% versus 4.2%. This makes sense: If you think that DoggyTreats is more likely, a priori (before seeing the $Y = 4$ survey), then you should also think that DoggyTreats is more likely a posteriori (after seeing the $Y = 4$ survey) as well. This point is demonstrated further in Example 13.3.

TABLE 13.4

Informative Prior Distribution
on the Discrete Parameter θ

θ	$p(\theta)$
BankTen	0.00
DoggyTreats	0.50
CraftyCrafts	0.00
AgBus	0.50
InternetSavvy	0.00
Total	1.00

TABLE 13.5

Posterior Distribution of θ, Given $Y = 4$ and an Informative Prior

θ	$p(\theta)$	$p(4\mid\theta)$	$p(4\mid\theta)\,p(\theta)$	$p(\theta\mid4)$
BankTen	0.00	0.196	$0.196 \times 0.00 = 0.0000$	$0.0000/0.1315 = 0.000$
DoggyTreats	0.50	0.025	$0.025 \times 0.50 = 0.0125$	$0.0125/0.1315 = 0.095$
CraftyCrafts	0.00	0.085	$0.085 \times 0.00 = 0.0000$	$0.0000/0.1315 = 0.000$
AgBus	0.50	0.238	$0.238 \times 0.50 = 0.1190$	$0.1190/0.1315 = 0.905$
InternetSavvy	0.00	0.045	$0.045 \times 0.00 = 0.0000$	$0.0000/0.1315 = 0.000$
Total	1.00	—	0.1315	1.00

Notice that if the prior probability is zero, then the posterior is also zero. This also makes sense: If an event is impossible, it simply can't happen. It doesn't matter what the data say. It is dangerous to use such priors, since doing so means that you are blindly dogmatic in your attitude. If you use a prior that places either 0% probability or 100% probability on particular states of Nature, then no amount of evidence will ever change your mind. Holding a dogmatic prior is therefore irrational behavior. In Example 13.2, we used prior probabilities of 0.00 for illustration purposes only. In reality, you can't be absolutely 100% certain, so you should place at least a tiny bit of prior probability, such as 0.0001, on the unlikely states of Nature.

Example 13.3: Partial Information (Informative Prior) Case, Continued

Suppose you knew, before looking at the unknown survey with $Y = 4$, that most of the people who filled out surveys were at the DoggyTreats company. That would change everything! Specifically, suppose 96% of the survey takers were from DoggyTreats and the rest evenly distributed among the other companies, so your prior distribution is as shown in Table 13.6.

Which company do you think the unknown survey came from now? Without checking the contents of the survey, you would reasonably suppose that it most likely came from DoggyTreats, simply because the responses were overwhelmingly from that company. Your posterior distribution is as shown in Table 13.7.

Notice that your posterior probability for DoggyTreats (0.810) is less than your prior probability (0.96), but it is still much higher than your posterior probabilities found when using the other priors (0.042 and 0.095).

As the examples show, your uncertainty about the unknown parameters (θ), as measured by your posterior distribution on θ, involves both the likelihood function $L(\theta\mid y) = p(y\mid\theta)$ and your prior distribution $p(\theta)$.

TABLE 13.6

Distribution of Survey Takers

θ	$p(\theta)$
BankTen	0.01
DoggyTreats	0.96
CraftyCrafts	0.01
AgBus	0.01
InternetSavvy	0.01
Total	1.00

TABLE 13.7

Posterior Distribution of θ, Given $Y = 4$ and an Informative Prior

θ	$p(\theta)$	$p(4\mid\theta)$	$p(4\mid\theta)\,p(\theta)$	$p(\theta\mid 4)$
BankTen	0.01	0.196	$0.196 \times 0.01 = 0.00196$	$0.00196/0.02964 = 0.066$
DoggyTreats	0.96	0.025	$0.025 \times 0.96 = 0.02400$	$0.02400/0.02964 = 0.810$
CraftyCrafts	0.01	0.085	$0.085 \times 0.01 = 0.00085$	$0.00085/0.02964 = 0.029$
AgBus	0.01	0.238	$0.238 \times 0.01 = 0.00238$	$0.00238/0.02964 = 0.080$
InternetSavvy	0.01	0.045	$0.045 \times 0.01 = 0.00045$	$0.00045/0.02964 = 0.015$
Total	1.00	—	0.02964	1.00

13.4 Bayesian Statistics: The Overview

Use what you know to predict what you don't know. You saw this mantric phrase in Chapter 5, in the discussion of conditional distributions, and it applies perfectly to Bayesian analysis. You don't know the parameter θ, because the *model has unknown parameters.* But you have observed some data, so you do know the data. Bayesian statistics is simply about using what you know (the data) to predict what you don't know (the parameter or parameters θ).

As shown in the examples earlier in Section 13.3, the essence of the Bayesian paradigm is simply to find the conditional distribution of θ, given the data. By Bayes' theorem, the posterior distribution of θ, given data $Y = (y_1, y_2, \ldots, y_n)$, is proportional to the probability distribution of Y given θ, times the prior distribution of θ. More succinctly

$$p(\theta \mid y_1, y_2, \ldots, y_n) \propto p(y_1, y_2, \ldots, y_n \mid \theta) \times p(\theta)$$

Since the pdf $p(y_1, y_2, \ldots, y_n \mid \theta)$ and the likelihood function $L(\theta \mid y_1, y_2, \ldots, y_n)$ are identical, you can rewrite Bayes' theorem as follows.

Bayes' Theorem for Statistical Analysis

$$p(\theta \mid y_1, y_2, \ldots, y_n) \propto L(\theta \mid y_1, y_2, \ldots, y_n) \times p(\theta) \qquad (13.1)$$

Equation 13.1 shows that the likelihood function $L(\theta \mid y_1, y_2, \ldots, y_n)$ is *almost* a pdf for θ. All you need to do to make $L(\theta \mid y_1, y_2, \ldots, y_n)$ a pdf is to multiply it by your prior $p(\theta)$ and make the resulting function of θ a pdf by choosing the right constant of proportionality.

The constant of proportionality makes the pdf sum to 1.0, in the case of a discrete parameter space as shown in the examples earlier, or integrate to 1.0, in the more common case of a continuous parameter space. Specifically, the constant is

$$c = \frac{1}{\int_{\Theta} L(\theta|y_1, y_2, \ldots, y_n) \times p(\theta) d\theta}$$

Once you have your posterior distribution $p(\theta|y_1, y_2, \ldots, y_n)$, you can get busy doing all the science you want to do.

Applications of the Posterior Distribution for θ

- Estimate the parameter θ using the mean of the posterior distribution $p(\theta|y_1, y_2, \ldots, y_n)$; this is called the **posterior mean**. You also have the flexibility to estimate the parameter using the **posterior median** if you wish.
- Obtain the standard deviation of the posterior distribution $p(\theta|y_1, y_2, \ldots, y_n)$; along with the mean of the posterior distribution, you can use this **Bayesian standard error** to describe your range of uncertainty about the parameter θ.
- Find an interval range (l, u) so that the true θ lies in the range (l, u) with specified probability (such as 90%); this is called a **Bayesian credible interval**.
- Find the probability that the parameter is greater than a constant of interest; for example, in regression analysis, one wants to know what is the chance that the slope parameter β_1 is more than zero; this is called **Bayesian hypothesis testing**.
- Generate plausible values of the parameter θ^* from the posterior distribution $p(\theta|y_1, y_2, \ldots, y_n)$ and then produce Y^* from $p(y|\theta^*)$ to predict the real Y values that will be seen in Nature; the values Y^* are samples from the **Bayesian predictive distribution**.
- Make decisions in the presence of uncertain states of nature (the value of the parameter θ); this is called **Bayesian decision analysis**.

13.5 Bayesian Analysis of the Bernoulli Parameter

In Hans' thumbtack and coin tossings, the parameter π represents the probability of a success, whether it is the thumbtack landing point up or the coin landing on heads. The parameter space is $\{\pi; 0 \leq \pi \leq 1\}$ in either case. Recall that the likelihood function from Hans' data was $L(\pi|2 \text{ successes and } 8 \text{ failures}) = \pi^2(1-\pi)^8$ in either case as well.

Example 13.4: Prior and Posterior Distributions, Thumbtack Example

In the case of the thumbtack toss, you have no prior information about the value of π. You may plead ignorance! If you feel that all values of π are equally likely, then your prior is the familiar uniform distribution over the range (0, 1), expressed by $p(\pi) = 1.0$, for $0 \leq \pi \leq 1$, and $p(\pi) = 0.0$ otherwise.

Then you can find your posterior distribution as follows:

$p(\pi \mid \text{data}) \propto L(\pi \mid \text{data}) \times p(\pi)$	(By Bayes' theorem [Equation 13.1])
$\Rightarrow p(\pi \mid \text{data}) \propto \pi^2(1-\pi)^8 \times p(\pi)$	(By substituting the likelihood function)
$\Rightarrow p(\pi \mid \text{data}) \propto \pi^2(1-\pi)^8 \times (1.0)$	(By substituting your ignorance-based prior for π)
$\Rightarrow p(\pi \mid \text{data}) = c \times \pi^2(1-\pi)^8 \times (1.0),$ for $0 \leq \pi \leq 1$, $p(\pi \mid \text{data}) = 0$ otherwise	(By definition of proportionality, $\propto$, and including the parameter constraints)

Since this example has a continuous parameter space, you can't list the values of π in a table and then sum up the values $\pi^2(1-\pi)^8$ to get c as shown in Section 13.3. Instead, you need to find the c that makes the area under the curve $c \times \pi^2(1-\pi)^8$ equal to 1.0; specifically:

$$c = \frac{1}{\int_0^1 \pi^2(1-\pi)^8 \, d\pi}$$

While this calculus problem is solvable, there is an easier trick involving the **kernel** of a distribution, a term we introduced in Chapter 2. Recall that the kernel of the distribution is the multiplicative part of the distribution that depends on the variable. For example, recall the normal distribution:

$$p(y) = \left\{ \frac{1}{(\sigma\sqrt{2 \times 3.14159\ldots})} \right\} \exp\left\{ -0.5 \frac{(y-\mu)^2}{\sigma^2} \right\}$$

Here the variable is y, and hence, the kernel is $\exp\{-0.5(y-\mu)^2/\sigma^2\}$. If you can recognize the kernel of a distribution, then you can recognize the entire distribution because the rest of the terms are simply what is needed to make the function integrate (or sum) to 1.0. Any pdf is completely determined by its kernel.

In the case of the posterior distribution $p(\pi \mid \text{data}) = c \times \pi^2(1-\pi)^8$, the variable is π, and the kernel is simply $\pi^2(1-\pi)^8$. So if you recognize the distribution for that kernel, then you know the posterior distribution of π.

The **beta distribution** is used to model variables that lie in the [0,1] range. The kernel of the beta distribution is $y^{\alpha-1}(1-y)^{\beta-1}$, when viewed as a function of a variable y. There are two parameters of the beta distribution, α and β. Note that when $\alpha = \beta = 1$, the beta distribution is the usual uniform distribution, so the uniform distribution is a special case of the beta distribution. You can find the constant c of the beta (α, β) distribution by looking it up on the Internet—go ahead, have a look! In cases where the parameters α and β are integers, $c = (\alpha + \beta - 1)!/\{(\alpha - 1)! \times (\beta - 1)!\}$. When the parameters are not integers, the constant can be found from the *gamma function*, which generalizes the *factorial function*. We won't use the gamma function, but it is interesting and useful; see other sources for more details.

You can see that the function $\pi^2(1-\pi)^8$. is the kernel of the beta distribution with $\alpha = 3$ and $\beta = 9$ because $\pi^2(1-\pi)^8 = \pi^{3-1}(1-\pi)^{9-1}$. Hence, the constant is $c = (\alpha + \beta - 1)!/\{(\alpha - 1)! \times (\beta - 1)!\} = 11!/(2! \times 8!) = 495$, and your posterior distribution in the thumbtack example is

$p(\pi \mid 2 \text{ "Point Up"}, 8 \text{ "Point Down"}) = 495\pi^2(1-\pi)^8$, for $0 \leq \pi \leq 1; = 0$ otherwise.

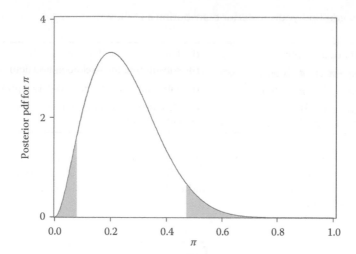

FIGURE 13.3
Posterior distribution of π in the thumbtack example, along with 90% equal-tailed credible interval (0.079, 0.470).

For practical purposes, you really don't need to know the constant $c = 495$. You are done once you recognize that $\pi^2(1-\pi)^8$ is the kernel of a beta(3, 9) distribution. You can access this distribution directly just by specifying the parameters $\alpha = 3$ and $\beta = 9$ in various software, including Microsoft Excel, and you can use the software directly to learn all you need to know about π through its posterior distribution.

Figure 13.3 shows a graph of your posterior distribution of π, $p(\pi \,|\, 2 \text{ point up, 8 point down}) = 495\pi^2(1-\pi)^8$, in the thumbtack example.

Please have a look at Figures 13.1 and 13.2 again—Figure 13.3 looks almost identical to these two! In fact, the curves are exactly proportional to one another; the only difference is that in Figure 13.3, the vertical axis is changed so that the area under the curve is 1.0. It is now a bona fide probability distribution! Since the prior is $p(\pi) = 1.0$ in this example, the posterior is simply the scaled likelihood. In particular, the area under the curves in Figures 13.1 and 13.2 is 1/495; dividing the likelihoods by this value gives you the posterior shown in Figure 13.3.

So what on earth do you do with this posterior distribution? Good question! One thing you can do with it is to report a **credible interval**, that is, a range of values of the unknown parameter π that has a prespecified probability. We attempted this in Chapter 12 using the likelihood function, but it was rather vague—we simply looked for a place where the function seemed to die out in both tails, and we reported the resulting range of values. We also used the Wald interval, but that was also crude because it assumed that the likelihood function was approximately proportional to a normal distribution.

You can be more precise. Instead of just eyeballing the likelihood function, or of using the approximate Wald interval, it would be better to specifically identify a range of values that has a precise probability such as 90%. One method for doing this is called the **equal-tailed credible interval**, shown in Figure 13.3.

The Equal-Tailed Credible Interval

Suppose $p(\theta \,|\, \text{data})$ is a continuous pdf for a one-dimensional parameter θ. Let l be the $\alpha/2$ quantile and let u be the $1 - \alpha/2$ quantile of $p(\theta \,|\, \text{data})$. Then $[l, u]$ is a $100(1 - \alpha)\%$ equal-tailed credible interval for θ.

To get a 90% credible interval using the thumbtack data and uniform prior, the value l is the 0.05 quantile of the beta(3, 9) distribution. You can access that value directly via Excel as BETA.INV(0.05, 3,9), which gives $l = 0.079$. The value u is the 0.95 quantile, accessible in Excel as BETA.INV(0.95, 3,9), which gives $u = 0.470$. Thus, you may conclude as follows:

"Given the observed data with two point up results and eight point down ones, along with my assumption of a uniform prior on possible values of the true probability, I deduce that there is a 90% probability that $0.079 \leq \pi \leq 0.470$, where π is the probability the thumbtack will land point up."

There is $100(1 - \alpha)\%$ probability that the interval covers θ, which you can see as follows:

$\Pr(l \leq \theta \leq u \,|\, \text{data})$

$= \Pr(\theta \leq u \,|\, \text{data}) - \Pr(\theta < l \,|\, \text{data})$ (Because the probability between two points is the difference between the two cumulative probabilities)

$= \Pr(\theta \leq u \,|\, \text{data}) - \Pr(\theta \leq l \,|\, \text{data})$ (Because the distribution is continuous)

$= (1 - \alpha/2) - \alpha/2$ (Since u and l are the $1 - \alpha/2$ and $\alpha/2$ quantiles, respectively)

$= 1 - \alpha$ (By algebra)

Unlike the Wald interval discussed in Chapter 12, the Bayesian credible interval is often nonsymmetric about the parameter estimate. In other words, the distance from the lower limit, l, to the estimate $\hat{\theta}$ is not same as the distance from the upper limit u to the estimate. In the case of the Wald interval $\hat{\theta} \pm c\hat{\sigma}$, both distances are the same, namely, $c\hat{\sigma}$. The asymmetry of the Bayesian interval is a good thing! With a skewed likelihood function or posterior distribution, such as shown in Figure 13.3, the interval should extend farther in one direction than the other. This asymmetry is an advantage of the Bayesian credible interval over the Wald interval.

While the equal-tailed credible interval is easy to compute, it lacks a certain common-sense element. The interval should *exclude* values in either tail that are *equally unlikely*. However, it is clear from Figure 13.3 that some of the excluded values in the lower tail are *more likely* than even the most likely excluded values in the upper tail. To solve this problem, you can use the **highest posterior density interval**, or the HPD interval for short. This interval contains all parameter values whose posterior density is larger than a constant, and for which $\Pr(l \leq \pi \leq u \,|\, \text{data}) = 1 - \alpha$. It is harder to compute than the equal-tailed credible interval, but computer software will do it for you. Figure 13.4 shows the 90% HPD interval for the thumbtack data, which is calculated as (0.056, 0.434). Still, you can say that $\Pr(0.056 \leq \pi \leq 0.434 \,|\, \text{data}) = 0.90$, as with the equal-tailed credible interval. However, the values that are excluded in the lower tail are now just as unusual as those in the upper tail.

No matter whether you prefer the equal-tailed credible interval or the HPD credible interval, the range of values you report will depend greatly on the probability that you choose to use. For example, if you choose 95% instead of 90%, then both the equal-tailed interval and the HPD interval extend to the right of 0.50 in Figures 13.3 and 13.4. That means you cannot state whether or not the probability is less than 0.5, if you need to be 95% sure.

In statistics, there are always tradeoffs. If you want more precision—that is, if you want a shorter interval range—then you have to sacrifice probability that the interval is correct. For instance, if you want to claim a more precise (i.e., shorter) interval for π based on the data 2/10, you might use a 50% credible interval instead of a 90% credible interval.

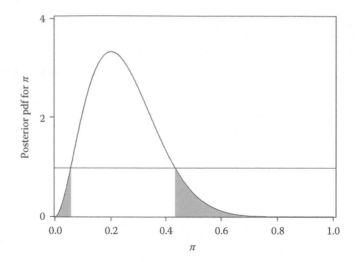

FIGURE 13.4
Posterior distribution of π in the thumbtack example, along with 90% HPD credible interval (0.056, 0.434).

If you only have to cover 50% of the center of the distribution, then you can see from either Figure 13.3 or 13.4 that your interval will be much shorter. Your 50% equal-tailed interval for π is $0.169 \leq \pi \leq 0.326$, suggesting much more precision than your 90% equal-tailed interval $0.079 \leq \pi \leq 0.470$. But increased precision comes at a cost: You are much more likely to be wrong! For this reason, you should report intervals that you believe to be correct. If you report a 50% interval, then you are as likely to be wrong as you are to be right. If you report a 90% interval, then you are probably right. If you report a 95% interval, then you are more likely still to be right. This calls for an ugly rule of thumb.

Ugly Rule of Thumb 13.1

Credible intervals having probability 90% have both sufficiently high probability and sufficiently narrow width.

Example 13.5: Prior and Posterior Distributions, Coin Toss Example

In the case of Hans' thumbtack toss, you had very little prior information about π. In the case of Hans' coin toss, on the other hand, you have *strong prior information* about the value of π. You feel that values of π close to 0.5 are very likely, and values of π far from 0.5 are very unlikely.

We are going to be inside your head for the rest of this example. Don't worry; we'll only stay there for a little while!

You feel that the probability of heads for the U.S. quarter, or π, is within ± 0.01 of 0.5 (between 0.49 and 0.51) with extremely high probability. You know that the mean ± 3 standard deviation range includes very high coverage probability, and you know that the beta distribution is appropriate for variables on the [0, 1], so you assume a beta prior with mean 0.5 and standard deviation 0.00333333. The mean of the beta(α, β) is $\alpha/(\alpha + \beta)$; hence, if your prior mean for π is 0.5, then the parameters of your prior satisfy $\alpha/(\alpha + \beta) = 0.5$ or $\alpha = \beta$.

The variance of the beta(α, β) distribution is $\alpha\beta/\{(\alpha + \beta)^2(\alpha + \beta + 1)\}$. Since your prior assumes $\alpha = \beta$, your variance is $\alpha^2/\{4\alpha^2(2\alpha + 1)\} = 1/(8\alpha + 4)$. Solving $1/(8\alpha + 4) = (0.00333333)^2$ gives you $\alpha = 11{,}250$; hence, $\beta = 11{,}250$ as well.

Thus, your prior distribution for the π of Hans' coin toss is the beta(11,250, 11,250) distribution given by

$$p(\pi) = c\pi^{11,250-1}(1-\pi)^{11,250-1}, \quad \text{for } 0 \leq \pi \leq 1; \ p(\pi) = 0 \text{ otherwise}$$

The constant c is just another constant of proportionality to make the function integrate to 1.0; you don't need to worry about it since it is not part of the kernel of the function.

Then you can find your posterior distribution as follows:

$p(\pi \mid 2 \text{ heads and } 8 \text{ tails})$

$\propto L(\pi \mid 2 \text{ heads and } 8 \text{ tails}) \, p(\pi)$ (By Bayes' theorem [Equation 13.1])

$\Rightarrow p(\pi \mid 2 \text{ heads and } 8 \text{ tails})$

$\propto \{ \pi^2(1-\pi)^8 \} \{ \pi^{11,250-1}(1-\pi)^{11,250-1} \}$ (By substituting your likelihood and prior distributions)

$\Rightarrow p(\pi \mid 2 \text{ heads and } 8 \text{ tails})$

$\propto \pi^{11,252-1}(1-\pi)^{11,258-1}$ (By algebra of exponents)

Again, you may recognize the kernel of the beta distribution here but with parameters $\alpha = 11,252$ and $\beta = 11,258$. Figure 13.5 shows both your prior and the posterior distributions for Hans' coin toss data. There is no discernable difference at the broad $0 \leq \pi \leq 1$ scale, so a zoomed-in version focusing on the range $0.49 \leq \pi \leq 0.51$ is also shown.

In the thumbtack example, your prior was the U(0, 1) distribution with a flat graph, and your posterior distributions were the beta distributions shown in Figures 13.3 and 13.4 with curved graphs. Thus, in the thumbtack case, your prior and posterior knowledge was very different because the data modified your opinion greatly. By contrast, in the coin toss example, your prior and posterior distributions are quite similar as shown in Figure 13.5.

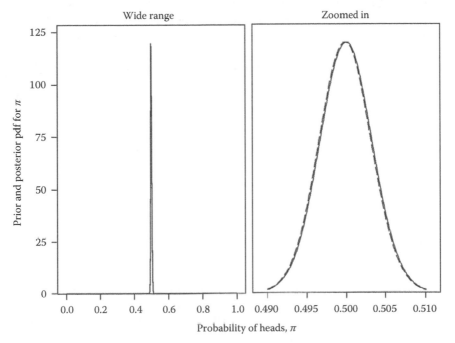

FIGURE 13.5
Prior (solid) and posterior (dashed) distributions for π with the coin toss data.

This happens because, in contrast to the *non-informative* prior you assumed for the thumb-tack case, your prior in the coin toss case was *highly informative*. That is, your prior expressed a strong opinion that the heads probability is between 0.49 and 0.51. With such a strong prior, it would take a lot of coin tosses to sway your opinion that Hans' coin has a probability that is much different from 0.5.

Again you can find the range of plausible values for π via an equal-tailed credible interval using the BETA.INV function: You believe in a 5% posterior probability that π is less than BETA.INV(0.05, 11,252, 11,258) = 0.4944, and you believe in a 95% chance that π is less than BETA.INV(0.95, 11,252, 11,258) = 0.5054. Thus, you have a 90% posterior probability (given two successes in 10 tries) that the probability of success (π) lies between 0.4944 and 0.5054. In proper symbols, Pr(0.4944 $\leq \pi \leq$ 0.5054|2 heads and 8 tails) = 0.90.

Thus, the data have had little effect on your conclusions in the coin toss case: A priori, you thought that the value of π was close to 0.5, and a posteriori, you still think the value of π is close to 0.5. Following the calculations earlier using your prior values $\alpha = \beta = 11{,}250$, your prior uncertainty about π is expressed using the prior (before seeing the data) credible interval Pr(0.4945 $\leq \pi \leq$ 0.5055) = 0.90. Your prior 90% credible interval (0.4945 $\leq \pi \leq$ 0.5055) differs very little from your posterior 90% credible interval (0.4944 $\leq \pi \leq$ 0.5054); hence, the data barely changed your mind about the value of π. This is a sensible interpretation, given your belief that the coin is nearly fair.

Let's reiterate the difference in the 90% posterior credible intervals in the thumbtack toss example and the coin toss example: Your intervals that express your uncertainty about π are (0.079 $\leq \pi \leq$ 0.470) and (0.4944 $\leq \pi \leq$ 0.5054), even though the data (two successes out of 10 trials) are exactly the same in both cases!

To interpret data properly, you must incorporate your prior information.

There is a general formula that you can see from Examples 13.4 and 13.5. Suppose your data are produced as independent and identically distributed (iid) Bernoulli(π) and you have observed s successes (cases where $y_i = 1$) and $n - s$ failures (cases where $y_i = 0$). If your prior for π is the beta(α, β) prior, then your posterior distribution is also a beta distribution.

Posterior Distribution for the Bernoulli Parameter π When Using a Beta(α, β) Prior

$$\pi \mid \{s \text{ successes}, n - s \text{ failures}\} \sim \text{beta}(\alpha + s, \beta + n - s) \qquad (13.2)$$

In the thumbtack toss case, you used the beta(1, 1) or uniform prior distribution. Applying expression (13.2), your posterior distribution is beta(1 + 2, 1 + 8) or beta(3, 9). In the coin toss case, you used the beta(11,250, 11,250) prior distribution. Applying (13.2), your posterior distribution is beta(11,250 + 2, 11,250 + 8) or beta(11,252, 11,258).

13.6 Bayesian Analysis Using Simulation

In the earlier examples with the thumbtack and coin tosses, you were able to find the posterior distributions precisely. These distributions were examples of the *beta distribution*, a distribution so well-known that it is even available in Excel. Of course you can find it in any dedicated statistical software as well.

However, the formula $p(\theta \mid y) \propto L(\theta \mid y) \times p(\theta)$ usually does not produce a nice named distribution, such as the normal, beta, or gamma distribution. The earlier examples with the beta prior for the Bernoulli parameter π are examples of **conjugate priors**. A conjugate prior is one that has the same mathematical form as the likelihood and allows for convenient Bayesian analysis in that the posterior distribution has a named form that you can then analyze using standard software.

Your selection of a prior distribution should be guided by your prior opinion and not by mathematical convenience. In the past, Bayesian analysis was very difficult due to computer limitations, and conjugate priors were necessary. Today, software can perform Bayesian analysis for priors that are not necessarily conjugate, that is, for priors that truly express your prior knowledge.

Even with a conjugate prior, however, Bayesian analysis can be quite complex in multiparameter models, just as the likelihood-based analyses presented in Chapter 12 are more complex in multiparameter models. Historically, with conjugate priors in multiparameter models, you would have to use complicated multivariable calculus to understand the analysis of your data. On the other hand, the currently available software makes it easy to understand even the analysis of multiparameter models, conjugate priors or not, without requiring calculus.

What is this wonderful method? It's yet another application of simulations! The current approach to Bayesian analysis is to *simulate* parameters from the posterior distribution and then base inferences on the statistical summaries of the posterior simulations. A recurring theme of this book is that with enough simulated DATA* from a distribution, you can recover everything you need to know about that distribution: The sample average of the DATA* will get closer to the mean of the distribution, and the histogram of the DATA* will look more and more like the distribution itself. The posterior distribution $p(\theta \,|\, \text{data})$ tells you all you need to know about θ, and if you can simulate enough DATA* from $p(\theta \,|\, \text{data})$, then you know everything you need to know about the posterior distribution $p(\theta \,|\, \text{data})$.

To be more specific, suppose, in the thumbtack example, that you could not figure out the name of the distribution $p(\pi) \propto \pi^2(1-\pi)^8$, which is actually the beta(3, 9) distribution. Your statistical software is smart enough to know how to simulate data from such distributions without knowing their names—the software only needs the proportional form $\pi^2(1-\pi)^8$. With enough simulated DATA* from this distribution, you can still calculate the 90% equal-tailed credible interval for the thumbtack case by simulating many, say a million, π^* values from the distribution $p(\pi) \propto \pi^2(1-\pi)^8$ and then taking the lower limit and upper limit of the 90% credible interval to be the 5th and 95th percentiles of the 1,000,000 simulated π^* values.

For example, with 1,000,000 simulated π^* values from $p(\pi) \propto \pi^2(1-\pi)^8$, we found 5th and 95th percentiles 0.079 and 0.471, leading to a 90% credible interval ($0.079 \leq \pi \leq 0.471$). Using the precise beta(3, 9) distribution, the credible interval is ($0.079 \leq \pi \leq 0.470$), and you can see that the simulation-based approximation is quite good. The approximation will become even better with more than 1,000,000 samples, but this level of precision seems unnecessary for the problem at hand, particularly given your uncertainty about your prior specification.

There are many different methods that statisticians have developed for simulating from posterior distributions. They all start with the basic premise that your posterior is proportional to the likelihood multiplied by your prior, expression (13.1) earlier. All you need to do to apply these methods is supply to your software the model $p(y\,|\,\theta)$ for your data, your prior distribution $p(\theta)$, and your data set. The software will do the rest, simulating as many samples as you desire from the posterior distribution $p(\theta \,|\, y)$.

A common method used by many software products for such posterior simulation is **Markov chain Monte Carlo** (MCMC for short). Figures 13.6 and 13.7 show the simulated

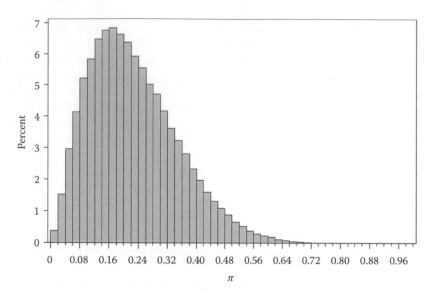

FIGURE 13.6
Simulated posterior distribution of π in the thumbtack example.

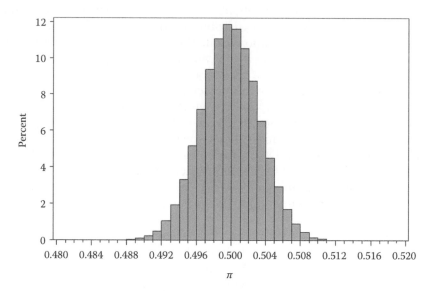

FIGURE 13.7
Simulated posterior distribution of π in the coin toss example.

posterior distributions, using MCMC, of π for the thumbtack and coin toss cases. The distributions are nearly identical to those shown in Figures 13.4 and 13.5.

As the coin toss example showed, your prior can have a dominating effect on your interpretation of data. Selecting a prior distribution $p(\theta)$ can be tricky, both from the standpoint of deciding "What do I know about θ?" and from the standpoint of deciding how to input your $p(\theta)$ into the computer software. In most empirical studies, you do not have such strong prior knowledge as in the coin toss. (If you do have such strong prior knowledge, then why are you bothering with collecting data?) The most common approach to Bayesian analysis is, therefore, to choose a **vague prior**—one that imposes little, if any, prior knowledge. The uniform U(0, 1) prior for π is an example of a vague prior.

When the parameter has an infinite range, as is most common (e.g., μ, σ, β_0, β_1 are parameters with infinite ranges), there is no bona fide uniform prior distribution because the area under the curve would be infinity, not 1.0. Nevertheless, uniform priors are still used in such cases; they are called **improper priors** because they are not really probability distributions at all. For example, the so-called uniform prior distribution for the mean of the normal distribution, $p(\mu) = 1$, for $-\infty < \mu < \infty$, is not a probability distribution at all because the area under the curve is $\int(1)d\mu = \infty$, which is a long way from the required value 1.0 for area under the curve of a pdf. Still, you can use improper priors as long as the posterior is a valid pdf. If you use the prior $p(\mu) = 1$, then the posterior is simply

$$p(\mu|\text{data}) \propto L(\mu|\text{data}) \times 1$$

The posterior distribution is thus the scaled likelihood function in this case, and it is a valid distribution when the integral of the likelihood function is finite.

Some purists do not like the idea of improper priors because they aren't distributions and therefore cannot model your prior thoughts correctly. Instead, they will choose a distribution with extremely large variance as a vague prior—such a prior is similar to a uniform prior, in that both state ambivalence, a priori, as to whether the parameter is a small number or a large number.

There are many vague priors; some are proper and some aren't. The prior $p(\sigma^2) = 1/\sigma^2$ is commonly used for variances; this prior corresponds to a uniform improper prior on $\ln(\sigma^2)$. The *Jeffreys prior* is another example of a vague prior.

A common feature of vague priors is that they "let the data talk for themselves." Such priors have little effect when they are multiplied by the likelihood function so that your posterior distribution is relatively unaffected by your prior. When you use a vague prior, you do not incorporate your subjective biases into the analysis of your data.

13.7 What Good Is Bayes?

Most software packages that do Bayesian analysis use vague priors by default. But if the resulting analysis is essentially a likelihood-based analysis, why should you care about Bayes? Why not just use likelihood-based analysis? In many cases, there really is not much difference. But whenever you need to *select* values of the parameters θ that are consistent with the data, then you need to use Bayes. You saw one example of this type of analysis way back in Example 1.9. There, the parameters θ were probabilities π_i of electoral wins in various U.S. states, and you saw how to select such values that were consistent with polling data. This analysis in turn allowed you to make projections as to the result of the election based on plausible values of the electoral college total, given the polling data.

In this chapter, you will see additional examples like this. Hopefully, the Bayesian logic will seem natural to you. If so, you may end up being surprised to learn in later chapters when we discuss classical (non-Bayesian) statistical methods that, with classical methods, you simply cannot perform the types of analyses shown in this section and in Example 1.9. And that's the answer to the question, "What good is Bayes?"

**Example 13.6: Using Bayesian Statistics to Quantify
Uncertainty in Logistic Regression Estimates**

In Chapter 12, we used the logistic regression model to estimate the probability of success in a workplace task as a function of job experience. The probability function was $\Pr(\text{success}|\text{experience} = x) = \exp(\beta_0 + \beta_1 x)/\{1 + \exp(\beta_0 + \beta_1 x)\}$. Figures 12.21 and 12.22 showed the likelihood function $L(\beta_0, \beta_1 | \text{data})$, and we used this likelihood function to show the uncertainty in the probability function: We selected plausible combinations (β_0, β_1) from the joint likelihood function; plugged them into the estimated success probability, $\exp(\beta_0 + \beta_1 x)/\{1 + \exp(\beta_0, \beta_1 x)\}$; and graphed the results in Figure 12.23.

The analysis shown in Figure 12.23 should have seemed strange to you. How can you pick values from the likelihood function? Which ones are the most likely ones you should pick? How many should you pick? Bayesian posterior simulation provides the answers to these questions, and you can do it using the following steps.

Selecting and Using Logistic Regression Parameter Values That Are Consistent with the Data

- Specify the logistic regression model using your software; this will allow your software to find the likelihood function $L(\beta_0, \beta_1 | \text{data})$.
- Select a vague prior for (β_0, β_1), unless you have strong prior knowledge. This might mean doing nothing at all, if you use the software defaults.
- Have your software generate a sample of pairs (β_0^*, β_1^*) from the posterior distribution $p(\beta_0, \beta_1 | \text{data})$ and draw a scatterplot.
- Draw the graphs shown in Figure 12.23, using the resulting sample, to see your uncertainty about the probability function $\exp(\beta_0 + \beta_1 x)/\{1 + \exp(\beta_0 + \beta_1 x)\}$.

Figure 13.8 shows a scatterplot of the resulting sample of 10,000 pairs (β_0^*, β_1^*) from the posterior distribution.

Notice how well Figure 13.8 matches the contour plot of Figure 12.22. Since the prior is vague, the values in Figure 13.8 are quite similar to samples from the scaled likelihood function, and the correspondence is therefore to be expected.

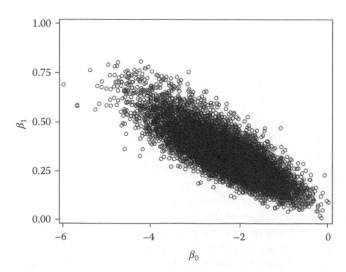

FIGURE 13.8
Scatterplot of 10,000 pairs (β_0^*, β_1^*) sampled from the posterior distribution of (β_0, β_1) in a logistic regression model.

Using these samples from the posterior distribution, you can refine the ad hoc analysis shown in Figure 12.23 to assess uncertainty about your probability function. Instead of picking a few values (β_0, β_1) that seem plausible as shown in Figure 12.23, a more systematic approach is to *generate* values (β_0, β_1) that *are* plausible, as shown in Figure 13.8, then use these to construct probability functions that are consistent with the observed data. Figure 13.9 shows probability functions of the form

$$\text{Pr*} (Y = 1 | x, \text{data}) = \frac{\exp(\beta_0^* + \beta_1^* x)}{1 + \exp(\beta_0^* + \beta_1^* x)}$$

These were calculated using 1s00 samples (β_0^*, β_1^*) from the posterior distribution $p(\beta_0, \beta_1 | \text{data})$.

The general upward trend is confirmed using the samples from the posterior, but the probability of success when $X = 20$, or $\exp\{\beta_0 + \beta_1(20)\}/[1 + \exp\{\beta_0 + \beta_1(20)\}]$, is clearly very uncertain, as seen in Figure 13.9 from the vertical variability in the "spaghetti" at the upper right of the plot.

You can easily construct a credible interval for this probability. Just calculate the following:

$$\pi_{20}^* = \frac{\exp\{\beta_0^* + \beta_1^* (20)\}}{1 + \exp\{\beta_0^* + \beta_1^* (20)\}}$$

Repeat for each of the sampled pairs (β_0^*, β_1^*) from the posterior distribution, and then take the 5th and 95th percentiles of the resulting π_{20}^* values to obtain an equal-tailed 90% credible interval for the true π_{20}. Using 10,000 sampled pairs (β_0^*, β_1^*) from the posterior distribution, the 5th and 95th percentiles of the π_{20}^* values are (in our simulation) 0.861 and 0.999, so a 90% credible interval for the true probability of success at a task for employees with 20 years of experience is $0.861 \le \pi_{20} \le 0.999$. Compare this interval with the vertical variability in the spaghetti at the upper right of Figure 13.9: You can see that the vertical range of the spaghetti at $x = 20$ extends beyond the limits $0.861 \le \pi_{20} \le 0.999$, particularly on the low side. That's okay—in statistics you don't need to be 100% sure about anything. If you are 90% sure or 95% sure, that's often good enough.

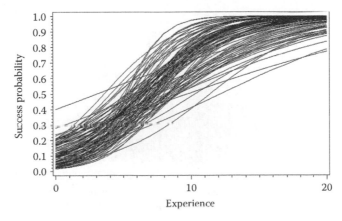

FIGURE 13.9
Probability functions that are consistent with the observed data, using 100 pairs (β_0^*, β_1^*) sampled from the posterior distribution.

Notice the chain of modeling steps you have to use for Bayesian analysis: First, you specify a model—a two-parameter logistic regression model—along with a prior distribution. Then, you generate thousands of parameter vectors (β_0^*, β_1^*) from the posterior distribution, and for each one, you construct an estimated model relating experience to probability of success. The estimated models are random because the (β_0^*, β_1^*) pairs are random. If this seems confusing to you, go back and reread the previous paragraphs. Better yet, do it using some software! We'll wait.

Ready to go on to another example? Okay!

Example 13.7: Using Bayesian Statistics to Quantify Uncertainty in the Estimates of Mean and Standard Deviation of Student Ages

In Chapter 12, we used the normal distribution to construct the likelihood function for (μ, σ) of the age data for students in a class and came up with maximum likelihood estimates $\hat{\mu} = 26.81$ years and $\hat{\sigma} = 4.79$ years. The likelihood function shows the ranges of uncertainty about the true values (μ, σ), but it is even easier to show the uncertainty by sampling from the posterior distribution. Using software, with a default (vague) prior, Figure 13.10 shows a scatterplot of 10,000 pairs (μ^*, σ^*) sampled from the posterior distribution.

While the scatterplot of Figure 13.10 suggests much uncertainty about the process mean and standard deviation, what you see in the graph are the outliers. The equal-tailed 90% credible intervals using the data graphed in Figure 13.10 show much less variability; they are $(24.66 \leq \mu \leq 28.99)$ and $(3.48 \leq \sigma \leq 5.98)$, which correspond to the 5th and 95th percentiles of the simulated μ and σ values.

Let's revisit the Mantra as it applies to Example 13.7 as follows:

Model produces data: The model that produces the 16 age values is assumed to be the iid normal model;

Model has unknown parameters: The parameters of the normal distribution are μ and σ; these are unknown.

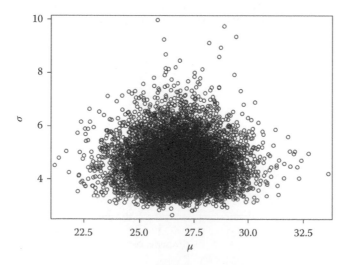

FIGURE 13.10

Scatterplot of 10,000 pairs (μ^*, σ^*) sampled from the posterior distribution of (μ, σ) of the age distribution, assuming the distribution is normal and using a vague prior.

Data reduce the uncertainty about the parameters: The credible intervals for μ and σ show the reduction in uncertainty.

We can now make the "reduction in uncertainty" concept more explicit. Before collecting data, you are uncertain about the parameters. (If you weren't uncertain about the parameters, you wouldn't need to collect any data at all!) You express your uncertainty before seeing the data using your prior distribution. After seeing the data your uncertainty is reduced, as shown by your posterior distribution, which is usually much narrower than your prior.

Bayesian methods are ideal for learning about process parameters when you have a small data set. The following example shows how.

Example 13.8: Bayesian Estimation of Value at Risk Using a Small Sample

Value at Risk (or VaR) in finance is the 95th percentile of the distribution of loss on an investment. Technically, since the loss or earnings occur in the future, you can interpret VaR as follows: "Among all potential futures, in 95% of them my loss will be less than the VaR." Since no one has access to data from the future, you have to estimate VaR from current relevant data and hope they make a reasonable proxy for the potential future data that you might see.

Suppose you have collected loss data from $n = 14$ current ventures where loss data are available. The numbers are −12, 5, −22, −1, −6, 1, −32, −6, −6, −7, −6, −3, −20, −14, all in millions of Swiss francs. The negative losses are gains; so, unlike most financial applications, the negative numbers here mean that somebody *made* some money rather than *lost* money.

What is the 95th percentile? It is the value that puts 95%, or 19 out of 20 of the data values to its left, when viewed on a number line. But here, with only 14 numbers, it is hard to understand what "19 out of 20" means.

If you assume that a normal distribution $N(\mu, \sigma^2)$ produced the loss data, then the 95th percentile is $\mu + 1.6445\sigma$, where 1.6445 is the 0.95 quantile of the standard normal (or $N(0, 1)$) distribution. You can estimate this 95th percentile as $\hat{\mu} + 1.6445\hat{\sigma} = -9.214 + 1.6445 \times 9.955 = 7.157$, but how certain can you be of this estimate? After all, 100% of your sampled losses are below this number, so you know that 7.157 can't be the true 95th percentile of the loss distribution.

Bayesian statistics to the rescue! It's pretty easy: Just sample many (μ^*, σ^*) from the posterior distribution of (μ, σ); given the $n = 14$ observations, calculate the plausible VaR values $\mu^* + 1.6445\sigma^*$ for each sample, and find the 90% credible interval from the resulting simulated data. Figure 13.11 shows the histogram of the plausible VaR values using the default prior of a statistical software package.

The 5th percentile and the 95th percentile of the data graphed in Figure 13.11 are −0.24 and 12.57, respectively therefore, the Bayesian 90% equal-tailed credible interval for VaR is estimated to be (−0.24, 12.57). Remember, VaR is defined as the 95th percentile of the loss distribution, so the graph in Figure 13.11 is a graph of plausible 95th percentiles, given the observed data.

There are quite a few assumptions in effect here, all of which are embedded in the following interpretation:

Assuming that these current loss data are an iid sample from potential future losses, and assuming these losses are produced from a normal distribution, *and assuming that you have no prior knowledge about the mean and standard deviation* (μ, σ) *of the earnings process*, then there is a 90% probability that the VaR of your future investment lies in the range (−0.24, 12.57).

So, based on past data and your assumptions, you should be prepared to lose up to 12.57 million Swiss francs.

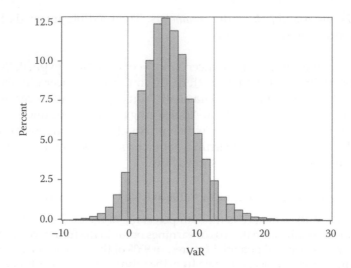

FIGURE 13.11
Histogram of plausible VaR values, given observed earnings data ($n = 14$), with 90% equal-tailed credible interval shown by vertical lines.

Example 13.9: Producing Plausible Potential Future Stock Price Trajectories Using Bayesian Analysis

All the way back in Chapter 1, in Example 1.8, we tried to convince you that the probabilistic *model produces data* concept was very useful to answer questions such as, "Should I buy a stock and hold it, or should I trade it based on previous price movements?" We did this by producing 1000 potential futures using the model $Y_t = Y_{t-1}(1 + R_t)$, where Y_t is the stock price on day t and R_t is the return from day $t - 1$ to day t. The numbers R_t were produced as iid from the N(0.001, 0.03²) model; that is, we assumed that the day-to-day returns come from a normal distribution with mean $\mu = 0.001$ and standard deviation $\sigma = 0.03$.

But we didn't really know μ and σ—nobody does, not even all the Nobel laureates in economics combined—so we suggested that sensitivity analysis was prudent, wherein you choose different μ and σ and redo the analysis, looking for consistency in your conclusions.

But how can you select the different (μ, σ) pairs to use in the sensitivity analysis? Bayesian posterior simulation provides the simple answer: Choose them from the posterior distribution of (μ, σ) given historical data! The question then arises, *which* historical data? You can use all historical data on returns going back as far in time as data exist, but performance far in the past may not be as relevant for future performance, so this is not such a good idea. If you want your posterior distribution to be meaningful for potential future trajectories of the stock price, you'll have to assume that the returns you sample are iid from a process $p(r)$ *and that the future returns are sampled from this same process*. This assumption is more palatable if the data you select are from the recent past rather than ancient history.

We picked 1 year of daily returns on the Dow Jones Industrial Average (DJIA), from April 1, 2010 to March 1, 2011, for our sample. Using these data, there were $n = 230$ returns r_t, which we modeled as iid from N(μ, σ^2). Using software that performs Bayesian posterior simulation, with the (vague) Jeffreys prior, we simulated a sample of 1000 pairs (μ^*, σ^*) from the posterior distribution and plotted them as a scatterplot in Figure 13.12.

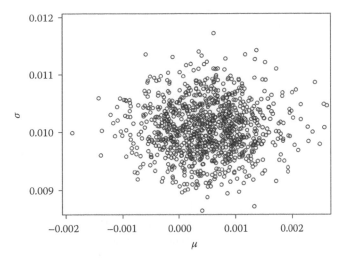

FIGURE 13.12
Scatterplot of 1000 pairs (μ^*, σ^*) sampled from the posterior DJIA return distribution, given data from April 1, 2010 to March 1, 2011.

Based on the scatterplot, it is not obvious that the mean return μ is necessarily positive, since a large proportion of the points graphed have a μ^* that is less than 0. This is bad news for investors! If the mean return really is negative, then putting your money in the stock is just like gambling at the casino: You expect to lose in the long run. However, the Bayesian analysis shown in Figure 13.12 used a vague prior. A financial analyst will likely have a more informative prior that will change the posterior. If the analyst thinks we are in a bad economic period, his or her prior will move the plausible mean returns to the left, farther into the negative territory of Figure 13.12. If the analyst thinks we are in a good period, his or her prior will move the plausible mean returns to the right, farther into positive territory of Figure 13.12.

You can generate a "future" plausible trajectory of the Dow Jones Index by taking a single pair (μ^*, σ^*) from the posterior (i.e., one of the points in Figure 13.12), generating "future" return values R_t^* as iid $N(\mu^*, (\sigma^*)^2)$, then generating "future" values of the Dow Jones Y_t^* as $Y_t^* = Y_{t-1}^*(1 + R_t^*)$. Of course, there are many plausible future trajectories Y_t^*, so you should repeat this procedure many times to see the range of plausible futures. This is precisely the analysis shown in Chapter 1, in the spreadsheet analysis, except that now each of the columns A through J is a random sample from a different normal distribution. Each column corresponds to a different (μ^*, σ^*) sampled from the posterior; that is, column A might be returns sampled from $N(0.00073, (0.0102)^2)$, column B might be returns sampled from $N(-0.00040, (0.0099)^2)$, etc. Otherwise, the plausible future trajectories are calculated exactly as before, only now using plausible values of (μ^*, σ^*) that are consistent with the historical data rather than using the setting $(\mu, \sigma) = (0.001, 0.030)$, which we chose simply as convenient values to illustrate the general idea.

The word *"future"* is in quotes because all this happens to be in the past by the time you are reading this. We'll start the "future" trajectories with $Y_0 = 12{,}058.02$, the value of the DJIA index on March 1, 2011. Figure 13.13 shows "future" trajectories of the DJIA for 30 trading days immediately after March 1, 2011, with the actual values of the DJIA for those 30 days indicated in bold dotted line.

From Figure 13.13, you can see that the Bayesian method works well to project the range of potential futures and that the actual "future" (now the actual past) falls mostly within the range of possibilities projected by the model.

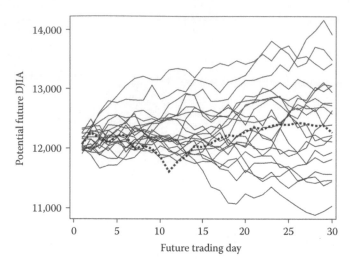

FIGURE 13.13
Projections of 20 potential future trajectories of the DJIA over 30 trading days, beginning March 1, 2011, and using Bayesian posterior simulation from the predictive distribution. The actual trajectory of the DJIA during this period is shown as a bold dotted line.

Example 13.10: Deciding Whether to Continue with Product Development

Suppose you wish to market a new drug to cure a disease. Based on your most recent study, 67 out of 98 diseased patients were cured using your drug, whereas 66 out of 101 diseased patients were cured using the current therapy that is on the market. The results seem nearly identical; should your company continue to develop the product or should they stop development?

To answer the question, let's make some simplifying assumptions. The first is that the cost of continued development will be 10 million (say, millions of euros). The second is that your company's gross profit will be 10 times the difference between the cure percentages between your drug and the standard therapy—that is, if your cure rate is 70% and the competitor's is 65%, then your profit will be 50. If the percentages are 75% and 65%, then your profit will be 100, reflecting increased demand. If your cure rate is less than the competitors', then your profit will be zero, and you will not recoup the development costs. Thus, your company's net profit, should you continue development, is given as

$$\text{Net Profit} = 10(100\pi_1 - 100\pi_0) - 10, \quad \text{if } \pi_1 > \pi_0$$

$$\text{Net Profit} = -10, \quad \text{if } \pi_1 \leq \pi_0$$

From the data, the estimated cure rates are $\hat{\pi}_1 = 67/98 = 0.684$ and $\hat{\pi}_0 = 66/101 = 0.653$, giving an estimated net profit of $10(68.4 - 65.3) - 10 = 21.0$ million euros. This sounds good, but what if the true difference $\pi_1 - \pi_0$ is negative? What is your risk if you continue with development? To answer, you can simulate thousands of plausible values π_1^* and π_0^* that are consistent with your observed data, then use these to construct plausible values of your net profit, given your observed data.

You can use Bayesian software to do this, but the solution is simple enough so that you can do it in Microsoft Excel. Consider expression (13.2), which states that the posterior

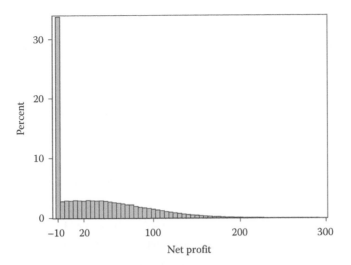

FIGURE 13.14
Histogram of potential future net profit based on an initial study.

distribution of the Bernoulli parameter is $\pi \mid \{s \text{ successes}, n - s \text{ failures}\} \sim \text{beta}(\alpha + s, \beta + n - s)$. If you assume independent uniform (beta(1, 1)) priors for both π_1 and π_0, then your posteriors are $\pi_1 \sim \text{beta}(68, 32)$ and $\pi_0 \sim \text{beta}(67, 36)$. You can simulate values π_1^* and π_0^* in Excel by generating two columns of U(0, 1) variables, say in columns A and B, and then using the inverse cumulative distribution function (cdf) method along with the BETA.INV function. Create π_1^* values in column C as BETA.INV(A1, 68,32), and create π_0^* values in column D as BETA.INV(B1, 67,36). Then create your net profit in column E as Net Profit $= 10(100 \, \pi_1^* - 100 \, \pi_0^*) - 10$, if $\pi_1^* > \pi_0^*$, and Net Profit $= -10$ otherwise.

Figure 13.14 shows the result of 100,000 such simulated values.

The average net profit over the 100,000 simulated scenarios is 34.3 million euros. Since this is a positive number, it suggests that your company ought to proceed with development. However, there is also a large probability that your company will lose the entire 10 million euros if they continue, as shown by the large spike having probability approximately 33% in Figure 13.14. The decision process at this point will depend on your company's product portfolio: If this is one of many in the pipeline, then it makes sense to continue. On the other hand, if the company simply cannot afford the 10 million loss, then it may be prudent to shift resources to other more promising products.

Again, what good is Bayes? The answer is that you cannot select parameter values θ^* any other way and so can't do the analyses shown in this section any other way. In the following chapters, we will present the alternative, classical methods, also called *frequentist* methods. The disadvantage of frequentist methods is that they are limited—you cannot perform the analyses shown in this section using frequentist methods. On the flipside, an advantage of using frequentist methods is that you do not have to make as many assumptions. Another advantage is that frequentist methods provide simple, intuitive ways for you to assess the effects of chance variation on your data. Also, from a pragmatic view, you are likely to see many statistical analyses that have been done using the classical frequentist approach, so you should understand frequentist methods, even if you prefer to use Bayesian methods.

Vocabulary and Formula Summaries

Vocabulary

Prior information	What you know before conducting a study.
Prior distribution	The probability distribution of θ, denoted $p(\theta)$, which currently exists in your mind, before observing the next data values.
Posterior distribution	The probability distribution of θ, denoted $p(\theta \mid \text{data})$, that exists in your mind after observing some new data.
Posterior mean	The mean of your posterior distribution.
Posterior median	The median of your posterior distribution.
Bayesian standard error	The standard deviation of your posterior distribution.
Bayesian credible interval	See credible interval below.
Bayesian hypothesis testing	Calculating the probability that a parameter is in some range of interest, using your posterior distribution.
Bayesian predictive distribution	The distribution of plausible future data Y, produced from $p(y \mid \theta)$, where θ itself is sampled from the posterior distribution $p(\theta \mid \text{data})$.
Bayesian decision analysis	Use of the posterior distribution $p(\theta \mid \text{data})$ to make optimal decisions.
Kernel	The part of a probability distribution that depends on the variable.
Beta distribution	A two-parameter continuous probability distribution used to model variables that lie in the interval from 0 to 1.
Credible interval	A range of values within which the parameter θ is believed to fall with a specified probability.
Equal-tailed credible interval	An interval that puts equal probability of noncoverage in each tail of the posterior distribution.
HPD interval	A credible interval containing values whose posterior density is larger than a constant, and for which $\Pr(l \leq \theta \leq u \mid \text{data}) = 1 - \alpha$.
Conjugate prior	A prior distribution that has the same mathematical form as the posterior distribution formed from it.
MCMC	A computer-based method of simulating from posterior distributions.
Vague prior	A prior distribution used to model the case where you have little prior information about θ.
Improper prior	A function used as a prior that is not a true probability distribution because the total area under the curve is not 1.

Key Formulas and Descriptions

$p(\theta \mid y) \propto p(y \mid \theta)\, p(\theta)$	Bayes' theorem, with one y data value.
$p(\theta \mid y_1, y_2, ..., y_n) \propto p(y_1, y_2, ..., y_n \mid \theta)\, p(\theta)$	Bayes' theorem, with a sample of n data values.
$p(\theta \mid y_1, y_2, ..., y_n) \propto L(\theta \mid y_1, y_2, ..., y_n)\, p(\theta)$	Bayes' theorem stated in terms of the likelihood function.
$c = 1/\{\Sigma_\Theta\, L(\theta \mid y_1, y_2, ..., y_n) \times p(\theta)\}$	The constant of proportionality for the posterior distribution of a discrete parameter θ.
$c = 1/\{\int_\Theta L(\theta \mid y_1, y_2, ..., y_n) \times p(\theta)\, d\theta\}$	The constant of proportionality for the posterior distribution of a continuous parameter θ.
$\alpha/(\alpha + \beta)$	The mean of the beta(α, β) distribution.
$\alpha\beta/\{(\alpha + \beta)^2(\alpha + \beta + 1)\}$	The variance of a beta(α, β) distribution.
$\pi \mid \{s$ successes, $n - s$ failures$\} \sim \beta(\alpha + s, \beta + n - s)$	The posterior distribution of the Bernoulli parameter, assuming a beta(α, β) prior and an iid sample with s successes and $n - s$ failures.
$\Pr(l \le \theta \le u \mid \text{data}) = 1 - \alpha$	The interpretation of a $100(1 - \alpha)\%$ credible interval for θ.
$p(\mu) = 1$, for $-\infty < \mu < \infty$	A uniform, improper prior for the mean μ.
$p(\sigma^2) = 1/\sigma^2$, for $\sigma^2 > 0$	An improper prior for the variance σ^2.
$\mu + 1.6445\sigma$	The 95th percentile of the $N(\mu, \sigma^2)$ distribution.

Exercises

13.1 Consider the data from Exercise 12.1.

    A. Find and graph your posterior distribution for π when you assume a uniform prior. How does this graph differ from the likelihood function?

    B. Graph the beta(9, 1) prior for π. When would you use this prior?

    C. Find and graph your posterior distribution for π using the beta(9, 1) prior. Compare the result with the graph in Exercise 13.1A and comment on the effect of your prior.

13.2 Consider the data from Exercise 12.5.

    A. Suppose you use the prior $p(\lambda) = 0.01\exp(-0.01\lambda)$. State your posterior distribution for λ, without finding the constant of proportionality, c.

    B. Graph the function in Exercise 13.2A and compare it to the likelihood function. Does this prior have much effect?

    C. Repeat Exercise 13.2A and B using the prior $p(\lambda) = 1000\exp(-1000\lambda)$. Which prior is more informative?

13.3   Consider the case of the unknown survey but that there were $n = 10$ surveys that got separated from the pile, with responses 4, 4, 4, 5, 4, 2, 4, 4, 4, 4. You may assume that the data come from an iid sample. Find the posterior distribution of θ = company, assuming a uniform prior.

13.4   Suppose that $Y|\theta \sim N(\theta, 1)$ and that $p(\theta) = 1$ for $-\infty < \theta < \infty$.

A.   Why is this prior called improper?

B.   Find the distribution of $\theta|Y = y$.

13.5   Example 12.4 showed the likelihood function $\pi_1^{392}\pi_2^{401}(1 - \pi_1 - \pi_2)^{209}$. Suppose the prior is $p(\pi_1, \pi_2) = 1$.

A.   What is the posterior distribution of (π_1, π_2)? (You don't need to find the constant of proportionality.)

B.   How is the posterior distribution in Exercise 13.5A similar to Figure 12.5? How is it different from Figure 12.5?

13.6   The maximum likelihood estimate is equal to the mode of the posterior distribution, assuming an improper uniform prior. Explain why this is true.

13.7   The stoplight green signal duration example of Chapter 4 specified a U(0, θ) distribution for Y = the amount of time the light stays green. Suppose you observe one value $Y = 0.6$.

A.   Assuming a U(0, 5) prior distribution for θ, find the posterior distribution of θ using calculus.

B.   Find the 95% equal-tailed interval for θ using the posterior distribution in Exercise 13.7A.

C.   Find the 95% HPD interval for θ using the posterior distribution in Exercise 13.7A.

D.   Suppose you observe a sample of values, assumed to be produced as iid from U(0, θ): $y_1 = 0.6$, $y_2 = 0.8$, $y_3 = 0.1$, $y_4 = 0.1$, $y_5 = 0.4$, $y_6 = 0.3$, and $y_7 = 0.3$. Repeat Exercise 13.7A through C and compare.

13.8   You can use the central limit theorem to perform approximate Bayesian analysis. Often, estimators are approximately normally distributed so that $\hat{\theta}|\theta \sim N(\theta, \hat{\sigma}^2)$. You can turn this around to state that $\theta|\hat{\theta} \sim N(\theta, \hat{\sigma}^2)$, where $\hat{\sigma}$ is the standard error of the estimate, when you assume a vague prior on θ. In Example 10.2, the mean and standard error of the DJIA return are given.

A.   State and graph the approximate posterior distribution of the mean DJIA return, μ, assuming a vague prior.

B.   Use the distribution in Exercise 13.8A to compute the approximate posterior probability $Pr(\mu < 0|data)$.

C.   Why is the probability in Exercise 13.8B interesting to investors?

13.9   Use the data in Exercise 12.8. Use software that simulates β_0 and β_1 from the posterior distribution, assuming a vague prior.

A.   Estimate the posterior probability $Pr(\beta_1 > 0|data)$ using your data sampled from the posterior distribution. Use this result to answer the question in Exercise 12.8C.

B.   Find the equal-tailed 90% credible interval for $Pr(Y = 1|X = 5.0)$ using your data sampled from the posterior distribution.

13.10 Use the data and the heteroscedastic model of Exercise 12.9. Use software that can simulate β_0, β_1, and σ from the posterior distribution, assuming a vague prior.

    A. Estimate $\Pr(\beta_1 > 0 | \text{data})$.

    B. Estimate $E(Y|X = 10) = \beta_0 + \beta_1(10)$ using the posterior mean.

    C. Find a 90% credible interval for $E(Y|X = 10)$.

    D. Estimate the probability that $E(Y|X = 10)$ is greater than 10.

13.11 See Exercise 12.11. Use calculus and a uniform prior.

    A. Find the mean and the median of the posterior distribution for θ.

    B. Compare the estimates in Exercise 13.11A with the MLE you found in Exercise 12.11B. In what way are the estimates in Exercise 13.11A more sensible than the MLE?

13.12 Use the data and the case study in Exercise 12.12. Draw a graph a function that is proportional to the posterior distribution of θ assuming a uniform prior.

13.13 Use the data and scenario of Exercise 12.13. Use software that simulates from the posterior distribution, along with different priors. How does the quasi-separation problem appear in terms of the simulated values of β_1?

13.14 Use the data and model from Exercise 12.14, and use software that simulates from the posterior distribution using a vague prior.

    A. Estimate $\Pr(\beta_1 > 0 | \text{data})$.

    B. Find a 90% credible interval for the mean number in the shortest line when $X = 2$.

    C. Suppose there are $X = 2$ lines. Estimate the predictive distribution of Y as follows: (i) Simulate a (β_0, β_1) combination from the posterior distribution, (ii) simulate a Y^* from the appropriate Poisson distribution, (iii) repeat (i) and (ii) 10,000 times, and (iv) summarize the data and interpret the results for the supermarket's manager.

14

Frequentist Statistical Methods

14.1 Introduction

Hans flips a fair coin, launches it high, and lets it fall where you can't see it. Then he steps on it. It's already flipped, but you can't see the result. He asks you, "What is the probability that the coin has landed with heads up?" Intuitively, you would answer as follows:

> Hans, I think there is a 50 percent probability that your coin has landed with heads up.

But you will not answer that way if you are a **frequentist**! If you are a frequentist, you will answer as follows:

> Hans, since you have already flipped the coin, its outcome is no longer random. The probability is either 100% or 0%, depending upon whether your coin landed heads up or tails up. I know that in the long run, over many repetitions of you flipping the coin and stepping on it, the coin will land heads up 50% of the time. Therefore, Hans, I am 50% confident that your coin has landed heads up on this particular occasion.

Sounds like a long way to go to describe what is essentially a 50% probability! Further, the two interpretations seem hardly different: One statement uses the word **probability**, and the other uses the word **confidence**. Yet the distinction between *probability* and *confidence* is very important to those who adhere to the frequentist camp of statistics. For frequentists, once the data are observed (i.e., once DATA turn into data) nothing is random, and no probability statements can be made.

Bayesians also agree that once the data are observed they are no longer random. But the Bayesian describes his or her uncertainty about the unknown parameter(s) θ using probability, because the Bayesian thinks the parameter's potential values are described by a posterior probability distribution.

On the other hand, once the data are collected, the frequentist does not describe his or her uncertainty about parameters using any probability distribution. In fact, a frequentist would argue that the entire previous chapter on Bayesian statistics is meaningless, because you cannot assign probability distributions to the unknown parameters θ. To the frequentist, the parameters are fixed—perhaps fixed at 0.5 for a fair coin, perhaps fixed at some other unknown π in the case of a bent coin, but fixed in either case. Figure 14.1 shows the difference between how a Bayesian and frequentist conceive of the value of the unknown probability π that a bent coin lands with heads facing up.

The Bayesian perhaps looks at the coin, sees the way it is bent, and thinks that heads will land somewhat less often. She then gives her prior as shown in Figure 14.1, a distribution

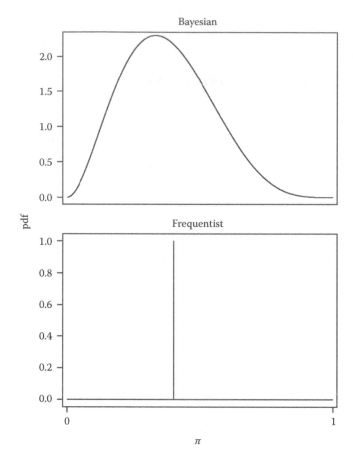

FIGURE 14.1
Bayesian and frequentist conceptions of an unknown parameter π.

that reflects her opinion that heads will occur less often than in 50% of flips; but she is not absolutely sure, so she allows some prior opinion that it could also be more than 50%.

The frequentist, on the other hand, will do nothing of the sort! Even after looking at the coin, she states simply that π is some fixed number whose value she does not know. She is not willing to give a formal guess as to whether π is more than 0.50 or less than 0.50. In Figure 14.1, we showed her π to be less than 0.50, but she actually thinks π might be anywhere in the 0–1 range. She thinks that π is some fixed number and that π has 100% probability of being equal to whatever value π is in fact equal to.

So, from the frequentist side of the fence, after observing the data, there is nothing random: The data are not random, and the parameters are not random. So if you are a frequentist, how can you state your uncertainty about the parameters following the observation of data? You do this in the same way as described earlier with Hans' coin toss: You envision repeated samples from the process. If, in repeated samples of tossing the coin and stepping on it, the coin turns up heads in 50% of the repeated samples, then you feel "50% confident" that the coin under the shoe will show heads on this particular occasion. By the same token, if your interval for θ is correct in 95% of repeated samples from the process, then you feel "95% confident" that it is also correct in your given sample. In this sense, you can be "95% confident" in the results, but as a frequentist, you will *not* say that the result is "true with 95% probability." Instead, you will interpret a 95% interval such as $56.2 \leq \theta \leq 76.9$ as follows:

> In 95% of repeated samples of size n from the same process, similarly constructed intervals will give different upper and lower limits, because every sample produces a different data set. However, 95% of these intervals will capture the true θ, so I am 95% confident that the interval $56.2 \leq \theta \leq 76.9$ is correct. In other words, I am 95% confident that the process parameter θ lies between 56.2 and 76.9.

While frequentist statistical methods might sound a little weird, they have historically been the standard methods. If you took a statistics class before this one, most likely you learned statistics from a frequentist viewpoint. And, in many ways, the frequentist methods are great.

Advantages of Frequentist Methods over Bayesian Methods

- You do not have to assume any priors $p(\theta)$.
- It is easier to use generic distributions $p(y)$ that make no assumption of distribution form (normal, Poisson, etc.).
- It is easier to understand frequentist methods using simulation.
- It is easier to validate frequentist methods using simulation.

The following sections illustrate these concepts.

14.2 Large-Sample Approximate Frequentist Confidence Interval for the Process Mean

Return to the data on ages of students (in years) from Example 11.5. The data are $y_1 = 36$, $y_2 = 23$, $y_3 = 22$, $y_4 = 27$, $y_5 = 26$, $y_6 = 24$, $y_7 = 28$, $y_8 = 23$, $y_9 = 30$, $y_{10} = 25$, $y_{11} = 22$, $y_{12} = 26$, $y_{13} = 22$, $y_{14} = 35$, $y_{15} = 24$, and $y_{16} = 36$. Suppose you want to use these data to construct an interval range around the parameter μ, the mean of the process that produced these data, but you don't want to make the questionable normality assumption that you made in Chapters 12 and 13. Instead, you'd rather assume that the distribution is generic, having the form shown in Table 14.1.

The parameter vector, θ, in this example is the list of the unknown probabilities:

$$\theta = \{\dots, \pi_{21}, \pi_{22}, \pi_{23}, \pi_{24}, \pi_{25}, \pi_{26}, \dots\}$$

You should always be skeptical about assumptions. The generic distribution shown in Table 14.1 does not assume that the π_y follow some restrictive distribution such as Poisson or discretized normal.

The mean of the distribution is $\mu = \Sigma_{\text{all } y}\, y \times \pi_y$. You want to estimate μ using the data, and you want to provide a range of uncertainty around your estimate.

While the generic distribution shown in Table 14.1 is more believable than, say, a normal distribution, you are not out of the woods yet. There is no free lunch! You always have to

TABLE 14.1

Generic Distribution for Age of
Students in a Graduate Class

Age, y	$p(y \mid \theta)$
...	...
21	π_{21}
22	π_{22}
23	π_{23}
24	π_{24}
25	π_{25}
26	π_{26}
...	...
Total	1.0

make some assumptions. Here, you'll have to assume that the observations are produced according to some probabilistic process, so go ahead and assume that the data are produced as independent and identically distributed (iid) from the distribution $p(y \mid \theta)$ shown in Table 14.1. But realize, as discussed in Chapter 7, that the iid assumption can and should be questioned. If the assumption is grossly violated—for example, if everyone decided to get together and report the same age for some reason (violating the independence assumption)—then stop reading now! If the iid assumption is violated, then the logic, conclusions, and methods for the remainder of this section are simply wrong.

So, before you have the data $y_1 = 36$, $y_2 = 23$, ..., $y_{16} = 36$, you assume the DATA $Y_1, Y_2, ..., Y_{16}$ will be produced as an iid sample from the pdf $p(y \mid \theta)$ shown in Table 14.1. The mean of this pdf is $\mu = \Sigma_{\text{all } y}\, y \times \pi_y$, and its variance is $\sigma^2 = \Sigma_{\text{all } y}\, (y - \mu)^2 \times \pi_y$. Thus, the future $\bar{Y} = (Y_1 + Y_2 + \cdots + Y_{16})/16$ that you will calculate is a random variable, and its distribution has mean μ and variance $\sigma^2/16$, as discussed in Chapter 10. These conclusions depend heavily on the iid assumption: Chapter 10 showed you how to trace the mathematical logic precisely from the iid assumption to the factual conclusions $E(\bar{Y}) = \mu$ and $\text{Var}(\bar{Y}) = \sigma^2/n$. There are logical steps along the way that require both independence and identical distributions, and if either assumption is violated, then these conclusions are simply wrong.

Yet another logical consequence of the iid assumption is that the distribution of $\bar{Y}$ is approximately a normal distribution by the central limit theorem (CLT). Putting it all together: *If the data are an iid sample from $p(y \mid \theta)$, then you can conclude that the distribution of the random variable $\bar{Y}$ is approximately a normal distribution with mean μ and variance $\sigma^2/16$.* In symbols:

$$\text{If } Y_1, Y_2, ..., Y_{16} \sim_{iid} p(y \mid \theta), \quad \text{then } \bar{Y} \dot{\sim} N(\mu, \sigma^2/16)$$

By the 68–95–99.7 Rule, approximately 95% of the random $\bar{Y}$ values will fall within two standard deviations of the mean. But here is a point of possible confusion: *Which* standard deviation? There is the σ that refers to the standard deviation of the age distribution—the distribution of Y—and there is a standard deviation of the possible values of $\bar{Y}$ that will be observed; this one is $(\sigma^2/16)^{1/2} = \sigma/\sqrt{16}$.

There is a big difference between σ and $\sigma/\sqrt{16}$. If the sample size was much larger, say 10,000, there would be an even bigger difference: σ versus $\sigma/\sqrt{10000}$. So it is important to get this one right! To understand the difference clearly, consider the 68–95–99.7 Rule in terms of some normally distributed but otherwise generic random variable W. You can state that 95% of the values of W will be within the range $E(W) \pm 2\text{StdDev}(W)$. Other than being normally distributed, it doesn't matter what is W, this statement is true. So in particular, substituting $\bar{Y}$ for W, you can state that 95% of the values of $\bar{Y}$ will be within the range $E(\bar{Y}) \pm 2\text{StdDev}(\bar{Y})$. Now it should be clear: Approximately 95% of the values of $\bar{Y}$, not Y, will be within the range $\mu \pm 2\sigma/\sqrt{16}$.

Another way to understand the distinction between σ and $\sigma/\sqrt{n}$ is to consider again what happens with a much larger n: If $n = 10{,}000$, then you know that the $\bar{Y}$ values that are calculated from those 10,000 observations will all be very close to the mean μ, by the law of large numbers (LLN). Then it makes sense that 95% of the values of $\bar{Y}$ will be within the narrow range $\mu \pm 2\sigma/\sqrt{10000}$ for such a large sample size n. It makes no sense at all that 95% of the Y values will be in the range $\mu \pm 2\sigma/\sqrt{10000}$: If the range of Y values is uniform from 0 to 100, for example, it makes no difference how many Y values you might observe, their range is still 0–100, not some narrow range like 49.4–50.6.

The CLT is a remarkably useful theorem. Even though the real age distribution shown in Table 14.1 is obviously non-normal, mainly due to right skewness induced by the occasional older, nontraditional student, you can count on the fact that the distribution of the average age $\bar{Y}$ will be approximately normal, and you can therefore count on the fact that this average, $\bar{Y}$, will be in the range $\mu \pm 2\sigma/\sqrt{n}$ for approximately 95% of the observable samples $Y_1, Y_2, \ldots, Y_{16}$.

Since this phrase "95% of the observable samples $Y_1, Y_2, \ldots, Y_{16}$" is key to understanding frequentist inferences, it needs more explanation. At the risk of making the notation more cumbersome, let's introduce a superscript to indicate the sample. Note that each random sample is actually a data set; the term *sample* here refers to the entire data set of $n = 16$ people, not to an individually sampled person within the data set.

How the Repeated Samples Look

Random Sample 1: $\left\{ Y_1^{(1)}, Y_2^{(1)}, \ldots, Y_{16}^{(1)} \right\}$, giving sample average $\bar{Y}^{(1)} = (1/16) \sum_{i=1}^{16} Y_i^{(1)}$.

Random Sample 2: $\left\{ Y_1^{(2)}, Y_2^{(2)}, \ldots, Y_{16}^{(2)} \right\}$, giving sample average $\bar{Y}^{(2)} = (1/16) \sum_{i=1}^{16} Y_i^{(2)}$.

Random Sample 3: $\left\{ Y_1^{(3)}, Y_2^{(3)}, \ldots, Y_{16}^{(3)} \right\}$, giving sample average $\bar{Y}^{(3)} = (1/16) \sum_{i=1}^{16} Y_i^{(3)}$.

...

Random Sample 10,000: $\left\{ Y_1^{(10,000)}, Y_2^{(10,000)}, \ldots, Y_{16}^{(10,000)} \right\}$, giving sample average $\bar{Y}^{(10,000)} = (1/16) \sum_{i=1}^{16} Y_i^{(10,000)}$.

It is easy to imagine simulating (using the computer of course) these 10,000 samples—if you knew the distribution $p(y)$. The phrase "95% of the observable samples Y_1, $Y_2, \ldots, Y_{16}$" refers to "$\cong 95\%$ of these 10,000 samples," or equivalently "$\cong 9{,}500$ of these 10,000 samples."

Now, if $\bar{Y}$ is within $\pm 2\sigma/\sqrt{n}$ of μ, then μ is also within $\pm 2\sigma/\sqrt{n}$ of $\bar{Y}$. This sounds like a rather simple statement, but it is actually deep, and it is important. As an analogy, if a mountain lion is less than 20 km from town, then it follows that the town is less than 20 km from the mountain lion. Like the mountain lion, the location of the average $\bar{Y}$ is variable. Like the town, the location of the mean μ is fixed. Continuing the analogy, the statement "95% of the time, $\bar{Y}$ is within $\pm 2\sigma/\sqrt{n}$ of μ" is analogous to the statement "95% of the time, the mountain lion is within 20 km of the town." Likewise, the statement "95%

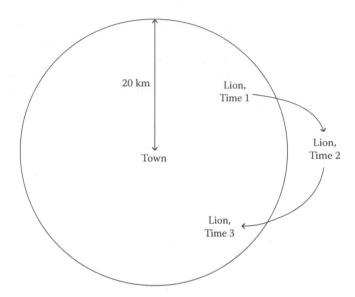

FIGURE 14.2
A mountain lion wandering nearby the town.

of the time, μ is within $\pm 2\sigma/\sqrt{n}$ of $\bar{Y}$" is analogous to the statement "95% of the time, the town is within 20 km of the mountain lion" (see Figure 14.2): Two out of the three times, the mountain lion is within 20 km of the town. In those two times, the town is also within 20 km of the mountain lion.

Using the actual age data sampled from the students, the sample average is observed to be $\bar{y}$ = 26.81 years. That's analogous to the location of the mountain lion, at a particular point in time. At a different point in time, the mountain lion will be somewhere else; see Figure 14.2. Likewise, for a different sampled data set, $\bar{y}$ will be also somewhere else, different from 26.81.

The town is analogous to the process mean μ. At a different point in time, the town is still where it was before. A different sample from the same process $p(y)$ will give you a different $\bar{y}$, but the process mean μ is still the same: It is still $\mu = \Sigma y\, p(y)$ in the discrete case, or $\mu = \int yp(y)\, dy$ in the continuous case.

Suppose that the town is within 20 km of the mountain lion 95% of the time. How about right now? You would guess that the town is probably within 20 km of the mountain lion, and you would probably be right.

Likewise, for 95% of the observable samples, μ lies within $\pm 2\sigma/\sqrt{n}$ of $\bar{Y}$. How about for the particular sample that gave you $\bar{y}$ = 26.81? You would guess that μ is within $\pm 2\sigma/\sqrt{16}$ of 26.81, and you would probably be right.

The range $\bar{y} \pm 2\sigma/\sqrt{n}$ is an *"approximate 95% confidence interval for μ."* In approximately 95% of samples from the process, this interval will contain the value μ. Keep the town/mountain lion analogy in your mind. The interval, like the mountain lion, is variable. Every sample from the process gives a different interval, since every sample gives a different $\bar{y}$. The value of μ, on the other hand, never changes. By your assumption that your data are produced as iid from the distribution $p(y)$, whose mean is μ, a different sample from the one you observed will give a different $\bar{y}$, yet the mean μ hasn't changed because it is still the mean of the same distribution $p(y)$ that produced the data.

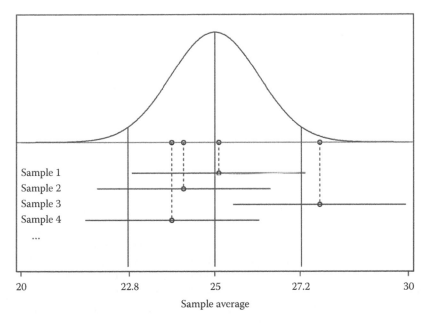

FIGURE 14.3
The results from four sampled data sets of $n = 20$ from a distribution with mean $\mu = 25.0$ and variance $\sigma^2 = 5^2$, each giving a sample average $\bar{Y}$ that is produced, approximately, from the N(25.0, 5^2/20) distribution that is shown. The mean $\mu = 25.0$ is within ±2.2 of the sample average whenever the sample average is within ±2.2 of the mean $\mu = 25.0$.

The town doesn't move. The mountain lion does move. The process mean, μ, doesn't move. The sample mean, $\bar{y}$, does move.

Figure 14.3 shows the idea using data instead of a mountain lion. Suppose your data are produced as iid from a distribution $p(y)$ whose mean is 25.0 and whose standard deviation is 5.0. Then a sample of $n = 20$ values will produce a sample average $\bar{Y}$ having a distribution whose mean is also 25.0 but whose standard deviation is StdDev($\bar{Y}$) = $5.0/\sqrt{20}$ = 1.118. Thus approximately 95% of the $\bar{Y}$ values will be within the range $25.0 \pm 2(1.118)$, or within ±2.2 of 25.0. Figure 14.3 shows four $\bar{Y}$ values, with one of them outside the range. (You expect only 1 in 20 to be outside the range.) There are four intervals shown under the curve in Figure 14.3, all centered at the values of the observed $\bar{Y}$. Notice that $\mu = 25.0$ is within the interval centered at the observed $\bar{Y}$, whenever the observed $\bar{Y}$ is within the interval centered at μ. In particular, the $\bar{Y}$ from Sample 3 that is more than two standard deviations from μ gives an interval that does not contain the value of μ.

While the range $\bar{y} \pm 2\sigma/\sqrt{n}$ can be called an approximate 95% interval, you may recall that the actual range leading to 95% of the normal distribution is not precisely ±2, but rather the ±1.96 standard deviation range. So it is more common to call the interval range $\bar{y} \pm 1.96\sigma/\sqrt{n}$ an "approximate 95% confidence interval for μ." It is still *approximate* because the distribution of $\bar{Y}$ is only *approximately* normal. If the distribution of $\bar{Y}$ were *exactly* normal, then you can call the interval $\bar{y} \pm 1.96\sigma/\sqrt{n}$ an "*exact* 95% *confidence interval for* μ."

An Approximate 95% Confidence Interval for μ, When σ Is Known

$$\bar{y} \pm 1.96\sigma/\sqrt{n}$$

TABLE 14.2

Confidence Levels, α, and Critical Values

Confidence Level (%)	α	$1 - \alpha/2$	Critical Value, $z_{1-\alpha/2}$
68	0.32	0.84	0.994
80	0.20	0.90	1.282
90	0.10	0.95	1.645
95	0.05	0.975	1.960
99	0.01	0.995	2.576
99.7	0.003	0.9985	2.968

If you desire different confidence levels, such as 90%, you can use an appropriate quantile of the standard normal distribution other than 1.96. Note that 1.96 is not the 0.95 quantile, it is the 0.975 quantile of the standard normal (or N(0, 1)) distribution, and −1.96 is the 0.025 quantile of the standard normal distribution. Thus, there is 95% probability between −1.96 and +1.96 under the standard normal distribution. In general, to capture $100(1 - \alpha)\%$ probability in the central portion of any normal distribution, you need $\pm z_{1-\alpha/2}$ standard deviations on either side of the mean, where $z_{1-\alpha/2}$ denotes the $1 - \alpha/2$ quantile of the standard normal distribution, also called the *critical value*. Table 14.2 gives some standard confidence levels and the corresponding critical values.

You'll notice the familiar 68–95–99.7 confidence levels are in Table 14.2, although the precise critical values are, as shown, ±0.994, ±1.960, and ±2.968 standard deviations, rather than ±1, ±2, and ±3 standard deviations. Still, you should memorize ±1, ±2, and ±3 standard deviations along with 68–95–99.7 instead because they are easier to memorize.

For the age data, you can say that $26.81 \pm 1.96\,\sigma/\sqrt{16}$ is an approximate 95% confidence interval for the process mean age, μ. What about σ? It is unknown, so you have to estimate it. As discussed in Chapter 11, either the plug-in estimator or the unbiased version is consistent, so if the sample size is large enough, then $\hat{\sigma} \cong \sigma$. And you've already admitted "approximate," so why not add another layer of approximation? Who will know the difference? Recall: *Approximation* is a weasel word! Thus, you can say, correctly, that the interval $26.81 \pm 1.96\,\hat{\sigma}/\sqrt{16}$ is also an approximate 95% confidence interval for μ. However, there is an additional layer of approximation here in that $\hat{\sigma} \cong \sigma$, so the approximation when using $\hat{\sigma}$ is not as good as the approximation when using σ. Recall also from Chapter 11 that the estimator $\hat{\sigma}$ is not very close to σ for smaller sample sizes such as the $n = 16$ used here, so the approximation is quite suspect indeed once you add the additional approximation $\hat{\sigma} \cong \sigma$.

An Approximate 95% Confidence Interval for μ, When σ Is Unknown

$$\bar{y} \pm 1.96\hat{\sigma}/\sqrt{n}$$

Even if the distribution of $\bar{Y}$ were *exactly* normal, you would *still* call the interval $\bar{Y} \pm 1.96\hat{\sigma}/\sqrt{n}$ an *approximate* 95% confidence interval for μ, simply because of the approximation $\hat{\sigma} \cong \sigma$. In the case where the distribution of $\bar{Y}$ is exactly normal, this weasely word *approximate* can be deleted if you use a critical value from Student's *t*-distribution, described in Chapter 16, instead of the 1.96 from the standard normal distribution.

Nevertheless, you now have an answer, if an approximate one. Since $\hat{\sigma} = 4.94$ (years) for the age data (why?), you can now say that you are approximately 95% confident that the process mean μ lies in the range $26.81 \pm 1.96 \times 4.94/\sqrt{16}$, or in the range 26.81 ± 2.42, giving the interval $24.39 \leq \mu \leq 29.23$. In the mountain lion analogy, you would say "I am approximately 95% confident that the town is within 2.42 kilometers of the lion." In the present context you would say "I am approximately 95% confident that μ is within 2.42 years of the sample average 26.81." If questioned about the *approximate* word, simply say, "I am a weasel, and weasels do not need to justify their assumptions and methods."

We're just kidding. Seriously, don't really say "I am a weasel." Instead, you can offer a more studied answer based on simulation analysis as shown in the next section.

14.3 What Does *Approximate* Really Mean for an Interval Range?

When you say "I am approximately 95% confident that μ lies in the range $24.39 \leq \mu \leq 29.23$ (years)" what is the approximation? One way to look at it is that the endpoints, 24.39 and 29.23, are only approximately correct, being based on the substitution of $\hat{\sigma}$ for σ. But you can never know σ in practice, so it is not relevant to judge the approximation by comparing it to an unknowable standard. And even if you knew σ, the interval would still be approximate because the distribution of $\bar{Y}$ is only *approximately* normal.

Rather than judge the degree of approximation by comparing the interval endpoints to some theoretically correct endpoints, you can instead judge whether the **true confidence level** is close to 95%. When you say "approximately 95%" you are admitting that your true confidence level is different from 95%. How much different is it? A simulation study will tell you.

To do any simulation study, you first need to define how you will generate the data (*model produces data*). Ideally, you would like to generate the data using the distribution $p(y)$ that produced the actual data; however, this distribution is unknown (*model has unknown parameters*) so you need to use one that is reasonably close instead. You should also perform sensitivity analysis by choosing other distributions that are reasonable, so as to ensure that your results are not completely specific to just one distribution that you happened to choose.

A good starting point is the bootstrap distribution $\hat{p}(y)$, which is a sensible estimate of the true process distribution $p(y)$. For the age data, the distribution $\hat{p}(y)$ places 1/16 probability on each of the observed data points, and since there are repeat observations on some of the ages, the probabilities are collated; see Table 14.3.

You can simulate repeated samples of size $n = 16$ iid observations from this distribution; these are called **bootstrap samples**. Bootstrap samples turn out to be **with replacement random samples** from the original data set of $n = 16$ observations. That is, you sample one observation at random from the $n = 16$ values, replace it, then sample another—which, because the sampling is "with replacement," could be an observation you have already sampled—and repeat this until you have $n = 16$ randomly sampled data values $y_1^*, y_2^*, ..., y_{16}^*$.

You can get bootstrap samples easily using various statistical software packages, including Microsoft Excel. Enter Table 14.3, then use the random number generator, selecting

TABLE 14.3

Bootstrap Distribution for the Age Data

Age, y	$\hat{p}(y)$
22	0.1875
23	0.1250
24	0.1250
25	0.0625
26	0.1250
27	0.0625
28	0.0625
30	0.0625
35	0.0625
36	0.1250
Total	1.000

the discrete distribution and selecting the table you just entered for value and probability input range. Specify 16 columns and 1000 rows, and click OK. Each resulting row is now a bootstrap sample of data values $y_1^*, y_2^*, \ldots, y_{16}^*$. For each of these samples (i.e., for each row), you can construct an approximate 95% confidence interval for the mean of the distribution by the formula $\bar{y}^* \pm 1.96\hat{\sigma}^*/\sqrt{16}$. Since you have defined the distribution yourself as in Table 14.3, you know that the true mean is in fact $\hat{\mu} = 26.8125$ (the plug-in estimate), but remember that you would not know the true mean μ in practice.

You can enter the confidence interval formula in Excel easily. The lower limit is:

$$= \text{AVERAGE}(A16 : P16) - 1.96 * \text{STDEV}(A16 : P16)/\text{sqrt}(16)$$

The upper limit is:

$$= \text{AVERAGE}(A16 : P16) + 1.96 * \text{STDEV}(A16 : P16)/\text{sqrt}(16)$$

The formulas assume the first row of your simulated data lies in row 16, columns A through P. Copying the function down for all 1000 data sets gives you 1000 confidence intervals. A few of these intervals are shown in Table 14.4.

The boldface interval in Table 14.4 shows an incorrect interval: The range is (28.2121, 33.4128), which does not include the true mean, 26.8125. A simple count shows that 95 out of the 1000 samples of $n = 16$ each give similarly incorrect intervals, in that 26.8125 is either below the lower endpoint or above the upper endpoint. Hence, the other 905 out of the of 1000 samples give a correct result in that the true mean lies between the lower and upper endpoints. This corresponds to an estimated true confidence level of 905/1000 = 0.905.

Now you know what the word *approximately* means in the phrase "I am approximately 95% confident." When you claimed your interval was "approximately 95%," you were acknowledging that the true confidence level was not precisely 95%. In fact, the true confidence level is closer to 90% than to 95%. Either way, you can be pretty confident in the result. If an interval is correct 90% of the time, then you can feel confident that the interval you observed from the original data, $24.39 \leq \mu \leq 29.23$, is likely correct (i.e., that it contains the true *unknown* mean μ), because similarly constructed intervals are correct 90% of the time. Still, your presentation smacks slightly of dishonesty, since your claim of 95% suggests more confidence than you really have.

TABLE 14.4

Confidence Intervals for the Mean $\hat{\mu}= 26.8125$ Using
Data Simulated from the Bootstrap Distribution
Constructed from the Student Age Data

Average	StdDev	Lower Limit	Upper Limit
26.2500	5.446712	23.58111134	28.91888866
26.8750	4.193249	24.82030821	28.92969179
27.9375	5.836309	25.07770871	30.79729129
29.0000	5.899152	26.10941528	31.89058472
27.3125	4.895151	24.91387618	29.71112382
27.7500	5.859465	24.87886201	30.62113799
26.3125	5.186119	23.77130155	28.85369845
25.3750	3.442383	23.68823238	27.06176762
25.8750	4.660114	23.59154392	28.15845608
26.1875	3.780983	24.33481827	28.04018173
27.1875	4.777988	24.84628602	29.52871398
25.9375	4.404070	23.77950566	28.09549434
30.8125	**5.306835**	**28.21215074**	**33.41284926**
25.5625	4.661455	23.27838684	27.84661316
...	...	...	...

The analysis with the bootstrap distribution does not provide the final answer, because the bootstrap distribution is not the distribution $p(y)$. You cannot find the exact true confidence level, even via simulation, because it would require that you simulate data from $p(y)$, which you do not know. So in any simulation study, sensitivity analysis is prudent. You can select another distribution, other than the bootstrap distribution, and compare results.

To pick another distribution for sensitivity analysis, you might use a shifted and rounded-off exponential distribution: The lower age limit seems to be around 21, so you might add 21 to an exponential random variable, making its mean $21 + 1/\lambda$ instead of $1/\lambda$. The mean and the standard deviation of the (non-shifted) exponential distribution are the same; since the standard deviation of the observed data is 4.94, you can pick $\lambda = 1/5$ so the mean of the shifted random variable is $21.0 + 5.0 = 26.0$ and its standard deviation is 5.0. Remember, this is sensitivity analysis: You are just trying out some other plausible distribution; you are not trying to pick the exactly correct distribution. Finally, to make the data look like actual age data, you can round them to the nearest integer. Here is one random sample of $n = 16$ data points generated in this fashion: 25, 22, 30, 24, 22, 21, 35, 26, 25, 23, 23, 26, 22, 22, 22, and 23. These look like age data for the class. With the understanding that good models produce data that look like data you will actually observe, this distribution appears to be a reasonable candidate for sensitivity analysis.

As described earlier, the true mean μ of the distribution of this shifted exponential random variable is 26.0 when the data are not rounded. Rounding makes the calculation of the mean more difficult, requiring some tedious calculus. But it is easy to get an approximately correct value by appealing to the LLN. Using 16,000 data values sampled from this shifted and rounded exponential distribution, we find an average of 26.05, and this will suffice as our proxy for the true mean μ.

Notice that in the first sample of $n = 16$ data values that gave 25, 22, 30, 24, 22, 21, 35, 26, 25, 23, 23, 26, 22, 22, 22, and 23, the sample mean and sample standard deviation are 24.438 and 3.614, respectively, leading to an approximate 95% confidence interval of $24.438 - 1.96(3.614/\sqrt{16}, 24.438 + 1.96(3.614/\sqrt{16},$ or $(22.667, 26.208)$. Since the true mean 26.05 lies within this range, the interval is correct.

Among 1000 intervals generated in this way, we find 891 are correct, leading to an estimated true confidence level of $891/1000 = 0.891$. Thus, the sensitivity analysis simulation study essentially confirmed the original simulation study: The true confidence level is closer to 90% than to 95%. Either way, confidence is reasonably high, but, either way, it smacks of dishonesty in that your claim of "approximately 95%" suggests more confidence than you can truly claim. After all, if you are wrong one in ten times (or 90% confident), then you are wrong twice as often than when you are wrong only one in twenty times (or 95% confident).

Closer inspection of the simulation results, using both the bootstrap distribution and the shifted exponential distribution, reveals an interesting fact: When the intervals are wrong, the true mean is most often above the upper endpoint (7.3% versus 2.2% when using the bootstrap distribution; 10.4% versus 0.5% when using the shifted exponential distribution). This suggests that when sampling from a skewed distribution, the confidence interval should really be asymmetric, as with the Bayesian intervals shown in Chapter 13. In this example involving age, both the lower and upper limits should be increased. You can use the bootstrap to construct sensibly asymmetric frequentist confidence intervals (see Chapter 19).

14.4 Comparing the Bayesian and Frequentist Paradigms

"Hey buddy, can you 'pare a dime?" is what you might hear from a wino with very dry lips. On the other hand, a **paradigm** is an overarching view of a subject area. Think of a paradigm as both a lens that you can use to view a subject matter and as a mouthpiece that you can use to interpret the subject matter. The Bayesian and frequentist paradigms have historically been at odds, with proponents on either side vigorously defending their turf.

However, a newer paradigm seems to have emerged where both paradigms are seen as useful. This new paradigm says that, if it works, use it! For some types of problems, Bayesian methods work very well, and for others, frequentist methods work very well. Problems involving smaller data sets, missing values, nonresponse, unobserved (latent) data, nonlinear functions of parameters, or predictions of future data are well suited to Bayesian methods. Problems involving large data sets, where assumptions are to be avoided at all costs, or where repeated-sampling validations are required are well suited to frequentist methods. But the divide is never perfectly neat, and these days you will see advanced statistical methods that often contain hybrids of frequentist and Bayesian methods, for both types of problems.

Whether you like the frequentist or the Bayesian approach, you have to start with the Mantra:

> *Model produces data.*
>
> *Model has unknown parameters.*
>
> *Data reduce the uncertainty about the unknown parameters.*

Both the Bayesian and frequentist approaches are identical with respect to the first two phrases of the Mantra. Where they differ is in the third phrase, *Data reduce uncertainty about the unknown parameters*. In the Bayesian approach, you express your uncertainty about the unknown parameters by your posterior distribution. In the frequentist approach, you express your uncertainty about the unknown parameters by envisioning your sample as one of many samples that could have been produced and interpreting your data analysis with reference to the long-run frequency of other possible outcomes in different samples.

For example, if you had used Bayesian methods to calculate the 95% interval $24.39 \le \mu \le 29.23$, you would say:

> Based on my posterior distribution for μ, there is 95% probability that μ is between 24.39 and 29.23.

If you had used frequentist methods, on the other hand, you would say:

> Since μ will lie within the upper and lower limits of similarly constructed intervals for 95% of the repeated samples, my sample is likely to be one of those samples where μ is within the upper and lower limits, and I am therefore 95% confident that μ is between 24.39 and 29.23.

As mentioned in the introduction to this chapter, the frequentist does not use the word *probability* to describe the interval (24.39, 29.23). From the frequentist standpoint the probability is either 1 or 0: Either μ is in the interval or μ is not in the interval. Since μ is fixed, nothing is random, and there can be no probability statement.

Bayesians agree that μ is fixed, but since it is unknown, a Bayesian is willing to model his or her uncertainty using a probability distribution. This is a mental model, and it is personal. So, when a Bayesian interprets an interval in terms of probability, it is his or her own probability. Someone else with a different prior will have a different probability statement.

This seeming lack of objectivity in the Bayesian approach is bothersome to frequentists. The frequentist interpretation, while cumbersome and wordy, is at least more objective in that different frequentists will always arrive at the same conclusion, provided they assume the same model $p(y|\theta)$. On the other hand, all frequentists acknowledge that their DATA are uncertain (random) and accept the initial subjective assumption that their DATA are produced by $p(y|\theta)$. The Bayesians argue, then, that if you have accepted the subjective assumption of random generation, and if you also have accepted the subjective assumption that $p(y|\theta)$ is the generator, then you have already "drunk the Kool-Aid," so why not accept one more subjective assumption? You know that there is uncertainty about the parameter(s) θ, so why not model this uncertainty via a prior distribution $p(\theta)$, just like you model your uncertainty about DATA via your model $p(y|\theta)$?

Neither approach can be classified as 100% right or 100% wrong. There are valid points of view on either side of the Bayesian/frequentist fence. While the two approaches differ philosophically, it is fortunate that they tend to provide similar results, especially when vague priors are used. In such cases, it does not matter which philosophy you adopt, as your conclusions will be essentially the same either way.

The next chapters will be primarily frequentist, although we give Bayesian perspectives as well. Good statistical analysis requires both paradigms.

Vocabulary and Formula Summaries

Vocabulary

Probability	A measure of degree of belief, a number between zero (impossibility) and one (perfect certainty).
Confidence	A measure of how often something happens in repeated samples, a number between zero (it never happens in repeated samples) and one (it always happens in repeated samples).
Bayesian	A person who analyzes data by using the posterior distribution of the parameter(s); an adjective describing the Bayesian's methods.
Frequentist	A person who analyzes data by considering the long-run frequency of outcomes of repeated samples from the process; a non-Bayesian; an adjective describing the frequentist's methods.
$100(1 - \alpha)\%$ confidence interval	A frequentist interval that will cover the parameter θ in $100(1 - \alpha)\%$ of repeated samples from the process.
Approximate $100(1 - \alpha)\%$ confidence interval	A frequentist interval that will cover the parameter θ in *approximately* $100(1 - \alpha)\%$ of repeated samples from the process.
Mountain lion	A large feline wild animal, carnivorous. Must move around to seek food.
Town	A smallish collection of homes and businesses. Unlike a mountain lion, it does not move around.
Critical value	The number of standard errors on each side of the estimate that defines the interval range; commonly a value near 2.0.
True confidence level	The long-run percentage of intervals, based on repeated samples from the process, that contain the parameter.
Bootstrap sample	An iid random sample from the bootstrap population distribution.
With replacement random sample	When used to sample from an existing data set, this gives a *bootstrap* sample.
Paradigm	An overarching view of a subject area, such as the *frequentist paradigm*, or the *Bayesian paradigm*.

Key Formulas and Descriptions

$\bar{y} \pm 1.96\sigma/\sqrt{n}$	An approximate 95% confidence interval for μ when σ is known (it is unlikely that σ is known).
$\bar{y} \pm 1.96\hat{\sigma}/\sqrt{n}$	An approximate 95% confidence interval for μ when σ is unknown (it is most likely that σ is unknown).

$z_{1-\alpha/2}$ The $1 - \alpha/2$ quantile of the standard normal distribution.

$\bar{y}^* \pm 1.96\hat{\sigma}^*/\sqrt{n}$ The approximate 95% confidence interval for μ based on simulated data.

Exercises

14.1 Hans puts a fair, six-sided die with numbers 1, 2, 3, 4, 5, and 6 inside a box. He shakes the box, and lets the die settle. You can't see the die. Is the die showing a 3?

    A. Give a frequentist answer.

    B. Give a Bayesian answer.

14.2 Hans puts a mangled, misshapen, six-sided die with numbers 1, 2, 3, 4, 5, and 6 inside a box, shakes the box, and lets the die settle. You can't see the die, and you know nothing about *how* the die is deformed, only that it *is* deformed. Is the die showing a 3?

    A. Give a frequentist answer.

    B. Give a Bayesian answer.

14.3 Draw a graph of the standard normal distribution. Locate the 90%, 95%, and 99% critical values on this graph. Explain from the graph why the 90% critical value is the 0.95 quantile rather than the 0.90 quantile.

14.4 Use the data from Exercise 12.9.

    A. Construct the approximate 90% confidence interval for μ_x.

    B. Is μ_x inside your interval from Exercise 14.4A? Explain carefully, like a good frequentist.

    C. Construct the approximate 99% confidence interval for μ_y.

    D. Is μ_y inside your interval from Exercise 14.4C? Explain carefully, like a good frequentist. Explain also how use of 99% rather than 90% affects your answer to the question.

    E. Use bootstrap sampling to estimate the true confidence level of your interval in Exercise 14.4A.

    F. Use bootstrap sampling to estimate the true confidence level of your interval in Exercise 14.4C.

14.5 Consider the data set 1, 0, 1, 1, 1, 1, 0, 0, 1, 0, 1, 0, 0, 1, 1, 1, 1, 1, 1, and 0 assumed produced as an iid sample of $n = 20$ observations.

    A. What distribution produced these data? Give the distribution in list form, as well as its mean, variance, and standard deviation.

    B. Give the bootstrap distribution for these data in list form, as well as its mean, variance, and standard deviation. Explain why the distribution in Exercise 14.5A is different from the bootstrap distribution.

    C. Give the approximate 95% interval for the mean of the distribution in Exercise 14.5A using the formula $\bar{y} \pm 1.96\hat{\sigma}/\sqrt{n}$ where $\hat{\sigma}$ is the plug-in estimate from Exercise 14.5B.

D. Statistics sources give the approximate 95% confidence interval for the Bernoulli parameter π as $\hat{\pi} \pm 1.96\sqrt{\hat{\pi}(1-\hat{\pi})/n}$. Show that your interval in Exercise 14.5C is identical to this Bernoulli confidence interval. (If you are mathematically inclined, prove they are the same for general data.)

E. Is π in your confidence interval of Exercise 14.5C? Explain carefully, like a good frequentist.

F. Generate 10,000 samples of $n = 20$ each from the Bernoulli(0.6) distribution to estimate the true confidence level of this procedure. Is it reasonably close to 95%? How about when $\pi = 0.01$? Comment on how Ugly Rule of Thumb 10.2 applies here.

14.6 Find the approximate 95% confidence interval for the mean return μ of Example 10.2. How does your interval compare with the Chebyshev analysis from Example 10.2? Explain how and why the intervals differ.

14.7 Consider the data from Exercise 12.5.

A. Construct the approximate 95% frequentist confidence interval for the mean μ.

B. Using software that generates posterior samples, assume a Poisson model with a vague prior and graph the posterior histogram for μ.

C. Obtain a 95% credible interval. Compare the asymmetry of the resulting interval to the symmetric interval of Exercise 14.7A, and explain using Exercise 14.7B why the asymmetric interval is preferred.

14.8 Consider the data from Exercise 12.11.

A. Construct the approximate 95% frequentist confidence interval for the mean μ.

B. Convert the interval in Exercise 14.8A to an interval for $\theta = 2\mu$.

C. Why is the interval in Exercise 14.8B so different from the interval for θ you found in Exercise 12.11C of Chapter 12?

15

Are Your Results Explainable by Chance Alone?

15.1 Introduction

Hans tosses a coin 10 times, and it turns up heads only twice. You are tempted to conclude that Hans is a trickster. That Hans! Always kidding around. But is the difference between Hans' 20% heads and the expected 50% heads result *explainable by chance alone*? If so, would you still conclude that Hans is a trickster?

In your collection of the data on the ages of students in your class, you notice that the average age of those seated in the back rows of the class is 27.5 years, while the average age of those seated in the front rows is 26.125 years. You are tempted to conclude that older students tend to sit in the back rows. But is the difference *explainable by chance alone*? If so, would you still conclude that older students tend to sit in the back rows?

You are a quality control supervisor. You notice that yesterday the morning shift produced 5% defective products, while the afternoon shift produced 8%. You are tempted to conclude that the employees on the afternoon shift need to be punished for their laziness and inattention to detail. But is the difference *explainable by chance alone*? If so, would you still decide to punish the afternoon shift?

You are a researcher studying the effect of a new AIDS drug on the viral load in patients. You find that in the group of patients receiving the new drug the average viral load is 2.13, whereas in the group of patients receiving the standard therapy the average viral load is higher, 3.66. You are excited and want to publish your findings! But is the difference *explainable by chance alone*? If so, would you still decide to tell the world that your new AIDS drug is effective?

You are trying to understand a q–q plot. You know that if the data come from a normal distribution, then the appearance of the plot is expected to be a straight line. You then notice that there are wiggles in the plot—the data values do not fall exactly on the line. You are therefore tempted to conclude that the data do not come from a normal distribution. But is the difference between the observed plot and what you would expect *explainable by chance alone*? If so, would you still conclude that the distribution is not normal?

You are studying the effect of ambiguity on business decisions. You develop a measure X = environmental ambiguity and another measure Y = quality of business decision. After collecting data, you estimate the correlation between X and Y to be −0.12. So you say, "Aha! Business decisions are worse when there is more environmental ambiguity!" But is the difference between the correlation −0.12 and 0 *explainable by chance alone*? If so, would you still shout "Aha!"?

The subject of **statistical significance testing**, also called **hypothesis testing**, gives you an answer to the question, "Are the results *explainable by chance alone*?" And that is the subject of this chapter.

Before proceeding, we must alert you to a seemingly subtle but actually huge difference in terminology. The phrase "the results are *explainable* by chance alone" means something completely different than the phrase "the results are *explained* by chance alone." The former states that chance is one *possible* explanatory mechanism, while the latter states that chance is the *only* explanatory mechanism. Once you understand this distinction clearly, you'll understand 90% of everything you really need to know about hypothesis testing.

The subject of hypothesis testing in statistics, for better or worse, is laden with an extraordinary number of special vocabulary terms that are unique to the subject. The plethora of such terms is a good indicator for the topic's perceived importance among scientists. In any scientific study, it is essential to rule out chance as an explanatory mechanism for your empirical results, so there is a great deal of interest across all scientific disciplines in hypothesis testing, and hence there is a great deal of associated vocabulary. So you will learn many new hypothesis-testing vocabulary terms in this chapter, as well as in the next chapters.

15.2 What Does *by Chance Alone* Mean?

Before addressing the question of whether the results are explainable by chance alone, you first need to come to grips with a particular concept: *Model produces data*. Have you heard that one before? Well, maybe. But do you remember the version of *"Model produces data"* that requires you to think about the concept of *by chance alone*? You might—go back and have a look at Sections 5.6 and 7.5, you'll see it there. The chance-only model is the fundamental concept in the subject of hypothesis testing; we'll review it and fill in many more details in this chapter.

In all of the cases considered in Section 15.1, the results of interest concerned *differences*: the difference between percentages, the difference between averages, difference between percentages again, the difference between averages again, the difference between actual and expected appearance of graphs, and the difference between a correlation coefficient and 0, respectively.

You can equate results of interest to differences found in your data. The phrase "the results are explainable by chance alone" means that the difference you observe in your data is within the expected range of results that you expect when there is no difference in reality. For example, if you flip a fair coin 10 times and get 40% heads, then flip it 10 times again and get 60% heads, the observed difference between 40% and 60% is in the range of chance variation when there is no difference in reality.

Since your model that produces data is supposed to mimic Nature, you can assess natural, chance variations by simulating DATA* from a model that mimics Nature *when there are truly no differences*. Using these DATA* you can calculate the difference statistics just as you calculated from your observed data, and then you can compare your observed difference based on your actual data to the differences based on your DATA*. Because your DATA* are generated from a model where there are truly no differences, the differences you see in DATA* are completely *explained* by chance alone. If the difference you observe in your actual data is well within the range of differences expected in the DATA* under the no-difference model, then your results are *explainable* by chance alone. They could be the result of chance, or not. It's ambiguous, because *explainable* means *"can be* explained," not *"is* explained."

Your model that mimics Nature in the no-difference case is called a **null model** and is given the symbol $p_0(y)$. The 0 in the subscript is meant to remind you that this is a model for Nature where there is, in reality, zero difference. For example, in the coin flip case above, the model for the first 10 and last 10 flips is the null model $p_0(y)$, which is simply the Bernoulli(π) model.

As in the case of any model $p(y)$ that you assume to produce the data, you usually cannot determine the null model $p_0(y)$ precisely. But you will see that there are logical types of null models $p_0(y)$ you can use, depending on the specific circumstances.

Example 15.1: A Null Model for Age and Seat Selection

The student age data are (still!) $y_1 = 36$, $y_2 = 23$, $y_3 = 22$, $y_4 = 27$, $y_5 = 26$, $y_6 = 24$, $y_7 = 28$, $y_8 = 23$, $y_9 = 30$, $y_{10} = 25$, $y_{11} = 22$, $y_{12} = 26$, $y_{13} = 22$, $y_{14} = 35$, $y_{15} = 24$, and $y_{16} = 36$. Suppose the data are arranged by row so that y_1 through y_8 are the ages of students in the front rows, while y_9 through y_{16} are the ages of students in the back rows. Then the average age for students in the back rows is $(30 + 25 + \cdots + 36)/8 = 27.5$, and the average age for students in the front rows is $(36 + 23 + \cdots + 23)/8 = 26.125$. The difference in the data is $27.5 - 26.125 = 1.375$ years. This statistic is an example of a **test statistic**, which is in general a function of your sampled data that you use to test a theory.

A null model would state there is no difference in the *process that produced these data* between ages of the students in the front and the back rows.

While there is no one precise model that is ever perfectly correct, you can often specify null models that make sense. A familiar model is that the data $Y_1, Y_2, \ldots, Y_{16}$ are an iid sample from $p_0(y)$; if this is how the data arise, then there is no difference between the distributions of *any* of the observations, let alone between the first eight and the last eight. Under the iid model, there is no preference for older people to sit in the front or the back—it's completely random.

The iid model is the most common null model, and we will use it later. It's a great null model, but still leaves open the problem that $p_0(y)$ is unknown, depending on unknown parameters, as shown in Table 14.1. To present this chapter's concepts as clearly as possible, we'll start with an even simpler model called the **randomization model**, where $p_0(y)$ does not depend on unknown parameters and is completely known.

Here is the model. Students with ages 36, 23, 22, 27, 26, 24, 28, 23, 30, 25, 22, 26, 22, 35, 24, and 36 wander into class the first day. The first student in selects a seat *at random* and sits down. The second student in selects a seat *at random* from the seats still available and sits down. The third student does the same, and so on. If there are only 16 seats, the last student who wanders in has no choice but to select the sole remaining seat.

In this model, the arrangement $Y_1, Y_2, \ldots, Y_{16}$ is random, by virtue of the students randomly selecting chairs. Also, it is clear that in this model there is no difference in the ages of people sitting in the front versus back seats, since the selections are random. The model $p_0(y_1, y_2, \ldots, y_{16})$ that produces the $Y_1, Y_2, \ldots, Y_{16}$ data is a *randomization model*, also called a **permutation model**: all possible permutations of the data values 36, 23, 22, 27, 26, 24, 28, 23, 30, 25, 22, 26, 22, 35, 24, and 36 are equally likely. You can think of this model exactly as you think of a well-shuffled deck of cards where all possible permutations of the card order are equally likely following a thorough shuffling. Each permutation, or shuffle, gives averages for the first eight (in the front rows) and the last eight (in the back rows), and the difference between these two is purely *explained by chance*, because, again, there is truly no difference between front and back seats in this model. By analogy, in a well-shuffled deck of cards, there is no systematic tendency for the lower-numbered cards to be in the top of the deck.

TABLE 15.1

Differences between Ages in Front and Back Rows That Result from 10 Random Permutations of the $n = 16$ Students' Ages

Randomization Scenario	Student Ages, from Front to Back of Class	Average Age in the Front of Class ($n_1 = 8$)	Average Age in the Back of Class ($n_2 = 8$)	Difference between Averages, Back Minus Front
1	27 24 22 23 24 22 36 26 36 28 23 26 25 30 35 22	25.5	28.125	2.625
2	24 25 23 28 35 36 23 26 36 30 27 24 26 22 22 22	27.5	26.125	−1.375
3	35 24 30 26 36 36 22 23 24 26 25 22 22 27 23 28	29.0	24.625	−4.375
4	22 23 36 30 24 26 24 22 23 27 25 36 22 28 26 35	25.875	27.75	1.875
5	25 28 26 24 23 30 36 22 27 36 26 22 24 23 35 22	26.75	26.875	0.125
6	36 23 30 24 22 28 36 26 26 22 23 35 27 25 24 22	28.125	25.5	−2.625
7	22 25 30 36 36 22 24 23 27 26 28 26 35 23 22 24	27.25	26.375	−0.875
8	22 25 26 22 36 36 24 27 35 23 22 28 24 26 23 30	27.25	26.375	−0.875
9	28 30 22 24 36 23 35 23 27 26 22 26 25 22 24 36	27.625	26.0	−1.625
10	26 23 24 22 23 22 27 28 35 26 30 22 25 24 36 36	24.375	29.25	4.875

Table 15.1 shows 10 such random permutations, along with the difference between averages—the test statistics—in the front and back rows. For example, in the first scenario shown in Table 15.1, the average age in the front of the class is $(27 + 24 + 22 + 23 + 24 + 22 + 36 + 26)/8 = 25.5$, and the average age in the back of the class is $(36 + 28 + 23 + 26 + 25 + 30 + 35 + 22)/8 = 28.125$.

The numbers in the final column are purely chance differences. Further, the actual average age difference you saw from the actual students, 1.375 years, appears well within the range of these differences shown in Table 15.1, which are again differences that are *explained by chance alone*. Thus, the difference you see in your data, 1.375, is easily *explainable by chance alone*.

But you *cannot* say that your observed difference, 1.375, is *explained* by chance alone. You have not proven that seat selection is independent of age in general. It could be that older students do tend to sit more often in the back rows. But it could also be that older students tend to sit more often in the front rows. All that you can conclude, based on the data and the randomization model results shown in Table 15.1, is that you simply don't know. Since chance alone is a possible explanation for the difference you saw, you can't argue, based on your data alone, that there is any relationship between age and position in the classroom.

What if the difference were −5.125 years? Would that difference be explainable by chance alone? From Table 15.1, the difference −5.125 seems outside the range of chance difference, but, on the other hand, Table 15.1 shows only the differences resulting from 10 possible random seat selections. Perhaps if you looked at a few more random shuffles, differences −5.125, or even more extreme, would appear?

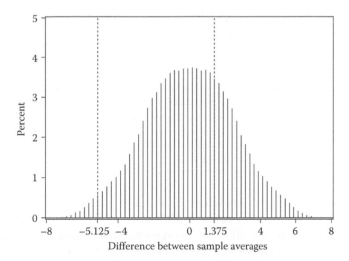

FIGURE 15.1
Permutation distribution of the difference between sample averages of student ages in back versus and front rows, with $n_1 = n_2 = 8$ in each group. The observed difference 1.375 is shown, as well as a hypothetical observed difference −5.125.

It is possible to enumerate all possible permutations, but, like shuffles of a deck of cards, there are a huge number of possibilities: with $n = 16$ data values, there are $16! = 16 \times 15 \times \cdots \times 2 \times 1 = 2.092 \times 10^{13}$ possible arrangements. Instead, it is common to simulate data from the permutation distribution: instead of sampling just 10 of the possible shuffles as shown in Table 15.1, you can sample, say, 1,000,000 of them using appropriate mathematical or statistical software. By the Law of Large Numbers, with many simulations, the resulting probability estimates—which are averages of Bernoulli data; see Section 8.5—will be very accurate.

Figure 15.1 shows the result of simulating 1,000,000 permutations, calculating the mean difference for each permutation, finding the frequency of each possible difference, and graphing the resulting estimate of the discrete distribution via a needle plot. The distribution shown in Figure 15.1 is called the **null distribution of the test statistic**.

In Figure 15.1, it is apparent that the observed difference, 1.375 years, is easily explainable by chance alone, since it is well within the range of the permutation distribution. Again, this doesn't prove that chance *is* the explanation for the difference. It just shows that the difference *can be* explained by chance alone. Since chance alone can explain the difference, you cannot argue, based on these data, that there is a systematic tendency for older students to sit toward the back.

On the other hand, if the observed difference were −5.125 years then chance is much less likely as an explanation, given its placement toward the tail of the distribution under the permutation model where seat selection is truly independent of age. If you observed the difference −5.125 years with these data, you could argue against chance as the reason for observing this extreme a difference, since the probability of seeing a difference this extreme, by chance alone, is small. In this case, you could make an argument for the existence of a systematic difference between seat selections of older versus younger students.

You can assume other null models instead of the permutation model. For example, you can assume the student data $Y_1, Y_2, \ldots, Y_{16}$ are produced as an iid sample from $p_0(y)$, for a $p_0(y)$ as shown in Table 14.1. The test statistic is still $T = \bar{Y}_2 - \bar{Y}_1$, where $\bar{Y}_1 = (Y_1 + \cdots + Y_8)/8$

and $\bar{Y}_2 = (Y_9 + \cdots + Y_{16})/8$. Using properties of expectation, and assuming $Y_1, Y_2, \ldots, Y_{16} \sim_{\text{iid}}$ $p_0(y)$, you can deduce the null distribution of $T = \bar{Y}_2 - \bar{Y}_1$ as follows:

$E(T) = E(\bar{Y}_2 - \bar{Y}_1)$	(By definition of T)
$\quad = E(\bar{Y}_2) - E(\bar{Y}_1)$	(By the linearity and additivity properties of expectation)
$\quad = \mu - \mu$	(Since the expected value of the sample average of data produced as iid from a process with mean μ is equal to μ; see Chapter 10)
$\quad = 0$	(By algebra)

So, the mean of the difference is zero when there really is no difference. This makes sense! The null model says there is no systematic difference in the process, and the calculation shown earlier shows that there is likewise no systematic difference in the data averages. Further, using properties of variance, you can deduce the following:

$\text{Var}(\bar{Y}_2 - \bar{Y}_1) = \text{Var}\{\bar{Y}_2 + (-1) \times \bar{Y}_1\}$	(By algebra)
$\quad = \text{Var}(\bar{Y}_2) + \text{Var}\{(-1) \times \bar{Y}_1\}$	(By the additivity property of variance for independent random variables; see Chapter 10)
$\quad = \text{Var}(\bar{Y}_2) + (-1)^2\text{Var}(\bar{Y}_1)$	(By the linearity property of variance; see Chapter 9)
$\quad = \text{Var}(\bar{Y}_2) + \text{Var}(\bar{Y}_1)$	(Since $(-1)^2 = +1$)
$\quad = \sigma^2/8 + \sigma^2/8$	(Since the variance of the sample average of data produced as iid from a process with variance σ^2 is equal to σ^2/n; see Chapter 10)
$\quad = \sigma^2/4$	(By algebra)

Also, since by the Central Limit Theorem $\bar{Y}_1$ and $\bar{Y}_2$ are approximately normally distributed, it follows that $T = \bar{Y}_2 - \bar{Y}_1$ is also approximately normally distributed—although the relevant linearity and additivity properties of the normal distribution are presented in Chapter 16.

Putting the pieces together, the null distribution of the test statistic $T = \bar{Y}_2 - \bar{Y}_1$ is approximately the normal distribution with mean 0 and variance $\sigma^2/4$. In summary, under the iid null model $Y_1, Y_2, \ldots, Y_{16} \sim_{\text{iid}} p_0(y)$:

$$\bar{Y}_2 - \bar{Y}_1 \dot{\sim} N\left(0, \frac{\sigma^2}{4}\right)$$

The variance σ^2 of $p_0(y)$ is unknown, but you can use this estimate:

$$\hat{\sigma}^2 = (1/n)\sum_i (y_i - \bar{y})^2 = \frac{1}{16}\left\{(36-26.81)^2 + (23-26.81)^2 + \cdots + (36-26.81)^2\right\} = 22.902$$

This is the bootstrap plug-in estimate given in Chapter 11. You can also use the $n - 1$ formula; both are just approximations and both become more accurate for larger n. Hence, an approximate null distribution for the test statistic $T = \bar{Y}_2 - \bar{Y}_1$ is the N(0, 22.902/4) distribution; i.e., the normal distribution with mean 0 and standard deviation $\sqrt{22.902/4} = 2.393$.

This distribution provides a good approximation to the null distribution based on the permutation model shown in Figure 15.1: using the normal approximation, the ±1, ±2, and ±3 standard deviation ranges for the difference $\bar{Y}_2 - \bar{Y}_1$ are (−2.393, 2.393), (−4.786, 4.786), and (−7.179, 7.179), respectively, appearing to capture roughly 68%, 95%, and 99.7% of the probabilities shown in Figure 15.1.

Which is the better null model, the iid model or the randomization model? Note that they provide similar null distributions for the test statistic in this example, so, fortunately, it makes little difference which model you use. However, the iid model gives a distribution that is only approximate, since you have to approximate the distributions of the averages using the normal distribution, and you have to approximate σ using $\hat{\sigma}$. On the other hand, the randomization approach is exact, requiring no approximation, and is preferable for that reason in this simple example.

While the discussion so far suggests that the randomization model is preferable, there are two sides to the story. The randomization model is somewhat restrictive and not available for more complex advanced statistical methods such as multiple regression models. In such cases, you'll have to use the iid and related approximate models instead of the simpler randomization model. The likelihood methods described in Chapter 12 give rise to a general class of testing methods called *likelihood ratio tests*, which allow you to test a wide variety of hypotheses in more complex statistical models (see Chapter 17).

15.3 The *p*-Value

The best way to understand, in the big picture sense, whether your statistical results are explainable by chance is to look at the big picture! That is, look at the big picture showing the null distribution of the test statistic, as shown in Figure 15.1. But there is a question of judgment: Where do you draw the line between values that are typical, and therefore explainable by chance, and values that are so unusual that chance alone can be ruled out, for all intents and purposes? The **p-value**, which we will abbreviate *pv* for short, is a specific measure of how easy it is to explain your results by chance alone.

You will see the symbol *p* for *p*-value in most other statistics sources, but there are already too many italic *p* symbols in this book, so we will use *pv* instead.

The term *p*-value is short for "probability value" and is indeed a certain probability; therefore, the *p*-value is always between zero and one. When the *p*-value is very small, you can confidently rule out chance as a plausible explanation for your results. Otherwise, you cannot easily rule out chance as a plausible explanation for your results. Here is its definition:

Definition of *p*-Value

The *p*-value is the probability of seeing a difference as extreme or more extreme than the difference that you observed, assuming your data come from a process where there is, in reality, no difference.

In other words, the *p*-value is the probability of seeing a result as extreme or more extreme than what you observed, by chance alone.

The phrase *as extreme or more extreme than* is cumbersome, and we will substitute the simpler *as extreme as* from now on.

Thus, with a smaller *p*-value, it is less likely to see a result as extreme as what you did see, by chance alone. Equivalently, "chance alone" is a less plausible explanation for your results when the *p*-value is small.

Notice the careful phrasings about *p*-values. They reflect conditional probabilities, and it is very easy to confuse conditional probabilities $Pr(A|B)$ and $Pr(B|A)$. A *p*-value is specifically:

$$pv = Pr(\text{the difference based on DATA is as extreme as the difference based on data}|$$

$$\text{no difference in the process})$$

The *pv* is *not* $Pr(\text{No difference in process}|\text{data})$. Such a probability would allow a more direct interpretation such as "there is a 92% probability that there is no difference in the process, given the observed data." Such a direct interpretation is desirable, but is only possible via the Bayesian calculation:

$$p(\text{difference in process}\,|\,\text{data}) \propto p(\text{data}\,|\,\text{difference in process})p(\text{difference in process})$$

You can use this Bayesian calculation to find the Bernoulli distribution of "difference in process," which can take the values yes or no. To perform the Bayesian calculation, you would have to assign prior probabilities to the different mechanisms that produced the data; for example, you might assign a 0.50 prior probability that chance alone is the mechanism that produced your data, and also a 0.50 prior probability that there are systematic differences in the process that produced your data. The notion of a *p*-value is strongly frequentist, though, and assumes no priors. For this reason, the interpretation of the *p*-value is necessarily stilted, requiring you to envision replications of your DATA under the null model $p_0(y)$ and then make a judgment as to whether your observed data are so unusual compared to those null DATA that you can rule out the chance mechanism $p_0(y)$, for all intents and purposes.

The *p*-value offers only indirect evidence about the null model. For example, assuming you get $pv = 0.03$ in your data analysis, your interpretation is as follows:

> Assuming there is no difference in the process, the probability of observing a difference as extreme as what I actually observed is 0.03. Therefore, it seems unlikely that the difference I observed is explained by chance alone.

Be careful interpreting *p*-values. Because people have a hard time understanding the difference between $Pr(A|B)$ and $Pr(B|A)$, there are as many incorrect interpretations of *p*-values in books and scientific literature as there are correct ones.

Example 15.2: Calculating the *p*-Value for the Age and Seat Selection Example

The *p*-value is the probability of seeing a difference as extreme as what you observed, by chance alone. Continuing Example 15.1, the difference is 1.375 (years), and the phrase "a difference as extreme as what you observed" translates to "a difference that is either 1.375 or higher, or −1.375 or lower." The lower tail is included because a difference of −1.375 is just as extreme as 1.375; it simply goes in the other direction. Figure 15.2 shows the *p*-value as shaded region; the sum of all the shaded probabilities is the *p*-value.

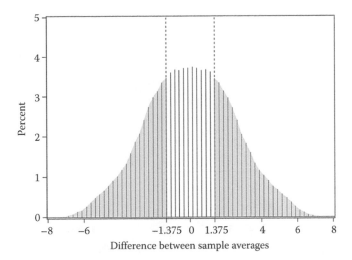

FIGURE 15.2
Permutation distribution of the difference between sample averages of student ages in back versus and front rows, with $n_1 = n_2 = 8$ in each group. The observed difference 1.375 is shown; the shaded values are differences that are "as extreme as" 1.375. The p-value is the sum of all the shaded probabilities.

The sum of the shaded probabilities in Figure 15.2 is (using the 1,000,000 random permutations) $pv = 0.632$. This probability is interpreted as the probability of seeing a sample average difference as extreme as 1.375, by chance alone. Since this probability is large, you cannot rule out chance as an explanatory mechanism.

Be careful that you do not say "there is a 0.632 probability that the mechanism is chance alone." Again, that interpretation is wrong because it conflates Pr(A|B) with Pr(B|A). The p-value is equal to Pr(difference is as extreme as 1.375|chance mechanism); the p-value is *not* equal to Pr(chance mechanism|difference is 1.375).

One more note about the word *extreme*: what counts as extreme is guided by your research goals, and you should take care to not be dogmatic about how you understand Nature. In this example we were open-minded about Nature: we wanted to see if there was a difference—any difference—in the average age of students sitting in the back row versus the front row. In this situation, any extreme difference (positive or negative) indicates that our results are not explainable by chance alone.

Suppose, though, that you were interested in determining whether older people sit in the back rows. If you ask the question narrowly this way, an extreme result happens only when the average difference is large and positive. If the difference is large (in absolute value) and negative, you would not declare it to be extreme, since you have decided *a priori* that only large positive differences matter. This would be an example of a **one-tailed test**; the analysis in Example 15.2 is a **two-tailed test**.

The p-value can be approximated using the distribution N(0, 22.902/4) found earlier, which assumes an iid null model, rather than a permutation null model. In EXCEL, recall that the NORM.DIST(y, mu, sigma, TRUE) returns the cdf, i.e., the lower tail probability. Thus, the probability greater than 1.375 is "– 1 – NORM.DIST(1.375, 0, sqrt(22.902/4),TRUE)", this returns 0.2828 for the area in the upper tail. Multiplying by 2 (since, by symmetry of the normal distribution, the probability greater than 1.375 is the same as the probability less than –1.375) gives the area in both tails as 0.565. The normal approximation p-value

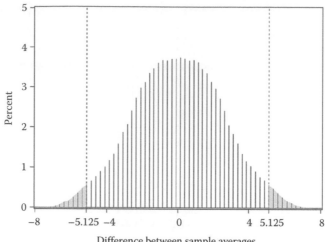

FIGURE 15.3

Permutation distribution of the difference between sample averages of student ages in back versus and front rows, with $n_1 = n_2 = 8$ in each group. The hypothetical observed difference −5.125 is shown; the shaded values are differences that are as extreme as −5.125. The p-value is the sum of all the shaded probabilities.

0.565 is uncomfortably far from the randomization p-value 0.632, but there is no essential difference in the conclusions: either way, you cannot rule out chance as a possible explanatory mechanism.

On the other hand, suppose that the difference between sample averages was −5.125 years. Again, the p-value is the probability of seeing a difference as extreme as what you observed, by chance alone. If the difference were −5.125 years, then the phrase "a difference as extreme as what you observed" translates to "a difference that is either −5.125 or lower, or 5.125 or higher." The upper tail is included because a difference of 5.125 is just as "extreme" as −5.125. Figure 15.3 shows the p-value as shaded region; the sum of all the shaded probabilities is the p-value.

The sum of the shaded probabilities in Figure 15.3 is $pv = 0.043$. This probability is interpreted as the probability of seeing a sample average difference as extreme as −5.125, by chance alone. Since this probability is relatively small, it seems less likely that the explanatory mechanism is purely chance. In the face of such a small probability, you have the logical choice that either (a) chance is the explanatory mechanism and a rare event has occurred or (b) the explanatory mechanism is not chance alone. Choice (b) is more logical when the probability of the event is small under the chance-only model.

Again, be careful that you do not say "there is a 0.043 probability that the mechanism is chance alone." That interpretation is wrong, again because it conflates Pr(chance mechanism|difference = −5.125) with Pr(difference is as extreme as −5.125|chance mechanism). The latter probability is equal to the p-value, not the former.

Using the distribution N(0, 22.902/4) found earlier, you can calculate the probability in the lower tail from −5.125 using Microsoft Excel as "= NORM.DIST(−5.125, 0, sqrt(22.902/4),TRUE)," which returns 0.0161 for the area in the lower tail. Multiplying by 2 gives the area in both tails to be 0.032. The normal approximation p-value 0.032 is different from the randomization p-value 0.043, but again there is no essential difference in the conclusions: either way, chance alone seems unlikely as a possible explanatory mechanism.

15.4 The Extremely Ugly "$pv \leq 0.05$" Rule of Thumb

In the example with age and seat selection, one *p*-value was 0.632, the other was 0.043. In the former case, we claimed that the results were explainable by chance alone and in the latter we claimed that chance was not as plausible as an explanatory mechanism. Where to draw the boundary between results that are explainable by chance alone and those that are, for all intents and purposes, not explainable by chance alone?

We have the famous historical figure Sir R.A. Fisher to blame for the $pv \leq 0.05$ rule of thumb, which we call *extremely* ugly because it is so over-used and abused by researchers.

Ugly Rule of Thumb 15.1

If the *p*-value is less than 0.05, then you can confidently rule out chance as the explanatory mechanism. If the *p*-value is more than 0.05, then the results can be classified as explainable by chance alone.

There is nothing magical about this ugly rule of thumb. Nature does not change mystically when the *p*-value moves from 0.051 to 0.049; yet unfortunately, researchers often think so. Fisher himself was not wed to the notion: his essential rationale was something to the effect of "it seems reasonable to me"; nothing more profound than that.

Nevertheless, the "$pv \leq 0.05$" rule of thumb is a permanent fixture on the statistical landscape. Journals routinely use it to gauge whether a researcher's results are interesting enough to publish, and international pharmaceutical regulatory agencies use it to assess whether a drug can be considered safe and/or effective.

One thing that you *can* say about the $pv \leq 0.05$ rule is this: If nothing but chance alone explains the observed differences in your study, then you will observe $pv \leq 0.05$ only in (approximately) 5% of similar studies. (This conclusion is explained further later.) So, by using the ugly $pv \leq 0.05$ rule, you can be assured that you will make the incorrect conclusion that the results are not explainable by chance alone only around 5% of the time, when the results are in fact explained by chance alone.

Sir R.A. Fisher's suggestion of 0.05 as a *p*-value threshold is not mandatory, nor for all its popularity is it universally adopted. If you want more assurance that the results are not due to chance, then you can use a lower threshold. In quality control, the threshold 0.003 is common, and in particle physics, thresholds of one in a million and less are sometimes used. But if you are more interested in making claims of discoveries even if you are wrong in making such claims, then you could use a higher threshold such as 0.10, or 0.20, or even 0.50.

The **level of significance** of your statistical procedure is the *p*-value threshold that you use to categorize a result as either explainable by chance alone or not explainable by chance alone. The level of significance is given the symbol α; commonly $\alpha = 0.05$, as per Fisher's recommendation.

No matter what level of significance α you pick, the conclusion is analogous to the $pv \leq 0.05$ rule: you will observe $pv \leq \alpha$ in (approximately) 100α% of studies where your data are produced by a null model. Stated as a probability, $\Pr(PV \leq \alpha \,|\,\text{no difference in process}) \cong \alpha$. This is the cumulative distribution of the *PV*; taking the derivative with respect to α gives the null probability distribution of *PV* as $p(pv) \cong (\partial/\partial\alpha)\alpha = 1$, for $0 < pv < 1$. This last statement gives you the following remarkable fact:

The null distribution of the *p*-value is (approximately) the uniform $U(0, 1)$ distribution.

To understand why this is true, we need to introduce more vocabulary. The **critical value** is the smallest value of the test statistic for which the p-value is less than α. In the example with age and seat selection, the probability that the difference between sample means (in absolute value) is 5.125 or more is 0.043, as shown earlier. The next smaller value of the test statistic is 4.875 (see Figure 15.1), and the probability that the difference between sample means (in absolute value) is 4.875 or more is 0.056. Hence, 5.125 is the $\alpha = 0.05$ critical value of the test statistic.

Now, the p-value is less than or equal to α if and only if the test statistic exceeds the α critical value. For example, see Figure 15.3. The p-value is less than or equal to 0.05 if and only if the test statistic $T = \bar{Y}_2 - \bar{Y}_1$ is either ≥ 5.125 or ≤ -5.125. Hence, the probability that the p-value is less than α is equal to the probability that the absolute test statistic exceeds the α critical value, which is approximately α by definition of the critical value. In the case of the age/seat selection case, the probability that the p-value is less than or equal to 0.05 is identical to the probability that the test statistic exceeds (in absolute value) 5.125, or 0.043.

The probability that the p-value is less than α is *approximately* α in the case of a discrete null distribution, rather than *exactly* α because the precise significance level α (e.g., 0.05), is usually not achievable. For example, achievable significance levels in the seat selection case are 0.043 and 0.056, but nothing in between. The probability is also *approximately* α when you use an approximate normal distribution for the null distribution of the test statistic. But there are some cases where the null distribution is continuous and known exactly, rather than approximately; these cases involve sampling from normal processes and the resulting t-tests, chi-squared tests, and F tests discussed in Chapter 16. In such cases, the null distribution of the p-value is *exactly* the uniform U(0, 1) distribution.

The fact the p-values are random variables may come as a surprise to you, if you have seen them before. Often, students learn (incorrectly) to interpret p-values as measures of absolute truth: they have learned to state that if the p-value is less than 0.05, then there is a systematic tendency, and otherwise nothing but chance is at work. These interpretations are clearly wrong when you understand that p-values are random variables. When there is nothing but chance at work, the p-value will be less than 0.05, 5% of the time.

Thus, if you are looking at thousands of p-values—for example in a genomics research project involving thousands of genes—you will easily see dozens, perhaps even hundreds, of p-values that are less than 0.05, many indicating nothing at all other than chance variation. To correct for this problem, you need to use **multiple testing methods**. Such methods use lower p-value thresholds than Fisher's 0.05 in order to screen out **false positives**.

Example 15.3: Are Stock Returns Independent of Previous Returns?

In Chapter 5, Figure 5.5, there are two graphs: one shows the distribution of the Dow Jones Industrial Average returns on days where the previous day's return was down, and the other shows the distribution of the Dow Jones Industrial Average returns on days where the previous day's return was up. We argued there that the difference was negligible, so that the independence model is therefore reasonable. But are the differences in the histograms shown in Figure 5.5 explainable by chance alone, or is there really some systematic difference?

To answer that question, you have to select a statistic to measure the difference. A statistic of interest to investors is the average return, so you could measure the difference between the histograms using difference between averages for the two scenarios, grouped by days where the return was down on the previous day versus days where the return was up on the previous day. When the previous day was down, the average historical return is -0.0004514. When the previous day was up, the average historical return is 0.00080521.

Very interesting! These data suggest that if you put money in the stock market following a down day, you should expect to *lose* money on the following day, but if you put money in the stock market following an up day, you should expect to *gain* money on the following day. The difference between these sample means is $0.00080521 - (-0.0004514) = 0.001257$ (or 0.1257%). But is this difference explainable by chance alone? You should be skeptical.

What does "by chance alone" mean here? It means that the financial processes that produce the returns when the previous day's return is down are identical to the financial processes that produce returns when the previous day's return is up. The permutation model used in the age/seat selection example is a possible model, although it makes less sense in the finance example than it does in the student example. While it is easily imaginable that a particular one of the $n = 16$ students might randomly select his or her seat, it makes no sense whatsoever that a particular one of the stock returns among the $n = 18,834$ consecutive trading days might randomly choose which of the 18,834 trading days in which to appear. Instead, the iid model for the null distribution is more reasonable: independence specifically states that today's return has no effect whatsoever on the distribution of tomorrow's return; thus if this model is true, then the 0.001257 difference is explained by chance alone.

So, suppose as a null model that all 18,834 returns are generated as an iid sample from a common distribution $p_0(y)$. In symbols your null model is

$$Y_1, Y_2, \ldots, Y_{18,834} \sim_{\text{iid}} p_0(y)$$

The distribution $p_0(y)$ is unknown, but you can estimate it using the bootstrap distribution $\hat{p}_0(y)$ that puts 1/18,834 probability on each of the observed returns; see Sections 8.7 and 14.3 for refreshers on the bootstrap distribution and on bootstrap sampling. You can then use this bootstrap distribution to generate an iid sample $Y_1^*, Y_2^*, \ldots, Y_{18,834}^*$, calculate the difference $\bar{Y}_2^* - \bar{Y}_1^*$ between simulated "days" where the previous "day" was up versus down, and repeat many, many times to arrive at an approximate null distribution for $T = \bar{Y}_2 - \bar{Y}_1$. For each simulated data set, even the sample sizes for "previous day up" and "previous day down" cases are random; the randomness in these sample sizes contributes to the randomness in the $\bar{Y}_2^* - \bar{Y}_1^*$ values and is captured correctly by the bootstrap sampling method.

Figure 15.4 shows the histogram of the 1000 bootstrapped values of $\bar{Y}_2^* - \bar{Y}_1^*$, along with the observed difference $\bar{y}_2 - \bar{y}_1 = 0.00080521 - (-0.0004514) = 0.001257$. Clearly, the observed difference is not explainable by chance alone! Over the history of the Dow Jones Industrial Average, there is a real difference in the distribution of returns following up days versus the distribution of returns following down days.

While the observed difference in stock returns is not attributable to chance variation, you still should be skeptical about whether the results are currently relevant because much of these data are from the distant past. In addition, the iid assumption for stock return data can be criticized on the basis that such data are dependent on previous data due to persistent volatility effects: When the market is highly variable—i.e., when $\text{Var}(R_t)$ is high—then points nearby in time are also highly variable. A model that accounts for such persistent volatility effects is the *autoregressive conditionally heteroscedastic*, or ARCH model, and your analysis would be improved if you use a null model where there are ARCH effects. So, before you rush out and invest money in stocks every day after the market goes up, you should do a little more research.

When a difference is not easy to explain by chance alone, it is called a **statistically significant** difference. Thus, there is a statistically significant difference between average returns in days following up days and days following down days. Be careful that you do not confuse **statistical significance** with **practical significance**. Results might be statistically significant, but of no practical interest whatsoever if the size of the difference

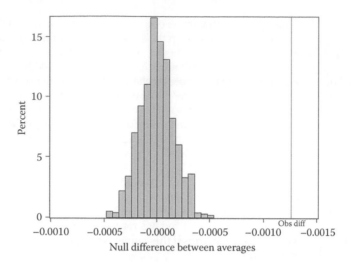

FIGURE 15.4
Null distribution of difference between averages after an up day versus averages after a down day. The observed difference, 0.001257, is shown with the vertical line to the right of the histogram.

is very small. Indeed, the 0.001257 difference shown in Figure 15.4 is statistically significant, but is so small that it may be of limited value to a stock trader, particularly when transaction costs are included in the calculation. Further, the difference is clearly small when compared to the day-to-day variability in returns: as shown in Figure 5.5, the typical range of returns far exceeds the difference between average returns. We concluded at the time that the returns were reasonably modeled as independent, since the two distributions are so similar. Here we conclude that there is statistically significant *dependence*; however, we'll stick with our original claim that independence is a reasonable model, because the difference in those distributions is so small and of questionable practical significance.

Results can be statistically significant but practically insignificant when the sample size is very large, as is the case here with $n = 18,834$ trading days. Under the null iid model, the difference $\bar{Y}_2^* - \bar{Y}_1^*$ has mean 0 and variance $\sigma^2/n_1 + \sigma^2/n_2$, where n_1 is the number of observations used to calculate $\bar{Y}_1^*$ and n_2 is the number of observations used to calculate $\bar{Y}_2^*$. A statistically significant difference is found when the difference $\bar{Y}_2 - \bar{Y}_1$ is more than 2 standard deviations from 0 or when the difference is outside the range $\pm 2\sqrt{\sigma^2/n_1 + \sigma^2/n_2}$. Therefore, with larger n_1 and n_2, you will be able to state that the difference between averages is statistically significant, even when the actual difference $\bar{y}_2 - \bar{y}_1$ is very small.

Example 15.4: Are Student Ages Produced by a Normal Distribution?

The answer to this question is, most resoundingly, "No!" The student age data-generating process has discreteness and skewness characteristics that make it obviously non-normal. Still, there is no harm in asking the question, "Are the deviations from normality *explainable* by chance alone?" If so, then you might think that the normal model (despite obviously being wrong) is a reasonable model.

Once again, the age data are $y_1 = 36$, $y_2 = 23$, $y_3 = 22$, $y_4 = 27$, $y_5 = 26$, $y_6 = 24$, $y_7 = 28$, $y_8 = 23$, $y_9 = 30$, $y_{10} = 25$, $y_{11} = 22$, $y_{12} = 26$, $y_{13} = 22$, $y_{14} = 35$, $y_{15} = 24$, and $y_{16} = 36$. Are these data produced by a normal distribution? Clearly not, since they are discrete. But suppose we had finer measurements such as $y_1 = 36.31$, $y_2 = 23.44$, $y_3 = 22.06$, $y_4 = 27.91$,

$y_5 = 26.33$, $y_6 = 24.32$, $y_7 = 28.25$, $y_8 = 23.56$, $y_9 = 30.80$, $y_{10} = 25.32$, $y_{11} = 22.29$, $y_{12} = 26.50$, $y_{13} = 22.89$, $y_{14} = 35.31$, $y_{15} = 24.33$, and $y_{16} = 36.99$. Is it believable that these data could have been produced by a normal distribution?

Figure 15.5 shows the quantile–quantile plot of these data. Recall that deviations from a straight line suggest non-normality.

There are deviations suggesting, as expected, positive skewness: the lowest age values are not as low as expected, had the age data come from a normal distribution, and the highest age values are higher than expected, had the age data come from a normal distribution.

But are these differences explainable by chance alone? In other words, if the data were really produced by a normal distribution, would it be possible to see deviations such as shown in Figure 15.5, purely by chance? Figure 15.6 shows a q–q plot of data that are produced by the $N(27.3, 5.0^2)$ distribution. The mean and standard deviation, 27.3 and 5.0, are the sample mean and standard deviation of the actual age data.

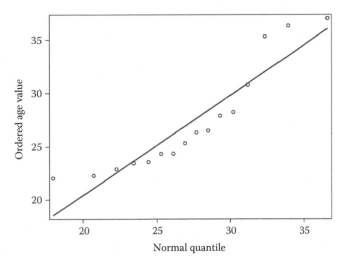

FIGURE 15.5
Quantile–quantile plot of student age data.

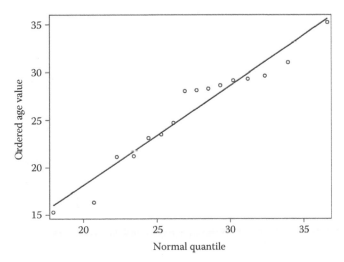

FIGURE 15.6
Quantile–quantile plot of data that are produced by a normal distribution.

As you can see, there are chance variations apparent in Figure 15.6: the data do not fall exactly on a line, even though the distribution that produced the data is *exactly* a normal distribution. Thus, the deviations in Figure 15.6 from the line are completely *explained* by chance alone.

Are the deviations in Figure 15.5 *explainable* by chance alone? There are two issues to consider. First, the graph in Figure 15.6 is just one sample; many more are needed to see all the possible differences that are explained by chance alone. Second, while an eyeball analysis is fine to understand the concept, it lacks the specific quantification you need to calculate a *p*-value. What you need is a measure of lack of fit to the line. The plug-in estimate of correlation given in Section 10.3 is one such measure—the smaller the correlation coefficient, the farther from the line are the data points.

With the original data in Figure 15.5, the correlation is 0.9286. With the simulated data graphed in Figure 15.6, the correlation is 0.9693. Thus, according to the correlation measure, the chance deviation in Figure 15.6 is smaller than the chance deviation shown in Figure 15.5. This suggests that the deviation in the original data is more than expected by chance alone, but many more samples from the normal distribution are needed to confirm such a statement.

Figure 15.7 shows the distribution of 50,000 correlation coefficients from *q–q* plots as shown in Figure 15.6. In other words, there are 50,000 graphs just like Figure 15.6, all samples from a normal distribution, and each has a different correlation because they are all random samples. (We won't show you all the graphs!)

As Figure 15.7 shows, it is in fact unusual to see a correlation as low as 0.9286 when the data are truly sampled from a normal distribution. The *p*-value is the probability of seeing a result as extreme as what is observed by chance alone; in this example, the "extreme" correlations are those 0.9286 and smaller because these suggest bigger differences between the observed data and the line. Among the samples represented by Figure 15.7, the correlation is less than 0.9286 (by chance alone) for only 2.16% of the 50,000 samples; hence, the *p*-value is estimated to be 0.0216. According to the ugly $pv \leq 0.05$ rule of thumb, you can rule out chance as an explanation for the discrepancy between the observed data and the expected line, shown in Figure 15.5, and conclude that the difference is statistically significant. Therefore, the distribution that produced the data is not a normal distribution.

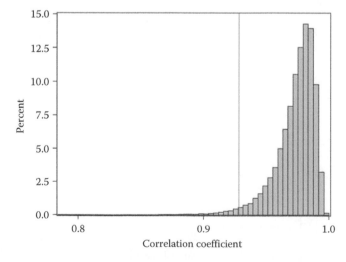

FIGURE 15.7
Distribution of correlation coefficients from 50,000 *q–q* plots of *n* = 16 observations sampled from a normal distribution. The observed correlation 0.9286 from the original data set is shown with a vertical line.

Notice, however, that even though you rejected the premise that the distribution is normal, you could be wrong! It is entirely possible (though unlikely) to observe coefficients as small as 0.9286 when the distribution is *exactly* normal, as shown in Figure 15.7. You are again, in essence, making a bet: you are wagering that your correlation in your original data is probably not one of the few odd ones you could get by chance, even when the distribution that produced the data is a normal distribution.

Had the result come out different, for example, with a *p*-value 0.34, you would conclude that the difference is *explainable* by chance alone. That is, the difference is within the expected range of variations that occur when the distribution is truly normal. What a mouthful! But it's very important to say all those words, because it underscores the difference between *explainable by chance alone* and *explained by chance alone*. In particular, you *cannot* conclude that the distribution is normal just because the deviation from the straight line in the *q–q* plot is within the chance deviations. Some other distribution could have produced the data as well.

Stating that a difference is *explainable by chance alone* is light-years apart from stating that the difference is *explained by chance alone*.

Vocabulary and Formula Summaries

Vocabulary

Statistical significance testing; hypothesis testing	A systematic approach to investigating whether the results of a study are explainable by chance alone.
Null model	A model that mimics Nature in the "no difference" or "no effect" case.
Test statistic	A function of the data used in statistical significance testing to measure a difference of interest.
Randomization (permutation) model	A null model constructed by considering all possible permutations of n data values as equally likely, conceptually identical to deals from a well-shuffled deck of cards.
Null distribution of the test statistic	The probability distribution of a test statistic when the null model is true.
p-value (pv)	The probability of seeing a result as extreme as your observed result, when the null model is true.
One-tailed test	A hypothesis test that examines differences in the process going in only one of two directions—either large positive or large (in absolute value) negative differences.

Two-tailed test

A hypothesis test that examines differences in both directions—both large positive and large (in absolute value) negative differences.

Level of significance

The threshold that you use to categorize a result as either explainable by chance alone or not explainable by chance alone; commonly denoted by α.

Critical value

The smallest positive value of the test statistic for which the p-value is less than the level of significance α.

Multiple testing methods

Methods that increase the burden of proof that a researcher must provide to declare evidence of a real difference when conducting several hypothesis tests; employed to screen out false positive results.

False positive

Declaring that a real difference in the process exists when there is actually no difference.

Statistically significant

A result that is not easily explainable by chance alone.

Statistically insignificant

A result that is easily explainable by chance alone.

Practically significant

When the true difference is large enough to be of practical importance to decision-makers.

Key Formulas and Descriptions

$p_0(y)$

A model that you assume to produce your data, in which there is no difference in reality regarding the quantity you are studying; a model that allows you to assess whether your results are explainable by chance alone.

$E(\bar{Y}_2 - \bar{Y}_1) = E(\bar{Y}_2) - E(\bar{Y}_1)$

The expected value of the difference of group averages.

$Var(\bar{Y}_2 - \bar{Y}_1) = Var(\bar{Y}_2) + Var(\bar{Y}_1)$

The variance of the difference of group averages assuming independence.

$pv = \Pr(\text{test statistic} \geq t \,|\, \text{no difference in process})$

The definition of p-value, assuming larger values of the test statistic are the extreme ones.

$pv \leq 0.05$	The extremely ugly rule of thumb for ruling out chance as a possible explanation for your observed difference.
$PV \sim U(0, 1)$	The distribution of the p-value is approximately the uniform distribution between 0 and 1 when the null model is true.
$p(pv) \cong 1$, for $0 < pv < 1$	The distribution of the p-value is approximately the uniform distribution between 0 and 1 when the null model is true.

Exercises

15.1 Suppose Hans flips a coin 100 times, getting 45 heads. You thought he should get 50 heads. Hans' data are Bernoulli, $y_1, y_2, ..., y_{100}$, where each y_i value is either 0 or 1 and where $\Sigma y_i = 45$. Answer the question, "Is the difference from 45 to 50 in Hans' data explainable by chance alone?" as follows:

A. State the relevant null model for the data $Y_1, Y_2, ..., Y_{100}$.

B. Find the approximate null distribution of the test statistic $T = \Sigma Y_i$ using the central limit theorem. Use your model in Exercise 15.1A to find the relevant mean and variance.

C. Use the distribution of T that you found in Exercise 15.1B to find the two-sided p-value for Hans' data, and then answer the question.

15.2 Ten adult dogs on a diet of Standard Dog Chow are observed for a day using video surveillance. The total length of time X_i, out of 24 hours, that the dogs appeared to be smiling, was recorded for each of dogs $i = 1, 2, ..., 10$. The next day, the dogs were all fed Happy Dog Chow. The video surveillance was repeated, and the total length of time Y_i, out of 24 hours, that the dogs appeared to be smiling, was recorded for each dog. The researchers then calculated the differences, $D_i = Y_i - X_i$, for each of dogs $i = 1, 2, ..., 10$, getting data 0.22, –1.79, 1.17, –1.46, 1.36, 1.20, 0.60, 0.62, –0.75, and –0.28. A null model is that the D_i are iid with $E(D_i) = 0$.

A. Find an approximate null distribution of the test statistic $\bar{D} = (1/10)\Sigma D_i$.

B. Use the distribution in Exercise 15.2B to find the p-value for the two-sided test. Are results explainable by chance alone?

C. You can create a bootstrap population distribution where the null is true by subtracting $\bar{d} = (1/10)\Sigma d_i$ from each of the data values d_i. Using this bootstrap population distribution produce the histogram of 1000 values of the test statistic and locate your observed test statistic on the graph.

D. Using the bootstrap data in Exercise 15.2C find the two-sided p-value and compare your results with Exercise 15.2B.

15.3   BankTen conducted a survey of 1250 customers' satisfaction on a 1, 2, 3, 4, 5 scale, with 1 being "Highly Dissatisfied" and 5 being "Highly Satisfied." The average response was a 3.05, with a standard deviation of 1.4. Further breakdown of the data into groups of $n_1 = 250$ loyal customers versus $n_2 = 1000$ newer customers shows averages 3.50 and 2.94, respectively. Answer the question, "Is the difference between these averages explainable by chance alone?" as follows:

A. State the iid null model. Explain the meaning of each unknown parameter of this model. (There are five unknown parameters and they add to 1.)

B. Find the approximate distribution of $\bar{Y}_1 - \bar{Y}_2$, assuming the iid null model of Exercise 15.3A is true.

C. Using the distribution in Exercise 15.3B find the p-value for the two-sided test and state your conclusion.

D. The approximation in Exercise 15.3B is suspect. Use a simulation model that produces iid Likert scale data to estimate the probability in Exercise 15.3C. Comment on whether the approximation in Exercise 15.3B is reasonable.

15.4   You are thinking of trying a new route to your job, class, the store, etc., in the morning. You decide to record the times it takes to get from your home to your destination going your current route for a few days, and then your time to get to your destination using the new route, also for a few days. There will be a difference in the average times.

A. Give a chance-only model that can explain this difference. Use an example from this chapter to help you describe this model.

B. Create a realistic null simulation model to explain the differences. Use it to generate a few differences between your average times, assuming $n = 10$ for both routes.

C. Based on your answer to Exercise 15.4C, if you found that the new route was faster by an average of 0.15 minutes, would you conclude that the new route is indeed faster?

15.5   A random variable Z has the N(0, 1) under the null model. Only small values of Z are called extreme; i.e., this is a lower-tailed test. Find the p-values for each of the following cases:

A. $z = -3.00$

B. $z = -2.00$

C. $z = -1.00$

D. $z = 0.00$

E. $z = 1.00$

F. $z = 2.00$

G. $z = 3.00$

15.6   In Example 15.3, the null model was stated to be $Y_1, Y_2, \ldots, Y_{18,834} \sim_{iid} p_0(y)$. Suppose the null model is actually true; in other words, suppose stock returns really are iid.

A. Why is the bootstrap distribution $\hat{p}_0(y)$ used in Example 15.3 different from $p_0(y)$?

B. If the null model is true, do you expect the bootstrap distribution $\hat{p}_0(y)$ to be close to $p_0(y)$? To answer, consider what each model tells you about $\Pr(Y \leq 0)$, for example.

15.7 Closing prices for 30 consecutive Dow Jones Industrial Average trading days in 2012 are as follows: 12,977.57, 12,962.81, 12,759.15, 12,837.33, 12,907.94, 12,922.02, 12,959.71, 13,177.68, 13,194.10, 13,252.76, 13,232.62, 13,239.13, 13,170.19, 13,124.62, 13,046.14, 13,080.73, 13,241.63, 13,197.73, 13,126.21, 13,145.82, 13,212.04, 13,264.49, 13,199.55, 13,074.75, 13,060.14, 12,929.59, 12,715.93, 12,805.39, 12,986.58, and 12,849.59. Answer the question, "Are these data produced by an iid process?" as follows:

A. State the null model.

B. Draw the lag scatter plot as shown in Figure 7.18.

C. Calculate the plug-in correlation estimate that represents the data in Exercise 15.7B.

D. Use bootstrap sampling to estimate the probability that, by chance alone, you will observe a correlation as large or larger than what you did observe in Exercise 15.7C. (This is a one-sided test.)

15.8 Students are asked two questions, "How often do you read online newspapers?" and "How often do you read print newspapers"? They answer on a 1, 2, 3, 4, 5 scale, where 1 = rarely and 5 = often. Call the former measure X and the latter Y. There is a question as to how X and Y are related: one theory is that readers read more of both, and nonreaders read less of both, leading to a positive correlation. Another theory is that by preference, some people prefer print to online, while others prefer online to print, leading to negative correlation. Observed data are (4 1), (5 1), (1 1), (1 2), (5 3), (3 3), (1 2), (4 2), (1 3), (5 1), (3 5), and (5 2).

A. Calculate the plug-in estimate of correlation.

B. Give the bootstrap population distribution of X.

C. Give the bootstrap population distribution of Y.

D. Simulate 1000 data sets, each with $n = 12$ iid pairs (X^*, Y^*), where X^* and Y^* are sampled independently from their respective bootstrap distributions. Calculate the 1000 correlation coefficients from these data sets and draw their histogram. (Since the X^* are independent of Y^*, all non-zero correlations are explained by chance alone.)

E. Find the percentage of correlations simulated in Exercise 15.8D that are as extreme or more extreme than the correlation you observed in Exercise 15.8A. What is the common name for this percentage?

15.9 Repeat Exercise 15.8A through E, but using the $n = 33$ data pairs shown in Example 5.2.

15.10 Consider Figure 15.1. Suppose the test statistic is the ratio of variances rather than the difference between means.

A. Find the ratio of the estimated variances, with students in the front rows comprising the numerator.

B. Construct the null permutation distribution of the variance ratio and graph it as shown in Figure 15.1.

C. Find the two-sided p-value. (Note: If the observed variance ratio is 1.5, then values more extreme are either ≥ 1.5 or $\leq 1/1.5$.)

15.11 Hans is desperately seeking significance. He tries an experiment and then another experiment, and finds a *p*-value more than 0.05 both times. But he is convinced that there is more than just chance at work in his study. So he considers churning out more studies, desperately trying to find one with $pv \leq 0.05$. Suppose that, sorry to say for Hans, there really is nothing but chance at work in his study.

   A. What is the distribution of one of Hans' *p*-values?

   B. Suppose Hans does 20 studies. What is the probability that he will erroneously find a significant result? In other words, what is the probability that one of his 20 *p*-values will be less than 0.05? To answer, simulate 20 *p*-values from the distribution you found in Exercise 15.11A and check to see if any are less than 0.05. Now, repeat that simulation of 20 *p*-values, 1000 times. Estimate the probability as a percentage of the 1000 simulations.

   C. Repeat Exercise 15.11B, but instead supposing Hans does 100 studies.

   D. The *Bonferroni method* is used to screen false positives. Instead of using 0.05 as a *p*-value threshold, you use $0.05/k$, where k is the number of *p*-values. Repeat Exercise 15.11B and C using the Bonferroni method. Is the Bonferroni method effective at screening false positives?

15.12 Use the data and the *q–q* plot from Exercise 4.9. Follow the method of Example 15.4 to calculate the *p*-value for assessing whether the deviations from a straight line in the *q–q* are explainable by chance alone, assuming the data really do come from an exponential distribution. **NOTE:** You can simulate all data from the exponential distribution with $\lambda = 1$, since the correlation coefficient is scale free. There is no need to re-compute the mean of the data for each simulated sample.

15.13 Use the data and the *q–q* plot from Exercise 4.10D. Follow the method of Example 15.4 to calculate the *p*-value for assessing whether the deviations from a straight line in the *q–q* are explainable by chance alone, assuming the data really do come from the U(0, 360) distribution.

15.14 When data are collected in time sequence, there is a question as to whether there is a drift, upward or downward, suggesting non-iid behavior. You can label the observations as they arrive with a variable *T* that indicates order of arrival: $T = 1, 2, \ldots$ The plug-in estimate of the correlation between your data *Y* and their order of arrival *T* is a measure of drift. Zero correlation in the process is consistent with no drift, but you will almost always see that the correlation between the (*T*, *Y*) observed data is nonzero, even when there is no drift in reality. Is the nonzero correlation explainable by chance alone? Suppose the data are, in order of appearance, 12.3, 16.5, 56.7, 45.3, 29.5, 23.8, 67.9, 100.1, 140.2, 86.6, 76.6, 188.7, and 146.0.

   A. Draw the time sequence plot of the data, *Y* on the vertical axis, and *T* on the horizontal axis.

   B. Calculate the plug-in estimate of correlation between *T* and *Y* and interpret it.

   C. Why did we say "you will almost always see that the correlation between the (*T*, *Y*) observed data is nonzero, even when there is no drift in reality"?

   D. Is the difference between 0 and the correlation you observed in Exercise 15.14B explainable by chance alone? Use the iid model, along with the bootstrap population distribution, to calculate the *p*-value. Interpret the *p*-value.

16

Chi-Squared, Student's t, and
F-Distributions, with Applications

16.1 Introduction

Too many weasel words! Approximation here, approximation there. Can you be less weasel-like? The answer is, "Yes, provided you are willing to make some assumptions." Statistics is full of trade-offs. There is no free lunch! If you want better results, you have to pay the price of making more assumptions.

You can stop using the annoying word *approximately* if you know the particular distribution that produced your data. The most common and also the most useful distribution is the normal distribution: Exact procedures are available for one-sample analysis, two-sample analysis, multi-sample analysis, and regression analysis when your data-producing model is the normal probability distribution function (pdf):

$$p(y \mid \mu, \sigma) = \frac{1}{\sqrt{2\pi}\sigma} \exp\left\{ -\frac{1}{2\sigma^2}(y-\mu)^2 \right\}$$

You don't have to know the mean μ and the standard deviation σ in order to believe that this is a good model. You just need to know that the data-generating model is the normal distribution for *some* μ and σ. The fact that μ and σ can be anything, and that you do not have to know their numerical values, improves the believability of this model enormously. However, normal still is normal, flexible parameters or not, and the normal model is never precisely correct in practice, due to discreteness, skewness, and other non-normal aspects of your DATA. Nevertheless, the normal model provides a useful *anchor*: If in fact the normal distribution were the data-generating process, then your procedures would be exact. Since your data-generating process is not precisely normal, your normality-assuming procedures are not exact, they are only approximate. The farther from normal is your data-generating process, the worse are the approximations of your normality-assuming statistical procedures.

In this chapter, we'll ask you to suspend your disbelief for a while and assume that your data are in fact produced by normal distributions. If so, then you can use all the standard statistical procedures, available in all statistical software, without having to say anything about approximations. These standard statistical methods include single-sample intervals, two-sample *t*-tests, ANOVA, and regression tests. All are based on distributions of statistics that are functions of independent and identically distributed (iid) normal random variables; these distributions include the chi-squared, Student's *t*, and *F*-distributions.

16.2 Linearity and Additivity Properties of the Normal Distribution

If your data come from a normal distribution, then one approximation disappears immediately. Recall from Chapter 10 that if your data $Y_1, Y_2, ..., Y_n$ are produced as an iid sample from *any* distribution $p(y)$ with finite variance, then the distribution of $\bar{Y} = (Y_1 + Y_2 + \cdots + Y_n)/n$ is *approximately* a normal distribution, by the central limit theorem (CLT), for *large n*. But if your data $Y_1, Y_2, ..., Y_n$ are produced as an iid sample from a *normal* distribution, then the distribution of $\bar{Y} = (Y_1 + Y_2 + \cdots + Y_n)/n$ is *exactly* a normal distribution, for *any n*.

This fact follows from the **linearity** and **additivity properties of the normal distribution**.

Linearity Property of the Normal Distribution

If $Y \sim N(\mu, \sigma^2)$ and $T = aY + b$, then $T \sim N(a\mu + b, a^2\sigma^2)$.

This property should look familiar: The mean and variance of T follow the standard linearity properties of mean and variance. The only thing new is that the distribution of T is a normal distribution. This is not an ugly rule of thumb or just a mere suggestion—it's a beautiful mathematical fact! In Chapter 9, Section 9.3, you saw the method for finding the distribution of a transformed continuous random variable. Here is how it works for $T = aY + b$ in the case $a > 0$; the case $a < 0$ is similar.

$P(t) = \Pr(T \leq t)$	(By definition of cumulative distribution function [cdf])
$\quad = \Pr(aY + b \leq t)$	(By substitution)
$\quad = \Pr\{Y \leq (t - b)/a\}$	(By algebra, since $a > 0$)
$\quad = P_Y\{(t - b)/a\}$	(By definition of the cdf of Y)

Now, $p(t) = P'(t)$, since the derivative of the cdf is the pdf (see Chapter 2). So let's find the derivative of the cumulative distribution function (cdf) of T and see what it looks like.

$P'(t) = (\partial/\partial t)P(t)$	(By definition of derivative)
$\quad = (\partial/\partial t)\, P_Y\{(t - b)/a\}$	(By substitution)
$\quad = P_Y'\{(t-b)/a\}(\partial/\partial t)\{(t-b)/a\}$	(By the chain rule, property D9 given in Section 2.5 of Chapter 2)
$\quad = P_Y'\{(t-b)/a\} \times 1/a$	(By the linearity and additivity properties of derivatives)

Now, $P'_Y(y) = p_Y(y) = \left(1/\sqrt{2\pi}\,\sigma\right)\exp\left\{-\left((y-\mu)^2/2\sigma^2\right)\right\}$, by the fact that the derivative of the cdf is equal to the pdf and by the assumption that $Y \sim N(\mu, \sigma^2)$. So, by substitution and algebra we get the following:

$$P'_Y\left\{\frac{(t-b)}{a}\right\} \times \frac{1}{a} = p_Y\left\{\frac{(t-b)}{a}\right\} \times \frac{1}{a}$$

$$= \frac{1}{\sqrt{2\pi}\sigma}\exp\left[-\frac{\{(t-b)/a - \mu\}^2}{2\sigma^2}\right] \times \frac{1}{a}$$

$$= \frac{1}{\sqrt{2\pi}a\sigma}\exp\left[-\frac{\{t-(a\mu+b)\}^2}{2a^2\sigma^2}\right]$$

This form you can recognize as the $N(a\mu + b, a^2\sigma^2)$ pdf for the random variable T.

In summary, a linear transformation of a normally distributed variable is also normally distributed. This is a remarkable property of the normal distribution, one that is not true for most other distributions.

Another beautiful mathematical fact, not an ugly rule of thumb, is the additivity property of the normal distribution.

Additivity Property of the Normal Distribution under Independence

Suppose $X \sim N(\mu_X, \sigma_X^2)$ and $Y \sim N(\mu_Y, \sigma_Y^2)$, and also suppose X and Y are independent random variables. Letting $T = X + Y$, it follows that $T \sim N(\mu_X + \mu_Y, \sigma_X^2 + \sigma_Y^2)$.

This property should also look familiar: The mean and variance of T follow the standard additivity properties of mean and variance, assuming independence. The only thing new, stated here without proof (because the mathematics is a little trickier than for the linearity property), is that the distribution of T is a normal distribution.

Like the linearity property, the additivity property is also a remarkable property of the normal distribution, one that is not true for most other distributions.

16.3 Effect of Using an Estimate of σ

Back to the subject of approximation. Another source of the "approximate" weasel word that you have seen is in the substitution of $\hat{\sigma}$ for σ. A procedure that is *exact* when you use σ becomes *approximate* when you use $\hat{\sigma}$, simply because $\hat{\sigma}$ and σ are different numbers. This is more of a problem with small n than with large n, because, with large n, the estimator $\hat{\sigma}$ is very close to σ (since $\hat{\sigma}$ is a consistent estimator, as discussed in Chapter 11). But what happens with small n? The case where $n = 2$ provides insight.

Example 16.1: Estimating Mean Failure Time Using Data with a Small Sample Size

Suppose that you are an engineer who would like to know the time until an item fails. This is a newly patented item, and it has not been tested much in practice. The only failure times that are available are 3.2 and 3.4 years. Using these data, what can you say about the mean failure time? Certainly, your best guess is $(3.2 + 3.4)/2 = 3.3$ years, but how accurate is this guess?

To answer, first you must assume that the data $y_1 = 3.2$ and $y_2 = 3.4$ are produced as an iid sample from some distribution $p(y)$. Suppose you are willing to assume further that the distribution $p(y)$ is a normal distribution, $p(y) = \left(1/\sqrt{2\pi}\sigma\right)\exp\left\{-\left(1/2\sigma^2\right)(y-\mu)^2\right\}$, where you don't know μ and σ. You want to assess your uncertainty about the process mean failure time μ based on only $n = 2$ observations, $y_1 = 3.2$ and $y_2 = 3.4$. Sounds like "Mission Impossible"! But it can be done.

Following the development in Chapter 14 (remember the mountain lion!), you start by assuming Y_1 and Y_2 are independent and identically distributed as $N(\mu, \sigma^2)$. Then $\bar{Y} = (1/2)(Y_1 + Y_2)$ has the $N(\mu, \sigma^2/2)$ distribution by the following logic:

$Y_1 + Y_2 \sim N(\mu + \mu, \sigma^2 + \sigma^2)$	(By the additivity property of the normal distribution under independence, and since Y_1 and Y_2 are produced by the same $N(\mu, \sigma^2)$ process)
$\Rightarrow Y_1 + Y_2 \sim N(2\mu, 2\sigma^2)$	(By algebra)
$\Rightarrow (1/2)(Y_1 + Y_2) \sim N((1/2)2\mu, (1/2)^2 2\sigma^2)$	(By the linearity property of the normal distribution)
$\Rightarrow (1/2)(Y_1 + Y_2) \sim N(\mu, \sigma^2/2)$	(By algebra)
$\Rightarrow \bar{Y} \sim N(\mu, \sigma^2/2)$	(By substitution)

Notice that there are no weasel words—so far. The distribution of $\bar{Y} = (1/2)(Y_1 + Y_2)$ is *exactly* a normal distribution, not *approximately* a normal distribution, under the assumption that the actual data Y_1 and Y_2 are produced as iid from a normal distribution. So you can say that the probability that $\bar{Y}$ will be within $\pm 1.96\sqrt{\sigma^2/2}$ of μ is *exactly* 0.95; that is, in the long run, for 95% of the repeated samples from an identical process, $\bar{Y}$ will be within $\pm 1.96\sqrt{\sigma^2/2}$ of μ.

Now here comes the weasel: You don't know σ^2 so you have to estimate it. Based on the sample of $n = 2$ observations, $\hat{\sigma}^2 = \{1/(2-1)\}\{(3.2 - 3.3)^2 + (3.4 - 3.3)^2\} = 0.02$. So your *approximate* 95% confidence interval is $3.3 \pm 1.96\sqrt{0.02/2}$, or $3.104 \le \mu \le 3.496$.

Recall from Chapter 14, Section 14.3, that the *true* confidence level is *not* 95% when you have an *approximate* 95% confidence interval; it could be more than 95% or it could be less than 95%. Also recall from Chapter 14 that you can evaluate the *true* confidence level by simulation. Here, you are assuming a normal distribution, so you will simulate data from a normal distribution to evaluate the true confidence level.

Algorithm to Evaluate True Confidence Level of a Confidence Interval from Normally Distributed Data Using an Estimated Standard Deviation

1. Simulate Y_1^*, Y_2^* from a particular $N(\mu, \sigma^2)$ distribution. (Pick any μ and pick any $\sigma^2 > 0$).
2. Calculate $\bar{Y}^* \pm 1.96\sqrt{\{\hat{\sigma}^*\}^2/2}$.
3. Note whether the μ you picked in Step 1 is within the range $\bar{Y}^* \pm 1.96\sqrt{\{\hat{\sigma}^*\}^2/2}$.
4. Repeat Steps 1 through 3 many times, say $NSIM = 1,000,000$, to estimate the true confidence level as the percentage of the $NSIM$ simulations where your μ was actually in the interval.

A little algebra simplifies the checking of whether μ is in the interval and also introduces the t-statistic.

μ is in the interval $\bar{Y} \pm 1.96\sqrt{\hat{\sigma}^2/2}$	
$\Leftrightarrow \bar{Y} - 1.96\sqrt{\hat{\sigma}^2/2} \le \mu \le \bar{Y} + 1.96\sqrt{\hat{\sigma}^2/2}$	(By definition)
$\Leftrightarrow -1.96\sqrt{\hat{\sigma}^2/2} \le \mu - \bar{Y} \le +1.96\sqrt{\hat{\sigma}^2/2}$	(By algebra)
$\Leftrightarrow -1.96 \le (\mu - \bar{Y})/\sqrt{\hat{\sigma}^2/2} \le +1.96$	(By algebra)
$\Leftrightarrow +1.96 \ge (\bar{Y} - \mu)/\sqrt{\hat{\sigma}^2/2} \ge -1.96$	(By algebra)
$\Leftrightarrow +1.96 \ge (\bar{Y} - \mu)/(\hat{\sigma}/\sqrt{2}) \ge -1.96$	(By algebra)
$\Leftrightarrow \left\| (\bar{Y} - \mu)/(\hat{\sigma}/\sqrt{2}) \right\| \le +1.96$	(By algebra)

	A	B	C	D	E	F	G
	Y1	Y2	Ybar	sigma-hat	T-stat	Check	Count
1	Y1	Y2	Ybar	sigma-hat	T-stat	Check	Count
2	2.879907	2.488927	2.68442	0.276465	-1.614317	TRUE	6931
3	3.097703	3.510589	3.30415	0.291955	1.473268	TRUE	
4	3.47934	3.693253	3.5863	0.151259	5.481633	FALSE	
5	2.126565	2.906328	2.51645	0.551375	-1.240259	TRUE	
6	3.438009	2.56532	3.00166	0.617085	0.003814	TRUE	
7	2.723918	2.323827	2.52387	0.282907	-2.380093	FALSE	
8	2.261236	2.608948	2.43509	0.24587	-3.249282	FALSE	
9	2.690597	2.152828	2.42171	0.380261	-2.150689	FALSE	
10	2.77283	2.838381	2.80561	0.046352	-5.931099	FALSE	
11	3.053941	2.853803	2.95387	0.141519	-0.460961	TRUE	
12	2.869204	2.851904	2.86055	0.012233	-16.12099	FALSE	
13	3.537057	2.965886	3.25147	0.403878	0.880548	TRUE	
14	2.925537	2.794717	2.86013	0.092504	-2.138406	FALSE	
15	3.788885	3.346269	3.56758	0.312976	2.56465	FALSE	
16	3.950262	2.738037	3.34415	0.857172	0.567798	TRUE	
17	3.664582	2.355041	3.00981	0.925986	0.014985	TRUE	
18	3.215579	3.360877	3.28823	0.102741	3.967425	FALSE	
19	3.767566	2.966193	3.36688	0.566656	0.915628	TRUE	
20	2.790482	3.270055	3.03027	0.33911	0.126232	TRUE	
21	2.84747	3.303045	3.07526	0.32214	0.330385	TRUE	

FIGURE 16.1

Excel screen shot showing how to estimate true confidence level of the confidence interval $\bar{Y} \pm 1.96 \hat{\sigma}/\sqrt{2}$, based on sample of $n = 2$ observations from a normal distribution..

So, μ is in the interval if and only if the **t-statistic** $T = (\bar{Y} - \mu)/(\hat{\sigma}/\sqrt{n})$ is no more than 1.96 in absolute value.

Figure 16.1 shows an Excel screen shot to illustrate. First, generate two columns, randomly from a normal distribution. It will not matter which mean μ you choose and which standard deviation σ you choose; the end results will be the same. But pick, say, $\mu = 3.0$ and $\sigma = 0.4$, just to get started.

In the Excel screenshot, columns A and B are 10,000 rows of numbers Y_1^*, Y_2^* sampled as iid from N(3.0, 0.4²). Column C contains the averages $\bar{Y}^* = (Y_1^* + Y_2^*)/2$, column D contains the estimated standard deviations $\hat{\sigma}^* = \sqrt{\{1/(2-1)\}\{(Y_1^* - \bar{Y}^*)^2 + (Y_2^* - \bar{Y}^*)^2\}}$, and column E contains the t-statistics $T^* = (\bar{Y}^* - \mu)/(\hat{\sigma}^*/\sqrt{2}) = (\bar{Y}^* - 3.0)/(\hat{\sigma}^*/\sqrt{2})$. Column F contains checks of whether the t-statistic is less than 1.96 in absolute value: "= ABS(E2)< = 1.96," and column G contains the count of how many of the 10,000 absolute t-statistics $|T^*|$ are smaller than 1.96: "= COUNTIF(F:F,TRUE)." Only 6931, or 69.31%, of them are less than 1.96, so the estimated confidence level of the method is 69.31%—a far cry from your desired 95%.

What has gone wrong here is that the estimate of σ can be very inaccurate with small n. See row 12 in Figure 16.1, for example: The estimated σ is 0.012, very far from the true value 0.400. For the data in row 12, the 95% confidence interval for μ is $2.86 \pm 1.96(0.012)/2^{1/2}$, or $2.84 \leq \mu \leq 2.88$. Since you know that μ is actually 3.0, this interval clearly misses the mark.

Typically, the intervals tend to be wrong when the estimate of σ is too small. In practice, you have no way of knowing whether your estimated σ is too small. However, you can correct for the problem by properly accounting for the variability in your estimate of σ; this is the purpose of using *Student's t-distribution* rather than the standard normal distribution. Student's t-distribution gives you a larger critical value than 1.96, one that explicitly accounts for the variability in your estimate of σ and will make your interval for μ *exactly* 95% when your data are produced as iid from a normal distribution.

A side note: "Student" is actually the famous historical figure William Gossett, who worked as a quality control technician at the Guinness brewery in Dublin, Ireland, in the early twentieth century. He found that the normal distribution was inadequate for his tests and figured out that the t-distribution was more appropriate. But his company didn't want him fooling around with academic publication—after all, there is plenty of work to be done making good beer! So, rather than risk his company's ire, Gossett published his work under the pseudonym "Student." The formal citation is as follows:

Student (1908), "The Probable Error of a Mean," *Biometrika*, 6, 1–25.

To get to Student's (actually, Gossett's) famous result, we need to discuss the variability in the estimator $\hat{\sigma}$, which is described in terms of the similarly famous **chi-squared distribution**.

16.4 Chi-Squared Distribution

The main points of the previous section are that $\hat{\sigma}$ is a random variable, that it is therefore different from σ, and that this difference implies that the confidence interval results are *approximate* rather than *exact*. To account for this randomness, you first need to know the distribution of $\hat{\sigma}$.

Since the estimated standard deviation involves squares, the story starts with the distribution of the sum of squares of iid standard normal random variables.

Definition of the Chi-Squared Distribution

Suppose $Z_1, Z_2, \ldots, Z_m \sim_{\text{iid}} N(0, 1)$, and let $V = Z_1^2 + Z_2^2 + \cdots + Z_m^2$. Then the distribution of V is the *chi-squared distribution with m degrees of freedom* or in shorthand $V \sim \chi_m^2$.

The term **degrees of freedom** may seem a little mysterious at this point, but we'll discuss it in more detail later. The actual function form of the chi-squared distribution is $p(v) \propto v^{m/2-1}e^{-v/2}$; to see the mathematical derivation of this function form of $p(v)$, you'll need to consult a more advanced statistics text than this one.

While this form of $p(v)$ may look slightly unfamiliar, the special case where $m = 2$ is one you have seen before: $p(v) \propto v^{2/2-1}e^{-v/2}$, $p(v) \propto v^0 e^{-v/2}$, or just $p(v) \propto e^{-v/2}$. Since $e^{-v/2}$ is the kernel of the exponential distribution with $\lambda = 1/2$, you now know that the chi-squared distribution with $df = 2$ is identical to the exponential distribution with parameter $\lambda = 1/2$; this distribution is graphed in the upper right panel of Figure 12.1 in Chapter 12.

One thing that should be not mysterious is that the mean of this distribution is equal to m, the degrees of freedom. This follows from the following result.

Expected Value of the Squared Standard Normal

If $Z \sim N(0, 1)$, then $E(Z^2) = 1$.

This you can deduce because $E(Z) = 0$ and $\text{Var}(Z) = 1$: By definition, $\text{Var}(Z) = E\{Z - E(Z)\}^2$. But since $E(Z) = 0$, $\text{Var}(Z) = E\{Z - 0\}^2 = E(Z^2) = 1$. This gives you the mean of the chi-squared distribution.

Expected Value of the Chi-Squared Random Variable

If $V \sim \chi_m^2$, then $E(V) = m$.

This is true by the additivity property of expectation, since $E(V) = E\left(Z_1^2 + Z_2^2 + \cdots + Z_m^2\right) = 1 + 1 + \cdots + 1 = m$.

You also know that the chi-squared distribution is approximately a normal distribution for large m, by the CLT. To see how the distribution looks precisely for different m, you can graph its kernel $v^{m/2-1} e^{-v/2}$. You can also simulate data $Z_1^*, Z_2^*, ..., Z_m^*$ as iid $N(0, 1)$, calculate $V^* = \left(Z_1^*\right)^2 + \left(Z_2^*\right)^2 + \cdots + \left(Z_m^*\right)^2$, repeat thousands of times, and graph the histogram of the resulting thousands of V^* values to see the approximate shape of the chi-squared distribution with m degrees of freedom. Either approach, a graph of $v^{m/2-1} e^{-v/2}$ or a histogram of many V^* values, will show the same shape. Figure 16.2 shows the histogram when $m = 5$, based on 100,000 simulated values V^*, and Figure 16.3 shows the graph of the kernel $v^{5/2-1}e^{-v/2}$.

Neither Figure 16.2 nor Figure 16.3 is actually the chi-squared pdf, since neither has area equal to 1.0. However, both are proportional (or approximately proportional, in the case of Figure 16.2) to the chi-squared pdf with degrees of freedom (or df) = 5. The main point is that both have the same appearance, with the same shape and the same range of plausible data values. If your software has the actual chi-squared pdf, you should graph that instead of the histogram or kernel as shown in Figures 16.2 and 16.3. The actual pdf will have the same appearance, just with different vertical axis numbers, and with area under the curve = 1.0.

So, now you know the distribution of the sum of squares of m iid standard random normal random variables. What is this good for? How often do you ever see $N(0, 1)$ data in practice? Well, as it turns out, you see data related to $N(0, 1)$ quite often. The $N(\mu, \sigma^2)$ model is often

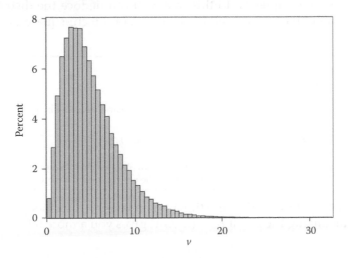

FIGURE 16.2
Histogram of 100,000 values $V^* = \left(Z_1^*\right)^2 + \left(Z_2^*\right)^2 + \cdots + \left(Z_5^*\right)^2$, where the Z values are produced as iid $N(0, 1)$.

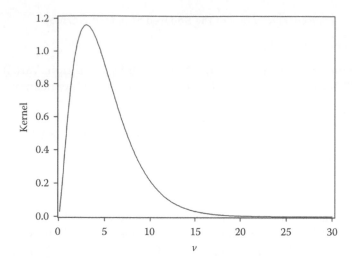

FIGURE 16.3
Graph of the kernel $v^{5/2-1}e^{-v/2}$ of the chi-squared distribution with $df = 5$.

a good one for your DATA, in which case the z-score defined in Chapter 9, $Z = (Y - \mu)/\sigma$, is distributed as N(0, 1), since $Z = (Y - \mu)/\sigma = (1/\sigma)Y - \mu/\sigma$. The linearity property of the normal distribution then implies that $Z \sim N\{(1/\sigma)\mu + (-\mu/\sigma), (1/\sigma)^2\sigma^2\}$, or, by algebra, that $Z \sim N(0, 1)$. Thus, z-scores calculated from normally distributed random variables have the standard normal or N(0, 1) distribution.

You can use the fact that $Z \sim N(0, 1)$ to find the distribution of the estimated variance, $\hat{\sigma}^2$. Suppose for a little while that you know the numerical value of the mean μ of your N(μ, σ^2). This is obviously false, but please play along for a bit. Then you can use the unbiased estimator of σ^2 as follows:

$$\hat{\sigma}^2 = \left(\frac{1}{n}\right)\sum (Y_i - \mu)^2$$

This was discussed in Chapter 11. In this case, you can deduce the distribution of $\hat{\sigma}^2$ as a simple application of the z-score and by definition of the chi-squared distribution. The logic is as follows:

$\hat{\sigma}^2 = (1/n)\sum (Y_i - \mu)^2$	(By assumption)
$= (1/n)\sigma^2 \sum (Y_i - \mu)^2/\sigma^2$	(By algebra, multiplying and dividing by σ^2)
$= (1/n)\sigma^2 \sum Z_i^2$	(By definition of z-score)
$\sim (\sigma^2/n)\, V$	(Where V has the chi-squared distribution with $df = n$, by definition of the chi-squared distribution)

Thus, the distribution of the variance estimator is related to the chi-squared distribution. Algebraically rearranging the result $\hat{\sigma}^2 \sim (\sigma^2/n)V$ gives you a more telling representation:

$$\frac{\hat{\sigma}^2}{\sigma^2} \sim \frac{\chi_n^2}{n} \tag{16.1}$$

Equation 16.1 provides excellent insights: Since χ_n^2 itself can be represented as a sum of squared standard normals, the ratio χ_n^2/n is their average. By the Law of Large Numbers (LLN), the average converges to the mean of the distribution of the squared standard normal, which is 1.0 as shown earlier. This also implies that $\hat{\sigma}^2/\sigma^2$ converges to 1.0 for large n. Conversely, with smaller sample sizes n, there is more variability in sample averages; hence, the ratio $\hat{\sigma}^2/\sigma^2$ tends to be farther from 1.0 with smaller n. In other words, $\hat{\sigma}^2$ tends to be farther from σ^2 with smaller n.

Of course, μ is unknown, so you can't use the variance estimator $\hat{\sigma}^2 = (1/n)\Sigma(Y_i - \mu)^2$. Instead, you use the unbiased estimator as follows:

$$\hat{\sigma}^2 = \left\{\frac{1}{(n-1)}\right\}\sum(Y_i - \bar{Y})^2$$

This was also discussed in Chapter 11. Recall that $\Sigma(Y_i - \bar{Y})^2$ is smaller than $\Sigma(Y_i - \mu)^2$ and that this is the reason for using the divisor $n - 1$ instead of n. This difference also shows up in the distributions: While $\Sigma(Y_i - \mu)^2/\sigma^2 \sim \chi_n^2$, the substitution of $\bar{Y}$ for μ changes the degrees of freedom, giving $\Sigma(Y_i - \bar{Y})^2/\sigma^2 \sim \chi_{n-1}^2$. This makes sense in the following way: You know that $\Sigma(Y_i - \bar{Y})^2$ is smaller than $\Sigma(Y_i - \mu)^2$, and you also know that the χ_{n-1}^2 distribution is shifted to the left of the χ_n^2 distribution, having mean $n - 1$ instead of n. While the formal proof of the loss of one degree of freedom is beyond the scope of this book, involving the algebra of n-dimensional vector spaces, there is an ugly rule of thumb associated with these degrees of freedom that appears over and over with advanced statistical methods:

Ugly Rule of Thumb 16.1

For every parameter you estimate, you lose a degree of freedom.

For example, you know that $\Sigma(Y_i - \mu)^2/\sigma^2 \sim \chi_n^2$. When you estimate the parameter μ, you lose a degree of freedom, giving you $n - 1$ instead of n.

The following example provides insight into the meaning of the term *degrees of freedom*.

Example 16.2: Degrees of Freedom and Dice Rolls

You roll an unbalanced six-sided die (hence μ is not 3.5). You get Y_1, which can be any number, 1, 2,..., or 6. There is no constraint. It is free to land however it wants. Thus, your outcome Y_1 has one degree of freedom. Roll it again, getting Y_2. That outcome is also free, unconstrained. It doesn't matter what the first roll showed, your second roll can still be any value 1, 2, ..., 6. Thus, there are two degrees of freedom in your two outcomes. Roll it again, three degrees of freedom; and so on. In general, for the numbers $Y_1, Y_2, ..., Y_n$ obtained as an iid sample, there are n degrees of freedom. The story does not change if you consider the transformed values $D_i = (Y_i - \mu)$. Because μ is a constant, the D_i are also iid and have n degrees of freedom, which corresponds to the n degrees of freedom for the chi-squared distribution of $\Sigma(Y_i - \mu)^2/\sigma^2$.

Now, since you don't know μ, you have to estimate it. Consider the transformed values $E_i = (Y_i - \bar{Y})$. These are no longer independent random variables: Since $\Sigma E_i = 0$, it follows that if you know $(n - 1)$ of the E_i, then you know the nth one. For instance, suppose that $n = 5$ and the deviations $e_i = (y_i - \bar{y})$ are $e_1 = 1.2$, $e_2 = 0.2$, $e_3 = 0.2$, and $e_4 = -1.8$. Then it follows that $e_5 = 0.2$. There are only four degrees of freedom in the E_i values when $n = 5$.

16.5 Frequentist Confidence Interval for σ

When you substitute $\bar{Y}$ for μ, you lose a degree of freedom; hence, $\Sigma(Y_i - \bar{Y})^2 / \sigma^2 \sim \chi^2_{n-1}$. Rearranging terms as in Equation 16.1 gives a following similar representation:

$$\frac{\hat{\sigma}^2}{\sigma^2} \sim \frac{\chi^2_{n-1}}{n-1} \qquad (16.2)$$

Here, $\hat{\sigma}^2$ is the unbiased estimator $\hat{\sigma}^2 = \{1/(n-1)\}\Sigma(Y_i - \bar{Y})^2$. Expression (16.2) provides similar insight as Equation 16.1: The estimate is more accurate with larger n. However, Expression (16.2) shows slightly less accuracy than Equation 16.1 since the average is comprised only of $n - 1$ terms rather than n terms.

The standard deviation is more interesting than the variance, for practical purposes, and Expression (16.2) also shows how close the estimator $\hat{\sigma}$ is to the estimand σ by simple square root transformation:

$$\frac{\hat{\sigma}}{\sigma} \sim \sqrt{\frac{\chi^2_{n-1}}{n-1}} \qquad (16.3)$$

Thus, the distribution of the ratio $\hat{\sigma}/\sigma$ is the same as the distribution of the square root of the average of $(n - 1)$ iid squared standard normal variables. You can do some fancy math to find this distribution, or you can just simulate it. Figure 16.4 shows

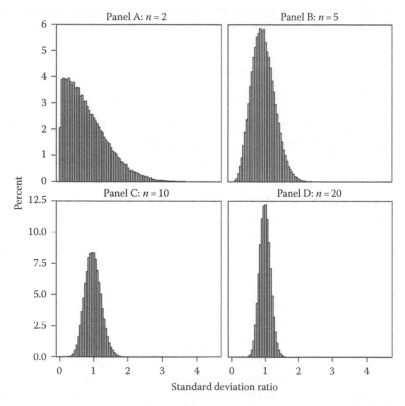

FIGURE 16.4
Simulated distributions of the ratio $\hat{\sigma}/\sigma$, for $n = 2, 5, 10$, and 20.

histograms of the square roots of the average of $(n - 1)$ iid squared standard normal variables, for different values of n.

Notice from Figure 16.4 that the distributions of the ratio $\hat{\sigma}/\sigma$ are centered near 1.0, a comforting result. However, the means are not exactly 1.0 by Jensen's inequality. Notice also that for larger n, the variability of the distributions becomes smaller, again a comforting result, because it tells you that the estimator $\hat{\sigma}$ tends to be closer to the estimand σ when n is larger. Finally, notice that the distributions are all right-skewed.

You can use the distributions shown in Figure 16.4 to construct a 95% confidence interval for σ. Suppose c_L is the 2.5th percentile of the distribution of $\hat{\sigma}/\sigma$ for a particular n, and suppose c_U is the 97.5th percentile. Then you can find a 95% confidence interval for σ using the following logic:

$\Pr(c_L \leq \hat{\sigma}/\sigma \leq c_U) = 0.95$	(By definition c_L and c_U as the 0.025 and 0.975 quantiles)
$\Rightarrow \Pr(1/c_L \geq \sigma/\hat{\sigma} \geq 1/c_U)$	(By algebra)
$\quad = 0.95$	
$\Rightarrow \Pr(\hat{\sigma}/c_L \geq \sigma \geq \hat{\sigma}/c_U) = 0.95$	(By algebra)
$\Rightarrow \Pr(\hat{\sigma}/c_U \leq \sigma \leq \hat{\sigma}/c_L) = 0.95$	(By rearrangement of terms so that the lower value is on the left and the upper value is on the right)

Notice that the *upper* quantile c_U of the distribution of $\hat{\sigma}/\sigma$ is in the denominator of the *lower* limit $\hat{\sigma}/c_U$ for σ and also that the *lower* quantile c_L of the distribution of $\hat{\sigma}/\sigma$ is in the denominator of the *upper* limit $\hat{\sigma}/c_L$ for σ.

You can get the values c_L and c_U by simulation; just pick the sample quantiles from the simulated data as shown in Figure 16.4. Or, you can use the chi squared distribution. In general, for a $100(1 - \alpha)\%$ interval

$$c_L = \sqrt{\frac{\chi^2_{n-1,\alpha/2}}{n-1}} \tag{16.4}$$

And

$$c_U = \sqrt{\frac{\chi^2_{n-1,1-\alpha/2}}{n-1}} \tag{16.5}$$

Here, $\chi^2_{df,p}$ denotes the p quantile of the chi-squared distribution. You can find these quantiles from many software packages. In SAS, for instance, the 0.025 quantile of the chi-squared distribution with nine degrees of freedom is quantile("chisquare", 0.025, 9), giving 2.70039; in Excel, you can use = CHISQ.INV(0.025,9) and get the same result.

$100(1 - \alpha)\%$ Interval for σ

Assuming $Y_1, Y_2, \ldots, Y_n \sim_{\text{iid}} N(\mu, \sigma^2)$, an exact $100(1 - \alpha)\%$ confidence interval for σ is $\hat{\sigma}/c_U \leq \sigma \leq \hat{\sigma}/c_L$, where c_L and c_U are given by Equations 16.4 and 16.5.

Example 16.3: Interval Estimation of the Standard Deviation of Failure Time with a Small Sample Size

In Example 16.1, the failure time data are $y_1 = 3.2$ and $y_2 = 3.4$. The estimated standard deviation is $\hat{\sigma} = \sqrt{\{1/(2-1)\}\{(3.2-3.3)^2 + (3.4-3.3)^2\}} = \sqrt{0.02} = 0.1414$. Assuming the data are produced as iid $N(\mu, \sigma^2)$, the 0.025 and 0.975 quantiles of the distribution of $\hat{\sigma}/\sigma$ are, using Equations 16.3 and 16.4:

$$c_L = \sqrt{\frac{\chi^2_{2-1,0.05/2}}{2-1}} = \sqrt{\frac{0.00098207}{1}} = 0.031338$$

And

$$c_U = \sqrt{\frac{\chi^2_{2-1,1-0.05/2}}{2-1}} = \sqrt{\frac{5.02388647}{1}} = 2.241403$$

Notice that 0.031338 and 2.241403 are the 0.025 and 0.975 quantiles of the distribution shown in panel A of Figure 16.4. On visual inspection of panel A, these quantiles appear to be correct. The exact 95% confidence interval for σ is then $\hat{\sigma}/c_U \le \sigma \le \hat{\sigma}/c_L$, 0.1414/2.241403 $\le \sigma \le$ 0.1414/0.031338, or 0.063 $\le \sigma \le$ 4.512.

The range of the confidence interval for σ is exceptionally wide; this happens because the sample size, $n = 2$, is so small. Note also that the range extends to a very large upper limit, 4.512, even though the estimate itself was small, 0.1414. The result makes sense when you consider that with a sample size of $n = 2$, it can easily happen that the two values just happened to be close together by chance alone, even when the actual variance is quite large. This type of occurrence is illustrated by panel A of Figure 16.4, where there is a large probability density for values of $\hat{\sigma}/\sigma$ that are near zero.

With small sample sizes, the Bayesian approach becomes more attractive. If you are the engineer tasked with determining how long a unit of the item can be expected to last before it fails, you may have prior knowledge about the possible range of variation in the failure times, and could incorporate this knowledge into a prior, then construct a posterior credible interval for σ that incorporates your prior knowledge. If your prior for σ places very low likelihood on values, say 2.0 and higher, then your posterior interval of σ will similarly have a lower upper limit, probably much smaller than 4.512, and closer to 2.0.

16.6 Student's *t*-Distribution

In Example 16.1, we noted that the *t*-statistic $T = (\bar{Y} - \mu)/(\hat{\sigma}/\sqrt{n})$ was less than 1.96 in absolute value only 69.3% of the time when $n = 2$, far from the expected 95% when the distribution of T is standard normal. The problem, as noted earlier, is simply that $\hat{\sigma}$ is random and not equal to the fixed σ. Also as noted earlier, the distribution of $\hat{\sigma}$ is related to the chi-squared distribution. The distribution of T is actually **Student's *t*-distribution** with $n - 1$ degrees of freedom; this distribution derives from both the standard normal and the chi-squared distributions.

Definition of Student's *t*-Distribution

Suppose $Z \sim N(0, 1)$, and suppose $V \sim \chi^2_m$, independent of Z. Then the distribution of $T = Z/\sqrt{V/m}$ is Student's *t*-distribution with m degrees of freedom. In shorthand, $T \sim T_m$.

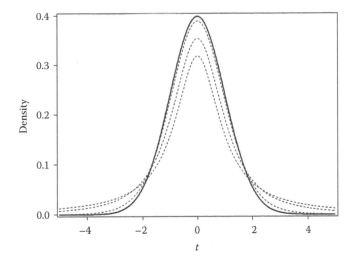

FIGURE 16.5
Student *t*-distributions (dotted curves) and the standard normal distribution (solid curve). Distributions are shown for *df* = 1 (the lowest peak), *df* = 2 (second lowest peak), and *df* = 10.

The derivation of the mathematical form of Student's *t*-distribution is beyond the scope of this book, but the mathematical form itself is not so complex: Its kernel is given by $p(t) \propto (1 + t^2/m)^{-(m+1)/2}$. Most statistical software packages have Student's *t*-distribution, and include the appropriate constant of proportionality that makes the area under $p(t)$ equal to 1.0. Figure 16.5 shows you how the *t*-distributions look for $m = 1, 2, 10$, and ∞; the case $m = \infty$ gives the standard normal distribution.

Compared to the standard normal distribution, the *t*-distribution has the same median (0.0) but with variance $df/(df - 2)$, which is larger than the standard normal's variance of 1.0. The variance is infinite when $df \leq 2$.

The connection of the *t*-distribution to real-world data is as follows.

Main Result for Student's *t*-Distribution

Suppose $Y_1, Y_2, ..., Y_n \sim_{\text{iid}} N(\mu, \sigma^2)$, and define $T = (\bar{Y} - \mu)/(\hat{\sigma}/\sqrt{n})$. Then $T \sim T_{n-1}$.

This result, along with Figure 16.5, explains why only 69.3% of the *t*-statistics were within the 0 ± 1.96 range in Example 16.1. Under the curve with $df = 1$ (the one with the shortest peak), only 69.3% of the area is between 0 ± 1.96. Using the T_1 cumulative distribution function rather than simulation, you can find this probability to be precisely $\Pr(-1.96 \leq T_1 \leq 1.96) = \Pr(T_1 \leq 1.96) - \Pr(T_1 \leq -1.96) = 0.84983 - 0.15017 = 0.6996$. On the other hand, there is precisely 95% of the area under the standard normal curve (the solid curve) between 0 ± 1.96.

It is mostly simple algebra to connect the main result with the definition of the *t*-distribution involving standard normals and chi-squares. But one result that requires higher math is this: If $Y_1, Y_2, ..., Y_n \sim_{\text{iid}} N(\mu, \sigma^2)$, then $\bar{Y}$ and $\hat{\sigma}$ are independent random variables. This is actually quite a remarkable fact, considering that both $\bar{Y}$ and $\hat{\sigma}$ are functions of the same data $Y_1, Y_2, ..., Y_n$. This result is another unique fact about the normal distribution. If the distribution $p(y)$ that produces the data Y_i is skewed, then $\bar{Y}$ and $\hat{\sigma}$ are dependent random variables.

Here is the logical connection between the main result where the t-distribution is defined in terms of data $Y_1, Y_2, \ldots, Y_n$ and the definition of the t-distribution in terms of standard normal and chi-squared random variables:

$$
\begin{aligned}
T &= \frac{\bar{Y} - \mu}{\hat{\sigma}/\sqrt{n}} & \text{(By definition)} \\[2mm]
&= \frac{\bar{Y} - \mu}{\sigma/\sqrt{n}} \times \frac{\sigma}{\hat{\sigma}} & \text{(By algebra)} \\[2mm]
&= Z \times \frac{\sigma}{\hat{\sigma}}, \text{where } Z \sim \mathrm{N}(0,1) & \text{(By linearity and additivity properties of normal} \\
& & \text{random variables)} \\[2mm]
&= \frac{Z}{\hat{\sigma}/\sigma} & \text{(By algebra)} \\[2mm]
&= \frac{Z}{\sqrt{V/(n-1)}}, \text{where } V \sim \chi^2_{n-1} & \text{(As shown in Section 16.5)} \\[2mm]
&= T, \text{where } T \sim T_{n-1} & \text{(By definition of the t-distribution, and using the} \\
& & \text{fact that $\bar{Y}$ and $\hat{\sigma}$ are independent for normally} \\
& & \text{distributed data)}
\end{aligned}
$$

You can use this result to construct an exact confidence interval for μ. The method is similar to that shown in Section 16.5 and goes as follows: Let $t_{df,p}$ be the p quantile of the T_{df} distribution, and define the $100(1 - \alpha)\%$ critical values $c_L = t_{n-1,\alpha/2}$ and $c_U = t_{n-1,1-\alpha/2}$. Figure 16.6 shows the 95% critical values for the case where $n = 10$ (or $df = 9$).

From Figure 16.6, you can see the following:

$$
\Pr\left\{ c_L \leq \frac{\bar{Y} - \mu}{\hat{\sigma}/\sqrt{n}} \leq c_U \right\} = 1 - \alpha \tag{16.6}
$$

Since the t-distribution is symmetric about 0, you can also see in Figure 16.6 that $c_L = -c_U$. So there is only one critical value needed for the t-distribution, unlike the chi-squared distribution. Call it "c" without any subscript:

$$
c = t_{n-1,1-\alpha/2} \tag{16.7}
$$

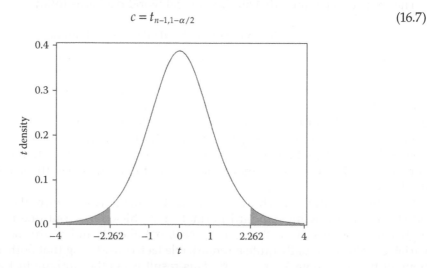

FIGURE 16.6
The T_9 distribution with 95% critical values: $c_L = -2.262$, and $c_U = 2.262$. The area in each shaded region is 0.025, and so the area in the center is 0.95.

Like the chi-squared quantiles, you can find these quantiles from many software packages. In SAS, the 0.975 quantile of the *t*-distribution with nine degrees of freedom is quantile('t', 0.975, 9), giving 2.262; in Excel you can use = T.INV(0.975,9) to get the same result.

Also, recall that the term $\hat{\sigma}/\sqrt{n}$ is called the standard error (or estimated standard deviation) of $\bar{Y}$, so let's abbreviate it as $s.e.(\bar{Y}) = \hat{\sigma}/\sqrt{n}$. Then you can rewrite Equation 16.6 as

$$\Pr\left\{-c \leq \frac{\bar{Y} - \mu}{s.e.(\bar{Y})} \leq c\right\} = 1 - \alpha \tag{16.8}$$

Rearranging the terms in Equation 16.8, you get

$$\Pr\left\{\bar{Y} - c \times s.e.(\bar{Y}) \leq \mu \leq \bar{Y} + c \times s.e.(\bar{Y})\right\} = 1 - \alpha \tag{16.9}$$

This gives you the interval for μ.

Exact Confidence Interval for μ Based on an iid Sample from $N(\mu, \sigma^2)$

$$\bar{Y} - c \times s.e.(\bar{Y}) \leq \mu \leq \bar{Y} + c \times s.e.(\bar{Y}) \tag{16.10}$$

Note the word *exact*. As long as the iid $N(\mu, \sigma^2)$ is valid, there are no weasely approximations.

You can identify the interval endpoints from the more compact formula $\bar{Y} \pm c \times s.e.(\bar{Y})$, a special case of a general formula that you see so often, it deserves "ugly rule of thumb" status.

Ugly Rule of Thumb 16.2

The endpoints of a confidence interval for a parameter are as follows:

(parameter estimate) ± (critical value) × (standard error of the parameter estimate)

For reasonably large sample sizes, the critical value is close to 2.0 from the 68-95-99.7 rule, giving the simpler approximate interval (parameter estimate) ± 2 × (standard error of the parameter estimate).

Example 16.4: Interval Estimation of the Mean Failure Time with a Small Sample Size

In Example 16.1, the data are $y_1 = 3.2$ and $y_2 = 3.4$, with an estimated standard deviation $\hat{\sigma} = 0.1414$. The 95% critical value from the *t*-distribution is $c = t_{n-1,1-\alpha/2} = t_{2-1,1-0.05/2} = t_{1,0.975} = 12.7062$, much larger than the 95% critical value 1.96 from the standard normal distribution that you would use if σ were known. Hence, assuming the data are produced as iid $N(\mu, \sigma^2)$, an *exact* 95% confidence interval for μ is given as follows:

$$3.3 - 12.7062\frac{0.1414}{\sqrt{2}} \leq \mu \leq 3.3 + 12.7062\frac{0.1414}{\sqrt{2}}$$

Or

$$2.03 \leq \mu \leq 4.57$$

Despite the small standard deviation, the interval is still very wide, again reflecting the problem with such a small sample size. Once again this is a good place to apply Bayesian statistics, particularly if you know, based on your prior knowledge, that the frequentist interval does not make sense.

16.7 Comparing Two Independent Samples Using a Confidence Interval

Recall the statistical science paradigm from Chapter 1: Nature → Design and Measurement → DATA. You learned in science classes that experimentation offers a valuable method to tap into Nature's workings. Statistical theory gives you the proper methods for analyzing the resulting data.

An **experiment** is a particular type of design and measurement scheme that allows you to assess cause-and-effect relationships in Nature. Many of the studies you will find in Internet searches that purport to assess the "effect of X on Y" come from experiments. Often, the X is binary. In a clinical trial, X might be a variable indicating whether a patient has received an active drug or a sugar pill (a *placebo*). In a study of students' ability to learn, X might be a variable indicating a specific educational software program that is used, either product A or product B. In general, the X variable in an experiment is called a **treatment** variable, and the individual values of the treatment variable are called **levels** of the treatment variable. For example, in the clinical trial, the two levels of X are (1) active drug and (2) placebo.

In a well-designed experiment, the **observational units**—be they people, plants, individual runs of a manufacturing process, animals, cities, whatever—are randomly assigned to particular X levels, and the outcomes Y are measured. If nothing differs in the way that the groups of observational units are treated, other than the difference between X values, then you can attribute differences in the Y outcome variable that are not explainable by chance alone to the causal effect of the X treatment.

Denote the measured data by Y_{ij}, where i indicates treatment group membership, either $i = 1$ or $i = 2$, and j indicates the observational unit in that particular group. For example, $Y_{2,6}$ denotes the sixth data value in the second treatment group. A reasonable model for such experimental data is as follows.

Statistical Model for Two-Group Experimental Data

$$Y_{ij} \sim_{\text{independent}} p_i(y), \quad \text{for } i = 1, 2; \quad \text{and} \quad j = 1, 2, \ldots, n_i$$

This model states that the data are all independent, but not necessarily identically distributed: The n_1 data values in group $i = 1$ are produced as iid from a distribution $p_1(y)$, and the n_2 data values in group $i = 2$ are produced as iid from a distribution $p_2(y)$. If the treatment X has no effect whatsoever, then these two distributions are identical: $p_1(y) = p_2(y)$. The case where $p_1(y) = p_2(y)$ is an example of the *no-effect model* discussed in Chapter 15. In this case, $Y_{ij} \sim_{\text{iid}} p_0(y)$, where $p_0(y)$ is the common distribution $p_1(y) = p_2(y) = p_0(y)$. If this null model is true, then any differences between data in groups $i = 1$ and $i = 2$ are *explained* by chance alone.

In Chapter 15, we presented both randomization models and bootstrap models that allowed you to avoid making any assumptions about the form (Poisson, exponential, normal, etc.) of the distributions $p_i(y)$. While those models were very useful, they have limitations: The bootstrap method is only approximate, and the randomization model cannot be easily extended to more complex statistical data. Methods based on normal assumptions, on the other hand, are exact (when the normality assumption is valid) and are easily extended to other, more advanced statistical methods such as multiple regression analysis and multivariate analysis.

The standard model for the **two-sample comparison** is $Y_{ij}\sim_{\text{independent}} N(\mu_i, \sigma^2)$. This model states that the data values are all independent and that the distributions $p_1(y)$ and $p_2(y)$ that produce the data in the two different groups are normal distributions, with possibly different means, but with common variances. The common variance assumption is also called the **homoscedasticity** assumption.

The *effect* of the treatment is $\delta = \mu_2 - \mu_1$. (**NOTE:** δ is the Greek lower case letter delta. Don't confuse it with the lowercase letter sigma, σ, and don't confuse it with the derivative operator ∂.) If $\delta > 0$, then treatment 2 causes generally higher values of the data to occur. If $\delta < 0$, then treatment 2 causes generally lower values of the data to occur. You have to be careful though: You can only use the term *causes* if you have a well-designed experiment. If not, then you have to say "associated with" rather than "causes"; for example, as in "If $\delta > 0$, then treatment 2 *is associated with* generally higher values of the data."

So a main interest in the experimental two-sample comparison is the estimation of the causal parameter δ. Following the development of Chapter 15 for the seat selection case, but assuming the normal, independent, homoscedastic model $Y_{ij}\sim_{\text{independent}} N(\mu_i, \sigma^2)$, you can construct an exact interval estimate for the causal effect δ as follows.

First, the obvious estimator of δ is $\hat{\delta} = \bar{Y}_2 - \bar{Y}_1$, where $\bar{Y}_1 = (Y_{11} + \cdots + Y_{1n_1})/n_1$ and $\bar{Y}_2 = (Y_{21} + \cdots + Y_{2n_2})/n_2$ are the within-group sample averages. You can see that $\hat{\delta}$ is an unbiased estimator, by using the linearity and additivity properties of expectation:

$E(\hat{\delta}) = E(\bar{Y}_2 - \bar{Y}_1)$	(By substitution)
$\quad = E(\bar{Y}_2) - E(\bar{Y}_1)$	(By the linearity and additivity properties of expectation)
$\quad = \mu_2 - \mu_1$	(Since the expected value of the sample average of data produced as iid from a process with mean μ is equal to μ; see Chapter 10)
$\quad = \delta$	(By definition)

You can also derive the variance of $\hat{\delta}$; this will allow you to see whether the observed difference is explainable by chance alone. It is described as follows:

$\text{Var}(\hat{\delta}) = \text{Var}(\bar{Y}_2 - \bar{Y}_1)$	(By substitution)
$\quad = \text{Var}(\bar{Y}_2) + \text{Var}(\bar{Y}_1)$	(By the linearity and additivity properties of variance for independent random variables; recall from Section 15.2 why the minus switches to a plus)
$\quad = \sigma^2/n_2 + \sigma^2/n_1$	(Since the variance of the sample average of data produced as iid from a process with variance σ^2 is equal to σ^2/n; see Chapter 10)
$\quad = \sigma^2(1/n_1 + 1/n_2)$	(By algebra)

Further, by the linearity and additivity properties of normally distributed random variables, the difference $\hat{\delta} = \bar{Y}_2 - \bar{Y}_1$ is *exactly* normally distributed. Putting the pieces together

$$\hat{\delta} \sim N\left\{\delta, \sigma^2\left(\frac{1}{n_1} + \frac{1}{n_2}\right)\right\} \tag{16.11}$$

If the model $Y_{ij} \sim_{\text{independent}} N(\mu_i, \sigma^2)$ is true, Formula (16.11) is an *exact* result, not an *approximate* result.

The variance σ^2 is unknown, but you have the two within-group unbiased estimators as follows:

$$\hat{\sigma}_1^2 = \left\{\frac{1}{(n_1-1)}\right\}\sum_j (Y_{1j} - \bar{Y}_1)^2$$

$$\hat{\sigma}_2^2 = \left\{\frac{1}{(n_2-1)}\right\}\sum_j (Y_{2j} - \bar{Y}_2)^2$$

From Section 16.2, you know that $\Sigma(Y_i - \bar{Y})^2/\sigma^2 \sim \chi^2_{n-1}$ when the data Y_i are iid $N(\mu, \sigma^2)$.
   Applying this result to the two groups separately, you also know

$$\frac{\sum\left(Y_{1j} - \bar{Y}_1\right)^2}{\sigma^2} \sim \chi^2_{n_1-1}$$

$$\frac{\sum\left(Y_{2j} - \bar{Y}_2\right)^2}{\sigma^2} \sim \chi^2_{n_2-1}$$

Now, the chi-squared distribution is the distribution of the sum of squared iid standard normals, so a chi-squared random variable with df_1 degrees of freedom plus an independent chi-squared random variable with df_2 degrees of freedom gives you another chi-squared random variable, one with $df_1 + df_2$ degrees of freedom. Since the data in the two groups are independent, it follows that:

$$\frac{\sum\left(Y_{1j} - \bar{Y}_1\right)^2}{\sigma^2} + \frac{\sum\left(Y_{2j} - \bar{Y}_2\right)^2}{\sigma^2} \sim \chi^2_{(n_1-1)+(n_2-1)}$$

Equivalently,

$$\frac{\{(n_1-1)\hat{\sigma}_1^2 + (n_2-1)\hat{\sigma}_2^2\}}{\sigma^2} \sim \chi^2_{(n_1-1)+(n_2-1)}$$

Recalling that variance estimates have distributions written generically as $\hat{\sigma}^2/\sigma^2 \sim \chi^2_{df}/df$, you can rearrange the terms to get the following:

$$\frac{\left\{\left((n_1-1)\hat{\sigma}_1^2 + (n_2-1)\hat{\sigma}_2^2\right)/\left((n_1-1)+(n_2-1)\right)\right\}}{\sigma^2} \sim \frac{\chi^2_{(n_1-1)+(n_2-1)}}{(n_1-1)+(n_2-1)} \tag{16.12}$$

This gives you the famous **pooled variance estimator**.

Pooled Variance Estimator

$$\hat{\sigma}^2_{pooled} = \frac{(n_1 - 1)\hat{\sigma}^2_1 + (n_2 - 1)\hat{\sigma}^2_2}{(n_1 - 1) + (n_2 - 1)}$$

Note that the pooled variance estimator is a weighted average of the individual within-group variances, where the within-group degrees of freedom are the weights. Since both $\hat{\sigma}^2_1$ and $\hat{\sigma}^2_2$ are unbiased estimators of σ^2, $\hat{\sigma}^2_{pooled}$ is also an unbiased estimator of σ^2 by the linearity and additivity properties of expectation. Plugging $\hat{\sigma}^2_{pooled}$ into Formula (16.12), the distribution of the pooled estimator is

$$\frac{\hat{\sigma}^2_{pooled}}{\sigma^2} \sim \frac{\chi^2_{\{(n_1 - 1) + (n_2 - 1)\}}}{(n_1 - 1) + (n_2 - 1)}$$

Note that the pooled estimator is more accurate than each estimator individually: The distribution of $\chi^2_{\{(n_1-1)+(n_2-1)\}}/((n_1 - 1) + (n_2 - 1))$ has less variability than that of either $\chi^2_{n_1-1}/(n_1 - 1)$ or $\chi^2_{n_2-1}/(n_2 - 1)$ since it is the average of more Z^2 terms.

Note also that in the total sample there are $n_1 + n_2$ free observations. By Ugly Rule of Thumb 16.1, you lose a degree of freedom for every parameter you estimate, and here you estimate two parameters, μ_1 and μ_2, via $\bar{Y}_1$ and $\bar{Y}_2$, respectively. So the degrees of freedom are $n_1 + n_2 - 2$, or $(n_1 - 1) + (n_2 - 1)$.

Recall that the variance of the parameter estimate $\hat{\delta}$ is $\sigma^2 (1/n_1 + 1/n_2)$; hence, its standard deviation is $\sigma\sqrt{1/n_1 + 1/n_2}$. Using the pooled variance estimate of σ^2, the standard error of the parameter estimate $\hat{\delta}$ is given as follows:

$$s.e.(\hat{\delta}) - \hat{\sigma}_{pooled}\sqrt{\frac{1}{n_1} + \frac{1}{n_2}}$$

Hang on, we're almost there! Similar to the main result for Student's *t*-Distribution given Section 16.6, it now follows logically that $T = (\hat{\delta} - \delta)/s.e.(\hat{\delta}) \sim T_{\{(n_1-1)+(n_2-1)\}}$, shown as follows:

$$T = \frac{\hat{\delta} - \delta}{\{\hat{\sigma}^2_{pooled}(1/n_1 + 1/n_2)\}^{1/2}}$$ (By definition)

$$= \frac{\hat{\delta} - \delta}{\{\sigma^2(1/n_1 + 1/n_2)\}^{1/2}} \times \frac{\sigma}{\hat{\sigma}_{pooled}}$$ (By algebra)

$$= Z \times \frac{\sigma}{\hat{\sigma}_{pooled}}, \text{where } Z \sim N(0,1)$$ (By the linearity and additivity properties of normally distributed random variables)

$$= \frac{Z}{\hat{\sigma}_{pooled}/\sigma}$$ (By algebra)

$$= \frac{Z}{\sqrt{V/\{(n_1 - 1) + (n_2 - 1)\}}}, \text{where } V \sim \chi^2_{\{(n_1-1)+(n_2-1)\}}$$ (As shown before)

$$\sim T_{\{(n_1-1)+(n_2-1)\}}$$ (By definition of the *t*-distribution, and using the fact that the $\bar{Y}$'s and $\hat{\sigma}_{pooled}$ are independent under normality)

So $T = (\hat{\delta} - \delta)/s.e.(\hat{\delta}) \sim T_{\{(n_1 - 1) + (n_2 - 1)\}}$. Now, define the critical value

$$c = t_{\{(n_1 - 1) + (n_2 - 1)\}, 1 - \alpha/2}$$

Then you have

$$\Pr\left\{ -c \leq \frac{(\hat{\delta} - \delta)}{s.e.(\hat{\delta})} \leq c \right\} = 1 - \alpha$$

Rearranging terms, it follows the following:

$$\Pr\{\hat{\delta} - c \times s.e.(\hat{\delta}) \leq \delta \leq \hat{\delta} + c \times s.e.(\hat{\delta})\} = 1 - \alpha \tag{16.13}$$

Equation 16.13 gives you the following famous result:

Exact Confidence Interval for the Difference between Means of Independent Homoscedastic Normal iid Samples

$$\hat{\delta} - c \times s.e.(\hat{\delta}) \leq \delta \leq \hat{\delta} + c \times s.e.(\hat{\delta})$$

Again, note the word *exact*. As long as your two-sample DATA are independently generated as samples from two normal distributions with common variance but possibly different means, there are no weasely approximations.

Example 16.5: Estimating the Average Age Difference for People in the Front and in the Back of the Classroom

The data in the seat selection example were $y_1 = 36$, $y_2 = 23$, $y_3 = 22$, $y_4 = 27$, $y_5 = 26$, $y_6 = 24$, $y_7 = 28$, $y_8 = 23$, $y_9 = 30$, $y_{10} = 25$, $y_{11} = 22$, $y_{12} = 26$, $y_{13} = 22$, $y_{14} = 35$, $y_{15} = 24$, and $y_{16} = 36$. The data are arranged by row so that y_1 through y_8 are ages of students in the front rows, while y_9 through y_{16} are ages of students in the back rows. In the double subscript format, $y_{1,1} = 36$, $y_{1,2} = 23$, $y_{1,3} = 22$, $y_{1,4} = 27$, $y_{1,5} = 26$, $y_{1,6} = 24$, $y_{1,7} = 28$, and $y_{1,8} = 23$ are the students in group $i = 1$, the front rows. In group $i = 2$, the back rows, $y_{2,1} = 30$, $y_{2,2} = 25$, $y_{2,3} = 22$, $y_{2,4} = 26$, $y_{2,5} = 22$, $y_{2,6} = 35$, $y_{2,7} = 24$, and $y_{2,8} = 36$. The sample averages are as before: $\bar{y}_1 = (36 + 23 + \cdots + 23)/8 = 26.125$, and $\bar{y}_2 = (30 + 25 + \cdots + 36)/8 = 27.5$. While in Chapter 15 we considered only the null model for data generation, let's now consider a broader perspective. Maybe there really *is* a difference in the age distributions that you find in the front rows versus the back rows. If so, the model $Y_{ij} \sim_{\text{independent}} p_i(y)$ is reasonable, as it allows that the age distribution in the front rows—namely, $p_1(y)$—possibly differs from the age distribution in the back rows, namely $p_2(y)$. This model allows approximate inferences via consistency and the CLT as shown in Chapter 15. If you want exact inferences, you'll have to make more assumptions.

So assume (for now) that the data are produced as $Y_{ij} \sim_{\text{independent}} N(\mu_i, \sigma^2)$, for $i = 1, 2$ and $j = 1, 2, \ldots, 8$ (here, $n_1 = n_2 = 8$). If this assumption were true—though of course it is not true; why not?—then the interval $\hat{\delta} \pm c\hat{\sigma}_{pooled}\sqrt{1/n_1 + 1/n_2}$ would be an *exact* $100(1 - \alpha)\%$ confidence interval for $\delta = \mu_2 - \mu_1$.

Here, $\hat{\delta} = \bar{y}_2 - \bar{y}_1 = 1.375$, as before. The critical value is $c = t_{[(8-1)+(8-1)], 1-\alpha/2}$; assuming $\alpha = 0.05$ for 95% confidence, the critical value is the 0.975 quantile of the t-distribution with 14 degrees of freedom or $c = t_{14, 0.975} = 2.145$. The pooled standard variance is

$\hat{\sigma}^2_{pooled} = \{(8-1)\hat{\sigma}^2_1 + (8-1)\hat{\sigma}^2_2\}/\{(8-1)+(8-1)\}$, which requires the within-group variance estimates $\hat{\sigma}^2_1, \hat{\sigma}^2_2$; you can calculate these as follows:

$$\hat{\sigma}^2_1 = \left(\frac{1}{7}\right)\{(36-26.125)^2 + (23-26.125)^2 + \cdots + (23-26.125)^2\} = 20.4107$$

$$\hat{\sigma}^2_2 = \left(\frac{1}{7}\right)\{(30-27.5)^2 + (25-27.5)^2 + \cdots + (36-27.5)^2\} = 30.8571$$

Further

$$\hat{\sigma}^2_{pooled} = \frac{\{(8-1)20.4107 + (8-1)30.8571\}}{\{(8-1)+(8-1)\}} = 25.6339$$

Notice that, in this example, the degrees of freedom are the same in each group, so the pooled variance estimate is a simple average of the two within-group variance estimates. Had the sample sizes differed in the two groups, the pooled variance estimate would be a weighted average of the two within-group variance estimates, with higher weight given to the variance from the group with the larger sample size.

The pooled standard deviation is thus $\hat{\sigma}_{pooled} = \sqrt{25.6339} = 5.063$, which is *not* an average (or even a weighted average) of the two within-group standard deviations—yet another application of Jensen's inequality! However, you can be assured that the pooled standard deviation will always be *between* the original two standard deviations (in this case 4.518 and 5.5549). The confidence interval for δ is $1.375 \pm 2.145(5.063)\sqrt{1/8+1/8}$, or 1.375 ± 5.430, or $-4.055 \le \delta \le 6.805$. Thus, the *process* difference between average ages in the back seats versus the front rows lies between −4.055 and 6.805 years.

The data show that the process mean age in the back rows is somewhere between 4.055 years *less than* the process mean age in the front rows, and 6.805 years *more than* the process mean age in the front rows, with 95% confidence. The logic for this interpretation is as follows:

$-4.055 \le \delta \le 6.805$	(This is your 95% confidence interval)
$\Rightarrow -4.055 \le \mu_2 - \mu_1 \le 6.805$	(By substitution)
$\Rightarrow \mu_1 - 4.055 \le \mu_2 \le \mu_1 + 6.805$	(By algebra)
$\Rightarrow$ (process mean age in front rows) − 4.055	(By definition)
$\le$ (process mean age in back rows)	
$\le$ (process mean age in front rows) + 6.805	

If $\mu_1 = \mu_2$, then there is no process mean difference between the front and back rows. The interval $-4.055 \le \delta \le 6.805$ admits the possibility that $\mu_1 = \mu_2$, since 0 is inside the interval for $\delta = \mu_2 - \mu_1$. Hence, chance alone is a plausible explanation for any differences seen in the data. In other words, the results are *explainable* by chance alone. But again, you cannot say that the results are *explained* by chance alone, since you have not proven that $\delta = 0$. There are many other plausible values of δ in the interval $-4.055 \le \delta \le 6.805$ other than $\delta = 0$; and if *any* of these non-zero values happens to be the true δ, then there is a systematic difference between the groups.

This result is similar to what we found in Chapter 15 for these data: The difference is explainable by chance alone. However, there are some differences in the methods in this chapter versus the methods of Chapter 15. Here, we assume normality, and we use a model that allows between-group differences; thus, the standard deviation estimate and critical values are a little different.

The advantage of the estimation-based approach given in this chapter over the testing-based approach given in Chapter 15 approaches is that, if there is a difference, you can quantify its size. Using the analyses of Chapter 15, all you can do is state whether the observed differences are explainable by chance alone. Here, you can state how big the difference might be, in addition to claiming whether the difference is explainable by chance alone. For instance, with the interval $2.12 \leq \delta \leq 11.91$, you could confidently state that $\delta \neq 0$ and hence that the differences are not explainable by the chance-only model where $\delta = 0$. But then you could go one step farther and state that the process mean age in the back rows is *at least* 2.12 years more than the process mean in the front rows. For this reason, some researchers eschew the entire null model testing-based approach presented in Chapter 15 and instead adopt the estimation-based approach shown in this chapter.

The exactness of these methods relies on assumptions, particularly normality, that are obviously false. Chapter 19 discusses this issue further. Why make the assumption at all? The answer is that the normality assumption, while always false, provides at least one case where the standard methods are, in some senses, optimal. The extent to which these methods remain good in cases where the normality assumption is violated depends on how badly violated the assumption is, in terms of extreme discreteness, skewness, and/or kurtosis.

16.8 Comparing Two Independent Homoscedastic Normal Samples via Hypothesis Testing

You can test for whether the results are explainable by chance in the context of the normal, independent, homoscedastic model by using the *two-sample t-test*. If you assume the model is $Y_{ij} \sim_{\text{independent}} N(\mu_i, \sigma^2)$, which allows possibly different means μ_1 and μ_2 for the two groups, then the chance-only model is one where $Y_{ij} \sim_{\text{iid}} N(\mu, \sigma^2)$ or one where $\mu_1 = \mu_2$. The statement $\mu_1 = \mu_2$ is a **hypothesis** about Nature. A hypothesis is a **constraint** you place on the types of models that you assume to produce your data.

The **null hypothesis** is a statement of no difference in the process, and the **alternative hypothesis** is a statement that there is a difference. Alternative hypotheses can be one-sided, such as $\mu_1 > \mu_2$ or $\mu_1 < \mu_2$, or two-sided, such as $\mu_1 \neq \mu_2$. We will mainly stick to two-sided hypotheses for all frequentist analyses in this book, for several reasons: One-sided hypotheses are too easily misused, two-sided tests are standard in many disciplines, and two-sided tests have a comfortable correspondence with confidence intervals.

Null and alternative hypotheses are often given the abbreviations "H_0" and "H_1," respectively, as in $H_0: \mu_1 = \mu_2$ and $H_1: \mu_1 \neq \mu_2$. These abbreviations are shorthand for the longer phrases, respectively as follows:

"If the null hypothesis is true, then $\mu_1 = \mu_2$."

And

"If the alternative hypothesis is true, then $\mu_1 \neq \mu_2$."

Which is the null and which is the alternative? Null hypotheses are always models in which the observed difference in the statistic of interest is explained by chance alone. For example, if H_0: $\mu_1 = \mu_2$ is true, then any difference between $\bar{y}_1$ and $\bar{y}_2$ is *explained* by chance alone. Sometimes null hypotheses are stated in terms of inequalities such as H_0: $\mu_1 \leq \mu_2$, but this also includes the chance-only model where $\mu_1 = \mu_2$. One thing that differentiates null and alternative hypotheses is that the chance alone model is always included within the null hypothesis, whereas the alternative hypothesis always states that there is a systematic difference.

Incidentally, the Bayesian viewpoint accommodates testing more naturally: Just calculate $\Pr(\text{Hypothesis is true}|\text{data})$. If that probability is sufficiently high (e.g., 90% or higher), then you can safely conclude that the hypothesis is true. There is no need to differentiate null and alternative hypotheses in the Bayesian framework, although you do have to think more carefully about prior probabilities when doing Bayesian hypothesis testing.

The *p*-value presented in Chapter 15 is distinctly non-Bayesian. It is a measure of whether your results are *explainable* by chance alone; that is, it is a measure of whether your results are *explainable* by the null hypothesis model. If the *pv* is small (e.g., less than 0.05), then you can essentially rule out chance as an explanation for the difference, and therefore **reject the null hypothesis**. On the other hand, if the *pv* is large (e.g., more than 0.05), then the observed differences are easily *explainable* by chance alone, and therefore you cannot rule out the null hypothesis. Some texts will say to **accept the null hypothesis** when the results are explainable by chance alone, but this is bad phrasing, because it suggests that the results are *explained* by chance alone, rather than simply *explainable* by chance alone. Better explanatory phrases are either "fail to reject H_0," or "the difference is statistically insignificant" when your results are explainable by chance alone.

If $Y_{ij} \sim_{\text{independent}} N(\mu_i, \sigma^2)$, recall from Section 16.7 that $T = (\hat{\delta} - \delta)/s.e.(\hat{\delta}) \sim T_{\{(n_1-1)+(n_2-1)\}}$. Under the chance-only (or null) model where H_0: $\mu_1 = \mu_2$ is true, $\delta = \mu_2 - \mu_1 = 0$; hence, when H_0 is true you know the following:

$$T = \frac{\hat{\delta}}{s.e.(\hat{\delta})} \sim T_{\{(n_1-1)+(n_2-1)\}}$$

This value T is called the **two-sample *t*-statistic**. It measures the size of the difference between sample means, or $\hat{\delta}$, relative to its standard error. This is the general form of a *t*-statistic seen in all computer outputs, not only for two-sample tests but also for tests involving more advanced models such as multiple regression, so we'll set it off as follows:

General Form of the *t*-Statistic in Computer Output

For a generic parameter θ, the *t*-statistic for testing H_0: $\theta = 0$ is given by $T = \hat{\theta}/s.e.(\hat{\theta})$.

Large values of the test statistic T, whether positive or negative, are extreme under the null hypothesis H_0: $\mu_1 = \mu_2$. Therefore, if the observed value of the test statistic is t, a positive number, then the *p*-value is equal to $\Pr(T \geq t|H_0) + \Pr(T \leq -t|H_0)$. Since the t distribution is symmetric, you can calculate this probability simply by doubling the tail probability beyond the absolute *t*-statistic: $pv = 2 \times \Pr(T \geq |t||H_0)$. This procedure is commonly known as the **two-sample *t*-test**.

Example 16.6: Testing the Average Age Difference for People in the Front and in the Back of the Classroom Using the Two-Sample *t*-Test

In Example 16.5, the student age versus seat selection example, the data and summary statistics are given. Using these summary statistics, you can calculate the two-sample *t*-statistic $t = \hat{\delta}/s.e.(\hat{\delta})$ as follows:

$$t = \frac{1.375}{5.063\sqrt{1/8 + 1/8}} = 0.543$$

The *p*-value is therefore $pv = 2 \times \Pr(T \geq 0.543)$, where T has the T_{14} distribution, and is calculated as $2 \times 0.2978 = 0.596$. This result is similar to the *p*-values calculated in Chapter 15 for this example using the randomization and iid null models—0.632 and 0.565, respectively. Since the *p*-value is quite a bit larger than 0.05, the observed difference is easily explainable by chance alone. Figure 16.7 shows the T_{14} distribution, along with the observed test statistic 0.543 and the critical value 2.145.

You can calculate *p*-values using any software that gives you cumulative probabilities for the standard types of distributions. In Excel, the *p*-value $pv = 2 \times \Pr(T_{14} \geq 0.543)$ is found as = 2*(1-T.DIST(0.543,14,TRUE)); in SAS you can access it as 2*(1 − cdf('t', 0.543,14)).

In this example, the 95% confidence interval for δ includes zero, meaning that the results are explainable by chance alone, and the *p*-value is greater than 0.05, also implying that the results are explainable by chance alone. This is no accident! The confidence interval for the mean difference and the two-sided *p*-value always give the same answer as to whether

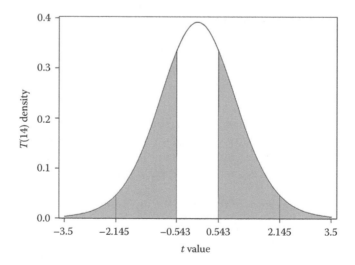

FIGURE 16.7
The null distribution of the two-sample *t*-statistic. The observed value $t = 0.543$ is shown, and the probability of values as extreme as $t = 0.543$ are shown in shaded area giving $pv = 2 \times 0.2978 = 0.596$. The critical value, 2.145, is the *t*-value giving $pv = 0.05$.

the results are explainable by chance alone. Here's why the test and the confidence interval provide similar information:

The confidence interval for δ includes 0

$$\Leftrightarrow \hat{\delta} - c \times s.e.(\hat{\delta}) \le 0 \le \hat{\delta} + c \times s.e.(\hat{\delta}) \qquad \text{(By definition)}$$

$$\Leftrightarrow -c \times s.e.(\hat{\delta}) \le -\hat{\delta} \le c \times s.e.(\hat{\delta}) \qquad \text{(By algebra)}$$

$$\Leftrightarrow -c \le \frac{-\hat{\delta}}{s.e.(\hat{\delta})} \le c \qquad \text{(By algebra)}$$

$$\Leftrightarrow c \ge \frac{\hat{\delta}}{s.e.(\hat{\delta})} \ge -c \qquad \text{(By algebra)}$$

$$\Leftrightarrow -c \le t \le c \qquad \text{(By algebra and the definition of the } t\text{-statistic)}$$

Now, the last expression indicates that the t-statistic will give a p-value greater than α, by definition of the critical value c.

Relationship between Hypothesis t-Tests and Confidence Intervals

The two-sided p-value for testing H_0: $\delta = 0$ is *greater than* α, *if and only if* the $100(1 - \alpha)\%$ confidence interval for δ *includes* 0.

Thus, the confidence interval for δ provides the same information as to whether the results are explainable by chance alone, but it gives you more than just that. It also gives the range of plausible values of the parameter, whether or not the results are explainable by chance alone. This correspondence means that the p-value is not necessary to establish whether results are explainable by chance alone—you can do it instead by using a confidence interval.

16.9 *F*-Distribution and ANOVA Test

You have seen that the statistically famous chi-squared and Student's t-distributions are distributions of functions iid standard normal random variables. Another famous distribution, known as the **F-distribution**, is defined similarly.

Definition of the F-Distribution

If $V_1 \sim \chi^2_{m_1}$, and if $V_2 \sim \chi^2_{m_2}$, independent of V_1, and if $F = (V_1/m_1)/(V_2/m_2)$, then $F \sim F_{m_1, m_2}$, which is called the *F-distribution with m_1 numerator degrees of freedom and m_2 denominator degrees of freedom.*

Here are some things you can see right away from the definition: Both the numerator and denominator have expectation 1.0, since the expected value of a chi-squared random variable is equal to its degrees of freedom. Hence, the center of the F-distribution is near 1.0, although 1.0 is not exactly the mean—Jensen's inequality again! Another thing you can see is that both the numerator and denominator are averages of squared standard normals.

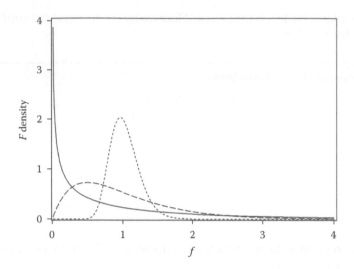

FIGURE 16.8
F-distribution with $(m_1, m_2) = (1, 10)$ (solid line); $(m_1, m_2) = (100, 100)$ (dotted line); and $(m_1, m_2) = (4, 101)$ (dashed line).

The more numbers in these averages, the closer they are to their true mean, which is 1.0 in this case. Therefore, not only is the *F*-distribution centered around the number 1.0, it also collapses to 1.0 when degrees of freedom increase in both the numerator and the denominator.

While its derivation is beyond the scope of this book, the kernel of the *F*-distribution has the following relatively simple form:

$$p(f) \propto f^{(m_1/2)-1}\left\{1+\left(\frac{m_1}{m_2}\right)f\right\}^{-(m_1+m_2)/2}$$

Standard software can graph the actual pdf including the constant *c* of proportionality; Figure 16.8 shows some *F*-distributions for different *df* combinations (m_1, m_2).

Notice in Figure 16.8 that the distributions all are centered around 1.0, and that the distribution collapses around 1.0 in the case with many degrees of freedom—large values of m_1 and m_2—in the numerator and denominator. Also notice that the solid line is nearly identical to the chi-squared distribution with one degree of freedom, since there is $\chi_1^2/1$ in the numerator and a random quantity that is approximately 1.0 in the denominator.

One application of the *F*-distribution that you can deduce is this: If a test statistic *T* has the T_m distribution, then $T^2 \sim F_{1,m}$. It is shown in the following:

$$T = \frac{Z}{\sqrt{V/m}}, \text{ where } V \sim \chi_m^2 \quad \text{(By definition of the } t\text{-distribution)}$$

$$\Rightarrow T^2 = \frac{Z^2/1}{V/m} \quad \text{(By algebra)}$$

$$\Rightarrow T^2 \sim F_{1,m} \quad \text{(By definition of the chi-squared random variable and of the } F\text{-distribution)}$$

This result explains why some software packages report *F*-statistics and some report *t*-statistics, even when the same models and hypotheses are considered. But either gives the same p-values, so it is not usually a concern whether the software reports F- or t-statistics. Here's why the p-values from the t-distribution and the F-distribution are identical.

$pv = \Pr(T_m \geq \lvert t \rvert \text{ or } T_m \leq -\lvert t \rvert)$	(By definition of the two-sided p-value)
$= \Pr\left(T_m^2 \geq t^2\right)$	(By algebra)
$= \Pr(F_{1,m} \geq t^2)$	(By the relationship between T^2 and F)

Thus, the *two-tailed* p-value from a t-test is equal to the *upper-tailed* p-value from the corresponding F-test, using t^2 as the test statistic.

The relationship between T^2 and F is not a good enough reason to have a completely separate distribution! The most common use of the F-distribution is for testing a **composite hypothesis**—that is, one that states that many hypotheses are simultaneously true.

Suppose that you have an experiment where you want to study the effects of three different treatments simultaneously. This occurs regularly in clinical trials: Some patients receive a standard treatment, some receive a new treatment formulation A, and others receive a different new treatment B. At the end of the study, there will be differences in the patient outcomes between the three groups, but are these differences explainable by chance alone?

The *analysis-of-variance model*, or **ANOVA model**, is used to analyze such multiple-group data.

Analysis-of-Variance Model

The standard analysis-of-variance (ANOVA) model is for data classified by group. The model states that the DATA are produced as $Y_{ij} \sim_{\text{independent}} N(\mu_i, \sigma^2)$, where $i = 1, 2, \ldots, g$ (the group label) and $j = 1, 2, \ldots, n_i$ (observation label within group i).

The ANOVA null hypothesis is, in the case of three groups, $H_0: \mu_1 = \mu_2 = \mu_3 = \mu$, which states that the process means for the three treatments are identical. If the null hypothesis is true, then any differences in the averages from the observed data are explained by chance alone.

The ANOVA null hypothesis is an example of a composite hypothesis. Whereas the two-sample t-test is a test of a simple hypothesis such as $H_0: \mu_1 = \mu_2$, the ANOVA hypothesis is a composite of both $H_0: \mu_1 = \mu_2$ *and* $H_0: \mu_1 = \mu_3$, because the ANOVA hypothesis states that there is no difference between any of the group means.

In general, with g groups and n_i observations per group, The F-statistic used to test the ANOVA hypothesis $H_0: \mu_1 = \mu_2 = \cdots = \mu_g = \mu$ is given as follows:

$$F = \frac{\sum_i n_i(\bar{Y}_i - \bar{Y})^2/(g-1)}{\sum_i \sum_j (Y_{ij} - \bar{Y}_i)^2/(n-g)} \tag{16.14}$$

Here, $\bar{Y}_i$ is the group i average (there are g of these), $n = n_1 + n_2 + \cdots + n_g$ is the total sample size, and $\bar{Y}$ is the ordinary sample average of all n observations.

The F-statistic is a measure of **variation between groups** relative to **variation within groups**. The numerator sum of squares $\sum_i n_i(\bar{Y}_i - \bar{Y})^2$ is larger when the group means differ greatly and hence measures variation between groups. The denominator sum of squares $\sum_i \sum_j (Y_{ij} - \bar{Y}_i)^2$ is larger when the data within groups differ greatly from their within-group means and hence measures variation within groups. The larger the

F-statistic, the less likely it is that the differences between sample means of the groups can be explained by chance alone.

Distribution of the F-Statistic

In the ANOVA model where H_0: $\mu_1 = \mu_2 = \cdots = \mu_g = \mu$ is true, the F-statistic in Equation 16.14 has the $F_{g-1, n-g}$ distribution.

The following lines show why the F-statistic has this null distribution. First, assume the model $Y_{ij} \sim_{\text{independent}} N(\mu_i, \sigma^2)$. Then

---

$$\bar{Y}_i \sim N\left(\mu_i, \frac{\sigma^2}{n_i}\right) \qquad \text{(By the linearity and additivity properties of the normal distribution)}$$

$$\Rightarrow \frac{n_i^{1/2}(\bar{Y}_i - \mu_i)}{\sigma} \sim N(0,1) \qquad \text{(By the linearity and additivity properties of the normal distribution)}$$

$$\Rightarrow \sum_i \left\{ \frac{n_i^{1/2}(\bar{Y}_i - \mu_i)}{\sigma} \right\}^2 \sim \chi_g^2 \quad \text{(By definition of the chi-squared distribution, and since the data are independent from group to group)}$$

$$\Rightarrow \sum_i \frac{n_i(\bar{Y}_i - \mu_i)^2}{\sigma^2} \sim \chi_g^2 \qquad \text{(By algebra)}$$

Hence, under the chance-only model where H_0: $\mu_1 = \mu_2 = \cdots = \mu_g = \mu$ is true:

$$\sum_i \frac{n_i(\bar{Y}_i - \mu)^2}{\sigma^2} \sim \chi_g^2$$

The value of μ is unknown, and you can substitute $\bar{Y}$, losing a degree of freedom according to Ugly Rule of Thumb 16.1 (which can also be proven rigorously using vector algebra):

$$\sum_i \frac{n_i(\bar{Y}_i - \bar{Y})^2}{\sigma^2} \sim \chi_{g-1}^2$$

That takes care of the numerator of the F-statistic. Now, consider the denominator: Note that

---

$$(Y_{ij} - \mu_i)/\sigma \sim_{\text{iid}} N(0, 1) \qquad \text{(By the linearity and additivity properties of the normal distribution, and since all data are independent)}$$

$$\Rightarrow \sum_i \sum_j \left\{ \frac{(Y_{ij} - \mu_i)}{\sigma} \right\}^2 \sim \chi_n^2 \qquad \text{(By definition of the chi-squared distribution, and because there are } n \text{ total summates)}$$

$$\Rightarrow \sum_i \sum_j \frac{(Y_{ij} - \mu_i)^2}{\sigma^2} \sim \chi_n^2 \qquad \text{(By algebra)}$$

Now, in the denominator of the F-statistic given by Equation 16.14, you see the μ_i are estimated using $\bar{Y}_i$. There are g of these estimators, so by Ugly Rule of Thumb 16.1 concerning degrees of freedom, you lose g degrees of freedom (again, this can be proven rigorously):

$$\sum_i \sum_j \frac{(Y_{ij} - \bar{Y}_i)^2}{\sigma^2} \sim \chi_{n-g}^2$$

Now you see the two chi-squares with $g - 1$ and $n - g$ degrees of freedom: They are $\sum_i n_i (\bar{Y}_i - \bar{Y})^2/\sigma^2$ and $\sum_i \sum_j (Y_{ij} - \bar{Y}_i)^2/\sigma^2$, respectively. One final piece of the puzzle, also requiring higher math, is the independence issue. It turns out that the numerator and denominator are indeed independent, for essentially the same reason that $\bar{Y}$ and $\hat{\sigma}$ are independent, assuming the normally distributed ANOVA model. Putting the pieces together we get the following:

$$\frac{\left\{\sum_i n_i(\bar{Y}_i - \bar{Y})^2/\sigma^2\right\}/(g-1)}{\left\{\sum_i \sum_j (Y_{ij} - \bar{Y}_i)^2/\sigma^2\right\}/(n-g)} \sim F_{g-1,n-g} \quad \text{(Using the aforementioned chi-squared distribution results, and by definition of the } F\text{-distribution)}$$

$$\Rightarrow \frac{\sum_i n_i(\bar{Y}_i - \bar{Y})^2/(g-1)}{\sum_i \sum_j (Y_{ij} - \bar{Y}_i)^2/(n-g)} \sim F_{g-1,n-g} \quad \text{(By canceling the } \sigma^2 \text{ terms)}$$

$$\Rightarrow F \sim F_{g-1,n-g} \quad \text{(By substitution from Equation 16.14)}$$

Again, this derivation assumes the chance-only, or null, model where $H_0: \mu_1 = \mu_2 = \cdots = \mu_g = \mu$ is true. Since large values of the F-statistic indicate differences that are less easily explained by chance alone, you should calculate the p-value from the upper tail only. If f is the observed value of the F-statistic (Equation 16.14), the p-value is

$$pv = \Pr\left(F_{g-1,n-g} \geq f\right)$$

As usual, if the p-value is less than your chosen significance level, α, then you can safely rule out chance as an explanation for the between-group differences. Equivalently, you can safely rule out chance if the F-statistic f is greater than the $1 - \alpha$ quantile of the $F_{g,n-g}$, or when $f \geq F_{g-1,n-g,1-\alpha}$. This quantile is called the *critical value* of the ANOVA F-test.

Example 16.7: Testing for Differences between Pharmaceutical Treatments

The time to develop a pharmaceutical product can be very long. First there is drug discovery, followed by preclinical safety testing, followed by a long process of testing in humans. Early drug trials involve few people and tend to be oriented more toward proof of concept and safety. Later trials are geared towards garnering approval from regulatory agencies. Suppose an early phase trial of five different treatments for Alzheimer's disease resulted in the data shown in Table 16.1.

TABLE 16.1

Summary Statistics for Study of Alzheimer's Disease

Treatment (i)	Sample Size n_i	Sample Mean $\bar{y}_i$	Standard Deviation $\hat{\sigma}_i$
1	13	61.00	7.69
2	12	59.75	8.29
3	10	45.20	8.88
4	16	83.00	11.23
5	6	84.00	10.58

There are differences between sample means; are the differences between sample means explainable by chance alone? The *F*-statistic and its associated *p*-value provide an answer.

Much of the information you need to calculate the *F*-statistic in Equation 16.14 is given directly in Table 16.1: $g = 5$, $n_1 = 13$, $n_2 = 12$, $n_3 = 10$, $n_4 = 16$, and $n_5 = 6$. Further, $n = n_1 + n_2 + n_3 + n_4 + n_5 = 57$, $\bar{y}_1 = 61.00$, $\bar{y}_2 = 59.75$, $\bar{y}_3 = 45.20$, $\bar{y}_4 = 83.00$, and $\bar{y}_5 = 84.00$. The overall mean $\bar{y}$ is the average of all 57 observations and is a weighted average of the individual group means $\bar{y}_i$:

$$
\begin{aligned}
\bar{y} &= \{(y_{1,1} + \cdots + y_{1,13}) + (y_{2,1} + \cdots + y_{2,12}) + (y_{3,1} + \cdots + y_{3,10}) && \text{(By definition of the sample average)}\\
&\quad + (y_{4,1} + \cdots + y_{4,16}) + (y_{5,1} + \cdots + y_{5,6})\}/57 \\
&= (13\bar{y}_1 + 12\bar{y}_2 + 10\bar{y}_3 + 16\bar{y}_4 + 6\bar{y}_5)/57 && \text{(By algebra)}\\
&= \{13(61.00) + 12(59.75) + 10(45.20) + 16(83.00) + 6(84.00)\}/57 && \text{(By arithmetic)}\\
&= 66.56
\end{aligned}
$$

The numerator sum of squares is given as follows:

$$
\sum_i n_i(\bar{y}_i - \bar{y})^2
$$

$$
\begin{aligned}
&= 13(61.00 - 66.56)^2 + 12(59.75 - 66.56)^2 + 10(45.20 - 66.56)^2 + 16(83.00 - 66.56)^2 \\
&\quad + 6(84.00 - 66.56)^2 \\
&= 11{,}670.9
\end{aligned}
$$

You can obtain the denominator sum of squares $\sum_i \sum_j (y_{ij} - \bar{y}_i)^2$ from the standard deviations as follows:

$$
\begin{aligned}
\sum_i \sum_j (y_{ij} - \bar{y}_i)^2 &= \sum_i \sum_j (n_i - 1)\{(y_{ij} - \bar{y}_i)^2/(n_i - 1)\} && \text{(By algebra)}\\
&= \sum_i \sum_j (n_i - 1)\hat{\sigma}_i^2 && \text{(By definition of the sample variance)}\\
&= \{(13 - 1)(7.69)^2 + (12 - 1)(8.29)^2 && \text{(By substitution and arithmetic)}\\
&\quad + (10 - 1)(8.88)^2 + (16 - 1)(11.23)^2 \\
&\quad + (6 - 1)(10.58)^2\} = 4626.66
\end{aligned}
$$

Dividing this sum of squares by $n - g = 57 - 5 = 52$ gives the pooled variance estimate $4626.66/52 = 88.97$ and the corresponding pooled standard deviation estimate $\sqrt{88.97} = 9.43$, which is comfortably in the middle of the five individual standard deviation estimates. Thus, the *F*-statistic is

$$
f = \frac{\sum_i n_i(\bar{y}_i - \bar{y})^2/(g - 1)}{\sum_i \sum_j (y_{ij} - \bar{y}_i)^2/(n - g)} = \frac{11670.2/4}{4626.66/52} = \frac{2917.5}{88.97} = 32.79
$$

Just by looking at this F-statistic, you should have the idea that the differences between means are too large to be explained by chance alone. Under the chance-only model, which states that the data arise as iid from $N(\mu, \sigma^2)$ regardless of group, the F-statistic has the $F_{4,52}$ distribution, and the F-distributions are centered around the number 1.0. To be more specific, the critical value of the test is the 0.95 quantile of the $F_{4,52}$ distribution, or $F_{4,52,0.95} = 2.55$. Since $32.79 > 2.55$, you can safely rule out chance as an explanation for the differences between sample means.

The probability of observing an F-statistic as extreme as $f = 32.79$, under the chance-only model, is $\Pr(F_{4,52} \geq 32.79) = 1.2 \times 10^{-13}$, infinitesimally small. Again, you can safely rule out chance as an explanation for the differences in group means and conclude that at least two of the group means differ.

The F-test simply tells you that *some* of the group means differ. To identify *which* of the group means differ, you should use multiple testing procedures; these go by names such as *Bonferroni's method* and *Tukey's method*, but are not discussed in this book.

Since the p-value $= 1.2 \times 10^{-13}$ is so small, and since many people are not familiar with exponential notation, researchers typically report such a result as $pv < 0.0001$. Historically, when access to computers was scarce, researchers used reference tables to determine significance, and only the 0.05 and 0.01 p-value thresholds were available. This has led to the practice of stating $pv \leq 0.05$ or $pv \leq 0.01$ when reporting significances. This is an ancient practice that is neither useful nor necessary. If the p-value is 0.023, simply report $pv = 0.023$. Do not report $pv < 0.05$, as the reader is then confused as to how small the p-value really is. Is it 0.023? Is it 1.2×10^{-13}? Even worse, sometimes you will see p-values reported as something like $pv < 0.023$! This type of incorrect reporting conveys blatant ignorance: The number 0.023 is not less than 0.023, it is equal to 0.023. A report such as $pv < 0.023$ is silly, but you will see examples of this obviously incorrect practice, even in the better research journals.

On the other hand, it is reasonable to report p-values that are below the minuscule 0.0001 threshold as simply $pv < 0.0001$, but even then, it is a good idea to report the precise value for the benefit of the reader.

16.10 *F*-Distribution and Comparing Variances of Two Independent Groups

Consider the two-group experiment. Can a treatment affect the variance of the distribution? Certainly. For example, suppose that a fuel additive improves efficiency only for a subset of cars. Then in an experiment involving cars and additive, the fuel additive group data will be a mixture of data that are just like the control group (the unaffected cars), with data whose mean is shifted (the affected cars). If the efficiency range for the control group is 10–15 (in some arbitrary efficiency units), and the range for the affected cars is 13–18, then in the additive group the range will be 10–18, reflecting larger variability than in the control group where the range is 10–15.

To estimate different variances, you need a model that allows different variances. Consider the normal, independent, *heteroscedastic* model $Y_{ij} \sim_{\text{independent}} N(\mu_i, \sigma_i^2)$. This model is called **heteroscedastic** because it allows non-equal variances σ_1^2 and σ_2^2; the homoscedastic model assumes the variances are identical or that $\sigma_1^2 = \sigma_2^2 = \sigma^2$. From Equation 16.2 of Section 16.5, you know that

$$\frac{\hat{\sigma}^2}{\sigma^2} \sim \frac{\chi_{n-1}^2}{n-1}$$

Applying this result to each group separately, you get

$$\frac{\hat{\sigma}_1^2}{\sigma_1^2} \sim \frac{\chi_{n_1-1}^2}{n_1-1} \quad \text{and} \quad \frac{\hat{\sigma}_2^2}{\sigma_2^2} \sim \frac{\chi_{n_2-1}^2}{n_2-1}$$

By assumption of independent samples, and by definition of the F-distribution, the ratio of ratios is thus distributed as F_{n_1-1,n_2-1}

$$\frac{\hat{\sigma}_1^2/\sigma_1^2}{\hat{\sigma}_2^2/\sigma_2^2} \sim \frac{\chi_{n_1-1}^2/(n_1-1)}{\chi_{n_2-1}^2/(n_2-1)} \sim F_{n_1-1,n_2-1}$$

Letting $\theta = \sigma_1^2/\sigma_2^2$ denote the variance ratio, you get a simpler and familiar-looking representation

$$\frac{\hat{\theta}}{\theta} \sim F_{n_1-1,n_2-1} \tag{16.15}$$

Since the F-distribution collapses to 1.0 as the numerator and denominator degrees of freedom increase, this distribution form shows that the ratio of the estimated variance ratio to the true variance ratio gets closer to 1.0 when the number of observations used to estimate each individual variance increases. This is hopefully an intuitively sensible result! If not, review the concept of *consistency* from Chapter 11.

You can use this result to find a confidence interval for the variance ratio θ, and also to test $H_0 : \theta = 1.0$. If H_0 is true, then any difference between $\hat{\theta}$ and 1.0 is explained by chance. Note that

$$\Pr\left(F_{n_1-1,n_2-1,\alpha/2} \le \frac{\hat{\theta}}{\theta} \le F_{n_1-1,n_2-1,1-\alpha/2} \right) = 1-\alpha \qquad \text{(By the definition of quantile)}$$

$$\Rightarrow \Pr\left(1/F_{n_1-1,n_2-1,\alpha/2} \ge \frac{\theta}{\hat{\theta}} \ge 1/F_{n_1-1,n_2-1,1-\alpha/2} \right) = 1-\alpha \qquad \text{(By algebra)}$$

$$\Rightarrow \Pr\left(\frac{\hat{\theta}}{F_{n_1-1,n_2-1,\alpha/2}} \ge \theta \ge \frac{\hat{\theta}}{F_{n_1-1,n_2-1,1-\alpha/2}} \right) = 1-\alpha \qquad \text{(By algebra)}$$

$$\Rightarrow \Pr\left(\frac{\hat{\theta}}{F_{n_1-1,n_2-1,1-\alpha/2}} \le \theta \le \frac{\hat{\theta}}{F_{n_1-1,n_2-1,\alpha/2}} \right) = 1-\alpha \qquad \text{(By algebra)}$$

This latter equation gives the $100(1-\alpha)\%$ confidence interval for the variance ratio $\theta = \sigma_1^2/\sigma_2^2$. To obtain the corresponding confidence interval for the (more relevant) ratio of standard deviations, simply take the square roots of the endpoints.

You can also use the distribution result of Expression (16.15) to test that $H_0: \theta = 1$ or, equivalently, that $H_0: \sigma_1^2 = \sigma_2^2$. If H_0 is true, then by Expression (16.15) you can see that $\hat{\theta} \sim F_{n_1-1, n_2-1}$. The *pv* is obtained by doubling the tail probability: If $\hat{\theta} > 1$, then $pv = 2 \times \Pr(F_{n_1-1, n_2-1} \geq \hat{\theta})$; if $\hat{\theta} < 1$, then $pv = 2 \times \Pr(F_{n_1-1, n_2-1} \leq \hat{\theta})$. This definition of *pv* provides the familiar comfortable correspondence with the confidence interval: The $100(1 - \alpha)\%$ confidence interval for θ *includes* 1.0 if and only if the *pv* is *greater than* α.

Example 16.8 Comparing Standard Deviations of Pharmaceutical Treatments

In Table 16.1, the standard deviations for groups 1 and 2 are $\hat{\sigma}_1 = 7.69$ and $\hat{\sigma}_2 = 8.29$, based on sample sizes $n_1 = 13$ and $n_2 = 12$. Is the difference in standard deviations explainable by chance? The variance ratio is $\hat{\theta} = (7.69)^2/(8.29)^2 = 0.8604$. The *p*-value is thus $pv = 2 \times \Pr(F_{13-1,12-1} \leq 0.8604) = 0.797$. The difference between sample standard deviations is easily *explainable* by chance alone.

The confidence interval provides additional information: $F_{13-1,12-1,0.025} = 0.3011$ and $F_{13-1,12-1,0.975} = 3.430$, so the interval for the standard deviation ratio is

$$\frac{(7.69/8.29)}{\sqrt{3.430}} \leq \frac{\sigma_1}{\sigma_2} \leq \frac{(7.69/8.29)}{\sqrt{0.3011}}$$

Or

$$0.50 \leq \frac{\sigma_1}{\sigma_2} \leq 1.69$$

The interval also shows that the chance-only model where $\sigma_1 = \sigma_2$ is plausible, but it also shows more: If there is a difference in standard deviations, the ratio is no more than 1.69, and it is no less than 0.50, with 95% confidence.

In the two-sample *t*-test of Section 16.8, the initial model was the homoscedastic model. If you are worried about the homoscedasticity assumption, you can use this method to test for equality of variances. Some might suggest using the *F*-test for variances as a precursor to deciding whether to use a homoscedastic model or a heteroscedastic model. This is a questionable practice. First, it conflates the phrase *explainable by chance* with the phrase *explained by chance*. The difference between variance estimates may be explainable by chance, but that does not mean the variances are truly equal. Another problem is that the test for variances is much less robust against violations of the assumptions, particularly normality, than is the two-sample *t*-test for means. The practice of testing the variances first has therefore been described as something akin to "putting out a rowboat to see if the water is safe for the Titanic." A third issue is that using the *F*-test as a precursor changes the operating characteristics of the overall procedure, since it is now a multi-step process involving random decisions along the way.

If you really want to use the *F*-test as a precursor to choosing homoscedastic or heteroscedastic models, do a literature search first to understand the issues more clearly. Or, better yet, do your own simulations to understand the issue. In Chapter 19, we show how to use such simulation studies to address the robustness of statistical procedures.

Vocabulary and Formula Summaries

Vocabulary

Linearity property of the normal distribution	The property that states that linear functions of normally distributed random variables are also normally distributed.
Additivity property of the normal distribution	The property that states that sums of independent normally distributed random variables are also normally distributed.
Degrees of freedom	The number of free variables; a parameter of the chi-squared distribution and the related Student's t-distribution.
Chi-squared distribution	The distribution of the sum of squared independent standard normal random variables, related to the distribution of the sample variance.
Student's t-distribution	The distribution of a standard normal random variable divided by the square root of an independent chi-squared random variable divided by its degrees of freedom; the distribution of the standardized average when the estimated standard deviation is used instead of the true standard deviation.
t-statistic	A statistic that has Student's t-distribution, or a parameter estimate divided by its standard error.
Experiment	A design and measurement strategy that allows you to assess causal effects; a study where you randomly assign experimental units to treatment groups.
Treatment	An experimental conditional that is assigned to a collection of observational units.
Treatment level	A particular treatment assignment; for example, *drug* is one level of a treatment, and *placebo* is another level.
Observational units	Items that data are collected from: people, plants, individual runs of a manufacturing process, animals, cities, etc.
No-effect model	A model that states that the experimental treatment has no effect on the data.

Two-sample comparison	The comparison of data values in two independent samples.
Homoscedasticity	A fancy word meaning equal process variances.
Pooled variance estimator	An estimate of a common (homoscedastic) variance that is the weighted average of the within-sample variance estimates, where the weights are the within-sample degrees of freedom.
Hypothesis	A constraint you place on the types of models that you assume to have produced your data.
Constraint	An assumed restriction, such as $\mu_1 = \mu_2$ or $\sigma_1 = \sigma_2$.
Null hypothesis	A hypothesis that states that there is no difference in the process with respect to the phenomenon studied.
Alternative hypothesis	A hypothesis that states that there is a difference in the process with respect to the phenomenon studied.
Reject the null hypothesis	When you decide to rule out chance as an explanation for the observed difference, for all intents and purposes.
Accept the null hypothesis	When you decide that you can't rule out chance as an explanation for the observed difference, for all intents and purposes. Don't use this phrase, as it sounds too much like you actually believe the null hypothesis to be true. Instead, just say that your observed difference is explainable by chance alone.
Two-sample *t*-statistic	The difference between sample means from two independent groups, divided by the estimate of the standard deviation of the difference.
Two-sample *t*-test	The determination of whether or not the difference between the sample means from two independent groups is explainable by chance alone.
Standard error of the estimate	The estimated standard deviation of the distribution of the parameter estimator.
***F*-distribution**	The null distribution of the ratio of variances estimated from independent normal samples.

Composite hypothesis	A single hypothesis that states that many individual null hypotheses are true simultaneously.
ANOVA model	A model used to analyze multiple-group data.
Variation between groups	Differences between group averages.
Variation within groups	Differences from data to the within-group averages.
Heteroscedasticity	A fancy word meaning unequal process variances.

Key Formulas and Descriptions

$T \sim N(a\mu + b, a^2\sigma^2)$	The linearity property of the normal distribution, where $T = aY + b$ and $Y \sim N(\mu, \sigma^2)$.
$T \sim N\left(\mu_X + \mu_Y, \sigma_X^2 + \sigma_Y^2\right)$	The additivity property of the normal distribution, where $T = X + Y$, with $X \sim N\left(\mu_X, \sigma_X^2\right)$ and $Y \sim N\left(\mu_Y, \sigma_Y^2\right)$, independent of X.
$Z_1^2 + Z_2^2 + \cdots + Z_m^2 \sim \chi_m^2$	The sum of squared independent $N(0, 1)$ random variables is distributed as chi-squared with m degrees of freedom.
$p(v) \propto v^{m/2-1}e^{-v/2}$	The kernel of the chi-squared distribution with m degrees of freedom.
$E\left(\chi_m^2\right) = m$	The expected value of a chi-square distributed random variable is equal to its degrees of freedom.
$\hat{\sigma}^2/\sigma^2 \sim \chi_{n-1}^2/(n-1)$	The ratio of the estimated variance to the true variance is distributed as chi-squared with $n - 1$ degrees of freedom, divided by $n - 1$.
$\hat{\sigma}/c_U \leq \sigma \leq \hat{\sigma}/c_L$	The confidence interval for σ.
$Z/\sqrt{V/m} \sim T_m$	A standard normal random variable divided by the square root of an independent chi-square over its degrees of freedom has Student's t-distribution.
$(\bar{Y} - \mu)/(\hat{\sigma}/\sqrt{n}) \sim T_{n-1}$	The standardized average has a t-distribution when the estimated standard deviation is used instead of the true standard deviation.
$\bar{Y} - c \times s.e.(\bar{Y}) \leq \mu \leq \bar{Y} + c \times s.e.(\bar{Y})$	The exact confidence interval for μ based on an iid sample from $N(\mu, \sigma^2)$.
$Y_{ij} \sim_{\text{independent}} p_i(y)$	A general model for data appearing in two or more groups.

$Y_{ij} \sim_{\text{independent}} N(\mu_i, \sigma^2)$

A more specific model for data appearing in two or more groups; the classic ANOVA model.

$\hat{\sigma}^2_{pooled} = \left((n_1 - 1)\hat{\sigma}^2_1 + (n_2 - 1)\hat{\sigma}^2_2\right)/\left((n_1 - 1) + (n_2 - 1)\right)$

The pooled variance estimator.

$\hat{\sigma}^2_{pooled}/\sigma^2 \sim \chi^2_{\{(n_1-1)+(n_2-1)\}}/\left((n_1 - 1) + (n_2 - 1)\right)$

The distribution of the pooled variance estimator.

$s.e.(\hat{\delta}) = \hat{\sigma}_{pooled}\sqrt{1/n_1 + 1/n_2}$

The standard error of the estimated difference between means of independent, homoscedastic normal samples.

$\hat{\delta} - c \times s.e.(\hat{\delta}) \le \delta \le \hat{\delta} + c \times s.e.(\hat{\delta})$

Confidence interval for difference between means.

$\hat{\delta}/s.e.(\hat{\delta}) \sim T_{\{(n_1-1)+(n_2-1)\}}$

The null distribution of the t-statistic for comparing means of independent, homoscedastic normal samples.

$T = \hat{\theta}/s.e.(\hat{\theta})$

The general form of a t-statistic.

$\left(\chi^2_{m_1}/m_1\right)/\left(\chi^2_{m_2}/m_2\right) \sim F_{m_1, m_2}$

The ratio of independent chi-squared random variables, divided by their degrees of freedom, has the F-distribution.

$p(f) \propto f^{(m_1/2)-1}\{1 + (m_1/m_2)f\}^{-(m_1+m_2)/2}$

The kernel of the F-distribution.

$H_0: \mu_1 = \mu_2 = \cdots = \mu_g = \mu$

The ANOVA null hypothesis.

$$F = \frac{\sum_i n_i(\bar{Y}_i - \bar{Y})^2/(g - 1)}{\sum_i \sum_j (Y_{ij} - \bar{Y}_i)^2/(n - g)}$$

The ANOVA F-statistic.

$\sum_i n_i(\bar{Y}_i - \bar{Y})^2/\sigma^2 \sim \chi^2_{g-1}$

The null distribution of the numerator of the ANOVA F-statistic.

$\sum_i \sum_j (Y_{ij} - \bar{Y}_i)^2/\sigma^2 \sim \chi^2_{n-g}$

The distribution of the denominator of the ANOVA F-statistic.

$F \sim F_{g-1, n-g}$

The null distribution of the ANOVA F-statistic.

$(\hat{\sigma}^2_1/\sigma^2_1)/(\hat{\sigma}^2_2/\sigma^2_2) \sim F_{n_1-1, n_2-1}$

The distribution of the ratio of sample variances computed from independent normal samples.

$\hat{\theta}/F_{n_1-1, n_2-1, 1-\alpha/2} \le \theta \le \hat{\theta}/F_{n_1-1, n_2-1, \alpha/2}$

The $100(1 - \alpha)\%$ confidence interval for the ratio of variances from independent normal samples.

Exercises

16.1 The linearity property of the normal distribution states that $Y \sim N(\mu, \sigma^2)$ and $T = aY + b$ implies $T \sim N(a\mu + b, a^2\sigma^2)$. This was shown to be true when $a > 0$ in Section 16.2. Follow the same method, and show that it is also true for $a < 0$. Pay attention to the inequalities!

16.2 Before coming to her 08:00 a.m. class, Olga has to do numerous things at home including eat, shower, feed and walk her dog, etc. Then she drives to school and walks to class. Suppose her time at home is normally distributed with mean 65 minutes and standard deviation 10 minutes, while her driving and walking time is independently distributed as normal with mean 30 minutes and standard deviation 3 minutes. Among days where Olga awakes at 06:15, how often is she late for class?

16.3 The additivity property of normality under independence states that sums of independent normal random variables are also normally distributed.

    A. Using simulation and a q–q plot, satisfy yourself that this is true for normally distributed random variables.

    B. Using simulation and a bar chart, satisfy yourself that this is *not* true for Bernoulli random variables. In other words, satisfy yourself that sums of independent Bernoulli distributed random variables are *not* distributed as Bernoulli random variables.

16.4 Suppose Z_1, Z_2, and Z_3 are independent $N(0, 1)$ random variables.

    A. Find the distribution of $(Z_1 + Z_2)/2$.

    B. Find the distribution of $(Z_1 + Z_2)/(2)^{1/2}$.

    C. Find the distribution of $\{(Z_1 + Z_2)/(2)^{1/2}\}/|Z_3|$. (Hint: Student's t-distribution. Why? What are the degrees of freedom?)

16.5 An Internet advertiser is comparing purchase amounts for two different banner ads. Behind the ads is a collection of identical products, but the ads themselves look different. For $n_1 = 123{,}410$ purchase totals arising from clicks on ad 1, the average purchase is 49.21, with a standard deviation of 56.10. For $n_2 = 156{,}770$ purchase totals arising from clicks on ad 2, the average purchase is 50.24, with a standard deviation of 69.04.

    A. Do you think the assumptions for the two-sample t-test are valid here? Look at the means and the standard deviations, think 68-95-99.7, and note that the purchase amounts cannot be negative.

    B. Assuming the two-sample t-test is valid, is the difference between the average purchase amounts explainable by chance alone? Answer by calculating the p-value.

    C. Construct a 95% confidence interval for the difference between average purchase amounts, and explain how this interval corroborates your answer of Exercise 16.5B.

    D. Based on your interval from Exercise 16.5C, do you think the difference between true means is practically significant?

16.6 See Exercise 15.8 of Chapter 15, the example concerning preferences for online and print newspapers. The data are (X_i, Y_i) for student i, where X_i indicates how often student i reads online newspapers, and where Y_i indicates how often student i reads print newspapers, with values (4 1), (5 1), (1 1), (1 2), (5 3), (3 3), (1 2), (4 2), (1 3), (5 1), (3 5), and (5 2).

A. Hans says that the 24 observations can be assumed to be independent. What is wrong with Hans' logic?

B. Construct the differences $d_i = x_i - y_i$. Is it reasonable to assume that these differences all come from a process that produces observations independently? Explain. How might the observations be dependent?

C. Is it reasonable to assume that the differences in Exercise 16.6B come from a $N(\mu, \sigma^2)$ distribution? Explain. There is no need to analyze the data, just look at it.

D. Assuming the differences in Exercise 6B *are* produced as iid from the $N(\mu, \sigma^2)$ distribution, construct and interpret the exact 95% confidence interval for μ.

E. Using your answer to Exercise 16.6D, is the observed difference between the sample averages explainable by chance alone?

16.7 Use the data from Exercise 16.6.

A. Construct a 90% confidence interval for the standard deviation of the D_i data.

B. Explain which assumptions needed for the interval in Exercise 16.7A are violated and which are reasonable. There is no need to calculate anything; just stare at the numbers.

16.8 Dow Jones Industrial Average (DJIA) returns for $n_1 = 27$ trading days prior to September 11, 2001, averaged −0.33%, with a standard deviation of 1.07%. On September 11, 2001, there was a terrorist attack, causing markets to close for a week. After the markets reopened on September 17, the closing Dow dropped 7.13% from its previous close on September 10. In the subsequent $n_2 = 32$ trading days, things stabilized, with an average return of 0.07% and a standard deviation of 1.70%. Assuming the $n_1 = 27$ and $n_2 = 32$ returns are produced independently from the $N(\mu_i, \sigma^2)$ distributions, where $i = 1$ denotes before and $i = 2$ denotes after, find and interpret the 95% confidence interval for $\mu_1 - \mu_2$.

16.9 Use the data from Exercise 16.8, assuming the model $Y_{ij} \sim_{independent} N(\mu_i, \sigma_i^2)$.

A. Construct a 95% confidence interval for the ratio of variances.

B. Calculate the p-value for testing equality of variances, and explain how the result corroborates the result of Exercise 16.9A.

16.10 Consider the wait time data from Exercise 4.9 in Chapter 4. Assume the first 20 observations are from call center 1, the second 20 observations are from call center 2, and the third 20 observations are from call center 3.

A. State the standard ANOVA model for how these data are generated.

B. Is the normality assumption of the ANOVA model valid? No data analysis necessary—just think about wait times; especially about the number zero and about the occasional case where there is heavier than expected call volume.

C. Is the independence assumption valid? No data analysis is necessary. What process elements could cause a violation of this assumption?

D. Is the homoscedasticity assumption valid? What process elements could cause nonconstant variance between the shifts? Again, no data analysis is necessary.

E. Now do some data analysis. Calculate the F-statistic and p-value for comparing group means. Interpret the result as if all assumptions were valid (even though they aren't).

16.11 Hans thinks that a fair coin should land heads 50% of the time. He flips the coin 100 times, getting 45% heads and is irritated. He repeats 100 flips, getting 52% heads. Now he is even more irritated! He repeats two more times, getting 53% and 49%. He thinks the universe is against him. Not only did he never get 50%, but he got a different percentage every time!

    A. Calculate the F-statistic and p-value from Hans' 400 binary outcomes to compare his four sets of flips ($g = 4$), and interpret the result in such a way to soothe Hans. (Hint: You can use the plug-in estimate of estimates of the mean and variance calculated from binary data are $\hat{\pi}$ and $\hat{\pi}(1 - \hat{\pi})$ respectively. Or you can create Hans' data set and feed it into a software package.)

    B. Without doing any data analysis, discuss the normality, independence, and homoscedasticity assumptions of the ANOVA model. Which assumption(s) are reasonable here, and which assumption(s) are not reasonable?

16.12 Hans is also irritated by the statement that the variance of the t-distribution is infinite when $df \leq 2$. "How can the variance be infinite?" he asks. "If you calculate a variance from a sample of data, you always get a number, you never get infinity." Solve Hans' mental dilemma by simulating $n = 10, 100, 1{,}000, 10{,}000, 100{,}000$, and $1{,}000{,}000$ values from the T_1 distribution, calculating the sample variance for each of those six samples, and drawing a graph of the six estimates with n on the horizontal axis, with both axes in logarithmic scale.

17

Likelihood Ratio Tests

17.1 Introduction

Just like you have many different ways to estimate parameters, you also have many different ways to test hypotheses. For example, to test whether a student's age is independent of their seat preference in a classroom you could use the *median age* of students in the front rows, minus the *median age* of students in the back rows as your test statistic. And if you think about it, there are lots of other test statistics you could use as well. Why not compare the *maximum age* in each group? And further, why does the F-statistic in Chapter 16, Equation 16.14, look like it does? Who decides these things anyway? How can you know what is the best test statistic to use?

Likelihood to the rescue! You saw in Chapter 12 that likelihood provides an automatic, usually highly efficient method to estimate parameters. It is similarly useful for testing hypotheses: The *likelihood ratio test* provides tests that are also usually highly efficient, in the sense of providing greatest ability to detect deviations from the chance-only (or null) model. The *power* of a test measures its ability to detect deviations from the null model, and will be discussed further in Chapter 18.

In mathematical statistics sources that are more advanced than this one, you will see that likelihood ratio tests are *optimal* in the sense of having the highest power among certain types of tests. We won't prove the mathematics, but you can find out more by searching the terms *Neyman-Pearson Lemma* and *optimality of likelihood ratio tests*.

Chapter 16 showed that you can use the F-statistic for comparing groups. It turns out that the F-statistic is a likelihood ratio statistic, as are the other test statistics presented in that chapter. This explains some of the mystery as to why the test statistics have their particular forms: It is because they are likelihood ratio test statistics, and are therefore optimal. If they *weren't* likelihood ratio tests, you wouldn't see them!

In addition to their optimality, likelihood ratio tests are useful because they give you a way to test hypotheses in *any* likelihood-based model, whether based on normal distributions, Poisson distributions, Bernoulli distributions, etc. They also provide a simple way to test hypotheses in very complex models—such as nonlinear structural equation models, to drop a fancy name—with relative ease. Thus, you will see applications of likelihood ratio tests everywhere in advanced statistical methods, including survival analysis, psychometrics, econometrics, and engineering design optimization, to name just a few. Applications discussed in this chapter include ANOVA, multiple regression, multiple logistic regression, goodness-of-fit tests, and contingency tables.

17.2 Likelihood Ratio Method for Constructing Test Statistics

The **likelihood ratio test** (LRT) is a test of a **full model** versus a **restricted model**. The restricted model is the null, or chance-only model. The better the data fit the full model relative to the restricted model, the less plausible is the null model—that is, the more difficult it is to explain your results by chance alone.

For example, the ANOVA full model is $Y_{ij} \sim_{\text{independent}} N(\mu_i, \sigma^2)$, $i = 1, 2, \ldots, g$; $j = 1, 2, \ldots, n_i$, and the restricted model is $Y_{ij} \sim_{\text{independent}} N(\mu, \sigma^2)$. In the restricted model, the restriction is that the means are all equal: $\mu_1 = \mu_2 = \cdots = \mu_g$. The F-test you saw in Chapter 16 tests the null hypothesis that the restricted model is true, and you can also use the LRT to test this hypothesis. As you will see, these two tests are really the same test.

To use the likelihood ratio method in general, you must specify a full model $p_1(y \mid \theta_1)$ and a restricted (null) model $p_0(y \mid \theta_0)$, which must be obtained by *restricting* the parameters in the full model $p_1(y \mid \theta_1)$. For example, you can obtain the ANOVA restricted model $Y_{ij} \sim_{\text{independent}} N(\mu, \sigma^2)$ from the full model $Y_{ij} \sim_{\text{independent}} N(\mu_i, \sigma^2)$ by restricting the parameters $\theta_1 = \{\mu_1, \mu_2, \ldots, \mu_g, \sigma^2\}$ so that all the μ_i are the same number, μ. In this case $\theta_0 = \{\mu, \sigma^2\}$.

The main idea behind the LRT is this: To test whether the restricted model is adequate compared to the unrestricted model, simply compare their likelihoods. If the maximized likelihood is much larger for the unrestricted model, then you should reject the restricted model.

Mathematical Fact about Model Comparison

> When using likelihood, unrestricted models always fit the data as well as or better than restricted models.

To see why this is true, let Θ_1 (that's the Greek upper-case letter theta with a 1 subscript) be the parameter space in the unrestricted model, and Θ_0 be the parameter space in the restricted model. The requirement that the null model $p_0(y \mid \theta_0)$ *must be obtained* by *restricting* the parameters in the full model $p_1(y \mid \theta_1)$ is equivalent to

$$\Theta_0 \subset \Theta_1$$

In other words, the parameter space of the restricted model must be a subset of the parameter space in the unrestricted model in order for you to be able to apply the likelihood ratio method.

You can then see that the unrestricted model fits better as follows:

$L(\hat{\theta}_1 \mid \text{data}) = \max_{\theta \in \Theta_1} L(\theta \mid \text{data})$	(By definition of maximum likelihood estimate)
$\geq \max_{\theta \in \Theta_0} L(\theta \mid \text{data})$	(Since the maximum over a restricted space can be no larger than the global maximum)
$= L(\hat{\theta}_0 \mid \text{data})$	(By definition of maximum likelihood estimate)

Figure 17.1 illustrates the idea. The largest point on the entire mountain, shown as a contour plot, is larger than the largest point on a constrained portion of the mountain—that is, the portion of the mountain where the "longitude" coordinate (the value on the horizontal axis) is fixed but the "latitude" coordinate (the value on the vertical axis) varies. In Figure 17.1, the unrestricted space is the entire graph, or $\Theta_1 = \{(\mu, \sigma); -\infty < \mu < \infty, 0 < \sigma < \infty\}$. The restricted space is $\Theta_0 = \{(\mu, \sigma); \mu = 310, 0 < \sigma < \infty\}$, a subset of Θ_1—that is, the vertical line is a subset of the entire graph.

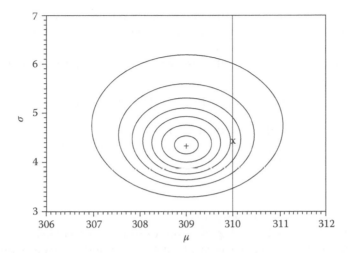

FIGURE 17.1
Contour plot of likelihood function of the parameter vector $\theta = \{\mu, \sigma\}$, showing slice where μ is constrained to be 310. The maximized likelihood over all Θ_1, located by the + mark in the center of the contours at $\hat{\theta}_1 = \{309.0, 4.344\}$, is higher than the maximized constrained likelihood over all Θ_0, located by the x mark on the vertical line at $\hat{\theta}_0 = \{310.0, 4.457\}$.

The fact that unrestricted models always fit the data better seems to suggest that you always should use unrestricted models. However, a better fit to the *data* does not necessarily mean better fit to the *process*, simply because *data* and *process* differ. The data are *random*, whereas the process is *fixed*. If you try to fit more and more wiggles and squiggles of the random data using models that are increasingly complex, your estimated model will get farther and farther from the process model. This is another application of the mantric phrase *nature favors continuity over discontinuity,* since more complex models tend to be less smooth than simpler ones. We'll return to this issue later in this chapter when we discuss *Akaike's Information Criterion.*

As discussed in Chapter 12, there are many reasons to use logarithms of likelihoods; the same is true for likelihood ratio tests. So let $LL_1 = \ln\{L(\hat{\theta}_1 | \text{data})\}$ and $LL_0 = \ln\{L(\hat{\theta}_0 | \text{data})\}$. A larger log likelihood indicates a better fit of the model to the data. The difference $LL_1 - LL_0$ is never negative because $L_1 \geq L_0$ as shown earlier, and because the ln function increases monotonically. In particular, $LL_1 \geq LL_0$ is true even when the true model is the restricted model $p_0(y | \theta_0)$. In this case, the difference between LL_1 and LL_0 is *explained by chance alone,* and the difference should be *small*. On the other hand, when $p_1(y | \theta_1)$ is the true model then its log-likelihood LL_1 should be *much larger* than LL_0.

Likelihood Ratio Method for Constructing Optimal Test Statistics

Let $\chi^2 = 2(LL_1 - LL_0)$, where LL_1 and LL_0 are the maximized log-likelihoods under the unrestricted and restricted models, respectively. Then χ^2 is the likelihood ratio test statistic, and you should reject the restricted model $p_0(y)$ when χ^2 is large.

The test is a likelihood *ratio* test because it is based on the *difference* of log likelihoods, which by property of logarithm is equal to the log of the *ratio* of likelihoods as follows:

$$LL_1 - LL_0 = \ln(L_1) - \ln(L_0) = \ln\left(\frac{L_1}{L_0}\right)$$

The *F*-statistic and others presented in Chapter 16 are not precisely equal to the χ^2 statistic, but they are closely related. To see how they are related, start with a simple example: If the χ^2 statistic were related to the *F*-statistic via $\chi^2 = 2F$, you would say that the statistics are equivalent, since larger values of χ^2 always correspond to larger values of *F*. Further, both would give the same *p*-value

$$\Pr(\chi^2 \geq v) \ = \ \Pr(2F \geq 2f) \ = \ \Pr(F \geq f)$$

This is true because $2F \geq 2f$, if and only if $F \geq f$.

Similarly, if the relationship were $\chi^2 = e^F$ you would also conclude that the statistics are equivalent, since they give the same *p*-value

$$\Pr(\chi^2 \geq v) \ = \ \Pr(e^F \geq e^f) = \ \Pr(F \geq f)$$

This is true because $e^F \geq e^f$, if and only if $F \geq f$.

These two examples, $\chi^2 = 2F$ and $\chi^2 = e^F$, are examples of *monotonically increasing* functions, where larger values of one variable always correspond to larger values of the other. If such a function is differentiable, then its derivative is always greater than zero.

Condition for a Monotonically Increasing Relationship

If $y = f(x)$ and $f'(x) > 0$ for all x, then $y = f(x)$ is a monotonically increasing relationship.

If the likelihood ratio statistic χ^2 statistic is a monotonically increasing function of another statistic *W*, then the test based on χ^2 and the upper-tail test based on *W* are equivalent, since both give the same *p*-value

$$\Pr(\chi^2 \geq v) \ = \ \Pr\{f(W) \geq f(w)\} \ = \ \Pr(W \geq w) \tag{17.1}$$

This is true because, for a monotonically increasing function f, $f(W) \geq f(w)$, if and only if $W \geq w$.

Thus, while the likelihood ratio statistics are not necessarily equal to standard test statistics such as presented in Chapter 16, they are often related monotonically. In such cases, the tests are equivalent since they give the same *p*-value.

Equation 17.1 is not true when the function is non-monotonic. For example, suppose $\chi^2 = f(T) = T^2$, a non-monotonic function. Suppose also that the observed value of χ^2 is $v = 4.0$. Then the *p*-value is as follows:

$$\Pr(\chi^2 \geq 4.0) \ = \ \Pr(T^2 \geq 4.0) \ \neq \ \Pr(T \geq 2.0)$$

Instead, $\Pr(\chi^2 \geq 4.0) = \Pr(T \geq 2.0) + \Pr(T \leq -2.0)$. So the test based on χ^2 statistic would not be equivalent to the upper-tail test based on the *T*-statistic in this case; instead it is equivalent to the upper-tail test based on the T^2 statistic. The following example establishes this connection in the case of the one-sample *t*-test.

Example 17.1: The One-Sample t^2-Statistic as a Likelihood Ratio Statistic

Suppose $Y_1, Y_2, \ldots, Y_n \sim_{\text{iid}} N(\mu, \sigma^2)$ is a reasonable model for how your data are produced, and that you are interested in whether the mean μ is equal to some particular constant value m_0. In Example 7.10 concerning widths of computer chips, the constant $m_0 = 310$ might be the desired target mean width for specification purposes. Thus, your null (restricted) model is

$$Y_1, Y_2, \ldots, Y_n \sim_{\text{iid}} N(m_0, \sigma^2)$$

Here, the parameter restriction is $\mu = m_0$ (for example, $\mu = m_0 = 310$); in the unrestricted model μ can be any number.

By the main result for Student's t-distribution presented in Chapter 16, Section 16.6, you know that if H_0: $\mu = m_0$ is true, then

$$T = \frac{\bar{Y} - m_0}{\hat{\sigma}/\sqrt{n}} \sim T_{n-1} \tag{17.2}$$

If $T \geq c$ or if $T \leq -c$, then you will reject H_0; from Equation 16.7, $c = t_{n-1,1-\alpha/2}$. Equivalently, you will reject H_0 if $T^2 \geq c^2$. Recall from Chapter 16, Section 16.9, that the distribution of T^2 is the $F_{1,n-1}$ distribution, so you can calculate p-value as follows:

$$pv = \Pr(F_{1,n-1} \geq t^2)$$

But is this method any good? If it is a likelihood ratio procedure, then it is not only *good*: It is *optimal*. So let's do the likelihood ratio analysis to check. The steps are as follows: (1) find the MLEs under the full and restricted models; (2) plug them into their respective log likelihood functions to get LL_1 and LL_0; and (3) compute $\chi^2 = 2(LL_1 - LL_0)$.

In Example 12.7 you saw that the MLEs of μ and σ^2 using a sample produced as iid from a $N(\mu, \sigma^2)$ were the ordinary sample mean and the plug-in variance estimate that uses n rather than $n - 1$, yielding

$$\hat{\mu}_1 = \frac{y_1 + y_2 + \cdots + y_n}{n} = \bar{y}$$

Also,

$$\hat{\sigma}_1^2 = \frac{(y_1 - \bar{y})^2 + (y_2 - \bar{y})^2 + \cdots + (y_n - \bar{y})^2}{n}$$

The 1 subscript on $\hat{\mu}_1$ and on $\hat{\sigma}_1^2$ denotes the unrestricted model $p_1(y | \theta_1)$.

Under the null model, the mean μ is no longer a free parameter but is a constant m_0 that you specify, such as $m_0 = 310$. So, in the null model, there is only one parameter, σ^2, rather than two parameters $\{\mu, \sigma^2\}$ as in the unrestricted model; that is, $\theta_1 = \{\mu, \sigma^2\}$ in the unrestricted model, and $\theta_0 = \{310, \sigma^2\}$ in the null model. The supposed mean $m_0 = 310$ is *not* a parameter. Recall the Mantra: *Model has unknown parameters.* If you are ever confused about what is or is not a parameter, just remember that parameters are values you have to *estimate* using the data.

You can show (see Exercise 17.1 at the end of this chapter) that the maximum likelihood estimate of σ^2 in the restricted model $N(m_0, \sigma^2)$ is

$$\hat{\sigma}_0^2 = \frac{(y_1 - m_0)^2 + (y_2 - m_0)^2 + \cdots + (y_n - m_0)^2}{n}$$

This estimate makes common sense: If you know that the mean is m_0, then a logical estimate of the expected squared deviation from Y to m_0—that is, of the variance—is just the average of the squared deviations from the y_i values to m_0. This estimate, when viewed as a function of random data Y_i, is unbiased under the null model since the mean is known–unlike the usual MLE for σ^2 which is biased, as discussed in Chapter 11, Section 11.3.

Plugging the appropriate MLEs into the likelihood function gives you the maximized values LL_1 and LL_0. In the unrestricted model, the likelihood function is given as follows:

$L(\theta_1 \mid y_1, y_2, \ldots, y_n)$
$= p_1(y_1 \mid \theta_1) \times p_1(y_2 \mid \theta_1) \times \cdots \times p_1(y_n \mid \theta_1)$ (Since the likelihood function of a sample produced as iid is equal to the product of the pdfs of the individual observations; see Chapter 12)

$$= \frac{1}{\sqrt{2\pi\sigma^2}} \exp\frac{-(y_1 - \mu)^2}{2\sigma^2}$$

$$\times \frac{1}{\sqrt{2\pi\sigma^2}} \exp\frac{-(y_2 - \mu)^2}{2\sigma^2} \times \cdots$$

$$\times \frac{1}{\sqrt{2\pi\sigma^2}} \exp\frac{-(y_n - \mu)^2}{2\sigma^2}$$ (By substitution)

$$= \left(\frac{1}{\sqrt{2\pi\sigma^2}}\right)^n \times \exp\left\{\frac{-(y_1 - \mu)^2 - (y_2 - \mu)^2 - \cdots - (y_n - \mu)^2}{2\sigma^2}\right\}$$ (By algebra and properties of exponents)

$$= \left(\frac{1}{\sqrt{2\pi}}\right)^n \left(\frac{1}{\sqrt{\sigma^2}}\right)^n \exp\left\{\frac{-\sum_i (y_i - \mu)^2}{2\sigma^2}\right\}$$ (By properties of exponents and definition of summation Σ)

$$= (2\pi)^{-n/2}(\sigma^2)^{-n/2} \exp\left\{\frac{-\sum_i (y_i - \mu)^2}{2\sigma^2}\right\}$$ (By properties of exponents)

Hence the log likelihood function for the unrestricted model is

---

$$LL_1(\mu, \sigma^2) = -(n/2)\ln(2\pi) - (n/2)\ln(\sigma^2) - \frac{\sum_i (y_i - \mu)^2}{2\sigma^2}$$ (By property of logarithms)

You get the maximized value of LL_1 by plugging the MLE $\hat{\theta}_1 = \{\hat{\mu}_1, \hat{\sigma}_1^2\}$ into the expression for $LL_1(\mu, \sigma^2)$. Since $\hat{\mu}_1 = \bar{y}$ and $\hat{\sigma}_1^2 = \{(y_1 - \bar{y})^2 + (y_2 - \bar{y})^2 + \cdots + (y_n - \bar{y})^2\}/n$, you get the following:

$$\frac{\sum_i (y_i - \hat{\mu}_1)^2}{2\hat{\sigma}_1^2} = \frac{\sum_i (y_i - \bar{y})^2}{2\sum_i (y_i - \bar{y})^2/n} = n/2$$ (By substitution and algebra)

Hence there is the following simple form for the maximized unrestricted likelihood:

$$LL_1 = -(n/2)\ln(2\pi) - (n/2)\ln\left\{\sum_i (y_i - \bar{y})^2/n\right\} - n/2$$ (By substitution of $\hat{\sigma}_1^2$ for σ^2 in the likelihood function)

$$= -(n/2)\ln(2\pi) - (n/2)\ln(\hat{\sigma}_1^2) - n/2$$ (By substitution)

Now, for the null model restricted likelihood. Following the same steps as shown for the unrestricted likelihood, you get

$$LL_0 = -\left(\frac{n}{2}\right)\ln(2\pi) - \left(\frac{n}{2}\right)\ln\left(\hat{\sigma}_0^2\right) - \left(\frac{n}{2}\right)$$

Therefore, the likelihood ratio chi-squared statistic has the following very simple form:

$\chi^2 = 2(LL_1 - LL_0)$	(By definition)
$= 2\left\{-(n/2)\ln(2\pi) - (n/2)\ln\left(\hat{\sigma}_1^2\right) - n/2 + (n/2)\ln(2\pi) + (n/2)\ln\left(\hat{\sigma}_0^2\right) + n/2\right\}$	(By substitution)
$= n\left\{\ln\left(\hat{\sigma}_0^2\right) - \ln\left(\hat{\sigma}_1^2\right)\right\}$	(By combining terms algebraically)
$= n\left\{\ln\left(\hat{\sigma}_0^2 / \hat{\sigma}_1^2\right)\right\}$	(By property of logarithms)

The likelihood ratio test procedure then will reject the restricted model $N(m_0, \sigma^2)$ if the observed value of $\chi^2 = n\left\{\ln\left(\hat{\sigma}_0^2/\hat{\sigma}_1^2\right)\right\}$ is too large. This makes sense: If the data are much farther from m_0 than they are from $\bar{y}$, then $\hat{\sigma}_0^2$ will be much larger than $\hat{\sigma}_1^2$, leading to a large value of the χ^2 statistic, and you should reject the notion that the true process mean is equal to m_0 in that case.

But is this really the same procedure as the one-sample t-test, where you reject H_0 for large values of t^2, where t is given by Equation 17.2? The answer is, yes! First note, by substitution, that

$$\chi^2 = n\left\{\ln\left(\frac{\hat{\sigma}_0^2}{\hat{\sigma}_1^2}\right)\right\} = n\ln\left\{\frac{\sum_i (y_i - m_0)^2/n}{\sum_i (y_i - \bar{y})^2/n}\right\}$$

There is a handy trick that allows you to simplify the ratio of variance estimates. You have seen it before in various places—it is the trick of adding and subtracting an appropriate constant. Here, it works like this to simplify the expression for $\hat{\sigma}_0^2$:

$\hat{\sigma}_0^2 = \sum_i (y_i - m_0)^2/n = \sum_i \{(y_i - \bar{y}) + (\bar{y} - m_0)\}^2/n$	(By adding and subtracting $\bar{y}$)
$= (1/n)\sum_i (y_i - \bar{y})^2 + (1/n)\sum_i (\bar{y} - m_0)^2 + (2/n)\sum_i (y_i - \bar{y})(\bar{y} - m_0)$	(By expanding the square and using summation properties)
$= \hat{\sigma}_1^2 + (\bar{y} - m_0)^2 + (2/n)(\bar{y} - m_0)\sum_i (y_i - \bar{y})$	(By substitution and noting that $(\bar{y} - m_0)$ is a constant with respect to the summation)
$= \hat{\sigma}_1^2 + (\bar{y} - m_0)^2$	(Since $\sum_i (y_i - \bar{y}) = 0$)

Thus you can see that the chi-squared statistic is as follows:

$$\chi^2 = n \times \ln\left(\frac{\hat{\sigma}_0^2}{\hat{\sigma}_1^2}\right) = n \times \ln\left[\frac{\hat{\sigma}_1^2 + (\bar{y} - m_0)^2}{\hat{\sigma}_1^2}\right] = n \times \ln\left\{1 + \frac{(\bar{y} - m_0)^2}{\hat{\sigma}_1^2}\right\} \qquad (17.3)$$

Now, the plug-in variance estimate $\hat{\sigma}_1^2$ (the n version), is related to the standard form $\hat{\sigma}^2$ (the $n - 1$ version) as follows:

$$\hat{\sigma}_1^2 = \frac{n-1}{n}\hat{\sigma}^2$$

Substituting this and Equation 17.2 into Equation 17.3, the likelihood-ratio chi-square statistic is then

$$\chi^2 = n \times \ln\left\{1 + \frac{n}{(n-1)}\frac{(\bar{y} - m_0)^2}{\hat{\sigma}^2}\right\} = n \times \ln\left\{1 + \frac{t^2}{(n-1)}\right\} \qquad (17.4)$$

Equation 17.4 shows that the likelihood ratio statistic is *monotonically related* to the t^2 statistic. You can see this because the function $f(x) = c_1 \times \ln(1 + c_2 x)$ has a positive derivative:

$$f'(x) = \frac{c_1 c_2}{(1 + c_2 x)} > 0$$

This is true for all $x > 0$ when c_1 and c_2 are positive. To relate this function to Equation 17.4, set $c_1 = n$, $c_2 = 1/(n - 1)$, and $x = t^2$.

Thus, if t^2 increases so does χ^2, and vice versa, implying that the test that rejects the null model when the likelihood ratio statistic is large equivalent to the test that rejects the null model when the t^2 statistic is large. This establishes that the two-sided t-test for H_0: $\mu = m_0$ is a likelihood ratio test, and is therefore an optimal test.

Assumptions are important! When the data-generating process is the iid $N(\mu, \sigma^2)$ model, Example 17.1 shows that the two-sided test based on the Student t-statistic is optimal. What if the distribution is non-normal? Well, different likelihoods imply different statistics, and this test is not optimal with non-normal distributions. The following example shows why, for the case of exponential distributions.

Example 17.2: The One-Sample t^2-Statistic is *Not* a Likelihood Ratio Statistic When the Distribution Is Exponential

Suppose the model $Y_1, Y_2, \ldots, Y_n \sim_{iid}$ Exponential(λ) is a reasonable model for how your data are produced, and that you want to know whether the mean $\mu = 1/\lambda$ is equal to some particular constant value m_0. For example if the data Y_i are customers' waiting times as introduced in Chapter 4, the constant $m_0 = 0.5$ minutes might be a desired target for the company's purposes. Thus, your null model is

$$Y_1, Y_2, \ldots, Y_n \sim_{iid} \text{Exponential}(2.0)$$

Recall from Example 12.2 that the likelihood function for the exponential sample is

$$LL = n\ln(\lambda) - n\lambda\bar{y} \tag{17.5}$$

Under the null restriction there is no parameter since $\lambda = 1/m_0$ in (17.5), and so there is no need to maximize anything. The null log likelihood is simply

$$LL_0 = n\ln\left(\frac{1}{m_0}\right) - \frac{n\bar{y}}{m_0}$$

Consider now the unrestricted log likelihood. As indicated in Example 12.6, the maximum likelihood estimate of λ is

$$\hat{\lambda} = \frac{1}{\bar{y}}$$

Plugging this into Equation 17.5, you get

$$LL_1 = n\ln\left(\frac{1}{\bar{y}}\right) - n$$

Hence the likelihood ratio test statistic is

$$\chi^2 = 2(LL_1 - LL_0) = 2n\ln\left(\frac{m_0}{\bar{y}}\right) - 2n\left(1 - \frac{\bar{y}}{m_0}\right)$$

For example, if $n = 10$ and $m_0 = 0.5$, then

$$\chi^2 = 20\ln\left(\frac{0.5}{\bar{y}}\right) - 20\left(1 - \frac{\bar{y}}{0.5}\right)$$

Notice that if $\bar{y} = 0.5$ then $\chi^2 = 0$, as is sensible since there would be no evidence against $\mu = 0.5$ in this case. Otherwise, the value of χ^2 becomes larger as $\bar{y}$ is either more or less than 0.5. But the function is not symmetric, as shown in Figure 17.2: The χ^2 statistic is larger than a critical value 5.0 if and only if the sample average is either less than 0.22 or greater than 0.94. The interval (0.22, 0.94) is an asymmetric interval around 0.5; therefore, the two-sided t-test is not optimal since it has a symmetric rejection region around m_0. You get different test statistics when you assume different distributions, so you should always be aware of your assumptions.

Example 17.3: The ANOVA F-Statistic Is a Likelihood Ratio Statistic

The ANOVA model introduced in Chapter 16 is for independent data in different groups, where the goal is to compare groups and identify differences. The standard model is $Y_{ij} \sim_{\text{independent}} N(\mu_i, \sigma^2)$, for groups $i = 1, 2, \ldots, g$ and observations $j = 1, 2, \ldots, n_i$, within group i. In other words, when you use this model you assume that the data within different groups are produced independently by normal distributions with possibly

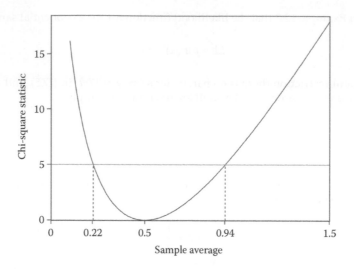

FIGURE 17.2
The likelihood ratio chi-squared statistic for testing an exponential mean, as a function of the sample average. The statistic is greater than 5.0 when the sample average is either less than 0.22 or greater than 0.94 (dashed vertical lines).

different means but common variance. The null model is $Y_{ij} \sim_{\text{independent}} N(\mu, \sigma^2)$, in which you assume that the same normal distribution produces *all* the data, regardless of group. This is a perfect example of the full-model versus restricted-model testing paradigm, and the likelihood ratio method is just dying to be employed for this purpose.

Again recall the following steps: (1) find the MLEs under the full and restricted models; (2) plug them in to their respective log likelihood functions to get LL_1 and LL_0; and (3) compute $\chi^2 = 2(LL_1 - LL_0)$.

First, the easy model. Since the null model simply states that all $n = n_1 + n_2 + \cdots + n_g$ observations are iid $N(\mu, \sigma^2)$, the maximum likelihood estimates are the familiar average and plug-in variance estimate, as applied to all the n data values, irrespective of group. Using double summation notation, these formulas give you the following:

$$\hat{\mu} = \sum_i \sum_j \frac{y_{ij}}{n}$$

$$\hat{\sigma}_0^2 = \sum_i \sum_j \frac{(y_{ij} - \hat{\mu})^2}{n}$$

The 0 subscript indicates estimated variance under the null model.

Now, the unrestricted model. The likelihood function is the product of the likelihoods for all the individual observations using the $N(\mu_i, \sigma^2)$ model:

$$L(\{\mu_1, \mu_2, \ldots, \mu_g, \sigma^2 | \text{data}) = \prod_i \prod_j \frac{1}{\sqrt{2\pi\sigma^2}} \exp\left\{ \frac{-(y_{ij} - \mu_i)^2}{2\sigma^2} \right\}$$

(By definition of likelihood function of a sample produced as independent observations)

$$= \left(\frac{1}{\sqrt{2\pi\sigma^2}} \right)^n \exp\left\{ \frac{-\sum_i \sum_j (y_{ij} - \mu_i)^2}{2\sigma^2} \right\}$$

(By algebra and properties of exponents)

No matter what σ is, the values of μ_i that *maximize* the likelihood have to *minimize* the numerator of the exponent, $\Sigma_i \Sigma_j (y_{ij} - \mu_i)^2$. Using calculus, you can find that

$$\hat{\mu}_i = \sum_j \frac{y_{ij}}{n_i}$$

In other words, the MLE of the within-group process mean is simply the within-group sample mean. This makes sense! Now, you can solve for the MLE of σ^2 by using calculus with the log-likelihood function, after plugging in the MLEs of the μ_i:

$LL(\{\hat{\mu}_1, \hat{\mu}_2, \ldots, \hat{\mu}_g, \sigma^2 \,|\, \text{data})$

$= -(n/2)\ln(2\pi) - (n/2)\ln(\sigma^2) - \Sigma_i \Sigma_j (y_{ij} - \hat{\mu}_i)^2 / 2\sigma^2$ (By substitution and properties of logarithms)

$\Rightarrow (\partial/\partial\sigma^2) LL(\{\hat{\mu}_1, \ldots, \hat{\mu}_g, \sigma^2 \,|\, \text{data})$

$= -(n/2)(1/\sigma^2) + \Sigma_i \Sigma_j (y_{ij} - \hat{\mu}_i)^2 / 2\sigma^4$ (By differentiation of terms with respect to σ^2; if this is confusing, just replace σ^2 with x and calculate $(\partial/\partial x)$)

Setting the derivative to zero and solving, you get

$$\hat{\sigma}_1^2 = \sum_i \sum_j \frac{(y_{ij} - \hat{\mu}_i)^2}{n}$$

Now you get the maximized log likelihoods for the unrestricted and restricted models by plugging the MLEs for the unrestricted and restricted models into their respective log likelihood functions. In either case the log likelihood function has the following form:

$$LL = -\left(\frac{n}{2}\right)\ln(2\pi) - \left(\frac{n}{2}\right)\ln(\sigma^2) - \sum_i \sum_j \frac{(y_{ij} - \mu_i)^2}{2\sigma^2}$$

In the case of the null model, just substitute μ for each μ_i. Plugging in the restricted model MLEs you get

$$LL_0 = -\left(\frac{n}{2}\right)\ln(2\pi) - \left(\frac{n}{2}\right)\ln\left(\hat{\sigma}_0^2\right) - \frac{n}{2}$$

Plugging in the unrestricted model MLEs you get

$$LL_1 = -\left(\frac{n}{2}\right)\ln(2\pi) - \left(\frac{n}{2}\right)\ln\left(\hat{\sigma}_1^2\right) - \frac{n}{2}$$

Hence the likelihood ratio test statistic is

$$2(LL_1 - LL_0) = n\left\{\ln\left(\frac{\hat{\sigma}_0^2}{\hat{\sigma}_1^2}\right)\right\}$$

This is a familiar-looking quantity: As in Example 17.1, the likelihood ratio statistic is related to the ratio of variances that are estimated under the restricted and unrestricted models. The *F*-statistic is also a ratio of variances, but not quite this one: Instead, it is a ratio of between-group to within-group variances. But you can see that the likelihood ratio procedure gives you the *F*-statistic using the same trick shown in Example 17.1, where you add and subtract the right numbers to simplify the expression for $\hat{\sigma}_0^2$.

$\hat{\sigma}_0^2 = \sum_i \sum_j \{(y_{ij} - \hat{\mu}_i) + (\hat{\mu}_i - \hat{\mu})\}^2 / n$ (By adding and subtracting $\hat{\mu}_i$)

$= \sum_i \sum_j (y_{ij} - \hat{\mu}_i)^2 / n + \sum_i \sum_j (\hat{\mu}_i - \hat{\mu})^2 / n$

$\quad + 2\sum_i \sum_j (\hat{\mu}_i - \hat{\mu})(y_{ij} - \hat{\mu}_i)/n$ (By expanding the square and using summation algebra)

$= \hat{\sigma}_1^2 + \sum_i \sum_j (\hat{\mu}_i - \hat{\mu})^2 / n$

$\quad + 2\sum_i (\hat{\mu}_i - \hat{\mu}) \sum_j (y_{ij} - \hat{\mu}_i)/n$ (By substitution and noting that $(\hat{\mu}_i - \hat{\mu})$ is a constant with respect to the summation index j)

$= \hat{\sigma}_1^2 + \sum_i \sum_j (\hat{\mu}_i - \hat{\mu})^2 / n$ (Since $\sum_j (y_{ij} - \hat{\mu}_i) = 0$)

Hence, by algebra:

$$2(LL_1 - LL_0) = n \times \ln\left(\frac{\hat{\sigma}_0^2}{\hat{\sigma}_1^2}\right)$$

$$= n \times \ln\left[\frac{\left\{\hat{\sigma}_1^2 + \sum_i \sum_j (\hat{\mu}_i - \hat{\mu})^2 / n\right\}}{\hat{\sigma}_1^2}\right] = n \times \ln\left\{1 + \frac{\sum_i \sum_j (\hat{\mu}_i - \hat{\mu})^2 / n}{\sum_i \sum_j (y_{ij} - \hat{\mu}_i)^2 / n}\right\}$$

Recall the *F*-statistic in Chapter 16:

$$f = \frac{\sum_i n_i (\bar{y}_i - \bar{y})^2 / (g-1)}{\sum_i \sum_j (y_{ij} - \bar{y}_i)^2 / (n-g)}$$

The likelihood ratio statistic is a simple function of the F-statistic as follows:

$$2(LL_1 - LL_0) = n \times \ln\left\{1 + \frac{\sum_i \sum_j (\hat{\mu}_i - \hat{\mu})^2 / n}{\sum_i \sum_j (y_{ij} - \hat{\mu}_i)^2 / n}\right\}$$ (Substituting results shown before)

$$= n \times \ln\left\{1 + \frac{\sum_i \sum_j (\bar{y}_i - \bar{y})^2 / n}{\sum_i \sum_j (y_{ij} - \bar{y}_i)^2 / n}\right\}$$ (By substitution)

$$= n \times \ln\left\{1 + \frac{\sum_i n_i (\bar{y}_i - \bar{y})^2}{\sum_i \sum_j (y_{ij} - \bar{y}_i)^2}\right\}$$ (Canceling the n, and noting that there are n_i identical terms in the numerator summation)

$$= n \times \ln\left\{1 + \frac{g-1}{n-g} \times \frac{\sum_i n_i (\bar{y}_i - \bar{y})^2 / (g-1)}{\sum_i \sum_j (y_{ij} - \bar{y}_i)^2 / (n-g)}\right\}$$ (Multiplying and dividing by $(g-1)/(n-g)$)

$$= n \times \ln\left\{1 + \frac{g-1}{n-g} f\right\}$$ (By substitution)

As shown earlier in Example 17.1, the function $f(x) = c_1 \ln(1 + c_2 x)$ is a monotonically increasing function. Thus, the likelihood ratio statistic is related one-to-one with the F-statistic since

$$\chi^2 = n \times \ln\left\{1 + \frac{g-1}{n-g} f\right\} \tag{17.6}$$

The fact that the likelihood ratio test is equivalent to the F-test justifies the particular mathematical form of the F statistic. You might have been tempted to use $\sum_i (\bar{y}_i - \bar{y})^2 / g$ in the numerator of the F-statistic, rather than $\sum_i n_i (\bar{y}_i - \bar{y})^2 / g$, since the former is a more obvious estimate of variance between groups than is the latter. However, the likelihood ratio shows that you are better off using the latter expression with the weighted sum of squares, because it provides you with the optimal test.

There is a comfortable intuition for using the weighted sum of squares in the numerator of the F-statistic. When the sample sizes are different, the F-statistic weights groups with larger sample sizes more heavily than groups with smaller sample sizes. This is good, because groups with larger sample sizes provide more accurate estimates $\bar{y}_i$. Despite its seemingly esoteric nature, the likelihood ratio procedure gives you common-sense results.

Example 17.4: The Multiple Regression R^2 Statistic and the Likelihood Ratio

Suppose you are thinking of buying a used car. The posted price depends on many factors, including age, condition, make and model, efficiency, size, safety features, and extras. Suppose you can quantify many of these factors, calling them X_1, X_2, ..., X_k. Vehicle price (your Y variable) is related to these X values, but the relationship is not deterministic: Two cars with exactly the same profile $\{x_1, x_2, ..., x_k\}$ can have different posted prices, simply because the seller can decide whatever number Y that he or she wants to post.

When you use the classic multiple regression model, you assume that the observations Y are produced independently from normal distributions $N(\beta_0 + \beta_1 x_1 + \beta_2 x_2 + \cdots + \beta_k x_k, \sigma^2)$, given observed values of the predictor variables $(X_1, X_2, ..., X_k) = (x_1, x_2, ..., x_k)$. The deterministic component of the model is the mean

$$E\{Y \mid (X_1, X_2, ..., X_k) = (x_1, x_2, ..., x_k)\} = \beta_0 + \beta_1 x_1 + \beta_2 x_2 + \cdots + \beta_k x_k$$

But the relationship is also probabilistic because there is an entire distribution of possible Y values, having variance σ^2, even among cases where the predictors are identical, with $(X_1, X_2, ..., X_k) = (x_1, x_2, ..., x_k)$.

Sometimes there is a question as to whether the process has any deterministic component whatsoever. For example, according to the financial theory of efficient markets, you cannot predict future stock returns using any publically available information. Imagine, then, that you create a list of attributes $x_1, x_2, ..., x_k$ of stocks, such as their current price, current trading volume, current price-to-earnings ratio, current company employee count, years in existence and so forth. Enamored with the possibility that advanced statistical machinery will allow you to predict future returns, and therefore allow you to make a profit, you proceed as follows.

You know that *model produces data*, so you assume the data-producer is the multiple regression model defined as

$$Y_i \mid (X_{i1}, X_{i2}, ..., X_{ik}) = (x_{i1}, x_{i2}, ..., x_{ik}) \sim_{\text{independent}} N(\beta_0 + \beta_1 x_{i1} + \beta_2 x_{i2} + \cdots + \beta_k x_{ik}, \sigma^2)$$

Model has unknown parameters, so you decide to estimate them via maximum likelihood, because you know that *data reduce the uncertainty about the unknown parameters*.

Then you decide to calculate the estimated deterministic component

$$\hat{E}\{Y \mid (X_1, X_2, ..., X_k) = (x_1, x_2, ..., x_k)\} = \hat{\beta}_0 + \hat{\beta}_1 x_1 + \hat{\beta}_2 x_2 + \cdots + \hat{\beta}_k x_k$$

And finally, you decide to use this estimate to make investment decisions.

Whoa, not so fast! First you had better answer the question of whether your estimated model is completely explainable by chance. If so, it would be financially unwise to use it!

The question as to whether an estimated regression model is completely explainable by chance alone falls nicely within the likelihood ratio testing paradigm. The null model is

$$Y_i \mid (X_{i1}, X_{i2}, ..., X_{ik}) = (x_{i1}, x_{i2}, ..., x_{ik}) \sim_{\text{independent}} N(\beta_0, \sigma^2)$$

In other words, the null model states that all the β's are zero except for the intercept β_0. In this model, the X values have no effect whatsoever on Y. Still, by chance alone, the MLEs $\hat{\beta}_1$ through $\hat{\beta}_4$ in the unrestricted model will be nonzero, even when the true values of β_1 through β_4 are all zero. The likelihood ratio test will tell you whether these nonzero estimates, as a group, are explainable by chance alone.

The regression null model is identical to the ANOVA null model. The only difference is that the term μ in the ANOVA null model is called β_0 in the regression null model. But in either case, the parameter is simply the process mean. Hence, as shown earlier in Example 17.3 for the ANOVA null model, you also know for the regression null model that

$$LL_0 = -\left(\frac{n}{2}\right)\ln(2\pi) - \left(\frac{n}{2}\right)\ln\left(\hat{\sigma}_0^2\right) - \frac{n}{2}$$

Using results of Example 17.3, $\hat{\beta}_0 = \sum_i y_i / n = \bar{y}$ and $\hat{\sigma}_0^2 = \sum_i (y_i - \bar{y})^2 / n$.

Now, the unrestricted model. The likelihood function is the product of the likelihoods for all the individual observations using the $N(\beta_0 + \beta_1 x_{i1} + \beta_2 x_{i2} + \cdots + \beta_k x_{ik}, \sigma^2)$ model, giving the following likelihood calculations

$L(\beta_0, \beta_1, \ldots, \beta_k, \sigma^2 \mid \text{data})$	(By definition of the likelihood function of a sample produced as independent observation)
$= \prod_i \dfrac{1}{\sqrt{2\pi\sigma^2}} \exp\left\{\dfrac{-\{y_i - (\beta_0 + \beta_1 x_{i1} + \cdots + \beta_k x_{ik})\}^2}{2\sigma^2}\right\}$	
$= \left(\dfrac{1}{\sqrt{2\pi\sigma^2}}\right)^n \exp\left\{\dfrac{-\sum_i \{y_i - (\beta_0 + \beta_1 x_{i1} + \cdots + \beta_k x_{ik})\}^2}{2\sigma^2}\right\}$	(By algebra and properties of exponents)

Notice there is a minus sign in the exponent. Thus, no matter what σ is, the values of the β_j that *maximize* the likelihood must *minimize* the expression $\sum_i \{y_i - (\beta_0 + \beta_1 x_{i1} + \cdots + \beta_k x_{ik})\}^2$. This implies that the MLEs are also **least squares estimates**; that is, estimates that minimize the sum of squared deviations from the observations y_i to the fitted values $\hat{y}_i = \hat{\beta}_0 + \hat{\beta}_1 x_{i1} + \cdots + \hat{\beta}_k x_{ik}$. For larger k, matrix algebra is needed to compute these $\hat{\beta}_j$, but the calculations are fairly simple with small k as shown in the end-of-chapter Exercise 17.16. In any case, the least squares estimates are readily available from any statistical software. The minimized sum of squared deviations, $\sum_i (y_i - \hat{y}_i)^2$, is called the **sum of squares for error**.

You can solve for the MLE of σ^2 by using calculus with the log-likelihood function, after plugging in the MLEs of the β's

$LL(\hat{\beta}_0, \hat{\beta}_1, \ldots, \hat{\beta}_k, \sigma^2 \mid \text{data})$	
$= -(n/2)\ln(2\pi) - (n/2)\ln(\sigma^2) - \sum_i (y_i - \hat{y}_i)^2 / 2\sigma^2$	(By substitution and properties of logarithms)
$\Rightarrow (\partial/\partial\sigma^2) LL(\{\hat{\beta}_0, \hat{\beta}_1, \ldots, \hat{\beta}_k, \sigma^2 \mid \text{data})$	(By differentiation of terms with respect to σ^2)
$= -(n/2)(1/\sigma^2) + \sum_i (y_i - \hat{y}_i)^2 / 2\sigma^4$	

Setting the derivative to zero and solving, you get

$$\hat{\sigma}_1^2 = \frac{\sum_i (y_i - \hat{y}_i)^2}{n}$$

As before in the ANOVA model, $LL_0 = (-n/2)\ln(2\pi) - (n/2)\ln\left(\hat{\sigma}_0^2\right) - n/2$. Plugging in the unrestricted model MLEs in the regression model you get, as in the ANOVA model, $LL_1 = (-n/2)\ln(2\pi) - (n/2)\ln\left(\hat{\sigma}_1^2\right) - n/2$. That means the likelihood ratio test statistic is

$$2(LL_1 - LL_0) = n\left\{\ln\left(\frac{\hat{\sigma}_0^2}{\hat{\sigma}_1^2}\right)\right\}$$

There it is again! That variance ratio. Once again, the likelihood ratio method tells you that if the estimated variance in the restricted (or null) model is much bigger than the estimated variance in the unrestricted model, then you should reject the null model. The only difference between the regression analysis shown here and the ANOVA analysis is that the form of $\hat{\sigma}_1^2$ differs.

To simplify the form of the likelihood ratio test statistic, we'll use the same "add and subtract" trick as before:

$$
\begin{aligned}
\hat{\sigma}_0^2 &= \sum_i \{(y_i - \hat{y}_i) + (\hat{y}_i - \overline{y})\}^2/n && \text{(By adding and subtracting } \hat{y}_i) \\
&= \sum_i (y_i - \hat{y}_i)^2/n + \sum_i (\hat{y}_i - \overline{y})^2/n \\
&\quad + 2\sum_i (\hat{y}_i - \overline{y})(y_i - \hat{y}_i)/n && \begin{array}{l}\text{(By expanding the square and} \\ \text{by summation algebra)}\end{array} \\
&= \hat{\sigma}_1^2 + \sum_i (\hat{y}_i - \overline{y})^2/n + 2\sum_i (\hat{y}_i - \overline{y})(y_i - \hat{y}_i)/n && \text{(By substitution)}
\end{aligned}
$$

There is a result from vector algebra that further simplifies this expression: Just as in the ANOVA case, the cross product disappears. Then you have

$$\hat{\sigma}_0^2 = \hat{\sigma}_1^2 + \sum_i \frac{(\hat{y}_i - \overline{y})^2}{n}$$

This latter form is famous in multiple regression. In words, it states

Total variance = Variance due to error + Variance due to model

This gives you the **R-Squared** statistic.

R^2 Statistic

$$R^2 = \frac{(\text{Variance due to model})}{(\text{Total variance})} = \frac{\sum_i (\hat{y}_i - \overline{y})^2/n}{\hat{\sigma}_0^2}$$

The R^2 statistic estimates the proportion of variance in Y that is explained by your X variables. Alternatively,

$$R^2 = 1 - \frac{\hat{\sigma}_1^2}{\hat{\sigma}_0^2}$$

From this form you can interpret R^2 as the proportional reduction in variance of Y, comparing the unconditional distribution of Y with the conditional distribution of Y given X.

Thus, the likelihood ratio is

$$
\begin{aligned}
2(LL_1 - LL_0) &= n\left\{\ln(\hat{\sigma}_0^2/\hat{\sigma}_1^2)\right\} \quad \text{(As shown before)} \\
&= n \times \ln\{1/(1 - R^2)\} \quad \text{(Since } R^2 = 1 - \hat{\sigma}_1^2/\hat{\sigma}_0^2\text{)}
\end{aligned}
$$

The function $n \times \ln\{1/(1 - R^2)\}$ is a monotonically increasing function of R^2; thus the likelihood ratio test is equivalent to a test based on the R^2 statistic. In other words, R^2 is an optimal statistic for detecting deviation from the chance-only model.

While the R^2 statistic is optimal for testing hypotheses, it is more commonly used to measure the goodness of a regression model. It is related to the plug-in estimate of the correlation coefficient $\hat{\rho}_{xy}$ given in Section 10.3; in fact, if there is just one X variable (i.e., if $k = 1$), then $R^2 = (\hat{\rho}_{xy})^2$. Thus, the R^2 statistic ranges from zero to one, with $R^2 = 0$ meaning no relationship, and $R^2 = 1$ meaning perfect relationship. You can use Ugly Rule of Thumb 10.1 for correlations to suggest to similar rules for R^2. For example, a correlation that is more than 0.7 corresponds to an R^2 that is more than 0.49.

17.3 Evaluating the Statistical Significance of Likelihood Ratio Test Statistics

The likelihood ratio procedure tells you the *form* of the test statistic you should use if you want an optimal test. However, just knowing that the optimal test statistic has the form $\chi^2 = 2(LL_1 - LL_0)$ doesn't tell you *when* to reject the restricted model. For that, you can calculate a *p*-value.

Larger values of χ^2 are the extreme ones that suggest the chance-only model is wrong; smaller values are consistent with the chance-only model. Hence, the *p*-value calculation for the likelihood ratio test is one-sided. Specifically, it is the probability greater than or equal to the observed value of χ^2, calculated under the chance-only, restricted model.

When you can calculate the exact *p*-value—for example, from the *t*-distribution or the *F*-distribution—you should do it that way. For other cases, the following result shows how you can compute an approximate *p*-value using a chi-squared approximation to the distribution of the likelihood ratio test statistic. You knew there had to be a chi-squared distribution here, right? Why else would we have called the statistic "χ^2"?

Approximate Distribution of the Likelihood Ratio Chi-Squared Statistic

Assume independence, large samples, and other regularity conditions such as that the MLEs occur in the interior of the parameter space. Then when the restricted model H_0 is true

$$
\chi^2 \sim \chi_{df}^2
$$

Here $df = \{$# of parameters in $p_1(y|\theta_1)\}$, minus $\{$# of parameters in $p_0(y|\theta_0)\}$. The approximation gets better as the sample size (n) gets larger.

Thus, if v is the observed value of $\chi^2 = 2(LL_1 - LL_0)$, then the probability of observing a value as extreme as v by chance alone is your p-value

$$pv \cong \Pr\left(\chi^2_{df} \geq v\right)$$

As usual, if $pv \leq \alpha$ then you can confidently rule out the notion that the difference between LL_1 and LL_0 is explained by the null model $p_0(y|\theta_0)$, that is, by chance alone.

Note that the chi-squared p-value is only an approximation. Yet another weasel! With smaller sample sizes these approximate p-values are more suspect. There can be no general ugly rule of thumb such as $n > 30$ to ensure the accuracy of these p-values, because it depends on the particulars of the model. However, we do give an ugly rule of thumb later in the chapter for how large the sample size should be when you have categorical data.

We will not prove the result that $\chi^2 \stackrel{.}{\sim} \chi^2_{df}$ under H_0; the proof requires mathematical statistics methods deeper than those of this book. But we will show that the result makes logical sense in some special cases from Section 17.2.

Example 17.5: The Chi-Squared Approximation to the Distribution of the One-Sample Likelihood Ratio Statistic

Example 17.1 presented the unrestricted model $Y_1, Y_2, \ldots, Y_n \sim_{\text{iid}} N(\mu, \sigma^2)$, and the null (restricted) model $Y_1, Y_2, \ldots, Y_n \sim_{\text{iid}} N(m_0, \sigma^2)$. The likelihood ratio test was shown to be related to the t^2 statistic in Equation 17.4 as follows:

$$\chi^2 = n \times \ln\left\{1 + \frac{T^2}{(n-1)}\right\}$$

According to the approximate distribution result, this χ^2 statistic is approximately distributed as χ^2_1 when the restricted model is true: Because there are two parameters (μ, σ^2) in the unrestricted model and one parameter in the restricted model (σ^2), the degrees of freedom for the chi-squared test is $2 - 1 = 1$.

Recall from Chapter 16 that a chi-squared random variable arises as a sum of squared standard normal random variables. The statistic $\chi^2 = n \times \ln\{1 + t^2/(n-1)\}$ doesn't look immediately like a sum of squared standard normals, but it is. There is a very useful mathematical approximation involving logarithms that you can use here to simplify the expression further.

$$\ln(1 + x) \cong x, \quad \text{for } x \text{ near } 0$$

Of course, *approximation* is (always and forever!) a weasel word, so you should ask, "How good is the approximation?" Figure 17.3 shows the answer.

As shown in Figure 17.3, the difference between $\ln(1 + x)$ and x is barely noticeable for $-0.1 < x < 0.1$, but becomes more noticeable when $|x| > 0.1$.

Now, when the null model is true the T^2 statistic has the F-distribution, meaning its values are in the vicinity of 1.0. Thus, $T^2/(n-1)$ will be near zero for large n, and the approximation $\ln(1 + x) \cong x$ will be good, implying

$$\chi^2 = n \times \ln\left\{1 + \frac{T^2}{(n-1)}\right\} \cong n \times \frac{T^2}{(n-1)}$$

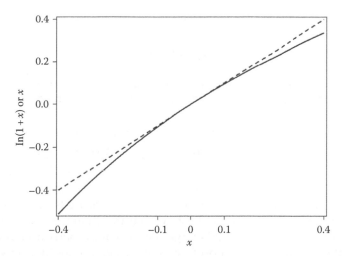

FIGURE 17.3
Graph of $f(x) = \ln(1 + x)$ (solid line) and $f(x) = x$ (dashed line).

For large n, $\{n/(n-1)\} \cong 1.0$. Further, from Chapter 16 you recall that T is approximately distributed as Z, where $Z \sim N(0, 1)$, when n is large. Thus

$$\chi^2 \cong Z^2$$

Since χ^2 is approximately the sum of just one squared standard normal variate, its distribution is approximately chi-squared with just one degree of freedom.

To apply the chi-squared likelihood ratio test to real data, Example 7.10 showed measurements of computer chip width: 311, 304, 316, 308, 312, 308, 314, 307, 302, 311, 308, 300, 316, 304, 316, 306, 314, 310, 311, 309, 311, 306, 311, 309, 311, 305, 304, 303, 307, and 316. Suppose the target is $m_0 = 310$, thus your null model is that these data are generated as an iid sample from $N(310, \sigma^2)$, with σ^2 being a free, unknown parameter. The MLE of the variance under the null (restricted) model is as follows

$$\hat{\sigma}_0^2 = \left(\frac{1}{30}\right)\left\{(311-310)^2 + (304-310)^2 + \cdots + (316-310)^2\right\} = 19.8667$$

The sample average of the data is the MLE of μ in the unrestricted model, or

$$\bar{y} = \left(\frac{1}{30}\right)(311 + 304 + \cdots + 316) = 309.0$$

So, the MLE of the variance under the unrestricted model is

$$\hat{\sigma}_1^2 = \left(\frac{1}{30}\right)\left\{(311-309)^2 + (304-309)^2 + \cdots + (316-309)^2\right\} = 18.8667$$

Thus, the likelihood ratio chi-squared statistic is

$$\chi^2 = n\left\{\ln\left(\frac{\hat{\sigma}_0^2}{\hat{\sigma}_1^2}\right)\right\} = 30\ln\left(\frac{19.8667}{18.8667}\right) = 1.549$$

The probability of seeing a chi-squared statistic as extreme as this one, by chance alone, is then

$$pv \cong \Pr\left(\chi_1^2 \geq 1.549\right) = 0.213$$

This probability is not small enough to rule out the null model—that is, the difference in log likelihoods is explainable by chance alone, where the true model is $N(310, \sigma^2)$.

In Chapter 16 you saw a different way to test this hypothesis using the *t*-statistic $T = (\bar{Y} - m_0)/\left(\hat{\sigma}/\sqrt{n}\right)$. Specifically: In the quality control example mentioned before, $t = (309 - 310)/(4.4178/30^{1/2}) = -1.240$, corresponding well with the χ^2 statistic since $(-1.240)^2 = 1.537 \cong 1.549 = \chi^2$. The *p*-value is $pv = 2 \times \Pr(T_{29} \leq -1.240) = 0.225$. This is an *exact* *p*-value under the iid normal assumption. While the results are the same in the sense that both the exact and approximate methods show that the chance-only null model is plausible, the *p*-value for the exact method is larger because it properly accounts for the variation inherent in the estimate $\hat{\sigma}$. This observation provides yet another ugly rule of thumb!

Ugly Rule of Thumb 17.1

The approximate likelihood ratio chi-squared test produces generally smaller *p*-values than corresponding exact tests based on *t*- or *F*-distributions (when such exact tests are available).

Since the exact tests are, well, *exact*, this ugly rule of thumb tells you that the approximate *p*-values of likelihood ratio chi-squared tests tend to be *too small* when they are calculated from the chi-squared distribution.

Example 17.6: The Chi-Squared Approximation to the Distribution of the Likelihood Ratio ANOVA Test Statistic

Consider Equation 17.6

$$2(LL_1 - LL_0) = n \times \ln\left\{1 + \frac{g-1}{n-g}f\right\}$$

Again the $\ln(1 + x)$ form appears. The "x" term will be small when the null model $N(\mu, \sigma^2)$ is true and n is large, since the ratio $(g - 1)/(n - g)$ will tend toward zero and f will be near 1.0. Thus, under the null model with large n

$$2(LL_1 - LL_0) \cong n\left\{\frac{g-1}{n-1}\right\}f \cong (g-1)f$$

This is true because $n/(n - g) \cong 1$ for large n.

Since the *F*-distribution is that of $\left\{\chi_{g-1}^2/(g-1)\right\}/\left\{\chi_{n-g}^2/(n-g)\right\}$, it follows that $(g - 1)F$ is approximately distributed as χ_{g-1}^2 for large n, since $\chi_{n-g}^2/(n-g) \cong 1$ in that case. Hence the ANOVA likelihood ratio χ^2 statistic is approximately distributed as χ_{g-1}^2 under the null model.

To see that there should be $(g - 1)$ degrees of freedom for the likelihood ratio test, note that there are $(g + 1)$ parameters in the unrestricted model (g μ_i terms and one σ^2 term), and two parameters in the restricted model (one μ and one σ^2). So the difference between the number of parameters in the full model and the number of parameters in the restricted model is $(g + 1) - 2 = g - 1$ degrees of freedom.

In Example 16.7 you saw that the difference between the groups was statistically significant via the F-test with $f = 32.79$ based on degrees of freedom 4 and 52, with p-value $pv = 1.2 \times 10^{-13}$. As shown earlier, the likelihood ratio statistic is a simple function of the F-statistic

$$2(LL_1 - LL_0) = n \times \ln\left\{1 + \frac{g-1}{n-g}f\right\} = 57 \times \ln\left\{1 + \frac{5-1}{57-5}(32.79)\right\} = 71.77$$

The p-value is the probability of observing a likelihood ratio statistic as large as 71.77 by chance alone under the chance-only model $N(\mu, \sigma^2)$. So, here, the approximate p-value is $pv \cong \Pr\left(\chi^2_{5-1} \geq 71.77\right) = 9.5 \times 10^{-15}$.

Once again, the approximate method provided a smaller p-value than the exact method, as suggested by Ugly Rule of Thumb 17.1. And, again, the preferred method is to use the F-distribution, simply because it is exact under the iid normal model while the calculation that uses the chi-squared distribution is only approximate.

Example 17.7: Evaluating the Significance of R^2 in Multiple Regression

Your esteemed colleague Hans is at it again. This time he says he can predict the future values of stock returns. He uses a company's current price, trading volume, price/earnings ratio, employee count, and years in existence. Using these variables he estimates a model based on $n = 40$ companies, finding $R^2 = 0.15$ for his model. According to the efficient markets theory of finance, the R^2 value should in theory be zero for such prediction models. Hans is very excited! His model explains 15% of the variation in stock returns! But is Hans' R^2 statistic explainable by chance alone?

The likelihood ratio statistic for Hans' data is $40 \times \ln\{1/(1 - 0.15)\} = 6.50$. Hans has used $k = 5$ predictor variables (or X variables) in his model, so the approximate distribution of the chi-squared statistic under the null model (where the X variables have no effect whatsoever on Y) is χ^2_5. Hence, the p-value is $pv \cong \Pr\left(\chi^2_5 \geq 6.50\right) = 0.261$. You now have the unpleasant task of telling Hans the disappointing news: His model does *not* demonstrate predictability of stock returns, because his results are explainable by a model where his predictor variables have no effect whatsoever on returns.

As in the case of the ANOVA model, the likelihood ratio is a function of a commonly used F-statistic. Recall that

$$2(LL_1 - LL_0) = n\left\{\ln\left(\frac{\hat{\sigma}_0^2}{\hat{\sigma}_1^2}\right)\right\} = n\ln\left\{1 + \frac{\sum_i (\hat{y}_i - \bar{y})^2/n}{\hat{\sigma}_1^2}\right\}$$

In sources that delve deeper into multiple regression analysis, you will see that there is an exact F-test for the hypothesis $H_0: \beta_1 = \beta_2 = \cdots = \beta_k = 0$. The statistic is called the *model F-statistic* because it is used to evaluate the significance of the overall model. The statistic is given by

$$f = \frac{\sum_i (\hat{y}_i - \bar{y})^2/k}{n\hat{\sigma}_1^2/(n-k-1)}$$

This F-statistic comes from the $F_{k,n-k-1}$ distribution when the restricted model is true. Thus, you should compute the p-value using this F-distribution, rather than the chi-squared distribution, because the F result is exact.

As in the case of the ANOVA F-test, the optimality of the likelihood ratio method motivates the form of the F-statistic. If the regression model F-statistic were not equivalent to a likelihood ratio statistic (see Exercise 17.7 at the end of this chapter), you wouldn't see it.

The normality-assuming ANOVA and regression models are cases where you shouldn't use the chi-squared distribution to calculate the p-value for likelihood ratio test; the F-distribution is better because it gives the exact p-value. In many cases there is no exact p-value calculation, in which case you need to calculate an approximate p-value from the chi-squared distribution. Logistic regression provides an example.

Example 17.8: Customer Scoring Using Multiple Logistic Regression

In marketing, you need to know who your customers are. If you want to sell baby diapers you probably don't want to waste your time trying to convince single men about how great your brand is. A great way to predict who *will* buy your product is to collect data on who already *has* bought your product or other products like it. Suppose you have a survey of consumers in your target market. The survey includes a binary Y variable, prior purchases (coded as 1 = purchase, 0 = no purchase), as well as demographic data on respondents' age (X_1), sex ($X_2 = 0$ for male, 1 for female), income (X_3), and education (X_4). A model for the respondent's likelihood of purchasing is the logistic regression model presented in Chapter 12, Example 12.8. Here it is again, but extended to multiple X variables

$$\Pr\left(Y = 1 \mid X_1 = x_1, X_2 = x_2, X_3 = x_3, X_4 = x_4\right)$$

$$= \pi(x_1, x_2, x_3, x_4) = \frac{\exp(\beta_0 + \beta_1 x_1 + \beta_2 x_2 + \beta_3 x_3 + \beta_4 x_4)}{1 + \exp(\beta_0 + \beta_1 x_1 + \beta_2 x_2 + \beta_3 x_3 + \beta_4 x_4)}$$

The logistic regression model for the entire sample of n observations is

$$Y_i \mid (X_{i1} = x_{i1}, X_{i2} = x_{i2}, X_{i3} = x_{i3}, X_{i4} = x_{i4}) \sim_{\text{independent}} \text{Bernoulli}(\pi(x_{i1}, x_{i2}, x_{i3}, x_{i4})).$$

You can estimate the parameters β_j using maximum likelihood, obtaining the estimated probability function as follows:

$$\hat{\pi}(x_1, x_2, x_3, x_4) = \frac{\exp(\hat{\beta}_0 + \hat{\beta}_1 x_1 + \hat{\beta}_2 x_2 + \hat{\beta}_3 x_3 + \hat{\beta}_4 x_4)}{1 + \exp(\hat{\beta}_0 + \hat{\beta}_1 x_1 + \hat{\beta}_2 x_2 + \hat{\beta}_3 x_3 + \hat{\beta}_4 x_4)} \tag{17.7}$$

This is called a **scoring function**, and it is used to assign a prospective customer a score (the probability), based on his or her demographic data (the x variables). Marketers then target the prospective customers with the highest scores. Those with low scores are not likely to buy the product, so your company may decide not to waste money targeting them.

While the sample sizes used to estimate the scoring function (17.7) typically tend to be very large, it may happen, for instance, that only a small survey sample size is available for a particular market region. In this case, before ranking prospective regional customers using the scoring function you would be wise to ask whether the coefficients in the model (the $\hat{\beta}_j$) are explainable by chance alone. A chance-only model would state that a given respondent's probability of purchasing your product ($\Pr(Y = 1)$) is constant for all x—that is, that $\beta_1 = \beta_2 = \beta_3 = \beta_4 = 0$. The null model therefore states that

$$\Pr\left(Y = 1 \mid X_1 = x_1, X_2 = x_2, X_3 = x_3, X_4 = x_4\right) = \pi = \frac{\exp(\beta_0)}{1 + \exp(\beta_0)}$$

Coupled with the independence assumption, the null model states that $Y_i|(X_{i1} = x_{i1},$ $X_{i2} = x_{i2}, X_{i3} = x_{i3}, X_{i4} = x_{i4}) \sim_{iid}$ Bernoulli(π), where π is a constant between 0 and 1 that is not affected by the X values. The MLE for π in this model is the intuitively obvious one: $\hat{\pi} = \sum_i y_i / n$ (see Exercise 17.8), which is simply the proportion of 1s (purchases) in the observed data. The restricted model log likelihood is therefore

$$LL_0 = \sum_{y_{i=1}} \ln(\hat{\pi}) + \sum_{y_{i=0}} \ln(1 - \hat{\pi})$$

The unrestricted model log likelihood does not have as simple a solution, but after finding the MLEs for the β values and plugging in to Equation 17.7, you get $\hat{\pi}_i = \hat{\pi} (x_{i1},$ $x_{i2}, x_{i3}, x_{i4})$, and then

$$LL_1 = \sum_{y_{i=1}} \ln(\hat{\pi}_i) + \sum_{y_{i=0}} \ln(1 - \hat{\pi}_i)$$

There is no simplified expression here, as there is in the case of the ANOVA and regression models, but still, under the chance-only null model where $Y_i|(X_{i1} = x_{i1}, X_{i2} = x_{i2},$ $X_{i3} = x_{i3}, X_{i4} = x_{i4}) \sim_{iid}$ Bernoulli(π), you know that

$$2(LL_1 - LL_0) \overset{\cdot}{\sim} \chi_4^2$$

The chi-squared distribution has four degrees of freedom because there are five parameters ($\beta_0, \beta_1, ..., \beta_4$) in the unrestricted model, and one parameter (β_0) in the restricted model.

We analyzed such a data set containing $n = 129$ such observations using the LOGISTIC procedure of the SAS/STAT software, yielding the screen shot shown in Figure 17.4.

Notice the values $-2 \ Log \ L$ are reported for the *Intercept Only* and for the *Intercept and Covariates* models. These are precisely the chance-only restricted model and the unrestricted model, respectively. By the way, a **covariate** is just another name for an X variable.

Thus

$$2(LL_1 - LL_0) = -2LL_0 - (-2LL_1) = 173.139 - 68.594 = 104.5450$$

You can see this reported as the likelihood ratio in the *Testing Global Null Hypothesis* portion of the output shown in Figure 17.4. The degrees of freedom are given as 4, as explained above, and the p-value is calculated as $pv = \ \Pr\left(\chi_4^2 \geq 104.5450\right)$, which is infinitesimally small in reality—recall that the mean of the chi-squared distribution is its degrees of freedom, or 4 in this case—but simply reported as <0.0001 in the output. Thus, the distribution of Y (purchasing behavior) depends on the collection of X variables: You can't easily attribute the difference in log likelihoods to chance alone.

```
                     Model Fit Statistics

                                             Intercept
                             Intercept             and
           Criterion              Only      Covariates

           AIC                 175.139          78.594
           OC                  177.999          92.893
           -2 Log L            173.139          68.594


           Testing Global Null Hypothesis: BETA=0

    Test                 Chi-Square      DF      Pr > ChiSq

    Likelihood Ratio       104.5450       4          <.0001
    Score                   79.0970       4          <.0001
    Wald                    32.2366       4          <.0001
```

FIGURE 17.4
Logistic regression output from the SAS/STAT software.

There are other test statistics reported in Figure 17.4 as well, the *score* test and the *Wald* test. The Wald test uses the Wald standard errors described in Chapter 12, and the score test is yet another method for testing the same chance-only hypothesis. The question always arises, "Which test is best to use?" While the likelihood ratio method is optimal in many situations, there is often no one test that is optimal for every possible situation. Simulation studies provide guidance for selecting the best test; see Chapter 19.

17.4 Likelihood Ratio Goodness-of-Fit Tests

Often data are nominal, or categorical with no ordering. For example, in Chapter 1 we offered a model for the choice of car color; red, green, or gray. These data are not numbers, but categories. Even if you recoded them as numbers, for example, red = 1, green = 2, and gray = 3, there still would be no ordering to the resulting values: Red is not less than green, and green is not less than gray.

With nominal data, the generic distribution of a single outcome Y is the *multinomial distribution* given in Table 17.1.

The C values in Table 17.1 are the categories that the nominal variable can take on, like car color. The model is familiar: $Y_1, Y_2, \ldots, Y_n \sim_{iid} p(y)$, but there is no Central Limit Theorem or Law of Large Numbers that applies directly to these data, simply because you can't take sums or averages of non-numeric data. On the other hand, the CLT and LLN do apply to numerical transformation of such data, such as a binary coding.

Example 17.9: Is the Die Fair?

Casinos worry about this kind of thing. If there is a systematic tendency for some values on a die to come up more often than others, then an observant gambler could make bets on the more common values and rake in the big bucks at the casino's expense.

In any set of n rolls of a die, there will not be exactly 16.666666% 1s, 16.666666% 2s, etc. Instead, you will observe percentages like 18.3% 1s, 13.9% 2s, etc. You can usually explain such differences purely by chance, but what if someone has rigged the die so that some outcomes are more frequent? How large do the differences have to be for you to rule out chance and conclude that the die was rigged?

You can answer this question using a likelihood ratio test. As before, you need a chance-only (or restricted) model, and an unrestricted model. The die categories are $C_1 = 1$, $C_2 = 2$, $\ldots$, $C_6 = 6$, and the general form of the multinomial distribution described above gives the model shown in Table 17.2.

TABLE 17.1

A Generic, Unrestricted Multinomial Distribution

y	$p(y)$
C_1	π_1
C_2	π_2
...	...
C_k	π_k
Total	1.00

TABLE 17.2

An Unrestricted Model
for Die Outcomes

y	p(y)
1	π_1
2	π_2
...	...
6	π_6
Total	1.00

TABLE 17.3

A Restricted Model
for Die Outcomes

y	p(y)
1	1/6
2	1/6
...	...
6	1/6
Total	1.00

In the restricted (or chance-only) model, the die outcomes all have probability 1/6, as shown in Table 17.3.

The observed data, from many rolls, are $y_1, y_2, ..., y_n$, where each y_i is either 1, 2, 3, 4, 5 or 6. In the restricted model, there are no parameters to estimate and plug in. The likelihood is simply

$$L_0 = \left(\frac{1}{6}\right) \times \left(\frac{1}{6}\right) \times \cdots \times \left(\frac{1}{6}\right) = \left(\frac{1}{6}\right)^n$$

The unrestricted model shown in Table 17.2 takes a little more work. Suppose the data are 3, 4, 3, 1, 2, 1, and 6. Then in the unrestricted model, the likelihood is

$$L(\pi_1, \pi_2, ..., \pi_6 | \text{data}) = \pi_3 \times \pi_4 \times \pi_3 \times \pi_1 \times \pi_2 \times \pi_1 \times \pi_6 = \pi_1^2 \times \pi_2^1 \times \pi_3^2 \times \pi_4^1 \times \pi_5^0 \times \pi_6^1$$

Notice that the exponents count occurrences of categories. Let f_1 denote the frequency (or count) of cases where $y_i = 1$, f_2 denote the frequency (or count) of cases where $y_i = 2$, etc. Then in the unrestricted model, the likelihood function is

$$L(\pi_1, \pi_2, ..., \pi_6 | y_1, y_2, ..., y_n) = \pi_1^{f_1} \times \pi_2^{f_2} \times \pi_3^{f_3} \times \pi_4^{f_4} \times \pi_5^{f_5} \times \pi_6^{f_6}$$

Further, the log-likelihood function is given as

$$LL(\pi_1, \pi_2, ..., \pi_6 | y_1, y_2, ..., y_n)$$
$$= f_1 \ln(\pi_1) + f_2 \ln(\pi_2) + f_3 \ln(\pi_3) + f_4 \ln(\pi_4) + f_5 \ln(\pi_5) + f_6 \ln(\pi_6)$$

Taking the derivatives of *LL* with respect to the π_j and setting them to zero, as shown in Equation 12.3 of Chapter 12, yields $f_j / \hat{\pi}_j = 0$, which has no solution for $\hat{\pi}_j$. Hmmm... something is not working right.

Aha! There really are not six degrees of freedom in the π_j; there are only five. Since the probabilities must add to 1.0, if you know any five of them, the sixth is completely determined.

Incorporating this constraint into the log likelihood function yields

$$LL(\pi_1, \pi_2, ..., \pi_5 \mid y_1, y_2, ..., y_n)$$

$$= f_1 \ln(\pi_1) + f_2 \ln(\pi_2) + f_3 \ln(\pi_3) + f_4 \ln(\pi_4) + f_5 \ln(\pi_5) + f_6 \ln(1 - \pi_1 - \pi_2 - \pi_3 - \pi_4 - \pi_5)$$

Now taking the derivatives with respect to π_j, and equating the result to zero as in Equation 12.3 gives

$$\left(\frac{\partial}{\partial \pi_j}\right) LL(\pi_1, \pi_2, ..., \pi_5 \mid y_1, y_2, ..., y_n) = \frac{f_j}{\pi_j} - \frac{f_6}{(1 - \pi_1 - \cdots - \pi_5)}, \quad \text{for } j = 1, 2, ..., 5$$

This result is an application of the chain rule in calculus (see Chapter 2, property D9). Setting these derivatives to zero at the MLEs $\hat{\pi}_j$ gives the intuitive result as follows:

$$\hat{\pi}_j = \frac{f_j}{n}$$

(See Exercise 17.9.) The likelihood ratio statistic is thus given as follows:

$$\chi^2 = 2(LL_1 - LL_0)$$

$$= 2\left\{\ln\left(\hat{\pi}_1^{f_1} \times \hat{\pi}_2^{f_2} \times \hat{\pi}_3^{f_3} \times \hat{\pi}_4^{f_4} \times \hat{\pi}_5^{f_5} \times \hat{\pi}_6^{f_6}\right) - \ln(1/6)^n\right\} \qquad \text{(By definition and substitution)}$$

$$= 2\{f_1 \ln(\hat{\pi}_1) + \cdots + f_6 \ln(\hat{\pi}_6) - n\ln(1/6)\} \qquad \text{(By the properties of logarithms)}$$

$$= 2\{f_1 \ln(f_1/n) + \cdots + f_6 \ln(f_6/n) - (f_1\ln(1/6) + \cdots + f_6\ln(1/6)\} \qquad \begin{array}{l}\text{(By substitution and the fact}\\ \text{that } \Sigma f_j = n)\end{array}$$

It can happen that an observed frequency f_i is zero. If so, there is no data in that category (category C_i), so there is no contribution to the likelihood function. Thus, any term $f_i \ln(f_i/n)$ is defined to be zero when $f_i = 0$, although the computer will give you an error message if you try to calculate $0 \times \ln(0/n)$, since $\ln(0)$ is undefined.

Now, there is a concept of **observed frequencies** and *expected frequencies* in contingency tables. The *observed* frequencies are the f_js—that is, they are what you actually observe. The *expected frequencies* are the estimates of frequencies you would expect, if the null model were true. Some sources call the expected frequencies *fitted frequencies* or *fitted values*, which are perhaps a better names for them, since in many cases they are not truly expected values but are instead just estimates.

In this example, the null model is that the probabilities are all (1/6), so you expect $e_j = n/6$ in each category. Noting that $e_j/n = 1/6$, you can re-write the likelihood ratio statistic given before completely in terms of observed and expected frequencies as follows:

$$2(LL_1 - LL_0)$$

$$= 2[f_1 \ln(f_1/n) + \cdots + f_6 \ln(f_6/n) - \{f_1 \ln(e_1/n) + \cdots + f_6 \ln(e_6/n)\}] \qquad \text{(By substitution)}$$

$$= 2[f_1\{\ln(f_1/n) - \ln(e_1/n)\} + \cdots + f_6\{\ln(f_6/n) - \ln(e_6/n)\}] \qquad \text{(By algebra)}$$

$$= 2\{f_1 \ln(f_1/e_1) + \cdots + f_6 \ln(f_6/e_6)\} \qquad \begin{array}{l}\text{(By properties of}\\ \text{logarithms and algebra)}\end{array}$$

There are five parameters in the unrestricted model and none in the restricted model, so the distribution of χ^2 is approximately χ_5^2 if the restricted model is true. As always, large likelihood ratio test statistics are unusual under the null hypothesis, so the p-value is the probability greater than the observed statistic, as calculated using the relevant chi-squared distribution.

Suppose you toss the die 74 times, resulting in 13 1s, 10 2s, 16 3s, 10 4s, 8 5s, and 17 6s. The expected frequencies are all $74/6 = 12.333$, so the likelihood ratio chi-squared statistic is

$$\chi^2 = 2 \left\{ 13 \times \ln\left(\frac{13}{12.333}\right) + 10 \times \ln\left(\frac{10}{12.333}\right) + 16 \times \ln\left(\frac{16}{12.333}\right) \right.$$

$$\left. + 10 \times \ln\left(\frac{10}{12.333}\right) + 8 \times \ln\left(\frac{8}{12.333}\right) + 17 \times \ln\left(\frac{17}{12.333}\right) \right\} = 5.30$$

The probability of seeing a statistic as extreme as this by chance alone is $pv \cong \Pr\left(\chi_5^2 \geq 5.30\right) = 0.381$, so the differences between the observed and expected frequencies are easily explainable by chance alone. There is no evidence that the die is loaded.

The weasel-like approximation appears again, in the calculation of the *p*-value. The approximation is good when the sample size is large, but it should be large enough so that there are sufficient numbers of observations in each of the cells of the table. This leads us to yet another ugly rule of thumb, one that you will see mentioned in statistical software printouts.

Ugly Rule of Thumb 17.2

The chi-squared approximation is adequate if at least 80% of the cells have expected frequencies of 5 or more.

This is the second time that this "5 or more" rule of thumb has appeared. The first time was in Ugly Rule of Thumb 10.2, for determining when the Bernoulli frequencies are approximately normally distributed. Rules of Thumb 10.2 and 17.2 have exactly the same purpose—to ensure that the continuous approximation to the discrete distribution is adequate.

In the example given before, 100% (six out of six) of the cells have expected frequencies of 5 or more, since $e_i = 12.333 \geq 5$, so the calculated *p*-value 0.381 is approximately correct, according to Ugly Rule of Thumb 17.2.

In general, the *multinomial goodness-of-fit test* is used for testing whether a generic multinomial distribution as shown in Table 17.1, with $k - 1$ unknown parameters, fits a particular multinomial distribution as shown in Table 17.4.

In the restricted model, the p_{0i} are probabilities that you specify, like the 1/6 in the die example. Thus there are no parameters to estimate, and there are no parameters in the restricted model. The expected frequencies in the restricted model are $e_i = np_{0i}$, and the likelihood ratio chi-squared statistic is

$$\chi^2 = 2 \sum f_i \ln\left(\frac{f_i}{e_i}\right) \tag{17.8}$$

In the ANOVA and regression models, the likelihood ratio procedure points you toward more commonly used *F*-statistics. While the *F*-statistic is not commonly used for testing

TABLE 17.4

A Restricted Multinomial Distribution

y	$p(y)$
C_1	p_{01}
C_2	p_{02}
...	...
C_k	p_{0k}
Total	1.00

goodness-of-fit in multinomial applications, there is another, more commonly used statistic known as the **Pearson chi-squared statistic**, which is used to test the same restricted model hypotheses and is a statistic that you are more likely to encounter. The Pearson chi-squared statistic is also defined in terms of observed and expected frequencies, as follows:

$$\chi^2 = \sum \frac{(f_i - e_i)^2}{e_i} \tag{17.9}$$

This statistic is also approximately distributed as χ^2_{k-1} under the null model of Table 17.4. The chi-squared form is more recognizable with (17.9) than with (17.8), since each summand of (17.9) can be viewed (loosely) as the square of an approximately N(0, 1) random variable, and also since the chi-squared distribution is the distribution of the sum of squared standard normals.

Despite the different-looking mathematical forms, the likelihood ratio chi-squared statistic (17.8) and the Pearson chi-squared statistic (17.9) are approximately equal under the null model. If this were not the case, you would never see the Pearson chi-squared test, because it would be sub-optimal.

To see why Equations 17.8 and 17.9 are approximately equal, recall the approximation $\ln(1 + x) \cong x$ that is graphed in Figure 17.3. We'll need a better approximation here; it is $\ln(1 + x) \cong x - x^2/2$ and is shown in Figure 17.5.

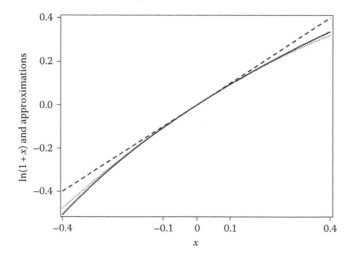

FIGURE 17.5

The function $f(x) = \ln(1 + x)$ (solid), the approximation $f(x) \cong x$ (dashed), and the improved approximation $f(x) \cong x - x^2/2$ (dotted).

Using this approximation, you can see why the Pearson chi-squared statistic is approximately equal to the likelihood ratio goodness of fit statistic as follows. Starting with the likelihood ratio chi-squared statistic

---

$$\chi^2 = 2 \sum_{i=1}^{k} f_i \ln(f_i / e_i)$$
(As shown before)

$$= 2 \sum_{i=1}^{k} \{e_i + (f_i - e_i)\} \ln\{1 + (f_i - e_i)/e_i\}$$
(By adding and subtracting e_i)

$$\cong 2 \left\{ \sum_{i=1}^{k} \{e_i + (f_i - e_i)\} \left\{ \frac{f_i - e_i}{e_i} - 0.5\left(\frac{f_i - e_i}{e_i}\right)^2 \right\} \right\}$$
(Since $\ln(1 + x) \cong x - x^2/2$)

$$= 2 \sum_{i=1}^{k} \left\{ (f_i - e_i) - 0.5\frac{(f_i - e_i)^2}{e_i} + \frac{(f_i - e_i)^2}{e_i} - 0.5\frac{(f_i - e_i)^3}{e_i^2} \right\}$$
(By multiplying the terms in braces algebraically)

$$= 2 \sum_{i=1}^{k} (f_i - e_i) + \sum_{i=1}^{k} \frac{(f_i - e_i)^2}{e_i} - \sum_{i=1}^{k} \frac{(f_i - e_i)^3}{e_i^2}$$
(By algebra)

Notice that the second summand in the final equation is exactly the Pearson chi-squared statistic shown in Equation 17.9. The first summand is 0, since $\Sigma f_i = n$, and since $\Sigma e_i = \Sigma n p_{0i} = n \Sigma p_{0i} = n \times 1 = n$ as well. That leaves the last summand, $\Sigma(f_i - e_i)^3 / e_i^2$. You can argue informally that this term should be small, and close to zero, as follows. Note that

$$\sum_{i=1}^{k} \frac{(f_i - e_i)^3}{e_i^2} = \sum_{i=1}^{k} \left\{ \frac{(f_i - e_i)^2}{e_i} \times \frac{(f_i - e_i)}{e_i} \right\}$$

Now

$$\frac{(f_i - e_i)}{e_i} = \frac{(n\hat{\pi}_i - n p_{0i})}{n p_{0i}} = \frac{(\hat{\pi}_i - p_{0i})}{p_{0i}}$$

Under the null model, $(\hat{\pi}_i - p_{0i})/p_{0i} \cong 0$ for large n, by the law of large numbers. Supposing n is so large that

$$\left| \frac{(\hat{\pi}_i - p_{0i})}{p_{0i}} \right| < \varepsilon$$

Here, ε is arbitrarily close to 0 for large n. Thus

$$\left| \sum_{i=1}^{k} \frac{(f_i - e_i)^3}{e_i^2} \right| < \varepsilon \sum_{i=1}^{k} \frac{(f_i - e_i)^2}{e_i} = \varepsilon \times (\text{an approximately } \chi^2 \text{ random variable})$$

So the last term in the summand is close to zero as well for large n under the null model. This concludes the demonstration that the likelihood ratio statistic shown in Equation 17.8 and the Pearson chi-squared statistic shown in Equation 17.9 are approximately equal to one another in the null case.

With the dice data there are again 13 1s, 10 2s, 16 3s, 10 4s, 8 5s, and 17 6s, and the Pearson chi-squared statistic is

$$\chi^2 = \sum \frac{(f_i - e_i)^2}{e_i} = \frac{(13 - 12.333)^2}{12.333} + \frac{(10 - 12.333)^2}{12.333} + \cdots + \frac{(17 - 12.333)^2}{12.333} = 5.30$$

This is nearly identical to the likelihood ratio chi-squared statistic, which is also 5.30 within rounding error.

You can also use the goodness-of-fit test to check whether the data are plausibly produced by particular distributions such as Poisson, normal, etc.; see Exercise 17.11 at the end of the chapter. If you have to estimate parameters in the null model, you should subtract a degree of freedom from the chi-squared distribution for each parameter you estimate, as indicated in the Ugly Rule of Thumb 16.1 of Chapter 16.

17.5 Cross-Classification Frequency Tables and Tests of Independence

In Chapter 6, we presented the following table, Table 6.14, showing observed joint frequencies of preference ratings for George H.W. Bush and Barbara Bush, given here again as Table 17.5.

As in the case of regression models, both classical and logistic, you can treat one of the variables as a response (Y) and one as a predictor (X). In the analysis that follows, it won't matter which is Y and which is X: According to the definition of independence, the variables are independent if the distribution of Y does not depend on $X = x$, and they are also independent if the distribution of X does not depend on $Y = y$.

An unrestricted model for the Barbara Bush rating (Y) and G. H. W. Bush rating (X) allows that the conditional probabilities $\pi_{j|i} = \Pr(Y = j | X = i)$ can be any nonnegative numbers that add to 1.0 for each i. Table 17.6 shows, as in Table 5.2, how these conditional probabilities look.

Assuming the 33 observations are independent, you can find the ML estimates of the $\pi_{j|i}$ by treating each row of data in Table 17.5 separately, and finding the MLEs as shown in Section 17.4. Thus you get the unrestricted MLEs as follows:

$$\hat{\pi}_{j|i} = \frac{f_{ij}}{f_i}.$$

These MLEs are intuitive: They are just the cell frequencies divided by the row totals.

TABLE 17.5

Cross-Classified Ratings Table

		Barbara Bush Rating					
		1	2	3	4	5	Total
George H.W.	1	5	1	0	0	0	6
Bush Rating	2	0	2	0	1	0	3
	3	1	1	3	1	1	7
	4	0	0	4	7	1	12
	5	0	1	0	1	3	5
	Total	6	5	7	10	5	33

TABLE 17.6

Unrestricted Conditional Distributions

		Y					
		1	2	3	4	5	Total
X	1	$\pi_{1\mid1}$	$\pi_{2\mid1}$	$\pi_{3\mid1}$	$\pi_{4\mid1}$	$\pi_{5\mid1}$	1.00
	2	$\pi_{1\mid2}$	$\pi_{2\mid2}$	$\pi_{3\mid2}$	$\pi_{4\mid2}$	$\pi_{5\mid2}$	1.00
	3	$\pi_{1\mid3}$	$\pi_{2\mid3}$	$\pi_{3\mid3}$	$\pi_{4\mid3}$	$\pi_{5\mid3}$	1.00
	4	$\pi_{1\mid4}$	$\pi_{2\mid4}$	$\pi_{3\mid4}$	$\pi_{4\mid4}$	$\pi_{5\mid4}$	1.00
	5	$\pi_{1\mid5}$	$\pi_{2\mid5}$	$\pi_{3\mid5}$	$\pi_{4\mid5}$	$\pi_{5\mid5}$	1.00

TABLE 17.7

Conditional Distributions under Independence

		Y					
		1	2	3	4	5	Total
X	1	π_1	π_2	π_3	π_4	π_5	1.00
	2	π_1	π_2	π_3	π_4	π_5	1.00
	3	π_1	π_2	π_3	π_4	π_5	1.00
	4	π_1	π_2	π_3	π_4	π_5	1.00
	5	π_1	π_2	π_3	π_4	π_5	1.00

If Y is independent of X, then the conditional probabilities $\pi_{j\mid i} = \Pr(Y = j \mid X = i)$ do not depend on i, meaning, as shown in Table 17.7 that

$$\pi_{j\mid i} = \pi_j$$

In the independence case, you can ignore the row ($X = i$) altogether, and obtain the MLEs as in Section 17.4 after collapsing the table to its column totals

$$\hat{\pi}_j = \frac{f_{\cdot j}}{n}$$

The MLEs under the restricted model are again intuitive: They are just the column totals divided by the total number of observations. In other words, they are the estimated marginal probabilities.

In either model, the likelihood function is the product of all the individual conditional likelihoods as shown in the following:

$$LL = \sum_{i=1}^{5} \sum_{j=1}^{5} f_{ij} \ln(\pi_{j\mid i})$$

In the unrestricted model, this gives you

$$LL_1 = \sum_{i=1}^{5} \sum_{j=1}^{5} f_{ij} \ln\left(\frac{f_{ij}}{f_{i\cdot}}\right)$$

While in the restricted model you get

$$LL_0 = \sum_{i=1}^{5} \sum_{j=1}^{5} f_{ij} \ln\left(\frac{f_j}{n}\right)$$

The likelihood ratio chi-squared statistic is then given as follows:

$$\chi^2 = 2(LL_1 - LL_0) = 2\left\{\sum\sum f_{ij} \ln\left(\frac{f_{ij}}{e_{ij}}\right)\right\} \tag{17.10}$$

Here, the expected frequencies are $e_{ij} = n\{(f_{i.}/n)\,(f_{.j}/n)\}$, exactly as shown in Table 6.16. The degrees of freedom for (17.10) are, in general, $(I - 1)(J - 1)$, since there are $IJ - I$ parameters in the unrestricted model—the $\pi_{j/i}$, losing one degree of freedom for each row because they add to 1.0—and $(J - 1)$ parameters in the restricted model—the marginal probabilities π_j subtracting one degree of freedom because they add to 1.0. The difference is $(IJ - I) - (J - 1) = (I - 1)(J - 1)$.

Example 17.10: Are the Trends in the Bush Likeability Data Explainable by Chance?

In the case of the Bush likeability data, the observed values are given in Table 17.5 and the expected values in Table 6.16, the likelihood ratio chi-squared statistic is given by

$$\chi^2 = 5 \times \ln\left(\frac{5}{1.09}\right) + 1 \times \ln\left(\frac{1}{0.91}\right) + 0 \times \ln\left(\frac{0}{1.27}\right) + \cdots + 3 \times \ln\left(\frac{3}{0.76}\right) = 43.11$$

(Recall as above that $0 \times \ln(0)$ is defined to be zero in the above expression since there is no contribution to the likelihood function for cells where there is no data.) The approximate p-value is $pv \cong \Pr\left(\chi^2_{16} \geq 43.11\right) = 0.0003$, indicating that the difference between observed and expected frequencies cannot easily be explained by chance alone.

However, according to Ugly Rule of Thumb 17.2, there is a problem. Here, there are 25 cells, and the expected frequencies e_{ij} are less than 5 in *all* cells, so the validity of the approximation is suspect. Methods based on resampling, such permutation and bootstrap models are appropriate in such cases; see Chapter 19 for a revisiting of this example.

You are more likely to see the Pearson chi-squared statistic for testing independence. As before it is defined as the sum of terms $(f - e)^2/e$; the only difference here is that the summation is over both rows and columns.

Pearson Chi-Squared Test of Independence in Two-Way Tables

Assuming that X and Y are independent, and that the observations (X_i, Y_i) are independent for $i = 1, 2, \ldots, n$, then

$$\text{Pearson } \chi^2 = \sum_{i=1}^{I} \sum_{j=1}^{J} \frac{(f_{ij} - e_{ij})^2}{e_{ij}}$$

is approximately distributed as $\chi^2_{(I-1)(J-1)}$, with the approximation becoming better with larger n.

Applied to the Bush data, the Pearson chi-squared statistic gives you $\chi^2 = 44.3$, again with $df = 16$, so the results are essentially the same regardless of whether you use the likelihood ratio or Pearson chi-squared statistic.

17.6 Comparing Non-Nested Models via the AIC Statistic

The likelihood ratio chi-squared test is applicable whenever the restricted model is just that—a restricted model. A restricted model is, by definition, a model that you get by constraining the parameters of the unrestricted model. Thus, you can say that the restricted model must be **nested** within the unrestricted model.

In all examples given so far you could get the null model by restricting the parameters of the unrestricted model. But what if one model is not nested within the other? For example, you might wish to test whether the normal distribution or the beta distribution is more plausible as the producer of your data. Or you might want to compare two regression models, one with a single predictor variable X_1 and another with a single predictor variable Z_1. In each of these examples neither model is a special case of the other, so you cannot apply the likelihood ratio test.

On the other hand, you can calculate the log likelihood for each model, and higher log likelihood indicates better fit. But you must be careful: As indicated earlier in this chapter, an unrestricted model has a higher log likelihood than a restricted model, even when the restricted model is the true model. So a simple comparison of log likelihoods is not appropriate—you have to account for the number of parameters in the model as well.

Akaike's information criterion (AIC) gives the commonly used *AIC statistic*, which penalizes the likelihood for the number of parameters that you estimate. One version of the AIC statistic is as follows:

$$AIC = LL - k \tag{17.11}$$

Here LL is the maximized log-likelihood of the model under question, and k is the number of parameters estimated in the model under question.

If you are comparing two models, say model 1 and model 2, then you will have two AIC statistics, $AIC_1 = LL_1 - k_1$ and $AIC_2 = LL_2 - k_2$. You can say that model 1 is better than model 2, according to the AIC statistic if $AIC_1 > AIC_2$. The form shown in (17.11) is therefore called a *larger is better* form, because larger values of (17.11) indicate better models. Depending on the software you use, you might also see a *smaller is better* form of AIC, where

$$AIC = -2LL + 2k. \tag{17.12}$$

It is easy to understand whether the form of AIC used by your software is *larger is better* or *smaller is better* by remembering two words: *Maximum likelihood*. In form (17.11), you want *higher AIC* because you want to maximize the likelihood. In form (17.12), you want *lower AIC*, again, because you want to maximize the likelihood.

It is instructive to consider how the AIC statistic works in the case of nested models, where the likelihood ratio test applies. Suppose model 1 is the unrestricted model and model 2 is the restricted model. Model 1 is favored by the AIC statistic if $AIC_1 > AIC_2$, or equivalently, if $2(LL_1 - LL_2) > 2(k_1 - k_2)$, or if $\chi^2 > 2df$. The rule is sensible in that the approximate expected value of the likelihood ratio χ^2 statistic is equal to df when model 2 is correct. Thus it is unlikely that the AIC statistic will choose model 1 when model 2 is correct, because it is unlikely that a chi-squared random variable will be more than twice its degrees of freedom.

Example 17.11: Comparing the Multinomial and Shifted Poisson Models for Dice

As shown above in Example 17.9, the multinomial model for the dice data gives $LL_1 = \Sigma f_j$ $\ln(f_j/n) = 13 \times \ln(13/74) + 10 \times \ln(10/74) + 16 \times \ln(16/74) + 10 \times \ln(10/74) + 8 \times \ln(8/74) + 17 \times \ln(17/74) = -129.94$.

Since the Poisson model predicts data 0, 1, 2, 3, … will occur, the dice data 1, 2, 3, 4, 5, 6 are clearly not Poisson as there are no zeros. However, you might suppose that the distribution of the dice is a shifted Poisson, that is, the distribution of $T = Y + 1$, where Y is Poisson, and where T is the outcome of the die roll. Then $\Pr(T = t) = \Pr(Y = t - 1) = \exp(-\lambda)\lambda^{(t-1)}/(t - 1)!$, for $t = 1, 2, 3, \ldots$. The MLE of λ in this model is $\hat{\lambda} = \bar{t} - 1$. Using the dice data, $\bar{t} = 3.554$, so $\hat{\lambda} = 2.554$.

Figure 17.6 shows a comparison of the estimated (unrestricted) multinomial model and the estimated shifted Poisson model.

The maximized log likelihood for the shifted Poisson model is $LL_2 = -n\hat{\lambda} + \Sigma\{(y_i - 1)\}\ln(\hat{\lambda}) - \Sigma \ln\{(y_i - 1)!\}$. Collecting terms where $y_i = 1, 2, \ldots, 6$, $LL_2 = -74(2.554) + \ln(2.554)\{13(1 - 1) + 10(2 - 1) + 16(3 - 1) + 10(4 - 1) + 8(5 - 1) + 17(6 - 1)\} - \{13\ln(0!) + 10\ln(1!) + 16\ln(2!) + 10\ln(3!) + 8\ln(4!) + 17\ln(5!)\} = -188.996 + 177.218 - 135.820 = -147.598$.

The multinomial model has a higher likelihood, but it also has more parameters than the Poisson model (five versus one). The AIC statistics are, respectively, $AIC_1 = -129.94 - 5 = -134.94$, and $AIC_2 = -147.60 - 1 = -148.60$. Even considering the extra parameters of the multinomial model, it is better as gauged by the AIC statistic than the shifted Poisson model. This conclusion only makes sense: As shown in Figure 17.6, the shifted Poisson model forces an up-then-down probability pattern, and also predicts that values greater than 6 can occur.

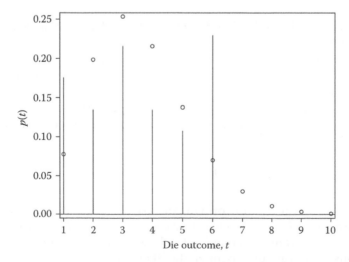

FIGURE 17.6
Comparison of estimated unrestricted multinomial model (needles) and the estimated shifted Poisson model (circles) for the dice data.

A final important point about AIC is that it is useful only as a *relative* measure of fit. There are no rules of thumb (not even ugly ones!) to tell you how large the AIC should be. The size of the AIC is data-dependent: AIC = –23.295 for a model may be very large or very small; you would need to fit other models on the same data, and compute their AIC statistics to judge which models are better.

Vocabulary and Formula Summaries

Vocabulary

Likelihood ratio	The ratio of the maximized likelihood functions under two scenarios, one in which the parameter space is unrestricted and the other in which the parameter space is restricted.
Full model versus restricted model tests	A general class of tests that compare the fit of the data under two different models, one in which the parameter space is unrestricted and the other in which the parameter space is restricted.
Likelihood ratio test	A type of full model versus restricted model test that uses the likelihood ratio as the test statistic.
Least squares estimates	Estimates of regression parameters that minimize the sum of squared deviations between the actual data values and the fitted values.
Sum of squares for error	The minimized sum of squared deviations between the actual data values and the fitted values in regression.
***R*-squared statistic**	A statistic that measures the proportion of variance in Y that is explained by X.
Scoring function	An estimate of probability of "success"; that is, of the probability that a person will buy a product, or will repay a loan.
Covariate	An X variable—that is, a variable that postulated to be deterministically related to another variable Y in a model.
Observed frequencies	In analysis of contingency tables, the frequencies of categories that you actually observed.
Expected frequencies	In analysis of contingency tables, estimates of the frequencies of categories you would expect if the null model were true.
Pearson chi-squared statistic	A statistic used to test goodness-of-fit for multinomial distributions.
Nested models	Unrestricted and restricted models; the restricted model is nested within the unrestricted model.

Akaike's Information Criterion (AIC)	A statistic that adjusts the fit of either nested or non-nested models by comparing their maximized log likelihoods while accounting for the number of parameters estimated in each model.

Key Formulas and Descriptions

Θ_0	The restricted parameter space.
Θ_1	The unrestricted parameter space.
$L(\hat{\theta}_1 \vert \text{data}) \geq L(\hat{\theta}_0 \vert \text{data})$	The unrestricted likelihood is always greater than or equal to the restricted likelihood.
$LL_0 = \max_{\theta \in \Theta_0} LL(\theta \vert \text{data})$	The maximized log-likelihood under the restricted model.
$LL_1 = \max_{\theta \in \Theta_1} LL(\theta \vert \text{data})$	The maximized log-likelihood under the unrestricted model.
$\chi^2 = 2(LL_1 - LL_0)$	The likelihood ratio chi-squared test statistic for testing whether the restricted model is valid.
$\chi^2 \,\dot{\sim}\, \chi^2_{df}$, where $df = \{\text{\# of parameters in } p_1(y \vert \theta_1)\} - \{\text{\# of parameters in } p_0(y \vert \theta_0)\}$	The approximate (large-sample) distribution of the likelihood ratio chi-squared statistic.
$\hat{\sigma}^2_0$	The estimated variance of Y using a restricted model.
$\hat{\sigma}^2_1$	The estimated variance of Y in an unrestricted model.
$2(LL_1 - LL_0) = n\left\{\ln\left(\hat{\sigma}^2_0 / \hat{\sigma}^2_1\right)\right\}$	The form of the likelihood ratio test statistic for testing whether the mean is equal to a constant, for testing equality of means in ANOVA, and for testing significance of regression coefficients.
$\ln(1 + x) \cong x$	The natural logarithm of $1 + x$ is close to x when x is close to zero.
$\ln(1 + x) \cong x - x^2/2$	The natural logarithm of $1 + x$ is even closer to $x - x^2/2$ when x is close to zero.
$Y_i \vert (X_{i1}, \ldots, X_{ik}) = (x_{i1}, \ldots, x_{ik}) \sim_{\text{independent}} \mathrm{N}(\beta_0 + \beta_1 x_{i1} + \cdots + \beta_k x_{ik}, \sigma^2)$	The unrestricted multiple regression model.
$Y_i \vert (X_{i1}, \ldots, X_{ik}) = (x_{i1}, \ldots, x_{ik}) \sim_{\text{independent}} \mathrm{N}(\beta_0, \sigma^2)$	The restricted multiple regression model.
$R^2 = 1 - \hat{\sigma}^2_1/\hat{\sigma}^2_0$	The R-squared statistic in multiple regression analysis.
$2(LL_1 - LL_0) = n \times \ln\{1/(1 - R^2)\}$	The likelihood ratio test in multiple regression.

$Y_i|(X_{i1}, ..., X_{ik}) =$
$(x_{i1}, ..., x_{ik})\sim_{\text{independent}} \text{Bernoulli}(\pi_i),$
where $\pi_i = \exp(\beta_0 + \beta_1 x_{i1} + \cdots + \beta_k x_{ik})/\{1 + \exp(\beta_0 + \beta_1 x_{i1} + \cdots + \beta_k x_{ik})\}$

The unrestricted multiple logistic regression model.

$Y_i|(X_{i1}, ..., X_{ik}) =$
$(x_{i1}, ..., x_{ik})\sim_{\text{independent}} \text{Bernoulli}(\pi_i),$
where $\pi_i = \exp(\beta_0)/\{1 + \exp(\beta_0)\}$

The restricted multiple logistic regression model.

$L(\pi_1, \pi_2, ..., \pi_k | y_1, y_2, ..., y_n) = \pi_1^{f_1} \times \pi_2^{f_2} \times \cdots \times \pi_k^{f_k}$

The likelihood function for the unrestricted multinomial model.

$\chi^2 = 2\sum f_i \ln(f_i/e_i)$

The likelihood ratio chi-squared statistic for a goodness-of-fit test, where f_i is the observed frequency and e_i is the expected frequency under the null model.

$\chi^2 = \sum (f_i - e_i)^2/e_i$

The Pearson chi-squared statistic for a goodness-of-fit test, where f_i is the observed frequency and e_i is the expected frequency under the null model.

$\chi^2 = 2\sum\sum f_{ij} \ln(f_{ij}/e_{ij})$

The likelihood ratio chi-squared statistic for testing independence in a contingency table, where f_{ij} is the observed frequency and e_{ij} is the expected frequency under independence.

$\chi^2 = \sum\sum (f_{ij} - e_{ij})^2/e_{ij}$

The Pearson chi-squared statistic for testing independence in a contingency table, where f_{ij} is the observed frequency and e_{ij} is the expected frequency under independence.

$AIC = LL - k$

The *larger is better* form of the AIC statistic.

$AIC = -2LL + 2k$

The *smaller is better* form of the AIC statistic.

Exercises

17.1 Refer to Example 17.1. Show that the maximum likelihood estimate of σ^2 for the $N(m_0, \sigma^2)$ restricted model is $\hat{\sigma}_0^2 = \{(y_1 - m_0)^2 + \{(y_2 - m_0)^2 + \cdots + \{(y_n - m_0)^2\}/n$, showing logical reasons for each step.

17.2 Explain how you get from Equation 17.3 to Equation 17.4.

17.3 Refer to Example 7.2. The likelihood ratio chi-squared statistic $\chi^2 = 2(LL_1 - LL_0) = 2n\ln(m_0/\bar{y}) - 2n(1 - \bar{y}/m_0)$.

    A. Rewrite the test statistic as $\chi^2 = 2(LL_1 - LL_0) = -2n\ln(\bar{y}/m_0) - 2n(1 - \bar{y}/m_0)$, and explain why you can do that.

B. Using the approximation $\ln(1 + x) \cong x - x^2/2$, show that $\chi^2 \cong n(\bar{y} - m_0)^2/m_0^2$. (Hint: First use an appropriate add-and-subtract trick to get logarithmic term into a $1 + x$ form.)

C. It is a fact that the mean and standard deviation of the exponential distribution are equal. Explain why the chi-squared statistic in Exercise 17.3B, when viewed as a function of random data, has a null distribution that is approximately a chi-squared distribution with one degree of freedom. Use the Central Limit Theorem in your answer, as well as the definition of the chi-squared distribution.

17.4 In Example 17.3, it is claimed that $\sum_i \sum_j (y_{ij} - \mu_i)^2$ is minimized by $\hat{\mu}_i = \sum_j y_{ij}/n_i$. Using calculus, show that this is true for the student age data in Example 16.5 where there are $g = 2$ groups.

17.5 Show that $\chi^2 = n \times \ln\{1/(1 - R^2)\}$ is a monotonically increasing function of the R^2 statistic.

17.6 Rewrite Ugly Rule of Thumb 10.1 for the R^2 statistic.

17.7 Show that the likelihood ratio statistic for the multiple regression model is a monotonic function of the model F-statistic for the multiple regression model.

17.8 Suppose $Y_i|(X_{i1} = x_{i1}, X_{i2} = x_{i2}, X_{i3} = x_{i3}, X_{i4} = x_{i4}) \sim_{\text{iid}}$ Bernoulli(π), for $i = 1, 2, \ldots, n$, where π is a constant between 0 and 1 that is not affected by the X variables. Show that the MLE for π in this model is $\hat{\pi} = \sum_i y_i/n$.

17.9 In Example 17.9, the derivatives of the log likelihood function are given as $(\partial/\partial\pi_j)LL(\pi_1, \pi_2, \ldots, \pi_5|y_1, y_2, \ldots, y_n) = f_{j}/\pi_j - f_6/(1 - \pi_1 - \cdots - \pi_5)$, for $j = 1, 2, \ldots, 5$. Set these five equations to zero, and solve them to get $\hat{\pi}_j = f_j/n$.

17.10 You are an instructor at a major university. One question on your exam is very hard, and you suspect that students are just randomly guessing. There are five possible answers to this multiple-choice question, and of the 670 students who took the test, the distribution of selected answers is as follows:

Answer	Count
A	120
B	130
C	150
D	160
E	110
Total	670

A. What are the data values $Y_1, Y_2, \ldots, Y_{670}$? Give a description, not actual values.

B. What do you assume about $Y_1, Y_2, \ldots, Y_{670}$ in the restricted model?

C. What do you assume about $Y_1, Y_2, \ldots, Y_{670}$ in the unrestricted model?

D. Compute likelihood ratio and Pearson chi-squared statistics and find their associated p-values.

17.11 Use the data from Exercise 2.12. Test whether the Poisson model could have produced the data. Estimate the Poisson parameter using a weighted average. Make the last category "4 or more" when calculating its expected value. Use both the likelihood ratio chi-squared test and the Pearson chi-squared test.

17.12 Show how to get the χ^2 statistic in Equation 17.10 from the expressions for LL_0 and LL_1, which are given in the lines immediately above Equation 17.10.

17.13 Consider the data from Exercise 12.1: 1, 1, 0, 0, 1, 1, 0, 0, and 0. Assume that they are sampled as iid from Bernoulli(π), and you want to test H_0: $\pi = 1/3$.

    A. Find LL_0 and LL_1.

    B. Compute $2(LL_1 - LL_0)$ using your answer to Exercise 17.13A. What is this statistic called?

    C. Apply Equation 17.8 to these data. You should get the same result as in Exercise 17.13B.

    D. Find the approximate p-value for the test. Explain why the degrees of freedom are 1.

    E. Why is the p-value in Exercise 17.13D *approximate* rather than *exact*?

    F. Calculate the Pearson chi-squared test and its approximate p-value.

17.14 Calculate the AIC statistic (smaller is better form) for the Poisson fit to the data Exercise 2.12. Compare it with the AIC statistic for the discrete uniform model for the data, where the last category is "4 or more."

17.15 What is the AIC statistic (smaller is better) for the unrestricted model regression analysis? For the restricted model? When does the restricted model have a lower AIC statistic?

17.16 Use the data from Exercise 12.9. Find the least squares estimates of β_0 and β_1 using calculus, and compare them to the MLEs in the classical model that were found in Exercise 12.9A.

17.17 Find the approximate probability that the AIC statistic will select the restricted model when the difference between the number of parameters in the restricted and unrestricted models is 1. Repeat for differences of 2, 3, ..., 20 and draw a graph of the results. Do you like what you see? Why?

18

Sample Size and Power

18.1 Introduction

You may recall the statistical science paradigm

$$\text{Nature} \rightarrow \text{design and measurement} \rightarrow \text{DATA}$$

And you may recall your model for this, as well. It looks like the following:

$$p(y) \rightarrow \text{DATA}$$

Or, in words, *model produces data*. If the necessity of this paradigm has not sunk in yet, this chapter should finally clinch it! In particular, simulation is often the best way to decide on a sample size. You don't have any data yet, so in order to see whether your data collection will result in a successful study, you can simulate plausible future data and check whether or not the analysis is successful. If not successful often enough, you should try a different design or sample size. Simulation is great for planning—there is no sense wasting money on a study that is likely to turn out badly!

Recall the related mantric phrase *data reduce the uncertainty about the unknown parameters*. As shown repeatedly in this book, a larger n provides more reduction in uncertainty. Statistically, larger n is always better. The only problem is cost: When n is too large, your study might be too costly.

There are two types of goals you might have for sample size selection. The first is that your estimates be *close enough* to the true parameter values. Opinion poll researchers have this kind of goal: If they claim "23% of respondents support candidate A," they would like their estimate 23%, to be close enough to $100\pi\%$, the process percentage. This is the simpler of the two types of goals, and is discussed in Section 18.2.

The other type of goal is to establish that chance alone cannot explain your results. For example, suppose that you are a physician who has developed a new procedure to cure a serious disease. Your procedure cures a proportion π_1 of the cases, while the standard procedure has a π_0 cure rate. Both π_1 and π_0 are unknown parameters. You can assume cases treated with your procedure are produced as independent and identically distributed (iid) Bernoulli(π_1), and you can assume cases treated with the standard procedure are produced as iid Bernoulli(π_0), but you can't assume that you know the values of π_1 and π_0.

People will gladly try almost any type of cure to improve their chances of survival, including nontraditional and even mystical cures. So if the chance of survival from the new treatment, π_1, is larger than the chance of survival from the standard, π_0, then the new therapy will certainly be interesting to a large group of people. If you knew that, on average, 30 out of 100 people (30%) are cured with the standard therapy and that 32 out of 100 people (32%) are cured with the new one, which one would you choose?

The statistical problem is that the true percentages are unknown. If you tried these two therapies on two sets of 100 patients, you might get 30 out of 100 and 32 out of 100, but the results would not replicate because every set of 100 patients is different from every other set of 100 patients. With *estimated* cure rates of 30% and 32%, you could certainly claim that your new method is worth looking into further—but if you claimed that it is in fact better, you would be guilty of medical, ethical, scientific, and legal fraud and could be prosecuted in a court of law.

Before making such claims, you first must establish that the results are not explainable by chance alone. A difference as small as 30 out of 100 versus 32 out of 100 is easily explainable by chance alone. To see why, just generate multiple sets of 100 observations from some Bernoulli distribution. Don't choose 0.30 or 0.32 for the π of the Bernoulli distribution because these are not the true parameters. The true parameter is unknown, so try $\pi = 0.35$ for now. Now, sum up the 100 Bernoulli(0.35) numbers for each set of 100 that you generate. These are the numbers of cured patients in different sets of 100 patients when the cure rate is identically 0.35 for both groups. Dividing these by 100 gives percentages; here are a few examples: (37.0%, 33.0%), (29.0%, 33.0%), (29.0%, 29.0%), (31.0%, 34.0%), and (32.0%, 38.0%). A process with the same cure rate in both groups produced all these pairs, yet the differences are large—many differences are more than 2%. A difference of 30% versus 32% is easily explainable by chance alone if there are 100 patients in each group.

If the seemingly arbitrary choice of 0.35 for the Bernoulli parameter bothers you, just try some other value, like 0.25, or 0.28, or 0.33. You'll see the same results: No matter what is the value of the unknown π—unless it's very close to 0.00 or 1.00—differences as large as 2% are easily explainable by chance alone when there are only 100 observations per group.

On the other hand, suppose that there are 100,000 people in each group, with the same 30%, 32% results—that is, 30,000 are cured in the standard therapy group and 32,000 are cured in the new therapy group. Is this difference explainable by chance alone? Not likely! Again, using a Bernoulli(0.35) model, some simulated percentages are (35.2%, 34.8%), (35.0%, 34.9%), (34.6%, 35.0%), (35.0%, 35.1%), and (34.9%, 34.8%). All of the differences are well under 2%. A difference of 2% is easily explainable by chance alone when there are 100 patients per group, but not when there are 100,000 patients per group.

If there really is a 2% difference between the two cure rates, you'll need a sample size that is somewhere between 100 and 100,000 per group to rule out chance. That's a pretty wide range! How many do you really need? A sample size of 100,000 people per group will work, but that would be too expensive. In this chapter, we'll help you narrow it down. The answer depends on the *power* of a statistical test, which we introduced in Chapter 17 and discuss in more detail in Section 18.3.

There is no one correct number for the sample size that you will need in your study. You have to make many subjective choices—some of which are essentially Bayesian in character—before deciding on a sample size. There are no choices that you can say are exactly correct, but many are clearly wrong, so you should think about them carefully, and be prepared to justify them to your boss, to reviewers of your research paper, or to reviewers of your grant proposal. We'll show you how to obtain reasonable sample sizes in this chapter via a series of examples, starting with Section 18.2 where we help you identify the sample size that you need to ensure that your estimates are sufficiently close to the true parameter values.

18.2 Choosing a Sample Size for a Prespecified Accuracy Margin

Example 7.1 of Chapter 7 showed how to use random sampling to estimate a mean inventory value. How many items should you randomly sample? It costs time and money to assess the value of each item sampled, so you don't want to sample more than necessary.

Example 18.1: Choosing a Sample Size for Estimating Mean Inventory Value

If you have a random sample, you know by the central limit theorem (CLT) that the sample average, $\bar{Y}$, has an approximately normal distribution with mean μ and variance σ^2/n. This is a remarkable and useful fact! It tells you that your data average $\bar{Y}$ will be within $\pm Z_{1-\alpha/2}\, \sigma/n^{1/2}$ of μ in approximately $100(1-\alpha)\%$ of your samples. Recall that each sample provides a different location for $\bar{Y}$ (the mountain lion), but the process mean μ (the town) doesn't move.

Let $a.m. = Z_{1-\alpha/2}\, \sigma/n^{1/2}$ be your **accuracy margin**. This is slightly different from the *margin of error*, which is calculated after collecting data and which uses $\hat{\sigma}$ instead of σ. You need to use *a.m.* instead of margin of error because you haven't collected any data yet, so you don't know $\hat{\sigma}$.

Choosing n is easy, from the standpoint of algebra. Solving $a.m. = Z_{1-\alpha/2}\, \sigma/n^{1/2}$ for n, you get a simple formula for determining n as

$$n = \frac{Z_{1-\alpha/2}^2 \sigma^2}{(a.m.)^2} \tag{18.1}$$

Applying (18.1) in practice is where all the difficulty lies. You have to decide what values to use for α, for *a.m.*, and for σ. Every choice can have a large effect on the sample size n, and there is no absolutely correct selection for any of them.

Let's start with *a.m.* How close do you want your estimated inventory value $\bar{Y}$ to be to the population average μ? Let's try some numbers. Is it acceptable that your data average is within $\pm\$200$ of the true average? You need local knowledge to help you decide: Perhaps you know, based on your company experience, that the inventory items are roughly \$300, on average. If this is true, then your estimate will be within a range, roughly, from \$100 to \$500. This interval range is clearly too wide because it means you could estimate your total value to be as much as 5/3 what it really is, or you could estimate your total value to be as little as 1/3 what it really is. Thus, you'll need to pick a smaller number for *a.m.* in this case. On the other hand, if you know that the inventory items are roughly \$300,000 on average, then your $\pm\$200$ range would be perfectly acceptable. By considering *relative* errors, you can narrow down the choice of *a.m.*

Still, that does not tell you precisely what number to pick for *a.m.* There is nothing more this textbook can tell you about it, either: You'll need to have a discussion about it in your boardroom, where you lay the numbers and choices on the table, and get some consensus among the stakeholders as to what *a.m.* you and your company would like to use. The choice is very important for the resulting sample size: If you choose *a.m.* = \$50, you'll get one sample size n from Equation 18.1; if you choose *a.m.* = \$25, you'll get a sample size n from Equation 18.1 that is *four times larger*, potentially costing your company four times as much.

There is no single number that is right for *a.m.* There are only numbers that you can suggest based on reason, using your inside knowledge about the process you are studying.

Things are no better for the other terms in Equation 18.1, α and σ. Consider α first.

The symbol α in (18.1) denotes the probability that your estimate $\bar{Y}$ will be farther than *a.m.* from μ. A larger α gives a smaller quantile $Z_{1-\alpha/2}$, hence a smaller n. For example, $\alpha = 0.50$ gives $Z_{1-\alpha/2} = Z_{0.75} = 0.67$, whereas $\alpha = 0.10$ gives $Z_{1-\alpha/2} = Z_{0.05} = 1.65$. Since $1.65^2/0.67^2 = 6.1$, your required sample size will be *six times larger* if you choose $\alpha = 0.10$ rather than $\alpha = 0.50$. So the choice of α is crucial. It is a personal choice that *you* must make. How willing are you to be wrong? If you want your estimated average inventory value to be within ±$50 of the population average, can you live with an estimate that will be within ±$50 only half the time? Probably not! If you have gone through the trouble to convince the boss and all the others in the boardroom that the estimate should be within ±$50 of the true average, you probably want this to be true with a high probability, such as 80%, 90%, 95%, or 99%. Still, there are big differences even among those critical values: The 80% probability corresponds to $Z_{1-\alpha/2} = Z_{0.90} = 1.28$, whereas the 99% probability corresponds to $Z_{1-\alpha/2} = Z_{0.995} = 2.58$. If you want to have a 99% probability that your estimate will be within the *a.m.* of the true value, rather than an 80% probability, then you'll need $(2.58/1.28)^2 = 4.06$ times as many observations. So, the choice of α is important and is a balance between how willing you are to be wrong versus how much money you have. If you are less willing to be wrong (smaller α), you'll need more money for the study, since your n will have to be higher.

So, there is no number that is "right" for α either, except that you want α to be small, but not so small that your study will wind up being too costly. What about the other term in Equation 18.1, namely, σ? Is there a definitive choice for σ?

Sadly, the answer is "No." The parameter σ is unknown. It's one of the unknown parameters in the Mantra *model produces data* and *model has unknown parameters*. You cannot know its value, but you may have some prior knowledge about it. For example, you can apply Chebyshev reasoning (since the distribution of inventory valuations is likely right-skewed, hence non-normal), to state that at least 75% of the individual item's valuations will be in the range $(\mu - 2\sigma, \mu + 2\sigma)$. So if you have a rough guess that the main range of most of the valuations spans $1000, then a very crude guess of σ is $s = 1000/4 = \$250$. But this is just a guess. And the guess matters: If you guess $500 for σ instead of $250, then you'll need a sample size that is four times higher, potentially costing four times as much.

The guess of σ not only affects sample size selection. It also affects your claim about how confident you are that the estimate will be within the *a.m.* Suppose, for example, that σ is really $500—unknown to you, of course—but you pick $s = \$250$ as your guess of σ, you state an *a.m.* of $50, and your desired error probability is $\alpha = 0.05$. What is the consequence of your choosing the wrong σ? First, based on your choices, you will apply Equation 18.1 and get $n = (1.96)^2(250)^2/(50)^2 = 96$ observations. "Not bad," says your boss. "That won't cost our company too much money." But what is the probability that your $\bar{Y}$ will be within ±$50 of μ? You'd like it be 95%, but it's not. Here's the real story:

$\Pr(\bar{Y}$ will be within ±$50 of $\mu)$

$= \Pr(\mu - 50 \leq \bar{Y} \leq \mu + 50)$ (By definition)

$= \Pr\{-50/(500/96^{1/2}) \leq (\bar{Y} - \mu)/(500/96^{1/2}) \leq 50/(500/96^{1/2})\}$ (By algebra)

$\cong \Pr(-0.98 \leq Z \leq 0.98)$, where $Z \sim N(0, 1)$ (Since $\sigma = 500$ and also by linearity and additivity properties of the normal distribution, by the CLT, and by arithmetic)

$= 0.67$ (By calculation using the N(0, 1) cumulative distribution function [cdf])

Thus, there is only a 67% chance that your estimate will be within ±$50 of μ if σ is really $500 but you pick $s = \$250$. In other words, in one out of three samples, your estimate, $\bar{Y}$, will be more than $50 from μ. To generalize, if you *underestimate* the standard deviation, your estimate is *less likely* to be within the *a.m.* Conversely, if you *overestimate* the standard deviation, your estimate will be *more likely* to be within the *a.m.*, but your sample size *n* will be needlessly large, wasting money.

While this discussion of Equation 18.1 might seem pessimistic, there is a positive message. You can use (18.1) to perform various what-if scenarios, selecting various reasonable values of *a.m.*, α, and σ. Present these options to the company's chief officers in the boardroom, discuss, and then offer your informed opinion about what sample size to choose. It's better than a wild stab in the dark.

When you want to estimate a proportion instead of an average, it is a little easier to estimate the variance, as the following example shows.

Example 18.2: Choosing a Sample Size for Estimating Burn Patient Mortality

People who have suffered severe burns often die. While the burn itself can be fatal, the risk of infection following the burn is high, and so burn patients often die of secondary causes. If the patient can survive for a year following burn, the patient is often classified as healed.

There are many burn and trauma centers worldwide. While most use standard practices, each has subtle differences in the physicians, facilities, and patient populations that can affect survival. Survival also depends strongly on the severity of the burn and on the patient's age. Suppose you wish to estimate the survival probability of a 20 year old patient with burns on 50% of the body. Since you probably won't see many patients with precisely these characteristics, suppose instead that you choose a neighborhood cohort including patients 15–25 years old with between 40% and 60% of their bodies burned. How many patients from this cohort will you have to sample to get a reasonably accurate estimate of the survival probability?

The answer starts, as always, with your model for how the data are produced. Here, you have to admit you really do like the idea of a *model* producing such data, rather than using the real, tragic human burn data. The data you will observe are $Y_1, Y_2, \ldots, Y_n$, and you can reasonably assume that they are produced as iid Bernoulli(π), where $Y = 1$ denotes survival. The parameter π is the unknown survival probability that you wish to estimate. You know by the law of large numbers discussed in Section 8.5 that the average of the Y values converges to π as n increases, that is, you know that $\lim_{n\to\infty} \bar{Y} = \pi$. So define $\hat{\pi} = \bar{Y}$. By the linearity and additivity properties of expectation and variance, you know that $E(\hat{\pi}) = \pi$ and that $Var(\hat{\pi}) = \sigma^2/n$, where $\sigma^2 = Var(Y)$. But for the Bernoulli distribution, $Var(Y) = \pi(1 - \pi)$, implying that $Var(\hat{\pi}) = \pi(1 - \pi)/n$. Further, because it is a sample average, the distribution of $\hat{\pi}$ is approximately normal by the CLT. Putting it all together, $\hat{\pi} \sim N\{\pi, \pi(1 - \pi)/n\}$, and hence, your estimate $\hat{\pi}$ will be within $\pm Z_{1-\alpha/2}\{\pi(1 - \pi)/n\}^{1/2}$ of π with probability approximately $1 - \alpha$.

Solving *a.m.* $= Z_{1-\alpha/2}\{\pi(1 - \pi)/n\}^{1/2}$ for *n*, the formula looks exactly as in Equation 18.1, except for the special form of the variance:

$$n = \frac{Z_{1-\alpha/2}^2 \pi(1 - \pi)}{(a.m.)^2} \tag{18.2}$$

Again you have to make some choices. As far as the *a.m.* goes, first note that you are estimating the proportion π, which is on the 0.00–1.00 scale rather than on the 0–100 percentage scale. So if you want a ±3% *a.m.*, you'll need to use *a.m.* = 0.03, not *a.m.* = 3.0. But what value to pick? This one is a little easier than the previous case because percentages are more familiar to everyone. Certainly you don't want an *a.m.* of 0.50; that would be worthless. The number 0.01 sounds great because ±1% is a natural thing to consider on the percentage scale, but that small, an *a.m.* may require too large a sample size. The value *a.m.* = 0.03 is often considered acceptable in opinion polls, but should be revised if the true proportion is very small, such as $\pi = 0.01$, because then an *a.m.* of ±0.03 would admit estimates that are quite far from 0.01. As in the case of the previous Example 18.1, you should consider the size of the *a.m.* relative to the value of the parameter you are estimating.

The choice of α has the same issues as before: There is no one right answer. You want α to be small, but not so small that your sample size is needlessly large.

The choice of σ is now replaced by a choice of π. Good news—this is relatively easy! In the case of burn patients, for example, historical data may suggest something around an 80% survival rate for this cohort, in which case you could substitute the guess 0.80 for π in the formula (18.2).

Using these values, along with $\alpha = 0.05$, you can use Equation 18.2 to get $n = 1.96^2(0.8)$ $(0.2)/(0.02)^2 = 1537$ patients. At this point, you'll probably want to step back and reevaluate your goals because it is doubtful that 1537 patients in that cohort will pass through your burn center any time soon.

One nice thing about Equation 18.2 is that it provides a worst-case scenario for the variance. When $\pi = 0.5$, the variance is maximized at $\pi(1 - \pi) = 0.25$. So you can use 0.5 in Equation 18.2 if you are clueless about π, and this will give you an upper bound for n. In the burn example, this gives you $1.96^2(0.5)(0.5)/(0.02)^2 = 2401$ patients. Even worse! It's better to have a guess at the value of π. But either way, you are going to have to accept more error in your estimate, such as *a.m.* = 0.03 or 0.04, if you want a manageable sample size for this study.

A final note: Example 18.2 is best analyzed using logistic regression. Choosing a sample size for a regression analysis is more complicated, but it gives you accurate estimates at the cohort level with many fewer observations in the cohort (perhaps none at all!), so it is worth the trouble (see Exercise 18.11).

18.3 Power

You can rule out chance when the *p*-value is less than your chosen significance level α. The chance-only model is the null hypothesis, H_0. When you rule out chance, you **reject H_0**. *Power* is simply the probability that you reject the null hypothesis correctly. In symbols, power is expressed as follows:

$$\text{Power} = \Pr(\text{Reject } H_0 | H_0 \text{ is false}) \tag{18.3}$$

Tables 18.1 and 18.2 help you to understand this concept.

Thus, if you incorrectly reject a null hypothesis, it is a **Type I error**. If you fail to reject a hypothesis that you should reject, it is a **Type II error**.

Power is defined as the probability of a correct decision when the null hypothesis is false, shown in Table 18.2.

TABLE 18.1

Hypothesis Testing Terminology

		Action Based on Data	
		Fail to Reject H_0	**Reject H_0**
State of Nature	H_0 **true**	Correct decision	Type I error
	H_0 **false**	Type II error	Correct decision

TABLE 18.2

Conditional Distributions of Actions, Given True States of Nature

		Action Based on Data		
		Fail to Reject H_0	**Reject H_0**	**Total**
State of Nature	H_0 **true**	$1 - \alpha$	α	1.0
	H_0 **false**	$1 - \text{Power}$	Power	1.0

There are two conditional distributions in Table 18.2, one in each row. When the null hypothesis is true, you *don't* want to reject it, and so you set the probability of a Type I error, α, to a small number like 0.05. On the other hand, when the null hypothesis is false, you *do* want to reject it, and so you want the power to be high in that case. Comparing Table 18.1 with Table 18.2, power is specifically the probability of making the correct decision when the null hypothesis is false and is equal to $1 - \Pr(\text{Type II Error})$.

You can use power to choose a sample size n, and you can use it to compare different procedures. When H_0 is false, larger n gives you higher power, so you can choose an n that is just large enough to ensure adequate power. When comparing two different test procedures—for example, a Pearson chi-squared test with a likelihood ratio chi-squared test—you would prefer to use the test with higher power. In Chapter 17, you learned that likelihood ratio tests are *optimal*; specifically, this means that they tend to be *more powerful* than other tests.

While the definition in Equation 18.3 seems simple enough, the devil is in the details. What does "H_0 is false" mean? It means that the chance-only model $p_0(y)$ is not the model. But there are infinitely many models other than $p_0(y)$, and to compute (18.3), you have to specify one particular model $p_1(y)$ that is different from $p_0(y)$. Which one to pick? Again there are many subjective choices you have to make. The following example provides details.

Example 18.3: The Power of a Test for Conformance with a Standard in Quality Control

In Chapter 7 and again in Chapter 17, we presented an example of comparing mean computer chip width to the standard $m_0 = 310$ based on a sample of n chips. Suppose that the quality control engineers are willing to assume that the process is in control when the width measurements Y_i are iid $N(310, 4.5^2)$.

If you sample $n = 10$ chips and compute the average $\bar{Y} = (Y_1 + Y_2 + \cdots + Y_{10})/10$, the natural and acceptable range of variation of the average width measurements $\bar{Y}$ is described by the $N(310, 4.5^2/10)$ distribution, by the linearity and additivity properties of the normal distribution. Figure 18.1 shows these two distributions. Notice that the range of acceptable values of Y_i is much wider than the range of acceptable values of $\bar{Y}$.

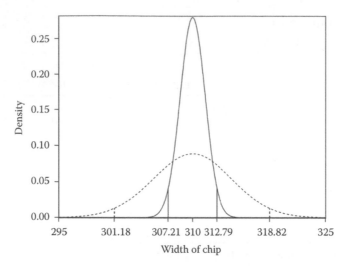

FIGURE 18.1
Distributions showing acceptable variation of individual chip widths (dotted line) and the average width of $n = 10$ chips (solid line), when the process is in control.

The vertical lines in Figure 18.1 are the ±1.96 standard deviation limits: For individual observations, the limits are $310 \pm 1.96(4.5)$, or $(301.18, 318.82)$; for averages of $n = 10$, the limits are $310 \pm 1.96(4.5)/10^{1/2}$, or $(307.21, 312.79)$. Since only 5% of the *individual* observations Y_i are outside the range $(301.18, 318.82)$ when the system is *in* control, an *individual* width observation Y_i that is outside that range is a good indication that the system is *not in* control.

As shown in Figure 18.1, there are different limits for averages. Since only 5% of the *averages* $\bar{Y}$ are outside the range $(307.21, 312.79)$ when the system is *in* control, an *average* $\bar{Y}$ that is outside that range is a good indication that the system is *not in* control. The narrower interval for the average simply reflects the fact that the average $\bar{Y}$ is a more accurate estimate of the process mean than is an individual observation Y_i; that is why averages are commonly used in quality control.

In the hypothesis testing jargon, you will reject H_0—that is, you will reject the null model $N(310, 4.5^2)$—when the average of $n = 10$ observations $\bar{Y}$ is outside the range $310 \pm 1.96(4.5)/10^{1/2}$, or outside the range $(307.21, 312.79)$. When the null model is true, you will incorrectly reject H_0 only with 5% of your samples. Specifically, suppose you take 100 samples of chips, with $n = 10$ chips in each sample, and calculate 100 $\bar{Y}$ values—one for each sample. If the individual chip process is in control—that is, described by the iid $N(310, 4.5^2)$ process—then about 5 of your 100 $\bar{Y}$ values will be outside the range $310 \pm 1.96(4.5)/10^{1/2}$.

Power, as given in Equation 18.3, is the probability of rejecting H_0 when H_0 is *false*. You already know that the rule "Reject H_0 when $\bar{Y}$ is outside the range $(307.21, 312.79)$" has 5% probability of rejecting H_0 when H_0 is true. What about when H_0 is false? Recall that the H_0 model here is $N(310, 4.5^2)$. If this model is false, the distribution could be anything, $N(311, 4.5^2)$, or $N(310, 8.2^2)$, or $N(308, 3.7^2)$, etc. There are infinitely many normal distributions other than $N(310, 4.5^2)$, and there are even more non-normal distributions!

Thus, there is no one number that you can identify as *the* power, and it is therefore impossible to compute *the* power of any test. Instead, power is a what-if calculation that gives different numbers for different scenarios. The distribution $N(311, 4.5^2)$ gives you one value of power; $N(310, 8.2^2)$ gives you another.

While the choice of a distribution $p_1(y)$ is nebulous, power itself is easy to find via simulation, once you do decide on a $p_1(y)$. Suppose you decide to find power when the distribution $p_1(y)$ of the widths is the $N(311, 4.5^2)$ distribution. You can find the power by simulating $n = 10$ observations $Y_1^*, Y_2^*, ..., Y_{10}^*$ from $N(311, 4.5^2)$, calculating the average

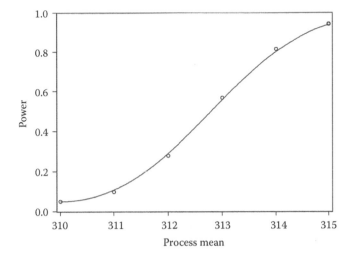

FIGURE 18.2
Power of the test of the N(310, 4.5²) process using a sample of $n = 10$ observations. The circles are based on 1000 simulated averages of $n = 10$ observations. The solid line is from theoretical calculations using the normal cdf.

$\bar{Y}^*$, and checking whether it is in the interval (307.21, 312.79) shown in Figure 18.1. Using Microsoft Excel with 1000 simulated sets of 10 observations, we found 98 of the 1000 resulting $\bar{y}^*$ values were outside the range (307.21, 312.79), leading to a power estimate of 98/1000, or 0.098. This is pretty low! It tells you and the engineers that if the process has slipped to the point where the average chip width is really 311 instead of 310, you will correctly identify that the process is out of control only with about 10% of your samples. For the remaining 90% of your samples, there will be no evidence of the problem, and you will incorrectly conclude that the process is in control.

Repeating this simulation with true means $m_1 = 311, 312, 313, 314, 315$ yields estimates 0.098, 0.278, 0.568, 0.816, and 0.943, as shown in Figure 18.2.

Notice in Figure 18.2 that the power of the test is higher with greater slippage of the process. This makes sense: The average $\bar{Y}$ is an estimator of the true process mean, so, for example, if the process mean has slipped to $m_1 = 315$, then the average $\bar{Y}$ will tend to be near 315 and hence likely outside the range (307.21, 312.79) that is predicted under the null model.

All the answers are in the graphs! To see why the power is 0.94 at $m_1 = 315$, in Figure 18.2, have a look at Figure 18.3. Two distributions are shown: the distribution of $\bar{Y}$ when the true mean is 310 (also shown as the solid curve in Figure 18.1 and the distribution of $\bar{Y}$ when the true mean is $m_1 = 315$. Power is the probability of rejecting the null model, that is, the probability of observing $\bar{Y}$ outside the range of (307.21, 312.79) when the alternative model is true, and is therefore the shaded area in Figure 18.3. (There is also a tiny bit of the probability to the left of 307.21 under the alternative distribution graph, but it is so small that it is negligible.)

You can calculate the power shown in Figure 18.3 exactly; simulation is not necessary. For example, in Microsoft Excel, use the formula

$$= \text{NORM.DIST}(307.21, 315, 4.5/\text{SQRT}(10), \text{TRUE})$$

$$+ (1 - \text{NORM.DIST}(312.79, 315, 4.5/\text{SQRT}(10), \text{TRUE}))$$

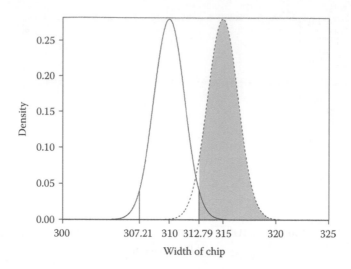

FIGURE 18.3
Null (solid line) and alternative (dotted line) distributions of $\bar{Y}$ (solid lines). Power of the test when $m_1 = 315$ is shown as the shaded area.

This gives you Power = 0.94. However, you should *think* about power in terms of simulation: If power is 94% for a particular alternative model, then if that model is true, you will correctly reject the null model with 94% of the potential data sets that you will sample. Simulation makes this interpretation clear because you can actually see what those potential future data sets look like and you can verify for yourself how often you will reject the null hypothesis with those data sets. In addition, simulation provides an easy way to answer more difficult questions, such as "What happens to the power of the test when the distribution is not a normal distribution?" We show you how to simulate power using non-normal distributions in Chapter 19, so you can assess the robustness of your methods when the normality assumption is false.

While simulation is great for understanding power, the analytic formula for calculating power gives you a simple formula that tells how power depends on all the various quantities. In the quality control example given earlier, you will reject the null hypothesis at the α level of significance if $\bar{Y} \geq m_0 + z_{1-\alpha/2}s_0/n^{1/2}$ or if $\bar{Y} \leq m_0 - z_{1-\alpha/2}s_0/n^{1/2}$, where $m_0 = 310$, $s_0 = 4.5$, $n = 10$, and $\alpha = 0.05$. If the alternative model $N(m_1, s_1^2)$ is true, then you can transform $\bar{Y}$ to a standard normal random variable via $Z = (\bar{Y} - m_1)/(s_1/n^{1/2})$. Hence, by algebra, the power is

$$\text{Power} = \Pr\left(Z \geq \frac{n^{1/2}(m_0 - m_1)}{s_1} + \frac{z_{1-\alpha/2}s_0}{s_1} \right) + \Pr\left(Z \leq \frac{n^{1/2}(m_0 - m_1)}{s_1} - \frac{z_{1-\alpha/2}s_0}{s_1} \right) \quad (18.4)$$

Equation 18.4 is useful particularly because it shows you how power depends on the sample size, n; on the α-level; and on the standard deviations.

In the case $m_0 = 310$, $m_1 = 315$, $s_0 = 4.5$, $s_1 = 4.5$, $n = 10$, and $\alpha = 0.05$, Equation 18.4 gives an alternative, but equivalent result as the Microsoft Excel calculation

$$\text{Power} = \Pr\left(Z \geq \frac{10^{1/2}(310 - 315)}{4.5} + 1.96 \right) + \Pr\left(Z \leq \frac{10^{1/2}(310 - 315)}{4.5} - 1.96 \right)$$
$$= \Pr(Z \geq -1.5536) + \Pr(Z \leq -5.4736)$$
$$= 0.9399 + 2 \times 10^{-8} = 0.94$$

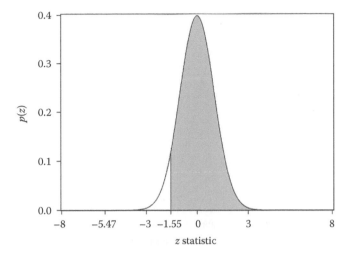

FIGURE 18.4
Distribution of standardized average in the quality control example, with power shown as shaded area.

Figure 18.4 shows this power calculation represented using the standard normal (or Z) distribution.

Suppose in Equation 18.4 that $m_1 > m_0$, as in the quality control example where $315 > 310$. Then the power of the test is directly related to the following term:

$$z = \frac{n^{1/2}(m_0 - m_1)}{s_1} + \frac{z_{1-\alpha/2}s_0}{s_1}$$

The smaller (further to the left in Figure 18.4) is z, the higher is the power, because there will be more area to the right of z as shown in Figure 18.4. Looking at how z depends on n, m_1, s_0, and α, you can then conclude the following.

What Affects the Power of a Test

- All else fixed, a larger n implies a higher power. If you take a larger sample size, you are more likely to reject the null hypothesis.
- All else fixed, a larger m_1 implies a higher power. The farther the alternative mean is from the null hypothesis mean, the more likely it is that you will reject the null hypothesis.
- All else fixed, a larger α implies a higher power. If you are willing to reject the null hypothesis more often, the interval range will be narrower, you will reject the null hypothesis more often, and power will be higher.
- All else fixed, a smaller s_0 implies a higher power. When standard deviation is lower, the interval range will be narrower, and you will reject the null hypothesis more often.

When $m_1 < m_0$, the graph of Figure 18.4 would show power as the area to the *left* of z, and all the same results hold when except for the second bullet, where smaller m_1 implies larger power. Either way, the main point is that when the alternative mean is farther from the null mean, power is higher.

Of special interest to design and measurement is the first bullet point. When all else is fixed, you can choose n to achieve whatever power you desire (presumably a high number).

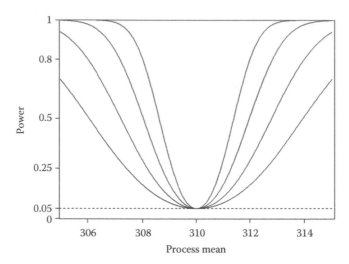

FIGURE 18.5
Power of the quality control tests when n = 5, 10, 20, and 40 (lowest to highest powers, respectively).

Figure 18.5 shows an expanded analysis of Figure 18.2, using different sample sizes n and with power calculations as given by Equation 18.4.

As indicated by Figure 18.5, you can choose n to make the power as large as you want, except in the null case where the mean is 310. In the null case, the probability of rejecting the null hypothesis is equal to the significance level (0.05 in Figure 18.5, indicated by the dotted line) for all n.

Example 18.4: The Power of a Test for Comparing Cure Rates

In the introduction, we discussed a case where you might wish to detect whether one treatment had a 2% higher cure rate than another. So if $\pi_1 - \pi_0 = 0.02$, how likely is it that you can rule out chance as the reason for the observed difference in the data? As always, you need to start with a model (*model produces data!*). Assume the data are iid Bernoulli(π_j) in groups $j = 0$ and $j = 1$ and that the groups are independent of each other. Then under the null hypothesis ($\pi_0 = \pi_1 = \pi$), the difference between sample means $\bar{Y}_1 - \bar{Y}_0 = \hat{\pi}_1 - \hat{\pi}_0$ is approximately normally distributed with mean zero and variance $\sigma^2(1/n_1 + 1/n_2)$, as derived in Chapter 15. For the Bernoulli distribution, the variance is $\sigma^2 = \pi(1 - \pi)$, and the overall average $\bar{Y} = \hat{\pi}$ is a consistent estimate (by the law of large numbers) of π under the null model, giving the following result:

$$Z = \frac{\hat{\pi}_1 - \hat{\pi}_0}{\sqrt{\hat{\pi}(1 - \hat{\pi})\{1/n_1 + 1/n_2\}}} \dot\sim N(0,1), \quad \text{under the iid Bernoulli}(\pi) \text{ null model}$$

$$(18.5)$$

Hence, the approximate p-value is $pv \cong 2 \times \Pr(Z \geq |z|)$, where $Z \sim N(0, 1)$, and the p-value is less than 0.05 whenever $|z| > 1.96$. So the power of the test is equal to the probability that the absolute value of the Z statistic in (18.5) exceeds 1.96.

Again, you can easily evaluate the power via simulation. Simply simulate many potential data sets having n_1 and n_0 patients, calculate the Z^*-statistic as in Equation 18.5 from the simulated data, and note whether $|Z^*| > 1.96$. For instance, suppose the true cure rates are 32% and 30% for the new and existing therapies and 100 patients in each group are sampled. You can use Microsoft Excel or other software that allows you to simulate

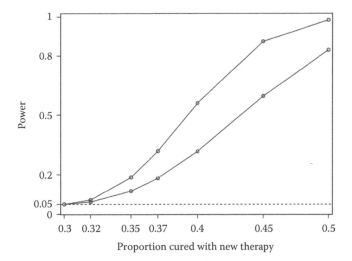

FIGURE 18.6
Power of the two-sample binomial test when there are 30% cured in the control group and either $n = 100$ patients per group (lower curve) or $n = 200$ patients per group (upper curve).

data, generating 100 Bernoulli random numbers with a 0.32 probability and another 100 Bernoulli random numbers with a 0.30 probability, calculating Z^*, and comparing it to 1.96. Repeating this process a large number of times, say 100,000, the power is the proportion of the 100,000 simulated studies yielding $|Z^*| > 1.96$.

Figure 18.6 shows the simulated power when the control therapy has a 30% cure rate and when the new therapy has a 30%, 32%, 35%, 40%, and 50% cure rate, when there are 100 patients per group. Also shown is what would happen if you had sampled 200 patients in each group.

It is unlikely that you will be able to rule out chance when the true cure proportions are 32% and 30%: Whether you have 100 per group or 200 per group, you will (incorrectly) conclude that the results are explainable by chance alone in more than 90% of your studies. In other words, you will commit a Type II error (see Table 18.1) in more than 90% of your studies, due to inadequate sample size. So you will need a much larger sample size in this case.

On the other hand, if the new therapy is much better than the old, say, with a 50% cure rate, Figure 18.6 shows that you will correctly rule out chance as a possible explanation of the difference you observe in your sample data more than 80% of the time, with just 100 patients per group.

18.4 Noncentral Distributions

Test statistics are often distributed as standard normal ($N(0, 1)$), student's t (T_{df}), chi-squared (χ^2_{df}), or F (F_{df_1, df_2}) when the null hypothesis is true. These distributions are centered at 0 for normal and t-distributions, at degrees of freedom (df) for the chi-squared distribution, and around 1.0 for the F-distribution. When used as null distributions for test statistics, these distributions are called **central distributions**.

When the null hypothesis is false, the distributions change: Their centers shift, resulting in **noncentral distributions**. The parameter that determines the extent of the shift is called the **noncentrality parameter**, abbreviated *ncp*, and denoted by the lowercase letter delta, δ. (This is not same number as the treatment effect δ defined earlier—sorry! There are only so many symbols.)

Here are the most famous noncentral distributions and their representations.

Noncentral Distributions

- *Noncentral Z-distribution*: The N(δ, 1) distribution. The parameter δ is the ncp.
- *Noncentral student's t-distribution*: The distribution of a noncentral Z (with parameter δ) divided by an independent $\sqrt{\chi^2_{df}/df}$ random variable.
- *Noncentral chi-squared distribution*: If $Y_i \sim_{\text{independent}} N(\mu_i, 1)$, $i = 1, 2, ..., k$, then ΣY_i^2 is distributed as noncentral chi-squared with k df and ncp $\delta = \Sigma \mu_i^2$.
- *Noncentral F-distribution*: The distribution of $\left\{ \chi^2_{df_1, \delta}/df_1 \right\}/\left\{ \chi^2_{df_2}/df_2 \right\}$, where the numerator and denominator are independent, is the noncentral F-distribution.

For example, in the quality control case, the distribution of the sample average $\bar{Y}$ under the null model $Y_1, Y_2, ..., Y_n \sim_{\text{iid}} N(310, 4.5^2)$ is $\bar{Y} \sim N(310, 4.5^2/n)$. Standardizing, you get $Z = (\bar{Y} - 310)/(4.5/n^{1/2}) \sim N(0, 1)$ under the null model; hence, you will reject the null model when $|Z| \geq 1.96$. Under an alternative model, the distribution of Z is no longer N(0, 1). For example, suppose that the data are produced as $Y_1, Y_2, ..., Y_n \sim_{\text{iid}} N(315, 4.5^2)$. Then the distribution of Z is given as follows:

$Z = (\bar{Y} - 310)/(4.5/n^{1/2})$	(By definition)
$= (\bar{Y} - 315 + 315 - 310)/(4.5/n^{1/2})$	(By adding and subtracting 315)
$= (\bar{Y} - 315)/(4.5/n^{1/2}) + (315 - 310)/(4.5/n^{1/2})$	(By algebra)
$= Z_0 + (315 - 310)/(4.5/n^{1/2})$, where $Z_0 \sim N(0, 1)$	(By the linearity and additivity properties of the normal distribution)

Hence, under the alternative model N(315, 4.5^2), the distribution of Z is N(δ, 1), where the ncp is given by $\delta = (315 - 310)/(4.5/n^{1/2})$. Figure 18.7 shows the central Z (or N(0, 1))

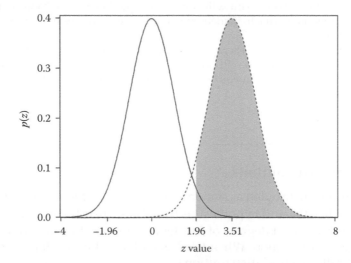

FIGURE 18.7

Null (central, solid line) and alternative (noncentral, dotted line) distributions of the Z statistic in the quality control example.

distribution as well as the noncentral distribution when $n = 10$. The rejection limits ± 1.96 are also shown, as well as the power of the test as a shaded region. Notice that the power of the test is expressed in terms of the ± 1.96 rejection limits from the null model and the probability distribution of the test statistic under the alternative model.

If your test statistic W has one of the standard forms—normal, student's t, chi-squared, or F—then the distribution of W under the alternative hypothesis has the same form but is noncentral. The ncp δ is equal to the value of W, but with all parameter *estimates* replaced by the corresponding parameter *values* under the alternative model.

In the quality control example, the test statistic is $Z = (\bar{Y} - 310)/(4.5/10^{1/2})$; replacing the estimate $\bar{Y}$ with the alternative mean $\mu_1 = 315$ gives $\delta = (315 - 310)/(4.5/10^{1/2}) = 3.51$ as shown in Figure 18.7.

With the noncentral student's t-distribution, the ncp is slightly smaller than the true mean because of Jensen's inequality. Figure 18.8 shows the noncentral student's t-distributions in the quality control example, with $df = 9$ and with ncp values $\delta = 10^{1/2}(m_1 - 310)/4.5$, where m_1 is either 310, 312, 315, or 320. The case $m_0 = 310$, where $\delta = 0$, is the ordinary central T_9 distribution. While the ncp of student's t-distribution is different from the mean, it is very close to its mean as you can see in Figure 18.8.

Noncentral distributions allow you to calculate power analytically—that is, without simulation—if your software computes the cdfs of these distributions. But you can always simulate power using spreadsheet or other software that allows you to simulate normal random variables.

In the quality control case with unknown variance, with $n = 10$ observations and the test statistic $T = (\bar{Y} - 310)/(\hat{\sigma}/n^{1/2})$, your p-value will be less than 0.05 whenever $|T| > T_{9,\,0.975} = 2.262$. So Power $= \Pr(|T| > 2.262)$, where T has the noncentral t-distribution with 9 df and ncp $\delta = (315 - 310)/(\sigma/n^{1/2})$. Figure 18.9 corresponds to Figure 18.7, where the null variance is known, but uses the model where the null variance is unknown, leading to a t-statistic. You get less power using the t-distribution than you do with the normal distribution: $\Pr(|T| > 2.262) = 0.877$, versus 0.94 for the normal distribution. While its power is less, the benefit of using student's t-test is that you do not have to prespecify σ. On the other hand, you do need to specify σ to calculate the power of the t-test; in Figure 18.9, it is specified as $\sigma = 4.5$.

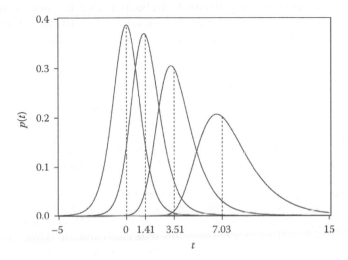

FIGURE 18.8
Graphs of central and noncentral T_9 distributions, with ncps indicated on the t axis with dotted lines.

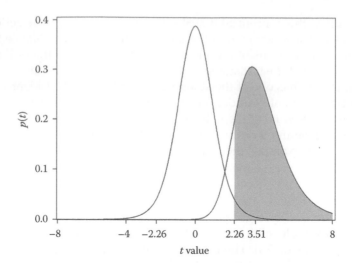

FIGURE 18.9
Central and noncentral t-distributions for the quality control example. The power of the test, 0.877, is indicated by the shaded area.

18.5 Choosing a Sample Size for Prespecified Power

If you were designing the study, what n of computer chips would you sample to determine that the chip manufacturing process is out of control? If you were designing the medical study to see whether the new therapy has a higher cure rate than the standard therapy, what n per group of patients would you choose to determine whether the new therapy is better? As Figures 18.5 and 18.6 show, a larger n will improve power, but the first question you should ask is, "How large a power do I need?" Also, as shown in Equation 18.4, power depends on more than just sample size, so you also need to ask "What values do I pick for α? For m_1? For s_1?" These are the same questions that you had to address when choosing a sample size for an acceptable *a.m.*, discussed in Section 18.2, but now you'll also have to prespecify the power of the test. And again, you need to think through the specifics of the study very carefully. We'll walk you through the details in the quality control case.

> **Example 18.5: Selecting a Sample Size for a Quality Control Test**
>
> Again suppose 310 is the target width. Presumably, if the machines were making chips whose widths differed on average from 310, you'd like to know about it, so you could fix the problem. That means you would like to rule out chance as the explanation for the difference between the sample average and 310, if in fact the process mean really differs from 310—as might happen, for example, if a machine has experienced sudden wear or breakage.
>
> But as shown in Figure 18.5, power can be very low when the process mean differs from 310 by just a little, even for large sample sizes. So the first thing that you and the quality control (QC) staff need to do is to decide on a **smallest meaningful difference** d. If the absolute difference between the true mean and 310 is more than d, then the QC staff must take corrective action; otherwise, no action is required. Clearly, a d as small as 0.0001 is not a concern: Who cares whether the process mean is 310.0001 or 310.0000? On the other hand, the engineers might tell you that if the process mean

slips to 315, then half of the product will be scrap. So the smallest meaningful difference is somewhere between 0.0001 and 5.0. What is it specifically? It's your choice. There is no one right answer. Like the *a.m.*, you can propose a few numbers, do some calculations, and take them to the boardroom to discuss with the quality control and engineering groups. A nice round number like $d = 1.0$ sounds like a good place to start, so let's use it for now.

As before, the α term is the error probability. In quality control applications, it is a very bad idea to stop the production process and check for problems needlessly, so the α is typically set much lower than 0.05 in QC applications. Instead, three standard deviations, or $\alpha = 0.003$, is a more common default. But there is no right answer. Instead, you should put some numbers on the table and have a conversation with the board about the pros and cons of the different options.

The σ parameter, again, is something that you simply have to guess. In the chip manufacturing application, perhaps historical data suggest that σ is near 4.5, in which case this could be your guess. But again there is no right answer here either.

If the process has slipped to a mean with the smallest meaningful difference of $d = 1.0$, then the mean is either 211 or 209, and you'd like your test to detect such slippage. Assuming the mean is 211—the end result will be the same if the mean has slipped to 209 because the tests are two-sided—you may be willing to assume that the data are produced as iid $N(211, 4.5^2)$. The ncp of the *t*-statistic is then $\delta = (211 - 210)/(4.5/n^{1/2})$, which you can use to choose n so that the power, which in this case is equal to $\Pr(|T| > T_{n-1,0.975})$, is sufficiently high.

One final choice: What is a "sufficiently high" power? You want to conclude that there is more than chance at work when the true mean slips to 311, but how certain do you wish to be that you will make this conclusion? Numbers like 90% and 95% seem generically reasonable, as these are the numbers used throughout this book to denote "high confidence" in the credible and confidence intervals. It's your call. However, one thing you should know is that if you want higher power, you'll need a larger n: Just look at Figure 18.5. In many circles—for example, for grant applications—it has become customary to use 80% as an acceptably high power figure. This means that if the process has slipped to a mean that differs from the target by the amount of the smallest meaningful difference, then you will conclude that the difference is more than can be explained by chance alone in four out of five samples, and in the remaining one out of five samples, you will conclude that the difference is explainable by chance alone.

If you have software that calculates the power of the noncentral *t*-distribution, you can evaluate Power = $\Pr(|T| > T_{n-1,0.975})$ for different n and choose the smallest n giving you 0.80 power. If you do not have access to such software, you will find free power calculators on the Internet to do this for you. Figure 18.10 shows the power of the quality control test for different n, using the smallest meaningful difference $d = 1.0$, along with a guess of 4.5 for σ, using $\alpha = 0.05$.

As Figure 18.10 shows, you'll need a sample size of $n = 161$ chips to reliably detect a significant difference when the process mean has slipped to 311 (or to 309). Perhaps this sample size is too high, because of the logistics of the production facility, to be implemented successfully. If so, you'll have to revise your inputs. The most logical course of action would be to choose a larger smallest meaningful difference, in consultation with the engineers. For example, if you change the smallest meaningful difference to 2.0, then you'll only need $n = 42$ observations. While that sounds great, the downside is that in cases where there is truly a one-unit slippage in the process mean, the difference between the sample mean and 310 will most likely be classified as "explainable by chance."

There are always trade-offs in statistics. When you understand these trade-offs, you'll know most of what you really need to know.

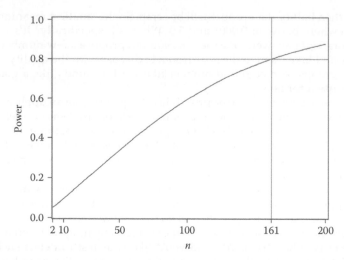

FIGURE 18.10
Power of the quality control test for detecting a smallest meaningful difference of 1.0 unit from the target 310.

18.6 Post Hoc Power: A Useless Statistic

After reading Sections 18.1 through 18.5 of *Understanding Advanced Statistical Methods*, Hans was despondent. He was hoping power calculation was easier, something you could simply let software give you from a data set to satisfy those pesky reviewers. So he downloaded the latest update of his favorite statistical software, *HappyStat® Version 1.2*, in hopes of finding something simpler. He was delighted to find a button labeled "Post-Hoc Power," which computes power from any given data set. Plug in the data set, the analysis method, and voilà! You get the power.

What on Earth is *HappyStat®* doing? As you have seen, power calculation requires that you specify values for unknown parameters, such as μ, σ, π, and $\mu_1 - \mu_2$. Since these values are always unknown (*model has unknown parameters*), you have to simply guess at these values and then perform sensitivity analysis because you know your guesses are wrong. Hans doesn't like ambiguity, and that's what bothered him about the previous sections of this chapter. *HappyStat®* computes post hoc power, also called *retrospective power*, which takes the guesswork out of these calculations by substituting the parameter estimates $\hat{\mu}$, $\hat{\sigma}$, $\hat{\pi}$, $\hat{\mu}_1 - \hat{\mu}_2$, etc., calculated from the data set, and proceeding as if they were the true values. While this might seem like a good idea, it's actually a useless and counterproductive practice.

To see why, consider the quality control example given earlier, where for simplicity's sake we assume an upper-tailed test with $\sigma = 4.5$. The rejection region for testing H_0: $\mu = 310$ is thus $\bar{Y} \geq 310 + 1.645(4.5)/n^{1/2}$ and Power $= \Pr(\bar{Y} \geq 310 + 1.645(4.5)/n^{1/2}\,|\,\mu)$, which depends on the true, unknown parameter μ, similar to what is shown in Figure 18.5. With post hoc power, you estimate the mean from the data, getting $\hat{\mu} = \bar{y}$, and proceed with the calculation. For example, suppose that $n = 10$ and $\bar{y} = 312.0$. Then you assume that $\bar{Y} \sim N(312, 4.5^2/10)$ and find that $\Pr(\bar{Y} \geq 312.3409) = 0.405$. "Aha!" says Hans. "The power, 0.405, is too low. No wonder I didn't reject the null hypothesis!"

A little algebra shows why this method is useless. Suppose the sample mean is $\bar{y}$. Then the post hoc power calculation gives

$$\text{Power} = \Pr\left(\bar{Y} \geq 310 + \frac{1.645(4.5)}{n^{1/2}} \middle| \mu = \bar{y}\right)$$

$$= \Pr\left\{\frac{(\bar{Y} - \bar{y})}{(4.5/n^{1/2})} \geq \frac{(310 - \bar{y})}{(4.5/n^{1/2})} + 1.645 \middle| \mu = \bar{y}\right\}$$

$$= \Pr\left\{Z_0 \geq 1.645 - \frac{(\bar{y} - 310)}{(4.5/n^{1/2})}\right\}$$

$$= \Pr(Z_0 \geq 1.645 - z_0) \tag{18.6}$$

Here, Z_0 is a $N(0, 1)$ random variable, and z_0 is the test statistic, $z_0 = (\bar{y} - 310)/(4.5/n^{1/2})$.

From Equation 18.6, you can see that power <0.5 when $z_0 < 1.645$; that is, post hoc power is less than 50% when you fail to reject the null hypothesis. Also from Equation 18.6, you can see that power >0.5 when $z_0 > 1.645$; that is, post hoc power is more than 50% when you reject the null hypothesis. So post hoc power provides no new information—it simply tells you whether or not you have rejected the null hypothesis, which you already knew anyway. It just gives you a repackaged test statistic (the post hoc power statistic) and a different critical value (50%).

Problems with the Post Hoc or Retrospective Power Statistic

- It is a monotonic function of the test statistic. As shown in Chapter 17, test statistics that are monotonically related provide equivalent information. So post hoc power gives you nothing new over and above the test statistic. Strike one.
- It conflates *parameter* with *parameter estimate*. Making the substitution of the estimate for the true value and ignoring the consequences provides misleading results and promotes confusion. Strike two.
- The interpretation of post hoc power is vacuous: It tells you what would happen if the parameter values were equal to their estimates. But the parameter values are never equal to their estimates, so the post hoc power calculation is irrelevant. Strike three.

On the other hand, it is a very good idea to use the parameter estimates from your current study to *suggest* values (although not the same as the estimates, since these are not the true values) that you can use to calculate power for *other* studies. And it is especially a good idea to use such suggested values along with different sample sizes, again for the purpose of evaluating other studies. But to use the existing data and sample size to calculate power for the current study—that is truly useless. It is also counterproductive because it promotes confusion about the subject of statistics. Don't promote confusion. Don't be a trained parrot.

But don't just take our word for it either. Read Hoenig and Heisey's paper "The Abuse of Power: The Pervasive Fallacy of Power Calculations for Data Analysis" in *The American Statistician* 55(1) (2001).

Vocabulary and Formula Summaries

Vocabulary

a.m.	The maximum difference (with a stated probability level) between your estimate and the estimand.
Reject H$_0$	To rule out the null hypothesis—that is, chance—as an explanation of observed results.
Type I error	An incorrect rejection of a true null hypothesis.
Type II error	An incorrect decision to fail to reject a false null hypothesis.
Significance level	The probability of committing a Type I error.
Power	The probability of correctly rejecting a false hypothesis or $1 - \text{Pr(Type II error)}$.
Noncentral distribution	The distribution of a test statistic when the alternative hypothesis is true.
Central distribution	The distribution of a test statistic when the null hypothesis is true.
ncp	The parameter that determines how far the noncentral distribution is from the central distribution.
Smallest meaningful difference	The smallest difference that is still important between a target value and the actual value in a process.
Post hoc power	Power calculated using parameter estimates from the current data set. Also called retrospective power. A useless statistic.

Key Formulas and Descriptions

$a.m. = Z_{1-\alpha/2}\, \sigma/n^{1/2}$	The maximum distance from $\overline{Y}$ to μ, with $1 - \alpha$ probability.
$n = Z_{1-\alpha/2}^2 \sigma^2/(a.m.)^2$	The sample size required for estimating a mean μ with *a.m.*
$\hat{\pi} \sim \text{N}\{\pi,\, \pi\,(1-\pi)/n\}$	The approximate distribution of the sample proportion when computed from an iid Bernoulli(π) sample.
$a.m. = Z_{1-\alpha/2}\{\pi\,(1-\pi)/n\}^{1/2}$	The maximum distance from $\hat{\pi}$ to π, with $1 - \alpha$ probability.
$n = Z_{1-\alpha/2}^2 \pi(1-\pi)/(a.m.)^2$	The sample size required for estimating a proportion π with *a.m.*

Power $= \Pr(Z \geq n^{1/2}(m_0 - m_1)/s_1 + z_{1-a/2}s_0/s_1)$
$\qquad + \Pr(Z \leq n^{1/2}(m_0 - m_1)/s_1 - z_{1-a/2}s_0/s_1)$

The power of a test for H_0: $\mu = m_0$, assuming alternative mean m_1 and null and alternative standard deviations s_0 and s_1.

$Z = (\hat{\pi}_1 - \hat{\pi}_0)/\sqrt{\hat{\pi}(1 - \hat{\pi})\{1/n_1 + 1/n_2\}} \sim N(0,1)$

The approximate null distribution of the standardized difference between sample proportions when sampling from the iid Bernoulli (π) model.

Exercises

18.1 Hans claims to have a solution to the fair coin dilemma. He'll flip it 10 times, and if the number of heads is 0, 1, 2, 8, 9, or 10, he'll call the coin biased. Use simulation to evaluate Hans' procedure, by generating 10,000 sets of 10 flips for various values of π from the Bernoulli distribution.

    A. Using $\pi = 0.5$, estimate the probability of a Type I error using your 10,000 simulations.

    B. Using $\pi = 0.6, 0.7, 0.8, 0.9$, and 1.0, estimate and graph the power function of Hans' procedure using 10,000 simulations for each π.

    C. How should Hans change his procedure with 100 flips, if he wants his Type I error rate to be the same as in this study? Use the result that $\hat{\pi} \sim N\{\pi, \pi(1 - \pi)/n\}$.

    D. Repeat Exercise 18.1B using the revised procedure in Exercise 18.1C, and comment on the effect of sample size on power.

18.2 You want to estimate the proportion of students who own cars.

    A. Choose and defend an *a.m.* As part of your answer, also choose an *a.m.* that you think is too large and explain why.

    B. Choose and defend an α level. As part of your answer, also choose an α that you think is too large and explain why.

    C. Choose and defend a guess as to the true value of the proportion. As part of your answer, also choose another value of the proportion that is obviously wrong and explain why.

    D. Use your guesses in Exercise 18.2A through C to decide on a sample size. Perform a sensitivity analysis to determine how sensitive your answer is to your initial guesses as well.

18.3 You want to estimate the average number of cups of coffee drunk by students in a week at a university.

    A. Choose and defend an *a.m.* As part of your answer, also choose an *a.m.* that you think is too large and explain why.

    B. Choose and defend an α level. As part of your answer, also choose an α that you think is too large and explain why.

C.   Choose and defend a guess as to the true value of the average. As part of your answer, also choose another value of the average that is obviously wrong and explain why.

D.   Use your guess of the average, as well as the relationship between the mean and variance of the Poisson distribution, to arrive at a guess of the variance of the number of cups of coffee drunk per week.

E.   Use your guesses in Exercise 18.3A through D to decide on a sample size. Perform a sensitivity analysis to determine how sensitive your answer is to your initial guesses as well.

18.4   Show that the ncp of the noncentral t-distribution is smaller than its mean using the representation $T_\delta = N(\delta, 1)/\sqrt{\chi^2_{df}/df}$. Use Jensen's inequality, as well as the product rule, for expectation of independent random variables.

18.5   Figure 18.5 illustrates how a result can be statistically significant but not practically significant.

A.   What will happen to the power graph as n gets even larger?

B.   What happens to the probability of rejecting H_0: $\mu = 310$ if the true mean is 310.001, for extremely large n?

18.6   Figure 18.5 is a good way to visualize effects of various terms, other than n, on power.

A.   Redraw Figure 18.5, with n fixed at 10, but with the various curves denoting different true standard deviations.

B.   Redraw Figure 18.5, with n fixed at 10 and with the standard deviation fixed at 4.5, but with the various curves denoting different values of α.

18.7   An agricultural researcher believes that a new fertilizer formulation increases cotton yield average by 10%. Current yields are 20 per plant with a standard deviation of 10. The researcher believes that if the mean increases by 10%, so does the standard deviation. The researcher is planning to perform the usual two-sample t-test on 100 plants in each of the two groups.

A.   Simulate 1000 studies, with each study having 100 in each group, assuming the researcher's claim of a 10% increase. For each of these 1000 studies, find the p-value for the two-sample t-test. Estimate the power of the test as the proportion of the 1000 studies where the $pv \leq 0.05$.

B.   Repeat 18.7A with other sample sizes until you get acceptable power, and write up your analysis, as if for the agricultural researcher's consideration.

18.8   You want to send out a survey via e-mail. One question on the survey asks whether people prefer online or in-store shopping, on a 1, 2, 3, 4, 5 scale where 1 denotes complete preference for online shopping and 5 denotes complete preference for in-store shopping. How many e-mails will you send? Write a report, applying the concepts of this chapter and accommodating the effect of nonresponse. Pay particular attention to the determination of σ as it relates to the 1–5 scale data.

18.9   Consider the infamous stoplight example, where the distribution of time the light stays green is $U(0, \theta)$. To test H_0: $\theta = 2.0$, you decide to measure one Y and to reject H_0 if $Y < 0.20$.

A.   Find the probability of making a Type I error.

B.   Find the probability of making a Type II error when $\theta = 1.0$.

C.   Find the power when $\theta = 1.0$.

D. Find the approximate 0.10 quantile of null distribution of $\bar{Y}$ based on an iid sample of 10 observations (use the CLT).

E. Find the approximate power of the test suggested by Exercise 18.9D when $\theta = 1.0$.

18.10 You are planning to compare species distributions on irrigated land versus nonirrigated land using the Pearson chi-squared test. The data you collect are $Y_{11}, Y_{12}, \ldots, Y_{1n_1}$, where n_1 is the number of animals identified in irrigated land and each Y is the species of the animal, either A, B, or C. Similarly, you will have data $Y_{21}, Y_{22}, \ldots, Y_{2n_2}$ from nonirrigated land. Based on past research, you think the distributions might look like this:

	Species A	Species B	Species C	Total
Irrigated	0.40	0.40	0.20	1.00
Nonirrigated	0.20	0.30	0.50	1.00

A. Simulate $n_i = 20$ observations from each group, so you have a total of 40 observations. Calculate the chi-squared test statistic, the critical value from the chi-squared distribution, and the p-value.

B. Using the $pv \le 0.05$ criterion, locate your decision of Exercise 18.10A in Table 18.1.

C. Repeat Exercise 18.10A 100 times, and estimate the power of your test using the 100 simulated data sets.

D. Repeat Exercise 18.10C but using 40 observations per group.

18.11 Suppose you plan to estimate burn patient mortality for patients with 50% burn using logistic regression. What sample size should you use? Suppose the model is $\Pr(\text{death} | \text{burn percentage} = x) = \pi(x) = \exp(\beta_0 + \beta_1 x)/\{1 + \exp(\beta_0 + \beta_1 x)\}$.

A. Suppose $\beta_0 = -4.3$ and $\beta_1 = 0.08$. Generate $n = 100$ X values from the U(0, 100) distribution, and for each $X = x$ so generated, generate a Y that is Bernoulli(π (x)). Then estimate the β values using maximum likelihood, and calculate $\hat{\pi}(50)$ $= \exp(\hat{\beta}_0 + \hat{\beta}_1(50))/\{1 + \exp(\hat{\beta}_0 + \hat{\beta}_1(50))\}$.

B. Repeat Exercise 18.11A many times (at least 100) and summarize the resulting estimates using a histogram. Based on the variation in the estimates, do you think $n = 100$ is an adequate sample size?

C. Repeat Exercise 18.11A and B using $n = 400$ observations. Does $n = 400$ provide more accurate estimates than $n = 100$? Do you think $n = 400$ is an adequate sample size?

D. The true parameter setting is not $(\beta_0, \beta_1) = (4.3, 0.08)$. Draw graphs of the logistic regression function $\pi(x) = \exp(\beta_0 + \beta_1 x)/\{1 + \exp(\beta_0 + \beta_1 x)\}$ for that parameter setting, as well as for several other parameter settings. Based on the graphs, pick two other parameter settings that seem plausible for predicting burn patient mortality, and perform sensitivity analysis for Exercise 18.11C using those settings.

E. Consider the U(0, 100) model for producing burn percentage data. Draw a graph of this distribution. When would this model be reasonable? When would it not be reasonable?

18.12 Consider Example 16.7 of Chapter 16, concerning the study of Alzheimer's drugs. Suppose also that you have not conducted the study yet, but you think that treatments 4 and 5 are the best and you want to select sample sizes to ensure that the F-test will reject the null hypothesis with high probability. Based loosely on prior studies, you decide to pick means $\mu_1 = \mu_2 = \mu_3 = 50$, $\mu_4 = \mu_5 = 60$, and $\sigma = 10$.

A. Simulate $n_i = 5$ observations from each group so that you have a total of 25 observations. Calculate the F-statistic, the critical value from the F-distribution, and the p-value.

B. Using the $pv \leq 0.05$ criterion, locate your decision from Exercise 18.12A in Table 18.1.

C. Repeat Exercise 18.12A 100 times, and estimate the power of your test using the 100 simulated data sets.

D. Repeat Exercise 18.12C but using $n_i = 10$ observations per group.

E. Using software that has the noncentral F-distribution, find the exact powers of the tests with 5 observations per group and then with 10 observations per group, and compare your answers to Exercises 18.12C and D.

19

Robustness and Nonparametric Methods

19.1 Introduction

Likelihood functions provide estimates and tests that are automatically efficient. You don't have to worry too much about how to proceed: Once you specify the model—or more likely, instruct the software about the model you want—these estimates and tests pop out automatically.

Likelihood-based methods are wonderful when your model $p(y|\theta)$ is correct. This model might involve normal, Poisson, Bernoulli, or other distributions. If your model includes a regression component, then it also involves function forms (e.g., linear, exponential, etc.) that relate your Y to your X.

Is your model correct? While there are many issues in statistics that have no firm resolution one way or another, you can count on this:

> All models are wrong.

That means *your* model $p(y|\theta)$ is wrong. This should not be a big surprise. Your DATA are not produced via an omniscient being using a random number generator; they are produced by the real process you are studying, as tapped through your design and measurement. Your model $p(y|\theta)$ is just a model, after all. A toy train is just a model for the train; it's not the real thing.

There are obvious ways in which your models are clearly wrong: No producer of real data is precisely a normal distribution, with all its infinite symmetries, just like no circular object existing in Nature is a mathematically perfect circle. The independence assumptions are questionable as well: Everything in Nature depends on everything else, although sometimes in a very minor way.

Nevertheless, we use models all the time, even though they are wrong. The rest of the phrase, attributed to the noted statistician George Box, is as follows:

> All models are wrong, but some are useful.

So, if your model is *wrong* and your estimates and tests are based on the model being *right*, what are the consequences? The concept called **robustness** answers this question.

Definition of Robustness

Robustness is the extent to which a statistical procedure is useful, despite failures of assumptions. Robustness is a question of *degree*.

To understand what is meant by *degree* of robustness, statistical fuzzy thinking helps. Throughout this book, you have seen many cases where there is no specific answer, like an ugly rule of thumb such as $n > 30$ but instead the more open-ended answer, "It depends." This answer applies especially well to the robustness of a statistical procedure: It depends on how badly violated are the assumptions. If you assume a normal distribution, and the distribution is not in fact normal but something close to normal in terms of its skewness, kurtosis, and discreteness characteristics, then your normality-assuming methods should be very robust. Conversely, if the true distribution is very far from normal in terms of skewness, kurtosis, and discreteness characteristics, then your methods will likely be non-robust, and the accuracy of your analysis will be questionable.

Non-robustness can manifest itself in two ways. First, your procedure might not work like you want it to: If your confidence target is 95% but the actual confidence level is 50%, then your procedure is clearly non-robust. Second, non-robustness can mean that an alternative procedure might work much better. For example, the sample mean is the best estimator of the center of a normal distribution, but the sample median is a much better measure to use when the distribution is symmetric with heavy tails, as discussed in Example 11.7.

There is no simple answer to the question "Is my method robust enough?" Not even an ugly rule of thumb! The answer does not lie in your observed data; it lies in the process that produced your data. (*Model produces data!*) And since you do not know the precise parameters of your data-generating model (*model has unknown parameters!*), you cannot provide a perfectly definitive answer to the question. Your data can help to answer the question (*data reduce the uncertainty about the unknown parameters!*)—for example, outliers provide hints of non-normality—but data are always incomplete and therefore cannot provide a definitive answer.

While you cannot provide a definitive answer to the question "Is my method robust enough?" you can nevertheless provide a reasoned answer involving various what-if scenarios by using computer simulation. For instance, if you assume that your data-generating process is $Y_1, Y_2, \ldots, Y_n \sim_{\text{iid}} N(\mu, \sigma^2)$, and arrive at the 95% confidence interval $\bar{Y} \pm T_{n-1, 0.975} \hat{\sigma}/\sqrt{n}$, you can assess how well the interval performs under alternative models $Y_1, Y_2, \ldots, Y_n \sim_{\text{iid}} p(y|\theta)$ via simulation. If the confidence level is close to 95% for a variety of models $p(y|\theta)$ that are reasonably close to your true data-generating process, then your method is robust enough in terms of confidence level.

Nonparametric methods are methods that do not require particular parametric models $p(y|\theta)$ such as the normal distribution, the Poisson distribution, etc. These methods work well across a variety of distributions $p(y)$ and therefore tend to be robust. They have the name *nonparametric* because the typical assumption is that the data are produced by a generic distribution $p(y)$, and parameters θ are usually not specified.

But the term *nonparametric* is somewhat of a misnomer. Even though the distribution is specified as $p(y)$, seemingly without parameters, the model still has unknown parameters such as $\theta_1 = E(Y)$ (the mean), $\theta_2 = Y_{0.5}$ (the median), etc. Ironically, *nonparametric* does not really mean "no parameters"; rather, it typically means "infinitely many parameters." For example, a continuous distribution $p(y)$ is determined by the infinity of parameters $\theta_y = p(y)$, one for each of the infinitely many y values in the continuous sample space. On the other hand, with a parametric model such as the normal model, all such $p(y)$ are functions of μ and σ^2, so there are only two unknown parameters of a normal model—not infinitely many.

In some cases, however, nonparametric methods involve distributions that are completely free of any unknown parameters; such methods are called **distribution-free**. An example is the randomization model used to create the null distribution in the age/seat selection example of Chapter 15, Example 15.1. Distribution-free models such as the randomization model are sometimes applicable for null (restricted) models, but not for alternative (unrestricted) models. You can never get away from parameters altogether!

While nonparametric methods are preferable in the sense that you do not have to make particular distribution assumptions, people often assume mistakenly that nonparametric models allow them to make no assumptions whatsoever. This is a false and dangerous notion! The independent and identically distributed (iid) assumption is even more prevalent in nonparametric methods than in parametric methods, and when the iid assumption is violated badly, nonparametric methods become as non-robust as parametric methods.

In this chapter, we introduce some commonly used nonparametric methods and show how to evaluate the robustness of statistical procedures when your assumptions are violated.

19.2 Nonparametric Tests Based on the Rank Transformation

Transformation is a common tool to improve robustness. With right-skewed data, the logarithmic transform often makes the distribution closer to normal, which implies that the usual normality-assuming methods will be robust when applied to the transformed data. But the logarithmic transform doesn't always work: Sometimes it can make the distribution even farther from a normal distribution than was the distribution of the original data. Other transformations might work, but the problem of identifying an appropriate transformation is tricky.

An easy, all-purpose solution is to use the **rank transformation**, as discussed in the paper "Rank transformations as a bridge between parametric and nonparametric statistics," published in the journal *The American Statistician* (Vol. 35, No. 3, August 1981) by W.J. Conover and R.L. Iman. To apply the rank transformation, you simply replace the data Y_i with their ranks R_i, assigned so that the smallest Y has $R = 1$, the second-smallest has $R = 2, \ldots$, and the largest has $R = n$. In case of multiple observations of the same Y, you can use the average rank for all the repeats of that Y.

For example, suppose the data are the student age data used repeatedly in earlier chapters. Table 19.1 shows the calculation of the ranks.

To understand Table 19.1, note that the smallest y value is 22, and this occurred three times. So the ranks of these values are 1, 2, and 3; taking the average gives $r = 2.0$ for the three cases where $y = 22$. The next smallest y value is 23, and this occurs twice, so the ranks of these values are 4 and 5; taking the average gives $r = 4.5$.

TABLE 19.1

Rank Transformation of Student Age Data

y	36	23	22	27	26	24	28	23	30	25	22	26	22	35	24	36
r	15.5	4.5	2	11	9.5	6.5	12	4.5	13	8	2	9.5	2	14	6.5	15.5

You can apply the rank transform in the same way that you would apply any other transformation: Simply perform the ordinary analysis on the ranks (the r values) instead of the original data (the y values).

Many nonparametric methods are based on the ranks. The **Wilcoxon rank sum test** is a classic nonparametric alternative to the two-sample t-test discussed in Chapter 16 for comparing two distributions. If you apply the two-sample t-test using the ranks r_i instead of the usual data, then the method is essentially equivalent to the classic Wilcoxon rank sum test.

Specifically, the two-sample rank-transformed test statistic is given by

$$T_r = \frac{\bar{R}_1 - \bar{R}_2}{\left\{ \hat{\sigma}^2_{r,pooled}(1/n_1 + 1/n_2) \right\}^{1/2}}$$

Here, the $\bar{R}_i$ are the averages of the ranks within the different groups, and $\hat{\sigma}^2_{r,pooled}$ is the pooled standard deviation calculated from the ranks. The two-sample rank test assumes that the data are $Y_{ij} \sim_{\text{independent}} p_i(y)$, and under the null model $p_1(y) = p_2(y)$, the distribution of T_r is approximately $T_{((n_1-1)+(n_2-1))}$, despite the fact that the ranks do not have a normal distribution (in fact, the distribution of the ranks is much closer to a uniform distribution than a normal distribution). Hence, the rejection rule $|T_r| \geq T_{((n_1-1)+(n_2-1)),1-\alpha/2}$ will give you a Type I error with probability $\cong \alpha$.

Example 19.1: The Two-Sample Rank Test Applied to Seat Selection

In Example 16.6, the two-sample t-statistic for comparing average ages in front and back of the room is given as $t = 1.375/\{(5.063)\sqrt{1/8 + 1/8}\} = 0.543$. This test statistic can be assumed to come from a T_{14} distribution when the null model $Y_i \sim_{\text{iid}} N(\mu, \sigma^2)$ is true. In this case, the normal model is clearly not true because the distribution of student age is right-skewed. In contrast, the rank transformation simply assumes the null model $Y_i \sim_{\text{iid}} p(y)$, for some unknown $p(y)$—a more reasonable model.

Applying the same calculation to the ranks shown in Table 19.1, where the first eight correspond to the students in the front and the last eight to the students in the back of the class, you get the two-sample rank transformation t-statistic:

$$T_r = \frac{\bar{R}_1 - \bar{R}_2}{\{\hat{\sigma}^2_{r,pooled}(1/n_1 + 1/n_2)\}^{1/2}} = \frac{8.8125 - 8.1875}{\{4.8876^2(1/8 + 1/8)\}^{1/2}} = 0.256$$

You get the p-value exactly as you did using the ordinary two-sample t-test in Example 16.6 using the T_{14} distribution:

$$pv = 2 \times \Pr(T \geq 0.256) = 0.802$$

Since the p-value is larger than any reasonable significance level α, there is no evidence to suggest that seat selection is anything other than random. This is the same conclusion we found with the parametric two-sample t-test, although you might have been concerned about that method's unreasonable assumption of normality.

The **Kruskal–Wallis test** is another famous nonparametric test that is an alternative to the usual normality-assuming analysis-of-variance (ANOVA) F-test. It is based on the ranks and is well approximated simply by calculating the F statistic from the rank-transformed data and by applying the usual rejection rule $F_r \geq F_{g-1, n-g, 1-\alpha}$.

Rank-based tests are, unfortunately, approximate, so the weasel still lurks. However, the approximations of rank-transformed tests usually outperform those of the parametric tests when the distribution assumptions (e.g., normality) are violated. To make this statement more precise, you can use simulations, as shown in Section 19.4. However, you can avoid the weasel altogether by using exact methods. These can be applied to any test statistics, whether those based on ranks, averages, or anything else. You've already seen them in Chapter 15—they involve the permutation distribution, where you shuffle the data like a deck of cards.

19.3 Randomization Tests

A randomization test is one where the permutation distribution is used to calculate the p-value; Example 15.1 provides a good illustration. The method works great for two-sample data and extends perfectly well to multi-sample (ANOVA) data.

The basis for the randomization test is—guess what?—*model produces data*. As with all advanced statistical methods, you start by specifying a model for how your data are produced. Suppose you feel that a reasonable model is $Y_{ij} \sim_{\text{independent}} p_i(y)$, where $i = 1, 2, ..., g$, and $j = 1, 2, ..., n_i$. You aren't stating that the distributions are normal or exponential or anything else, so the model seems very believable indeed. The chance-only (null) model states that all the distributions are the same, or that $p_i(y) = p(y)$, for all $i = 1, 2, ..., g$. In an experiment with g treated groups, this model states that the treatments do absolutely nothing—as if every person took the same sugar pill, regardless of which group they fell into. If the null model is true, then the data are in fact independent and identically distributed, regardless of which group i they are in: Specifically, $Y_{ij} \sim_{\text{iid}} p(y)$, for all $n = n_1 + n_2 + \cdots + n_g$ observations.

Suppose Hans told you that there were $n = 30$ data values coming from groups $i = 1, 2,$ and 3, where 10 of the values came from each group. Hans even gave you the data set: The values are 34.1, 37.1, 24.7, 34.0, ..., 41.1, but Hans didn't tell you which observations came from which groups. If the data were in fact iid from the same distribution $p(y)$, what would you think the average of the $n_1 = 10$ values in group $i = 1$ could be? Answer: The average of 10 randomly selected values from the list 34.1, 37.1, 24.7, 34.0, ..., 41.1. Right?

What about the other two averages? What values might they be? If you think about it, any shuffling of the data 34.1, 37.1, 24.7, 34.0, ..., 41.1 will give a possible collection of averages. Take the first 10 of the shuffled values and average them to get a possible average for the first group, take the second 10 of the shuffled values and average them to get a possible average for the second group, and take the last 10 of the shuffled values and average them to get a possible average for the last group.

Take another random shuffle, and this produces another collection of plausible averages for the three groups, under the null iid model.

So, under the null iid model, all shuffles of the data are equally likely when you don't know the group labels. This gives you the idea behind the permutation test.

The Permutation Testing Procedure for Testing $p_1(y) = p_2(y) = \cdots = p_g(y)$

1. Choose a test statistic T for which larger values of T suggest greater deviation from the null model. For example, with two-sample data, T could be the absolute t-statistic; with multiple-group ANOVA data, you might choose T to be the rank-transformed ANOVA F-statistic; with contingency table data, T could be the Pearson chi-squared test statistic.
2. Evaluate T on your data y_{ij}. Call the value t.
3. Randomly shuffle the data y_{ij}. Let the first n_1 of the shuffled values be $y_{11}^*, y_{12}^*, \ldots, y_{1m}^*$, the second n_2 shuffled values be $y_{21}^*, y_{22}^*, \ldots, y_{2n_2}^*$, ..., and the last n_g shuffled values be $y_{g1}^*, y_{g2}^*, \ldots, y_{gn_g}^*$.
4. Recompute your test statistic T using the shuffled data y_{ij}^*; call the result t^*. Check whether t^* is greater than or equal to your original statistic t.
5. Repeat steps 3 and 4 a large number (e.g., millions) of times.
6. The p-value is the probability of observing a statistic as extreme as t by chance alone; in other words, the proportion of the millions of permutations for which $t^* \geq t$.

If you are able to enumerate all possible permutations instead of just sample millions of them, the procedure is an *exact test*. Software packages routinely compute such exact tests; the most famous one is the *Fisher exact test* that is used for testing independence in 2×2 contingency tables. You can also compute exact tests for general contingency tables. The algorithm is as shown earlier, with the multiple groups being rows of the table and the data Y_{ij} being the multinomial outcomes.

Example 19.2: Testing for Independence in a Sparse Contingency Table

In Example 17.10, the analysis of the George H.W. Bush/Barbara Bush ratings data via likelihood ratio chi-squared test yielded $X^2 = 43.11$, based on 16 degrees of freedom and a p-value $pv \cong 0.0003$. This result suggested that the observed frequencies in the 5×5 contingency table are not easily explainable by chance alone, under a chance-only model where a person's rating of Barbara Bush is independent of their rating of George H.W. Bush.

However, according to Ugly Rule of Thumb 17.2, the approximation was suspect, since none of the 25 cells in the 5×5 table had expected frequencies of 5 or more. Exact tests to the rescue! Here's how it works in this case:

The distributions of interest are $p_1(y) = p(y|\text{GHWBush} = 1)$, $p_2(y) = p(y|\text{GHWBush} = 2)$, ..., $p_5(y) = p(y|\text{GHWBush} = 5)$, where Y denotes the Barbara Bush rating. The null hypothesis restriction is that $p_1(y) = p_2(y) = \cdots = p_5(y)$—that is, that the distributions of the Barbara Bush ratings are identical regardless of the George H.W. Bush rating. You could also do it the other way as well: Perform an analysis of the George H.W. Bush ratings distributions, conditional on the Barbara Bush rating. But it doesn't matter because the resulting p-value will be identical.

The first two steps in the algorithm are to choose and calculate a test statistic; the likelihood ratio chi-square is a good default choice, giving $t = 43.11$. The third and fourth steps are to shuffle the data and recompute the statistic t. From Table 6.9 (also shown in Section 17.5), the data values are as follows:

Group $i = 1$: $y_{11} = 1$, $y_{12} = 1$, $y_{13} = 1$, $y_{14} = 1$, $y_{15} = 1$, $y_{16} = 2$. Note that $n_1 = 6$.

Group $i = 2$: $y_{21} = 2$, $y_{22} = 2$, $y_{23} = 4$. Note that $n_2 = 3$.

Group $i = 3$: $y_{31} = 1$, $y_{32} = 1$, $y_{33} = 3$, $y_{34} = 3$, $y_{35} = 3$, $y_{36} = 4$, $y_{37} = 5$. Note that $n_3 = 7$.

Group $i = 4$: $y_{41} = 3$, $y_{42} = 3$, $y_{43} = 3$, $y_{44} = 3$, $y_{45} = 4$, $y_{46} = 4$, $y_{47} = 4$, $y_{48} = 4$, $y_{49} = 4$, $y_{4,10} = 4$, $y_{4,11} = 4$, $y_{4,12} = 5$. Note that $n_4 = 12$.

Group $i = 5$: $y_{51} = 2$, $y_{52} = 4$, $y_{53} = 5$, $y_{54} = 5$, $y_{55} = 5$. Note that $n_5 = 5$.

Now, if all $n = 33$ data values y_{ij} are from the same distribution, then any permutation (or shuffling) of these 33 values is equally likely to be observed. One such shuffling yields the following data:

Group $i = 1$: $y_{11}^* = 1$, $y_{12}^* = 4$, $y_{13}^* = 3$, $y_{14}^* = 1$, $y_{15}^* = 3$, $y_{16}^* = 2$. Again, $n_1 = 6$.

Group $i = 2$: $y_{21}^* = 3$, $y_{22}^* = 5$, $y_{23}^* = 5$. Again, $n_2 = 3$.

Group $i = 3$: $y_{31}^* = 1$, $y_{32}^* = 5$, $y_{33}^* = 4$, $y_{34}^* = 4$, $y_{35}^* = 3$, $y_{36}^* = 3$, $y_{37}^* = 1$. Again, $n_3 = 7$.

Group $i = 4$: $y_{41}^* = 3$, $y_{42}^* = 1$, $y_{43}^* = 2$, $y_{44}^* = 5$, $y_{45}^* = 4$, $y_{46}^* = 2$, $y_{47}^* = 1$, $y_{48}^* = 5$, $y_{49}^* = 4$, $y_{4,10}^* = 4$, $y_{4,11}^* = 3$, $y_{4,12}^* = 4$. Again, $n_4 = 12$.

Group $i = 5$: $y_{51}^* = 2$, $y_{52}^* = 2$, $y_{53}^* = 4$, $y_{54}^* = 4$, $y_{55}^* = 4$. Again, $n_5 = 5$.

Table 19.2 tabulates these data.

From the data in Table 19.2, the likelihood ratio chi-squared statistic is 20.898. Based on one sample from the chance-only model, it appears that the observed chi-square from the original data, 43.11, is larger than can be explained by chance alone, but you need more than just one sample to verify this. Table 19.3 shows another tabulation of a random shuffle.

From Table 19.3, the likelihood ratio chi-squared statistic is 10.854; again, it appears that the observed chi-square from the original data, 43.11, is larger than can be explained by chance alone.

Repeating the random shuffles 20,000 times, we obtained only 4 cases where the calculated likelihood ratio chi-squared statistic was greater than or equal to 43.11; hence, we estimate the exact p-value to be $4/20,000 = 0.0002$. While it may sound weird to say

TABLE 19.2

Cross-Classification of a Random Permutation of the Preference Data

		Barbara Bush Rating					
		1	2	3	4	5	Total
George H.W. Bush	1	2	1	2	1	0	6
Rating	2	0	0	1	0	2	3
	3	2	0	2	2	1	7
	4	2	2	2	4	2	12
	5	0	2	0	3	0	5
	Total	6	5	7	10	5	33

TABLE 19.3

Cross-Classification of Another Random Permutation of the Preference Data

		Barbara Bush Rating					
		1	2	3	4	5	Total
George H.W. Bush	1	2	0	2	1	1	6
Rating	2	0	1	1	1	0	3
	3	0	2	1	3	1	7
	4	3	1	2	4	2	12
	5	1	1	1	1	1	5
	Total	6	5	7	10	5	33

estimated and *exact* in the same breath, it is okay in the sense of the law of large numbers (LLN): If we simulate increasingly more than 20,000, our estimated p-value gets closer to the true, exact p-value.

Various software packages can calculate the exact p-value by enumerating all possible tables, rather than simulating from them. Using PROC FREQ in SAS/STAT, you get an exact p-value of 0.00018, very close to our simulation-based approximation of 0.0002. So, while the approximate p-value was 0.0003 based on the chi-squared approximation to the likelihood ratio statistic and while this was quite suspect according to Ugly Rule of Thumb 17.2, the exact p-value of 0.00018 was fairly close. The conclusion is the same; namely, that the observed frequency pattern is not easily explainable by a model where the ratings are independent. Thus, you can conclude that the responses to the two questions are related.

While the difference between the exact test and the approximate one was slight in the previous example, don't get complacent about the approximate methods. In some cases, they perform very badly. It is good practice to use exact methods whenever they are available.

19.4 Level and Power Robustness

There are two main types of statistical robustness. One is **level robustness**, which refers to how close the true significance level (or error rate) of the method is to the claimed significance level. The other is **power robustness**, which refers to how well the power of the method compares to that of competing methods. As it turns out, the usual methods based on the normality assumption are often very *level robust* when the distributions are non-normal, but are often greatly outperformed by transformed methods—for example, logarithm or rank—in terms of power when the data are produced by heavy-tailed distributions.

Simulation (*model produces data*) is the way to understand these concepts.

Example 19.3: Evaluating the Robustness of the Two-Sample *t*-Test with Discrete Ordinal Data

Suppose you wish to compare two medical therapies using the two-sample *t*-test. The measurements are the numbers 1, 2, 3, 4, 5, a measure of respiratory health in a patient with asthma. The coding is 1 = very poor health, 2 = poor health, 3 = medium health, 4 = good health, and 5 = very good health. A physician examines the patient and assigns an appropriate number.

The groups consist of patients treated with either one or another of two different inhaled corticosteroid therapies. The patients are randomly assigned to the two groups in a double-blind fashion. After the data are measured and collected, the blind is broken so that the analyst knows which patients were assigned to which formulation of corticosteroid. A two-sample *t*-test is then used to compare the groups to ascertain whether chance alone can explain the difference. Here, the sometimes abstract concept *chance alone* has a very concrete manifestation: Even if the corticosteroid is exactly the same for all patients, chance alone will cause there to be a difference between the two groups' averages, due to the random assignment of patients to groups.

TABLE 19.4

Generic Discrete Distribution

y	$p(y)$
1	π_1
2	π_2
3	π_3
4	π_4
5	π_5
Total	1.00

The model for the two-sample t-test is, from Chapter 16, $Y_{ij}\sim_{\text{independent}} N(\mu_i, \sigma^2)$. If this model is true, then the t-statistic $T = \hat{\delta}/\{\hat{\sigma}^2_{pooled}(1/n_1 + 1/n_2)\}^{1/2}$ is distributed as $T_{\{(n_1-1)+(n_2-1)\}}$ when $\mu_1 = \mu_2$ and is distributed as noncentral T otherwise. But, clearly, the data are not produced by normal distributions since they are highly discrete. So does that mean the two-sample t-test is useless? To answer that question, you can evaluate its level robustness and power robustness via simulation.

Table 19.4 shows the type of model $p(y)$ to use for the simulation.

Once you specify the values of the π_i, you can easily simulate data from the model with any of a variety of software, including a simple spreadsheet program. To pick the π_i, you should think like a Bayesian: What values of the π_i are reasonable for the asthma study? If you are really serious about this study, surely you have some inside information about the types of health ratings that are assigned by the physicians. Maybe some historical data can suggest a few nice, round numbers. Regardless, it is not essential that you get the values absolutely right, because you'll do some sensitivity analysis with other sets of π_i to see whether the essential conclusions change.

Now, you need to decide on sample sizes n_1 and n_2 to simulate from the two groups using your chosen model $p(y)$. This one is easy; just choose sample sizes that you would use in a typical study like this. If you are trying to decide whether the two-sample t-test is valid for a study that you intend to do with 100 patients per treatment group, then use $n_1 = n_2 = 100$. You can also vary the n_i to learn how sample size affects robustness.

Now, get to work! Simulate 100 observations Y^* from $p(y)$ and call those data *group 1*. Simulate another 100 observations Y^* from $p(y)$ and call those data *group 2*. Calculate the two-sample t-statistic $T^* = \hat{\delta}^*/\{\hat{\sigma}^{*2}_{pooled}(1/n_1 + 1/n_2)\}^{1/2}$, and check whether it is greater than $T_{\{(n_1-1)+(n_2-1)\},0.975}$ in absolute value. Repeat many times—the computer doesn't mind, so you can repeat the process 10,000 times or maybe even 1,000,000 times. You can estimate the true level of the test by using the proportion of simulated data sets for which you reject the null hypothesis. Your target here is 0.05, and the method is reasonably level robust if the simulated error rate is close to 0.05—say, between 0.03 and 0.07.

Table 19.5 contains the results of such a simulation, using several different parameter configurations, all estimated using 10,000 simulated data sets, each with 200 total observations. The Wilcoxon two-sample rank test is included as well.

Based on the analyses shown in Table 19.5, it seems that the normality assumption is not crucial in terms of the level robustness of the two-sample t-test. If you claim 5% Type I errors, then you'll get around 5% Type I errors. It does not seem to matter which discrete distribution you pick, although the results seem to drift from the target 0.05 in the pathological case shown in the bottom row of Table 19.5. Bayesian thinking again surfaces. The pathological case is not very likely *a priori*: Will the physicians really rate most of the people in excellent health? It is therefore of limited relevance to the study at hand, but it is comforting that the method works, essentially as advertised with the 5% error rate, even in the pathological case.

Because of the CLT and the LLN, you could have anticipated that the two-sample t-test would be approximately level robust. The two-sample t-statistic has an

TABLE 19.5

Estimated Type I Error Rates for the Two-Sample *t*-Test and the Wilcoxon Two-Sample Rank Test with a Discrete Data-Generating Process and $n_1 = n_2 = 100$ Observations per Group

Multinomial Probabilities					Estimated Type I Error Rate	
π_1	π_2	π_3	π_4	π_5	Two-Sample *t*-Test	Wilcoxon Two-Sample Rank Test
0.5	0.2	0.1	0.1	0.1	0.048	0.051
0.3	0.4	0.2	0.05	0.05	0.054	0.053
0.2	0.2	0.2	0.2	0.2	0.052	0.052
0.1	0.2	0.3	0.3	0.1	0.055	0.055
0.1	0.1	0.1	0.1	0.6	0.052	0.052
0.01	0.01	0.01	0.01	0.94	0.043	0.043

approximate N(0, 1) distribution, regardless of the form of the distribution $p(y)$, so as long as its variance is finite, and you know that the *t*-distribution and standard normal distribution are similar for large degrees of freedom. Nevertheless, it is appropriate to resort to simulation, because *approximately* is a weasel word (still!). Without the simulation, you won't know how good the approximation is. All you know is that larger sample sizes give better approximations.

While the *t*-statistic is robust for level, maybe an alternative test will be more robust for power? It is fine that the test rejects the null model only 5% of the time when you *don't* want to reject the null model. But what if the test only rejects 10% of the time when you *do* want to reject the null model, and another test rejects 90% of the time in such cases? Then the two-sample *t*-test would be called *non-robust for power*.

Both types of robustness are important, and you want your test to both have the right level and be powerful. As shown in Table 19.5, both the two-sample *t*-test and the Wilcoxon two-sample rank test are robust for level, but is one test more powerful than the other?

To assess power robustness, you need to specify the alternative (unrestricted) model where the distributions differ in the two groups. Both groups have 1, 2, 3, 4, 5 data, so the distributions must be as shown in Table 19.4. However, the probabilities π_i will differ for the two groups. Again, Bayesian thinking helps. How might the distributions look, based on your prior knowledge and subject matter expertise? Use that knowledge to pick plausible distributions for the two groups. Presumably, the distributions should not differ by too much, since most inhaled corticosteroids have similar modes of action to combat asthma.

Perhaps due to the screening requirements for patients' entry into the study, the ratings might be generally low, so you might think that the top row of Table 19.5, with probabilities 0.5, 0.2, 0.1, 0.1, 0.1, might be a reasonable model for one group. (Again, this would be a Bayesian-style subjective determination.) In the other group, if the new and improved corticosteroid formulation offers improved health, then the probabilities will be somewhat shifted to the right, higher up the discrete scale. Table 19.6 shows the results of a simulation study to compare the two-sample *t*-test with the Wilcoxon two-sample rank test for different, improved-health distributions in this group, again using 10,000 simulated data sets with 200 patients in each data set.

The two-sample *t*-test is power robust in this simulation study since it does not lose power relative to the Wilcoxon rank test; the powers of the two methods are actually very similar. Hence, relative to the Wilcoxon two-sample rank test, the two-sample *t*-test is level robust as well as power robust according to this study.

TABLE 19.6

Estimated Power for the Two-Sample t-Test versus the Wilcoxon Two-Sample Rank Test with a Discrete 1, 2, 3, 4, 5 Scale Data-Generating Process and $n_1 = n_2 = 100$ Observations per Group

	Multinomial Probabilities					Estimated Power	
	π_1	π_2	π_3	π_4	π_5	Two-Sample t-Test	Wilcoxon Two-Sample Rank Test
Group 1 model	0.50	0.20	0.10	0.10	0.10		
Group 2 models	0.40	0.21	0.11	0.13	0.15	0.345	0.350
	0.35	0.21	0.12	0.14	0.18	0.663	0.674
	0.31	0.20	0.12	0.15	0.22	0.887	0.891
	0.27	0.19	0.13	0.16	0.25	0.974	0.974

While the previous example suggests that you could use either method, you might be more comfortable using the Wilcoxon test because there is no assumption about the distributions. In addition, the power of the test is slightly better for most scenarios shown in Table 19.6, further favoring the Wilcoxon test. However, if you want to use a more complicated model, for example, to allow baseline covariates, then you'll need to use another model altogether, such as an *ordinal logistic regression* model, which you might see in your next statistics course.

Example 19.4: Evaluating the Robustness of the Two-Sample t-Test with Shifted Cauchy Data

The two-sample t-test is robust in terms of power and level for the discrete model shown in Example 19.3. However, a discrete distribution on the numbers 1, 2, 3, 4, and 5 produces no outliers. Things are different when you have a data-generating process that is prone to producing occasional data values that are extremely far from the rest of the pack. One example is with the shifted Cauchy distribution, defined as a random variable distributed as $T_\delta = \delta + T_1$, where T_1 has the central t-distribution with 1 degree of freedom. The t-distribution with 1 degree of freedom is also known as the Cauchy distribution, an outlier-prone distribution having median zero but no defined mean. Thus, the median of the shifted Cauchy distribution is δ, but it also has no defined mean. This distribution might be used as a model for financial returns that are occasionally very wild: Most of the time, the return is near zero, but occasionally, there are wild returns as a result of an extreme shock to the financial system.

Tables 19.7 and 19.8 essentially duplicate Tables 19.5 and 19.6 but use the standard Cauchy distribution to evaluate level robustness, and both the standard Cauchy and the shifted Cauchy distributions to evaluate power robustness.

In contrast with the case of the discrete distribution, the two-sample t-test is not level robust, having error rates usually much smaller than 5%, except for the extremely unbalanced (100,10) case, where the Type I error rate is unacceptably high. While Type I error rates less than 5% might seem acceptable on the surface—who wants to make errors?—they are not acceptable after all because they indicate a lack of power. If you state a rejection rate of 5%, then you are allowing that many false rejections. Level and power are related: If you reject less than 5% of the time, then your method is not rejecting null hypotheses often enough, and therefore your method will tend to have less power when the alternative is true.

On the other hand, the Wilcoxon test is quite level robust, as you can see from Table 19.7. The reason? Because there is no outlier problem with rank data. The largest rank,

TABLE 19.7

Estimated Type I Error Rates for the
Two-Sample *t*-Test and the Wilcoxon
Two-Sample Rank Test with a Cauchy
Data-Generating Process

Sample Sizes		Estimated Type I Error Rates	
n_1	n_2	Two-Sample *t*-Test	Wilcoxon Two-Sample Rank Test
100	100	0.020	0.050
50	50	0.021	0.048
100	50	0.028	0.047
10	10	0.020	0.044
100	10	0.082	0.048
200	200	0.021	0.050

TABLE 19.8

Estimated Power for the Two-Sample *t*-Test
versus the Wilcoxon Two-Sample Rank Test
with a Shifted Cauchy Data-Generating Process
and $n_1 = n_2 = 100$ Observations per Group

	Estimated Power	
δ	Two-Sample *t*-Test	Wilcoxon Two-Sample Rank Test
0.25	0.026	0.162
0.50	0.037	0.474
0.75	0.055	0.802
1.00	0.082	0.951
1.25	0.108	0.993
1.50	0.136	0.999

n, is just one unit away from the second-largest rank ($n - 1$), even if the largest y is billions of units away from the second-largest y.

How about power robustness? Here, the two-sample *t*-test fails miserably, as shown in Table 19.8.

Table 19.8 shows an example where the two-sample *t*-test is very non-robust for power. The problem is that the Cauchy distribution produces extreme outliers, which in turn inflate the variance estimate and make the usual two-sample *t*-statistic small.

19.5 Bootstrap Percentile-*t* Confidence Interval

Nonparametric methods are methods where you don't have to assume a particular parametric model $p(y|\theta)$, such as normal, Poisson, exponential, Bernoulli, etc. Instead, you just assume a generic distribution $p(y)$. The bootstrap, introduced already in this book several times, is such a method.

Recall that the bootstrap distribution does not assume any particular parametric form for $p(y)$, but instead is an estimate of the generic $p(y)$ that puts $1/n$ probability on each observed data value: $\hat{p}(y) = 1/n$ for $y = y_i$, assuming no repeats; otherwise, $\hat{p}(y) = \#y_i/n$ when $y = y_i$.

As shown in Section 19.4, rank-based methods, and, more generally, methods based on percentiles (such as medians) are useful for outlier-prone processes. However, there are cases where you really need to use the mean. For example, if you wish to know about the total of the y values, such as total cost, then you are more interested in the mean than the median, because the mean is directly related to the total and the median is not. The bootstrap is an all-purpose tool that can provide inferences about any parameter of a distribution, mean, median, variance, etc. The following example applies the bootstrap to estimate the mean of an outlier-prone process.

Example 19.5: Estimating Mean Days Lost Due to Back Injury

Warehouse workers often have to lift heavy objects as part of their job, and back injuries are common. Not only are such injuries excruciatingly painful and potentially debilitating, they are also incredibly costly: A large proportion of workers' compensation costs is for back injuries sustained on the job. Back injury can force a worker to miss one or more days of work, and the amount of workers' compensation is related to days lost.

A data set having number of days lost in a year due to back injuries for $n = 206$ warehouse workers is summarized in Table 19.9 in the form of a bootstrap distribution. Note that most workers lost no days, but one worker lost 152 days.

The data indicate a process that is very far from normal: Not only is there a pronounced discreteness component with the large percentage of zeros, but the bootstrap plug-in skewness and kurtosis values are quite extreme, with values 11.6 and 144.4, respectively. These numbers are well beyond the thresholds 2 and 3 indicated by Ugly Rules of Thumb 9.2 and 9.3; thus, they indicate extreme non-normality of the distribution.

Nevertheless, you may be interested in estimating the mean days lost, because it can help you predict how many days of work that workers will lose in the future. For example, if the mean days lost is 0.5, then you can predict 250 days lost among a future cohort of 500 warehouse workers over a year. On the other hand, the median is useless for predicting of total days lost.

TABLE 19.9

Bootstrap Distribution of Day Lost Data

Days Lost (y)	Frequency	$\hat{p}(y)$
0	190	0.9223
2	4	0.0194
3	3	0.0146
4	1	0.0049
5	2	0.0097
14	1	0.0049
15	1	0.0049
19	1	0.0049
27	1	0.0049
52	1	0.0049
152	1	0.0049
Totals	206	1.000

The sample mean from the $n = 206$ measurements is $\bar{y} = 1.505$ days, and the standard deviation is $\hat{\sigma} = 11.468$ days, leading to the normality-assuming 95% confidence frequentist confidence interval for μ given by $1.505 \pm T_{205,0.975}(11.468/206^{1/2})$, or $-0.070 \le \mu \le 3.080$ days. Hmmmm… this doesn't seem so good. Mean days lost can be negative? Obviously, there is a problem. The extreme discrepancy between the actual data-generating process and the normal distribution explains the problem.

In Chapter 16, you saw that the distribution of $T = (\bar{Y} - \mu)/(\hat{\sigma}/\sqrt{n})$ is T_{n-1} when the data Y_i are produced as iid $N(\mu, \sigma^2)$; this fact implies that $\Pr(\bar{Y} - t_{n-1,1-\alpha/2}\hat{\sigma}/\sqrt{n} \le \mu \le \bar{Y} + t_{n-1,1-\alpha/2}\hat{\sigma}/\sqrt{n}) = 1 - \alpha$. Here, the data-generating process is not close to $N(\mu, \sigma^2)$ with its pronounced discreteness, skewness, and kurtosis characteristics, so you can't assume that the interval has the coverage rate $1 - \alpha$ (0.95 in this example). You might be tempted to appeal to the CLT, arguing that $\bar{Y}$ should be approximately normally distributed for the sample size $n = 206$, which is often considered "large enough"—the horrifically ugly rule of thumb $n > 30$ is used in many naïve statistics sources. But, because there is extreme skewness in the data-generating process, you can't assume that $n = 206$ is adequate for the distribution of $\bar{Y}$ to be approximately normal.

You can use the bootstrap distribution to construct an alternative confidence interval, one that does not assume normality. The reasoning goes as follows: Suppose you knew the distribution $p(y)$ that produces the back injury data $Y_1, Y_2, …, Y_{206}$. Then you could find the distribution of $T = (\bar{Y} - \mu)/(\hat{\sigma}/\sqrt{n})$ by simulating data $Y_1^*, Y_2^*,…,Y_{206}^*$ as iid from $p(y)$, calculating $T^* = (\bar{Y}^* - \mu)/(\hat{\sigma}^*/\sqrt{206})$ (if you know $p(y)$ you also know μ), and repeating thousands (better, millions) of times. The histogram of the resulting thousands (or millions) of T^*'s is a very good estimate of the distribution of T, which you can then use to construct an exact confidence interval for μ: Letting c_L be the $\alpha/2$ quantile of the distribution of T, and letting c_U be the $1 - \alpha/2$ quantile, it follows that:

$\Pr(c_L \le T \le c_U)$	
$= 1 - \alpha$	(By the definition of quantile)
$\Rightarrow \Pr(c_L \le (\bar{Y} - \mu)/(\hat{\sigma}/\sqrt{n}) \le c_U) = 1 - \alpha$	(By substitution)
$\Rightarrow \Pr(c_L\hat{\sigma}/\sqrt{n} \le \bar{Y} - \mu \le c_U\hat{\sigma}/\sqrt{n}) = 1 - \alpha$	(By algebra: multiply through by $\hat{\sigma}/\sqrt{n}$)
$\Rightarrow \Pr(-c_L\hat{\sigma}/\sqrt{n} \ge \mu - \bar{Y} \ge -c_U\hat{\sigma}/\sqrt{n}) = 1 - \alpha$	(By algebra: multiply through by -1)
$\Rightarrow \Pr(\bar{Y} - c_L\hat{\sigma}/\sqrt{n} \ge \mu \ge \bar{Y} - c_U\hat{\sigma}/\sqrt{n}) = 1 - \alpha$	(By algebra: adding $\bar{Y}$ through)
$\Rightarrow \Pr(\bar{Y} - c_U\hat{\sigma}/\sqrt{n} \le \mu \le \bar{Y} - c_L\hat{\sigma}/\sqrt{n}) = 1 - \alpha$	(By rearrangement)

So, if you knew the distribution of T, an exact confidence interval for μ would be

$$\bar{Y} - c_U\hat{\sigma}/\sqrt{n}, \quad \bar{Y} - c_L\hat{\sigma}/\sqrt{n}$$

Here, c_L and c_U are the $\alpha/2$ and $1 - \alpha/2$ quantiles of the distribution of T. Notice that, as in the case of the confidence interval for the variance presented in Chapter 16, the upper quantile is part of the lower limit and the lower quantile is part of the upper limit. If this seems contradictory, retrace the algebraic steps to see what happened.

The problem with this approach is that you don't know the distribution of $T = (\bar{Y} - \mu)/(\hat{\sigma}/\sqrt{n})$, because you don't know the distribution $p(y)$ that produces the data. The bootstrap approach gives a simple solution: First, simulate the data from $\hat{p}(y)$ shown in Table 19.9 and find the values:

$$T^* = \frac{\bar{Y}^* - 1.505}{\hat{\sigma}^*/\sqrt{206}}$$

Recall that the mean of the bootstrap distribution is just the sample mean of the data, which is 1.505 here. Second, calculate estimated quantiles $\hat{c}_L$ and $\hat{c}_U$ using thousands (better, millions) of simulated T^* from the bootstrap distribution. The bootstrap confidence interval for μ is then:

$$\bar{Y} - \hat{c}_U \hat{\sigma}/\sqrt{n}, \;\; \bar{Y} - \hat{c}_L \hat{\sigma}/\sqrt{n}$$

This isn't really too hard. You can even generate bootstrap samples in spreadsheet software such as Microsoft Excel, calculate T^* for each sample, and repeat enough to get reasonable estimates of $\hat{c}_L$ and $\hat{c}_U$. Figure 19.1 shows the histogram of 100,000 thusly calculated T^* values. Notice that the distribution is quite different from the T_{205} distribution: It is highly asymmetric, and the range of values extends well outside the -3 to $+3$ range—in fact, some values of T^* were much less than -20.

Interestingly, the skewness of the T^* statistic is opposite that of the data: While the data have pronounced *positive* skewness, Figure 19.1 shows pronounced *negative* skewness. This phenomenon is explained by the correlation between $\bar{Y}^*$ and $\hat{\sigma}^*$. In samples where $\bar{Y}^*$ is higher than the mean 1.505, you will find the outlier upper values such as 52 and 152 (see Table 19.9), also resulting in an inflated $\hat{\sigma}^*$ and causing T^* to be small (but positive). But in samples where $\bar{Y}^*$ is lower than the mean 1.505, you will not find the outlier upper values such as 52 and 152, resulting in a deflated $\hat{\sigma}^*$ and causing T^* to be possibly very far in the lower tail.

The quantiles of the 100,000 T^* statistics depicted in Figure 19.1 are $\hat{c}_L = -8.100$ and $\hat{c}_U = 1.358$, both far cries from the corresponding T_{205} distribution quantiles -1.972 and $+1.972$. The resulting 95% bootstrap confidence interval for μ—the process mean number of days lost—is thus

$$1.505 - 1.358(11.468/206^{1/2}) \le \mu \le 1.505 - (-8.100)(11.468/206^{1/2})$$

or

$$0.420 \le \mu \le 7.977 \text{ days}$$

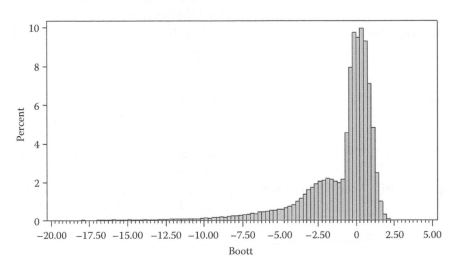

FIGURE 19.1
Bootstrap distribution of the *t*-statistic (labeled as "Boott") for the days lost data.

The bootstrap interval makes more sense than the interval based on the assumption of a normal distribution, or $-0.070 \leq \mu \leq 3.080$ days, because the lower limit of the bootstrap interval is more than 0, and because the bootstrap interval is quite asymmetric, like the data.

Vocabulary and Formula Summaries

Vocabulary

Robustness	The degree to which a statistical method is useful despite violations of its assumptions.
Nonparametric methods	Methods that do not require a specification of a particular parametric model $p(y \mid \theta)$.
Distribution-free methods	Nonparametric methods that are completely free of unknown parameters; typically permutation-based methods.
Rank transformation	The function that transforms the Y value to its rank order, a number 1, 2, ..., n, with the average rank usually used in the case where there are repeated Y values.
Wilcoxon rank sum test	A classic nonparametric alternative to the two-sample t-test for comparing two distributions. It uses the ranks of the observations in the two groups rather than the observations themselves.
Kruskal–Wallis test	A nonparametric alternative to the ANOVA F-test that uses the ranks of observations in the various groups rather than the observations themselves.
Randomization test	A distribution-free test for equal distributions that uses the shuffled data to determine the null distribution of the test statistic.
Level robustness	The degree to which the specified error rate (e.g., 5%) is maintained, despite failures of assumptions.
Power robustness	The degree to which a test procedure remains powerful, despite failures of assumptions.
Bootstrap percentile-t confidence interval	A confidence interval for the mean constructed using the bootstrap distribution of the T-statistic.

Key Formulas and Descriptions

$$T_r = (\bar{R}_1 - \bar{R}_2)/\{\hat{\sigma}^2_{r,pooled}(1/n_1 + 1/n_2)\}^{1/2}$$ Rank-transformed two-sample test statistic, where $\bar{R}_i$ represents the average of the ranks in group i, $i = 1, 2$.

$$\bar{Y} - \hat{c}_U \hat{\sigma}/\sqrt{n}, \quad \bar{Y} - \hat{c}_L \hat{\sigma}/\sqrt{n}$$ The bootstrap percentile-t interval

---

Exercises

19.1 Use the data from Exercise 15.7 in Chapter 15. Assume the first 15 trading days were prior to a major public financial announcement, and the remaining days followed the announcement.

   A. Compute the 29 returns, deleting the one that straddles the announcement, leaving $n = 28$ returns. Find the rank-transformed values of these 28 returns.

   B. Perform the two-sample t-test to compare mean returns before and after the announcement.

   C. Perform the rank-transformed two-sample t-test corresponding to Exercise 19.1B, and compare results.

   D. Why might a financial analyst prefer to use the analysis in Exercise 19.1C? Your answer should not refer to the specific set of 28 returns, but instead should refer to returns in general.

   E. Use the permutation-based method to find the exact p-value for Exercise 19.1C, using the absolute value of the two-sample rank t-statistic as the test statistic. The p-value will not really be exact unless you can enumerate all permutations. Use as many permutations as you can to get the p-value, and compare your results with Exercise 19.1C. Does the t-distribution provide an adequate approximation for the exact p-value for the rank-transformed two-sample t-test?

19.2 Use the data set in Exercise 15.2 of Chapter 15. Construct the 90% percentile-t bootstrap confidence interval for $E(D)$.

19.3 Consider Example 16.6 of Chapter 16, the study of Alzheimer's drugs. Suppose also you want to decide whether to perform an ordinary F-test or a rank-transformed F-test.

   A. Generate 10 observations per group from a normal distribution with mean 50 and standard deviation 10 in every group. Calculate the ordinary and rank-transformed F-statistics, and compare them to the appropriate critical value from the F-distribution. Which tests made the correct decision?

   B. Repeat Exercise 19.3A many times (at least 100). Do the tests appear to have the correct levels?

   C. Repeat Exercise 19.3B with means $\mu_1 = \mu_2 = \mu_3 = 50$ and $\mu_4 = \mu_5 = 60$, and $\sigma = 10$. Which test appears most powerful?

D. Repeat Exercise 19.3B except with shifted Cauchy distributions for all groups. Are the tests level robust?

E. Repeat Exercise 19.3B except with shifted Cauchy distributions using the means of Exercise 19.3C as shift parameters. Which test is more robust for power?

19.4 The following table lists the number of premium televisions sold in a given day during the last calendar year at a local electronics retailer:

y	Frequency	$\hat{p}(y)$
0	30	0.082
1	80	0.219
2	80	0.219
3	90	0.247
4	47	0.129
5	38	0.104

A. Specify a normal distribution model in terms of the unknown parameters, one that an unenlightened statistics student might assume to have produced these sales data. Then critique that model.

B. Find the normality-assuming 95% confidence interval for the mean number of sales per day. (Hint: Recall that the plug-in estimate of the mean is the same as the ordinary average but that the plug-in estimate of variance involves n rather than $n - 1$. So perform the appropriate correction to the plug-in estimate of the variance to get the $n - 1$ formula.)

C. Use the bootstrap percentile-t method to obtain a 95% confidence interval for the mean number of sales per day.

D. Other than normality, what assumptions are you making in your analysis for Exercise 19.4B? (You also make these assumptions in the analysis of Exercise 19.4C.) Critique these assumptions.

19.5 You have collected the following data on car color preferences from younger and older customers:

Age	Preference
Younger	Red, red, green, gray, red, red, red
Older	Gray, gray, gray, red, gray, green, green

A. Construct the contingency table, and compute the Pearson chi-squared test for independence of color choice and age.

B. Are the sample sizes adequately large? Find and apply the appropriate (ugly) rule of thumb.

C. Follow the method shown in this chapter to find an exact randomization-based p-value for the Pearson chi-squared statistic.

20

Final Words

The main learning outcome of this text is summed up by the Mantra *Model produces data, model has unknown parameters, data reduce the uncertainty about the unknown parameters.* If this is still a mystery, please read Chapter 1 again.

While this text is not intended to be a cookbook, you can find many recipes in the end-of-chapter summaries. But rather than apply statistics like a trained parrot, please aim for a deeper understanding. We hope that you understand, by now, that the subject of statistics is not a collection of arbitrary ugly rules of thumb. Rather, it is a science whose logical foundation is probability theory, and there are mathematical consequences of statistical assumptions that are 100% true—provided the assumptions are true. When the assumptions are wrong—and they *always* are—the conclusions cannot be entirely trusted. Then you have to deal with approximations. If we did our job with this book, you now have a healthy skepticism for any approximate, weasel-like method, as well as the tools (including simulations—*model produces data*) to judge when the approximations are good.

We also hope you have a better understanding of how statistical models represent natural processes. The sub-Mantra, *nature favors continuity over discontinuity*, is key to understanding how to model statistical relationships between variables and also provides an excellent reason why you should avoid using the *population* terminology when it comes to specifying these models.

Finally, we hope you have found this book to be practically useful. Our final Mantra, *use what you know to predict what you don't know*, tells you specifically how to use statistics—for research, profit, enlightenment, longevity, fun, and maybe even romance. Contrary to what you may have heard, statistics don't lie—only people do. Don't let Hans fool you!

Index